Communications in Computer and Information Science **2893**

Rationale
The CCIS series is devoted to the publication of proceedings of computer science conferences. Its aim is to efficiently disseminate original research results in informatics in printed and electronic form. While the focus is on publication of peer-reviewed full papers presenting mature work, inclusion of reviewed short papers reporting on work in progress is welcome, too. Besides globally relevant meetings with internationally representative program committees guaranteeing a strict peer-reviewing and paper selection process, conferences run by societies or of high regional or national relevance are also considered for publication.

Topics
The topical scope of CCIS spans the entire spectrum of informatics ranging from foundational topics in the theory of computing to information and communications science and technology and a broad variety of interdisciplinary application fields.

Information for Volume Editors and Authors
Publication in CCIS is free of charge. No royalties are paid, however, we offer registered conference participants temporary free access to the online version of the conference proceedings on SpringerLink (http://link.springer.com) by means of an http referrer from the conference website and/or a number of complimentary printed copies, as specified in the official acceptance email of the event.

CCIS proceedings can be published in time for distribution at conferences or as post-proceedings, and delivered in the form of printed books and/or electronically as USBs and/or e-content licenses for accessing proceedings at SpringerLink. Furthermore, CCIS proceedings are included in the CCIS electronic book series hosted in the SpringerLink digital library at http://link.springer.com/bookseries/7899. Conferences publishing in CCIS are allowed to use our online conference service (Meteor) for managing the whole proceedings lifecycle (from submission and reviewing to preparing for publication) free of charge.

Publication process
The language of publication is exclusively English. Authors publishing in CCIS have to sign the Springer CCIS copyright transfer form, however, they are free to use their material published in CCIS for substantially changed, more elaborate subsequent publications elsewhere. For the preparation of the camera-ready papers/files, authors have to strictly adhere to the Springer CCIS Authors' Instructions and are strongly encouraged to use the CCIS LaTeX style files or templates.

Abstracting/Indexing
CCIS is abstracted/indexed in DBLP, Google Scholar, EI-Compendex, Mathematical Reviews, SCImago, Scopus. CCIS volumes are also submitted for the inclusion in ISI Proceedings.

How to start
To start the evaluation of your proposal for inclusion in the CCIS series, please send an e-mail to ccis@springer.com

Hongxing Wei · Jianwei Niu
Editors

Collaborative Robot Technology and Application

5th International Conference on Collaborative Robot Technology and Application, COBOT 2025
Hangzhou, China, November 22, 2025
Proceedings

Springer

Editors
Hongxing Wei
Beihang University
Beijing, China

Jianwei Niu
Beihang University
Beijing, China

ISSN 1865-0929 ISSN 1865-0937 (electronic)
Communications in Computer and Information Science
ISBN 978-981-92-2073-1 ISBN 978-981-92-2074-8 (eBook)
https://doi.org/10.1007/978-981-92-2074-8

This Springer imprint is published by the registered company Springer Nature Singapore Pte Ltd.
The registered company address is: 152 Beach Road, #21-01/04 Gateway East, Singapore 189721, Singapore

Preface

This volume comprises the papers presented at the Collaborative Robot Technology and Applications Conference (Cobot 2025). The conference was sponsored by the Robotics Society of the Chinese Institute of Electronics and organized by the Embodied Intelligent Robotics Institute of Beihang University, AUBO (Beijing) Robotics Technology Co., Ltd., the Key Laboratory of Service Robot Technology, and Hangzhou Dianzi University.

Cobot 2025 was dedicated to fostering academic exchange among experts, scholars, engineers, and graduate students in the fields of robotics and collaborative robotics. It aimed to promote the research and application of cutting-edge theories, technologies, and systems in embodied intelligence, human-robot interaction, and intelligent perception and control.

The conference was held at Hangzhou Dianzi University (Xiasha Campus), Hangzhou, China, on November 22, 2025. It featured keynote speeches by leading experts in robotics and collaborative systems, along with extensive seminars on emerging topics and technologies. From human-robot collaboration and mobile robots to embodied intelligence and robotic applications across industrial, agricultural, service, educational, and medical domains, Cobot 2025 brought together the latest research trends and facilitated deep discussions among participants from both academia and industry.

The main topics of the conference included, but were not limited to, the following areas:

Collaborative Robots and Human-Robot Interaction Systems, Mobile and Composite Robots, Service Robots and Intelligent Perception Systems, Robot Navigation and Control, Embodied Intelligence Foundations, Embodied Large Models and Robot Integration, and Robotics Applications in various fields.

The conference received a total of 33 submissions. After a rigorous single-blind peer review process, 29 full papers were accepted for publication in this proceedings volume. Each paper was reviewed by three independent experts, who provided 2-3 detailed reviews. Authors were required to address all comments before final acceptance. The acceptance rate was 87.9%. No short papers or invited papers were included.

The editors of the proceedings extend their sincere gratitude to the Cobot 2025 Advisory Committee, especially Guangnan Ni of the Chinese Academy of Engineering, and the Conference Chairs, Hongxing Wei and Jianwei Niu of Beihang University. We also thank all authors, reviewers, and committee members for their valuable contributions. Special thanks are due to Springer for publishing the proceedings in the Communications in Computer and Information Science series, and to Taylor & Francis for supporting publication in the international journal Embodied Intelligent Robot.

November 2025

Hongxing Wei
Jianwei Niu

Organization

Conference Chairs

Hongxing Wei	Beihang University, China
Jianwei Niu	Beihang University, China

Advisory Committee

Guangnan Ni	Chinese Academy of Engineering, China

Program Committee

Huasong Min	Wuhan University of Science and Technology, China
Diansheng Chen	Beihang University, China
Guofeng Tong	Northeastern University, China
Shuo Wang	Institute of Automation, Chinese Academy of Sciences, China
Guanghua Yu	University of Science and Technology Beijing, China
Dangxiao Wang	Beihang University, China
Wei Wang	Beihang University, China
Mu Li	Beihang University, China
Jinchang Liu	National Natural Science Foundation of China, China
Jianhua Zhang	University of Science and Technology Beijing, China
Zhangguo Yu	Beijing Institute of Technology, China
Zhaoyao Shi	Beijing University of Technology, China
Geng Yang	Zhejiang University, China
Zheming Tong	Zhejiang University, China
Guangzhe Zhao	Beijing University of Civil Engineering and Architecture, China
Xiaohui Xiao	Wuhan University, China
Chi Guo	Wuhan University, China
Xianfeng Huang	Wuhan University, China
Ming Li	Wuhan University, China

Organizing Committee

Review Committee

Conference Secretaries-General

Zhikun Zhu	AUBO (Beijing) Robotics Technology Co., Ltd., China
Peng Wang	Beijing Institute of Petrochemical Technology, China
Ya Chen	Beijing Institute of Petrochemical Technology, China
Shuangyu Fan	AUBO (Beijing) Robotics Technology Co., Ltd., China
Huiwen Zhang	Green Ear Robot Technology (Beijing) Co., Ltd., China

Contents

Section I

Calibration of the In-Situ Integrated Torque Sensor in Harmonic Reducer of Robot Joint

Wenqiang Shen, Zhifeng Lou, Jiacheng Li, Ze An, and Xiaodong Wang(✉)

School of Mechanical Engineering, Dalian University of Technology, Dalian 116024, China
xdwang@dlut.edu.cn

Abstract. Torque sensor plays an important role in enabling functions such as collision detection and compliance control for collaborative robots. Integrating the torque sensor directly in the harmonic reducer is a preferable method, offering the advantage of preserving the structure stiffness and performance of the joint. However, this approach is susceptible to the interference caused by the squeezing action of the wave generator while rotating. To enhance the measurement accuracy, a strain rosette ring is employed as the sensitive unit of the torque sensor. Additionally, an adhesive bonding setup for attaching the developed ring structure is designed, and the bonding process is illustrated. The manufacturing process can be automatically operated, making mass production feasible. A calibration method based on the joint angle is proposed to deal with the interference during in-situ measurement. Moreover, a step-by-step compensation method is adopted to further suppress the influence of the ripple signal, reducing the peak-to-peak error of the ripple signal from 13.323 mV to 1.907 mV. Experimental results indicate that the developed torque sensor offers acceptable measurement accuracy.

Keywords: Torque Sensor · Harmonic Reducer · Collaborative Robot

1 Introduction

Over the past period, the development of robotic technology has evolved from its initial applications in assembly pipeline to a broad range of fields, encompassing healthcare, education, services and manufacturing. As tasks become increasingly intricate, the demand for robots capable of adapting to various environments and collaborating effectively with humans has risen significantly [1–3]. An important issue in human-robot collaboration is ensuring safety. To address this problem, robots are typically equipped with the torque sensor to measure the output torque. Based on the real time information, collision detection is performed, enabling robots to take actions to prevent accidental collisions [4–6].

An alternative method involves installing the torque sensor at the end of each joint. By detecting the deformation of the additional elastomer induced by external load, accurate torque information can be obtained. Wei [7] developed a novel joint torque sensor through optimization the cross-beam structure using finite element analysis. The optimized structure reduces interference from assembly errors and axial forces from the

H. Wei and J. Niu (Eds.): COBOT 2025, CCIS 2893, pp. 3–14, 2027.
https://doi.org/10.1007/978-981-92-2074-8_1

motor and maintains good performance in high-temperature environments. Min et al. [8] developed a monolithic joint torque sensor, which is a hexahedral block incorporating eight strain gauges and features a support structure that minimizes the moment load on the sensing block. By identifying the sensitive areas through FEM, the arrangement of the strain-sensitive units was optimized, improving the measurement accuracy. Kim et al. [9] developed a torque sensor utilizing the strain amplification mechanism comprising a trapezoidal beam, flexure hinges and frames. The operational principle is based on an initial displacement generated in the trapezoidal beam, followed by its amplification through the flexure structure.

However, this approach reduces joint stiffness, complicating high-precision position control [10]. Additionally, it increases the size and weight of joint. Considering that output torque is proportional to the current of the motor, researchers have proposed estimating output torque information based on the variation of current. This method is simple and cost-effective. However, due to the difficulty in accurately modeling non-linear factors such as friction, measurement accuracy of this approach is inherently constrained. Ma et al. [11] proposed a model for detecting collision indirectly based on the current of each joint motor. Demonstrating its capability to detect collision. Docherty et al. [12] introduced a compensated method to reduce the noise of the motor current, enhancing the accuracy of the estimated joint torque. Park et al. [13] developed two methods based on machine learning: Support Vector Machine and Convolutional Neural Network. These approaches utilize only the motor current, the dynamic model, and an observer, eliminating dependency on the friction model. The integration of machine learning enables the automatic adjustment of collision detection thresholds for individual joints.

Furthermore, researchers employed the six-axis force/torque sensor to monitor contact forces during operation, enabling precise control in tasks such as grinding and precision assembly. Yan et al. [14] proposed an orthogonally decoupled beam structure to achieve low coupling and high sensitivity in six-axis force/torque sensor. Yao et al. [15] introduced a redundant parallel six-component force/torque sensor with a spoke structure that combines parallel and flexible mechanisms, enhancing the measurement performance. However, this approach has significant limitations. As the torque sensor is typically mounted at the wrist of robot, its perception is confined to this local area, preventing the detection acting on other areas. Additionally, the cost associated with the six-axis force/torque sensor remains a practical concern.

Integrating the torque sensor directly into the harmonic reducer is a preferable method. This approach utilizes the flexspline, which is charactered by low torsional stiffness, as the elastomer of the torque sensor to preserve the performance of joints. However, it should be noted that this method presents challenges in accurately obtaining torque information. The output signal includes not only the target torque but also the interference of the squeezing action of the wave generator during rotation, known as the ripple signal [16, 17].

Taghirad et al. employed four symmetrical strain gauges to suppress the interference of the ripple signal, while a transparent film was used to enhance the positioning precision of the strain gauges. During the installation of the strain gauges, a microscope was employed. Tests were conducted on the developed torque sensor, the calibration results demonstrated good linearity and showed that the sensor was independent of the position of the flexspline. The developed harmonic reducer with the sensor presents a promising option for torque feedback applications. Additionally, a Kalman filter was applied for improving accuracy [18]. Jung et al. used an order tracking method to compensate for the ripple signal and combined the method with a notch filter to further reduce the amplitude of the ripple signal [19]. Furthermore, they installed the harmonic reducer on a robot and compared the results of the proposed signal processing method with those obtained from the current-based method. During loading, the maximum error of the current-based method was 8.2 N·m, while the proposed order tracking method reduced the maximum error to 4.8 N·m. Pan et al. [20] used vector addition principle to minimize the ripple signal and developed a torque sensor based on the three-vector cancellation method to suppress higher-order ripple signals. They also improved the accuracy of the torque sensor by using redundant strain gauges. And corresponding experiments were conducted. Results demonstrate that the developed sensor prototype possesses high sensitivity and accuracy. When the wave generator was fixed, the primary error was hysteresis, with a value of approximately ±0.54% F.S. Under the rotating of the wave generator, the error increased to ±2.3% F.S.

To address the ripple signal, a strain rosette ring is utilized as the sensitive unit of the torque sensor which is in-situ integrated in the harmonic reducer. This ring structure suppresses the ripple signal by measuring strain at multiple circumferential positions around flexspline flange. An adhesive bonding setup is designed to ensure precise attachment of the strain rosette ring. Also, the bonding process is also illustrated. Furthermore, a calibration method for the developed torque sensor based on joint position is proposed, and the performance of the torque sensor is tested.

2 Structure of the Strain Rosette Ring

Harmonic reducer primarily comprises the wave generator, the circular spline, and the flexspline. The hat-shaped harmonic reducer is studied, where the flexspline is fixed, the wave generator serves as the input, and the circular spline functions as the output. In the assembly state, the wave generator deforms the flexspline. This engages the external teeth of the flexspline near the major axis of the wave generator with the internal teeth of the circular spline, while the teeth near the minor axis fully disengage. As the wave generator rotates, the external teeth of flexspline sequentially engage with and disengage from the teeth of circular spline, thereby enabling power transmission.

When the wave generator is installed in the flexspline, the squeezing action causes notable initial uneven strain in the flexspline under no-load conditions. This squeezing action significantly decreases the in-situ measurement accuracy. In addition, the amplitude of the interference signal is higher than that of the external torque signal. To quantify the influence of various disturbances on measurement, a finite element analysis is performed to investigate the strain distribution on flexspline with the action of the wave generator and external torque. During the simulation, the deformation of the wave generator induces a corresponding deformation in the flexspline. The contact between the flexspline and the wave generator can be equivalently modeled as frictionless. In this case, this contact pair is defined as frictionless. The contact surface is set as the inner wall of the flexspline cylinder, and the target surface is the outer surface of the wave generator. The meshing between the circular spline and the flexspline involves pure sliding friction. And this contact pair is configured with friction. The contact surface is defined as the tooth surface of the flexspline and the target surface is the tooth surface of the circular spline, with a friction coefficient of 0.1 and a normal stiffness factor of 1. Subsequently, constraint conditions are applied to the harmonic reducer according to its actual state within the joint. Specifically, the wave generator is constrained to allow only rotational motion. A fixed support constraint is applied to the flexspline, restricting all translational degrees of freedom (X, Y, and Z directions). The circular spline is also permitted to undergo only rotational motion. Its translational degrees of freedom in the X, Y, and Z directions are constrained, while its rotational degree of freedom is kept unconstrained.

Results are presented in Fig. 1(a). It can be observed that the strain distribution is symmetrically distributed about the long axis of the wave generator. Subsequently, the strain values at different locations on the flexspline are extracted. Specifically, at the diameter of 47 mm on flexspline flange, strain at the circumferential positions of 0°, 30°, and 90° are measured as -3.29 με, 62.3 με, and -10.5 με, respectively. Under the application of external torque, strains at the same positions are 119.8 με, 140.2 με, and 125.2 με. The strain variations induced by the same torque differs significantly across the various locations. Consequently, it is difficult to accurately determine the external torque based on the strain from one single position under such conditions. And with the rotation of the wave generator, the strain at the same position still demonstrates different distribution.

To minimize this interference and ensure accurate measurement, research on flexspline strain variations under different operating conditions are conducted previously. And a strain rosette ring is developed as the sensitive unit of the torque sensor, as shown in Fig. 1(b) [21]. This strain rosette ring employs the Wheatstone full bridge circuit, which can effectively amplify the output signal and provides high measurement sensitivity. The configuration of the circuit is illustrated in Fig. 1(c). To simplify the wiring associated with the Wheatstone circuit, the wires are integrated directly onto the strain rosette ring. The dimensions of wires are significantly larger than those of the sensitive grids to minimize interference caused by deformation. In this case, the output of the developed torque sensor is related to the external torque and is insensitive to the position of the harmonic reducer.

Due to the large area and thin thickness, the strain rosette ring tends to exhibit curling without constraints, complicating the adhesive bonding process. Excessive curling also poses a risk of damaging the sensitive unit. Furthermore, positioning of the strain rosette ring is important for maintaining measurement accuracy through the analyzing of the strain distribution of flexspline. To address these issues, an adhesive bonding setup for the developed strain rosette ring is designed, as shown in Fig. 2. This setup comprises a top visual measurement module, a bottom visual measurement module, a plat-form, a vacuum suction module and a glue module, which together facilitate the pickup, positioning, release and adhesive bonding of the strain rosette ring. To enable high-quality integration of the strain rosette ring on the flexspline of the harmonic reducer, a bonding method based on capillary action is proposed. Specifically, this procedure involves placing the strain rosette ring on the flexspline. Adhesive is then injected into the gap between the strain rosette ring and the flexspline. Driven by capillary force between the two surfaces, the adhesive progressively flows, covering the entire area. The proposed scheme effectively eliminates the potential for air entrapment, ensuring a bubble-free adhesive layer after curing.

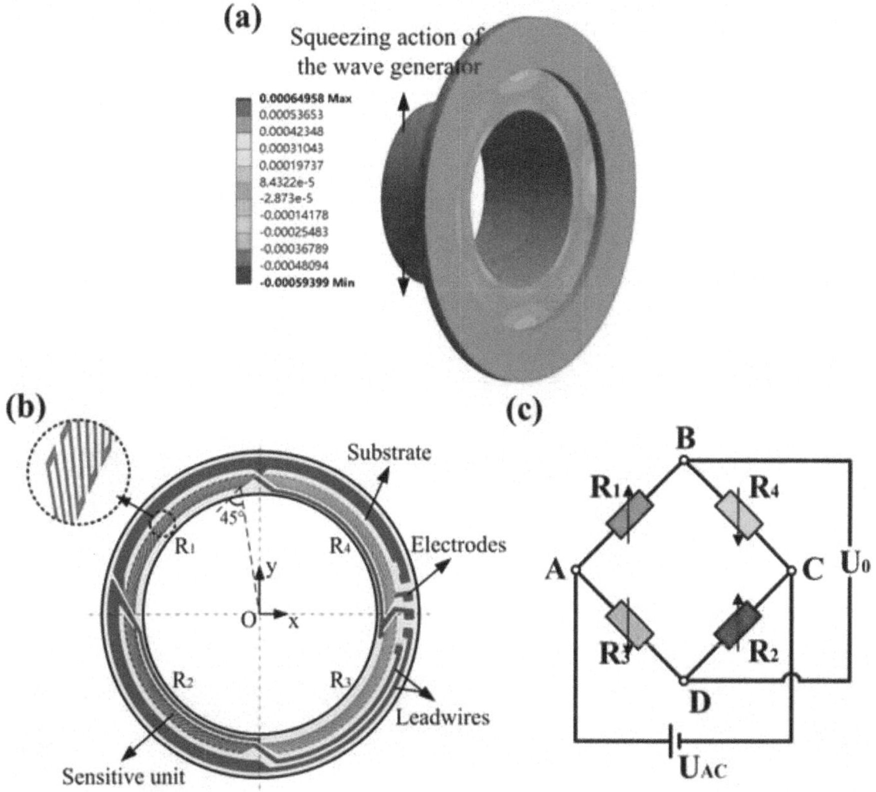

Fig. 1. (a) Initial uneven strain caused by the squeezing of the wave generator. (b) Structure of the torque sensor in-situ integrated into harmonic reducer. (c) Configuration of the strain rosette ring.

The adhesive bonding process is as follows:

(1) The top visual measurement module captures images of four positions around the circumference of the flexspline. Utilizing the image stitching and iterative fitting algorithms, the center position (X1, Y1) of the flexspline is determined. As the dimension of the flexspline exceeds the field of view of the visual measurement module, a sub-region image acquisition strategy is employed. Subsequently, a multi-region image fusion technique is applied to reconstruct the complete model of the flexspline. The discrete edge coordinates from this model are then iteratively fitted

to remove outliers and extract a precise profile. Simultaneously, the light intensity parameters need to be adjusted while acquiring the image.

(2) A circular cupule is designed and attached to the Z-axis guide rail through a magnetic connector. During positioning, negative pressure is applied to the circular cupule, allowing the strain rosette ring to be securely picked up through the evenly distributed holes at the bottom of the cupule. Herein, the bottom of the cupule is equipped with an array of holes. This configuration ensures the stable positioning of the strain rosette ring and prevents damage caused by uneven force distribution.

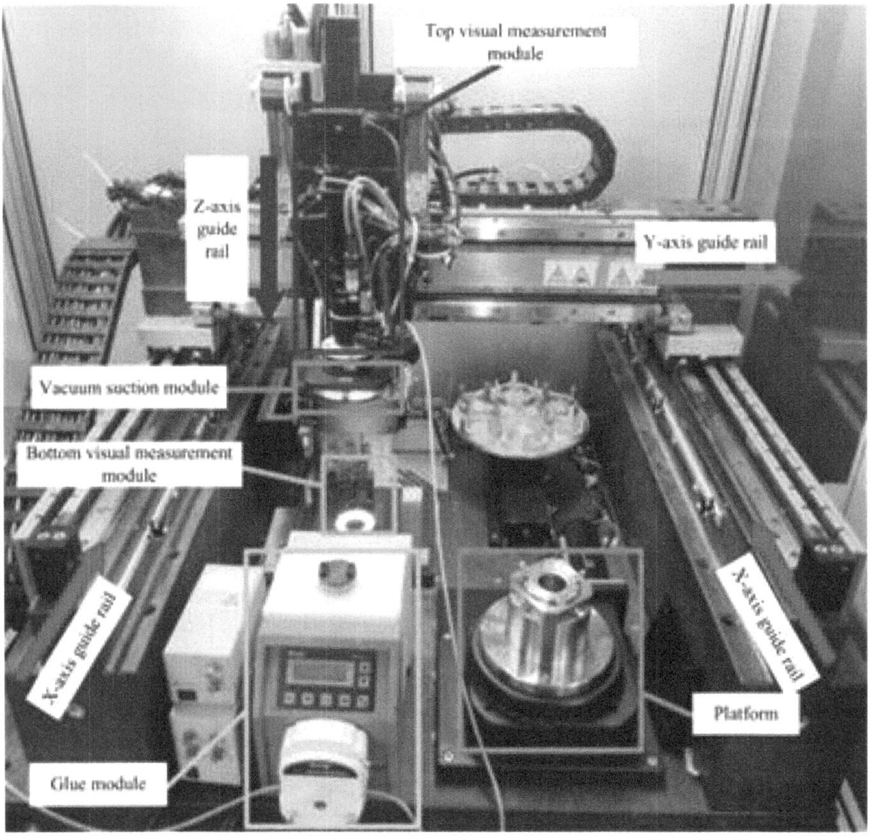

Fig. 2. Adhesive bonding setup of the strain rosette ring.

(1) Move the strain rosette ring to the bottom visual measurement module and capture images of the ring. These images are then stitched and fitted to obtain the center position (X2, Y2) of the strain rosette ring. Simulation analysis revealed that the results obtained by the developed torque sensor keep unaffected by the strain gradient at the bottom of flexspline flange when the eccentricity between the strain rosette ring and the flexspline is within 0.2 mm. Therefore, a dual-vision system is employed for

dimensional matching between the flexspline and the ring structure. Experimental results demonstrate that the maximum eccentricity between the flexspline and the strain rosette ring is 45 μm while positioning.

(2) By combining the positions of both the flexspline and the strain rosette ring, the strain rosette ring is then moved above the flexspline. Subsequently, the Z-axis guide rail is moved downward. Once the force sensor between the strain rosette ring and the flexspline reaches the threshold, turn off the vacuum and release the strain rosette ring.

(3) Position the needle on the Z-axis guide rail over the marked locations on the strain rosette ring for adhesive bonding. Utilizing the capillary force, the adhesive can completely fill the gap. After gluing, remove the cupule and transfer the flexspline to the oven for curing.

To obtain high positional accuracy, it is essential to compensate for the installation errors of the components of the developed setup and establish a compensation model. Furthermore, to facilitate operation and monitoring of the assembly process, the upper computer software is developed. The interface displays real-time parameters from the gantry guideways, rotary table, and force sensor, while also showing the progress and status of the assembly process. Additionally, images from both the main and bottom visual measurement modules can be provided.

To address issues such as insufficient glue volume and glue overflow, experiments are carried out to optimize the adhesive bonding parameters. An acrylic ring with the same dimensions as the strain rosette ring is fabricated to observe the glue flow and identify any bubble residue after bonding. During adhesive bonding the strain rosette ring, the filling rate is primarily influenced by the properties of the adhesive. While experiment, it is observed that an excessive dispensing rate leads to overflow. Conversely, an insufficient duration results in a total adhesive volume that is less than the gap, consequently causing substantial bubble. It is important to match the dispensing rate to the filling rate, preventing excessive dispensing speed. The dispensing rate is controlled by adjusting the rotational speed of the peristaltic pump in the developed setup. Based on multiple experiments, the peristaltic pump speed is set to 1 rpm, with an adhesive application time of 15 s at each position.

3 Results and Discussions

3.1 Calibration Setup

Due to the manufacturing error of the torque sensor, parameters of the sensitive units at different positions are inconsistent, resulting in residual signal fluctuations. There-fore, calibration is needed to suppress the residual ripple signal. Research has indicated that the output signal of the developed torque sensor is closely related to the angle of the wave generator. Also, with the rotation of the wave generator, the signal exhibits a sinusoidal pattern. In this case, a calibration method utilizing the position of the joint is proposed, combining the wave generator position with the output of the developed torque sensor. The calibration setup is shown in Fig. 3, consists of a joint, a reference torque sensor (Messtechnik 0261Due, accuracy class: ±0.1%F.S.), an output linkage, and a strain instrument. The reference torque sensor provides accurate torque information, while the output linkage applies external load to the setup by varying the mass. The strain instrument provides the circuit of the full-bridge Wheatstone bridge. While calibration, the harmonic reducer integrated the torque sensor is installed into the joint.

Fig. 3. Setup of the torque sensor integrated on the harmonic reducer.

3.2 Calibration

To suppress the residual ripple signal, the position of the wave generator is correlated with the output signal. The calibration involves rotating the wave generator every 6° until 360°. At each position, the angle of the wave generator and the voltage of the developed sensor are recorded. The relationship between the wave generator angle and the torque sensor output voltage is then fitted. The results are shown in Fig. 4(a). To accurately characterize the sinusoidal function relationship and reveal features of multiple frequency components, a second-order Fourier series is employed for fitting:

$$U' = a_0 + a_1 \cos(\omega\theta_{wg}) + b_1 \sin(\omega\theta_{wg}) + a_2 \cos(2\omega\theta_{wg}) + b_2 \sin(2\omega\theta_{wg}) \tag{1}$$

where, *U'* is the output voltage of the developed torque sensor, θwg is the angle of the wave generator. The coefficients *a0*, *a1*, *b1*, *a2*, *b2*, and ω are obtained by calibration. Values of these coefficients are listed in Table 1. The coefficient R^2 which determines the goodness of fit, is found to be 0.998. The second-order Fourier model can achieve high fitting accuracy with only a small number of parameters.

Table 1. Coefficients of the fitting curve.

Coefficients	Value
a0	−1.123
a1	14.561
b1	4.019
a2	−1.207
b2	0.869
w	0.036

As observed, the developed torque sensor effectively suppresses the interference with the proposed compensation method, reducing the maximum difference to 2.360 mV, decreasing the interference by 71.96%. And the root mean square (RMS) error after compensation is 0.208 mV. In addition, it is noted that the relationship between the wave generator angle and the voltage of the sensor varied at different positions during rotation, as illustrated by the black line in Fig. 4(b). This inconsistency is attributed to variations in the manufacturing and assembly of the harmonic reducer. To address this issue, a step-by-step compensation method is proposed. Specificality, this method involves uniformly segmenting data within the 0–360° range of the circular spline and calibrating each segment separately. The results with compensation are shown by the red line in Fig. 4(b). In this case, the peak-to-peak error of the residual ripple signal is reduced from 13.323 mV to 1.907 mV, and the RMS error is 0.237 mV. Importantly, the signal of the developed torque sensor is independent of the position of the harmonic reducer after compensation.

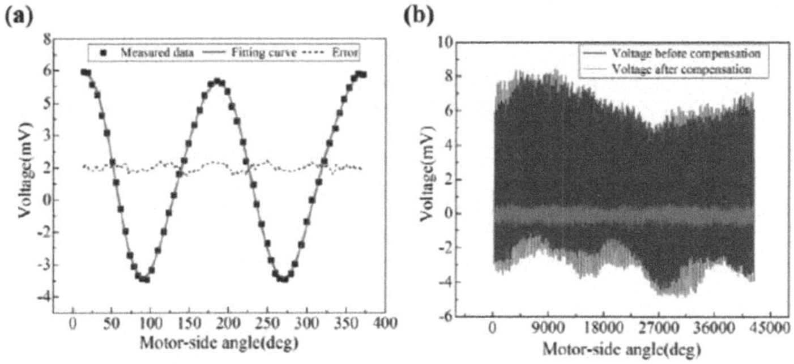

Fig. 4. (a) Relationship between the output signal and the angle of the wave generator. (b) The output signal while the circular spline rotates one cycle.

After suppressing the ripple signal, the next step is to calibrate the relationship between the developed torque sensor and the applied torque. Stepwise loads ranging from -15 to 15 N·m are applied to the output linkage. Meanwhile, the voltages from both the developed torque sensor and the reference sensor are recorded. To ensure the accuracy of the calibration, the external load is applied three times at each position. The data is then fitted using the least squares method, as shown in Fig. 5. After calibration, the relationship between the voltage of the developed torque sensor and the applied torque can be expressed as:

$$U - U' = K \cdot T + B \tag{2}$$

where, U is the voltage of the developed torque sensor, T is the output torque, K and b are obtained by calibration.

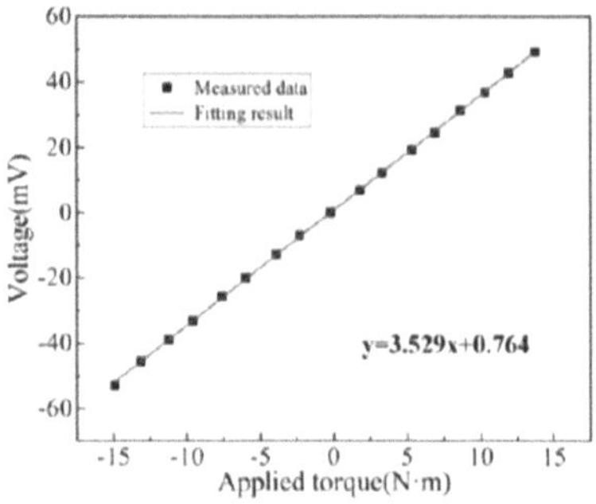

Fig. 5. Calibration of the developed torque sensor.

3.3 Experiment

In the first experiment, the wave generator is fixed while loads are applied to the linkage. Figure 6(a) displays the voltage of the torque sensor when various magnitudes and directions of loads are applied. In this case, the peak-to-peak error (Emax) between the developed torque sensor and the reference torque sensor is 0.708 N·m, with an RMS error of 0.151 N·m. Also, experiments with constant torque are conducted at different circular spline positions, and the corresponding results are shown in Table 2. Results show that the signal is unaffected by the position of the harmonic reducer. In the next experiment, the wave generator rotates continuously. Meanwhile, a mass is applied to the output linkage, inducing sinusoidal torque at the joint as the wave generator rotates. Figure 6(b) shows the outputs from both the developed torque sensor and the reference torque sensor. The outputs of the torque sensor and the reference torque sensor are consistent, with a peak-to-peak error of 1.239 N·m and an RMS error of 0.298 N·m.

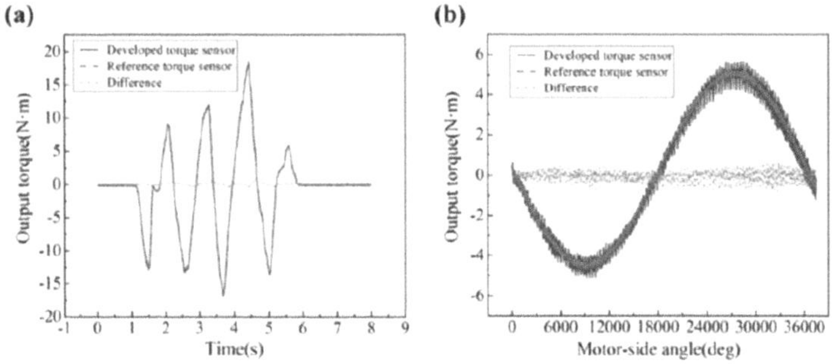

Fig. 6. (a) Output of the developed torque sensor while wave generator is fixed. (b) Output of the developed torque sensor while wave generator is rotating.

Table 2. Results at different positions of the harmonic reducer.

Positions (Circular spline)	Emax	RMSE
0	0.803	0.182
60	0.699	0.144
120	0.773	0.176
180	0.719	0.133
240	0.855	0.179
300	0.821	0.188

4 Conclusion

Torque sensor in-situ integrated in the harmonic reducer does not affect the original performance of the joint, making it an excellent scheme. However, the ripple signal exists while in-situ measurement. In this case, a strain rosette ring is employed as the sensitive unit for an in-situ torque sensor integrated in the harmonic reducer to suppress the interference. Also, an adhesive bonding setup is designed to ensure precise attachment of the strain rosette ring and the proposed adhesive bonding method is illustrated in detail. This setup facilitates the pickup, positioning and adhesive bonding of the strain rosette ring. The manufacturing process of the developed sensor can be automatically operated using the designed setup, making it suitable for mass production. To further address the interference induced by the squeezing action of the wave generator, a calibration method based on the joint angle information is proposed. Specificality, a step-by-step compensation approach is adopted, improving the measurement accuracy of the torque sensor and suppressing the nonlinearity of the developed sensor. According to the experimental results, the peak-to-peak error between the developed torque sensor and the reference torque sensor is 1.239 N·m, and the RMS error is 0.298 N·m as the wave generator rotates. As a continuation of the research, the influence of temperature needs to be discussed. Also, the performance of the developed sensor after long-term operation will be investigated.

Acknowledgments. This study was funded in part by the National Key Research and Development Program of China under Grant 2023YFB4704301 and in part by LiaoNing Revitalization Talents Program under Grant XLYC2002020.

Disclosure of Interests. The authors have no competing interests to declare that are relevant to the content of this article.

References

1. Avanzini, G., Ceriani, N., Zanchettin, A., et al.: Safety control of industrial robots based on a distributed distance sensor. IEEE Trans. Control Syst. Technol. **22**(6), 2127–2140 (2014)

2. Haddadin, S., Luca, A., Albu-Schäffer, A.: Robot collisions: a survey on detection, isolation, and identification. IEEE Trans. Rob. **33**(6), 1292–1312 (2017)
3. Huang, S., Gao, M., Liu, L., et al.: Collision detection for cobots: a back-input compensation approach. IEEE/ASME Trans. Mechatron. **27**(6), 4951–4962 (2022)
4. Wahrburg, A., Bös, J., Listmann, K., et al.: Motor-current-based estimation of cartesian contact forces and torques for robotic manipulators and its application to force control. IEEE Trans. Autom. Sci. Eng. **15**(2), 879–886 (2017)
5. Yu, W., Gong, J., Zhang, H., et al.: One-leg motion control of gecko robot based on three-dimensional force feedback. Chin. J. Sci. Instrum. **32**(12), 2674–2680 (2011)
6. Fu, B., Cai, G.: Design and calibration of a joint torque sensor for robot compliance control. IEEE Sens. J. **21**(19), 21378–21389 (2021)
7. Wei, N., Sun, H., Jia, Q., et al.: Analysis and design optimization of a compact and lightweight joint torque sensor for space manipulators. Adv. Mech. Eng. **5**, 111–114 (2013)
8. Min, J., Song, J.: Sensor block type joint torque sensor insensitive to crosstalk error. IEEE Sens. J. **20**(7), 3469–3475 (2020)
9. Kim, U., Maeng, C., Bak, G., et al.: High-stiffness torque sensor with a strain amplification mechanism for cooperative industrial manipulators. IEEE Trans. Industr. Electron. **69**(3), 3131–3141 (2021)
10. Nguyen, D., Kuo, C.: Design and optimization of a joint torque sensor for lightweight robots. IEEE Sens. J. **21**(8), 9788–9797 (2021)
11. Ma, J., Wu, H.: Sensorless collision detection for robot manipulator. J. Test Meas. Technol. **27**(3), 254–259 (2013)
12. Cao, F., Docherty, P., Ni, S., et al.: Contact force and torque sensing for serial manipulator based on an adaptive Kalman filter with variable time period. Robot. Comput.-Integr. Manuf. **72**, 102210 (2021)
13. Park, K., Kim, J., Park, J., et al.: Learning-based real-time detection of robot collisions without joint torque sensors. IEEE Robot. Autom. Lett. **6**(1), 103–110 (2020)
14. Yan, J., Jiang, Q., Zhang, Y., et al.: Self-uncoupling wrist six-axis force/torque sensor based on Fiber Bragg grating. Measurement **215**, 112827 (2023)
15. Yao, J., Zhang, H., Xiang, X., et al.: A 3-D printed redundant six-component force sensor with eight parallel limbs. Sens. Actuators, A **247**, 90–97 (2016)
16. Hashimoto, M., Kiyosawa, Y., Hirabayashi, H., et al.: A joint torque sensing technique for robots with harmonic drives. In: IEEE International Conference on Robotics and Automation, pp. 1034–1039 (1991)
17. Sensinger, W., Weir, R.: Improved torque fidelity in harmonic drive sensors through the union of two existing strategies. IEEE/ASME Trans. Mechatron. **11**(4), 457–461 (2006)
18. Taghirad, H., Belanger, P.: Intelligent built-in torque sensor for harmonic drive systems. IEEE Trans. Instrum. Meas. **48**(6), 1201–1207 (1999)
19. Jung, B., Kim, B., Koo, J., et al.: Joint torque sensor embedded in harmonic drive using order tracking method for robotic application. IEEE/ASME Trans. Mechatron. **22**(4), 1594–1599 (2017)
20. Pan, X., Wang, H., Jiang, Y.: Development of a built-in torque sensor in harmonic drive gears for robots. Chin. J. Sci. Instrum. **35**(1), 154–161 (2014)
21. Shen, W., Lou, Z., Du, Z., et al.: Torque sensor developed with a strain rosette ring for in situ integration in harmonic reducer. IEEE Sens. J. **24**(16), 25802–25813 (2024)

Application of Collaborative Robots in Automotive Lamp Production

Ning Li(✉)

AUBO (Beijing) Robotics Technology Co. Ltd., Beijing 102300, China
lining@aubo-robotics.cn

Abstract. Intelligent collaborative robots have become more and more important in the field of industrial automation. Having many advantages over the conventional industrial robots, the AUBO smart collaborative robots meet the development requirements of the domestic manufacturing industry. As is known, robots are widely applied in manufacturing the overall frameworks of automobiles. However, they are seldom used in auto-parts factories. To enhance the automation level in this field, the AUBO smart collaborative robots are applied to develop automatic production lines for a car-lamp factory. In the past several months, the automated production lines work well and yield remarkable economic and social benefits. These production lines provide good references for the transformation and upgrading of the car-lamp industry.

Keywords: Intelligent collaborative robots · Car-lamp · Automated production lines · Manufacturing · Transformation and upgrading

1 Introduction

As a typical industrial manufacturing industry, the automation production level of the automotive industry is far ahead of other industries. More than one-third of global industrial robots are used in automobile manufacturing, playing an irreplaceable role in welding, stamping, painting and other processes [1, 2]. However, most automotive parts companies are still labor-intensive, relying on manual operation of specialized equipment to produce small and lightweight non-core components. With the acceleration of automobile updates and replacements, the demand for flexibility in parts production has surged. Traditional heavy-duty non cooperative industrial robots are difficult to adapt to the requirements of quickly replacing tooling equipment due to their large size and lack of flexibility (Table 1).

Currently, most automobile parts companies are still labor-intensive, and in automobile manufacturing, many parts are mainly produced through commissioned processing. Compared to automobile manufacturers, the automation production level of most commissioned factories is still relatively low. As is well known, most automotive parts are non-core components that are small in size and light in weight. For this type of non-core components, traditional heavy-duty non cooperative industrial robots often have difficulty functioning, and their production and manufacturing generally rely mainly on

H. Wei and J. Niu (Eds.): COBOT 2025, CCIS 2893, pp. 15–27, 2027.
https://doi.org/10.1007/978-981-92-2074-8_2

Table 1. Core Issues Faced

Question Type	Specific Manifestations
Low Level of Automation	Commissioned processing factories rely on manual operations, resulting in low efficiency
Insufficient Flexible Production	Traditional robots are difficult to adapt to production lines with frequent tool changes
High Equipment Cost	Traditional robots have high procurement and maintenance costs
Potential Safety Hazard	Traditional robots in human-machine collaboration scenarios are prone to collision accidents

manual operation of specialized equipment. In addition, with the increasing frequency of automobile updates and replacements, the flexibility of automobile parts production is becoming higher and higher, and the replacement frequency of tooling equipment on the production line is also getting faster and faster. Therefore, traditional heavy-duty non cooperative industrial robots are often unable to meet the requirements of flexible production.

2 Development and Characteristics of Collaborative Robots

Collaborative industrial robots are the new generation of industrial robots. According to their working methods, industrial robots can be divided into two types: non cooperative and cooperative. Traditional industrial robots generally belong to non-cooperative types and have poor safety. When they collide with humans, they often cause injury accidents, so they generally need to work in isolated environments (such as guardrails). This situation seriously restricts the development and application of industrial robots, because in industrial production, besides sometimes using robots to replace humans to independently carry out work, in more situations, robots and workers need to cooperate and collaborate in the same workspace to achieve complementary advantages between humans and machines. For example, robots can carry out repetitive operations while humans can perform creative labor. In order to eliminate the above-mentioned drawbacks of traditional industrial robots, in the past decade, people have used high-precision force sensor technology and force control technology to design and manufacture a new generation of industrial robots that can work collaboratively with humans in the same environment, namely collaborative industrial robots [3–5]. Due to its collision detection and force control functions, it can automatically stop and achieve safety protection in the event of a collision. Collaborative industrial robots not only possess the speed and accuracy of traditional industrial robots, but also have outstanding advantages such as safety, reliability, plug and play, simple operation, convenient maintenance, and low cost that the latter do not have. Therefore, it has a broader application prospect than traditional non cooperative industrial robots and is an irreplaceable important automation equipment in industrial manufacturing. The internationally renowned investment bank

Barclays Bank predicts that by 2020 and 2025, the global market demand for collaborative industrial robots will reach 150000 and 700000 units respectively, with market values of $3 billion and $12 billion, respectively. Its industrial scale will far exceed that of traditional industrial robots.

Collaborative robots have gradually become an increasingly important development direction for industrial robots. A series of collaborative industrial robot brands have been launched both domestically and internationally, such as UR (Universal Robots) from Denmark [6], AUBO from China, Rethink from the United States, Franka from Germany, Elibot from China, and so on. Among the aforementioned brands, UR and AUBO have already achieved large-scale mass production. From a global market perspective, UR still dominates [7, 8]. However, in the Chinese Mainland market, AOBO's share has been basically equal to that of UR. Since the 2017 AUBO smart collaborative robots passed the EU CE, North American NRTL, and South Korean KCs technology safety certification, AUBO's sales in the international market have been continuously increasing, showing a strong development momentum.

As the first mass-produced and export qualified lightweight collaborative robot in China, the AUBO smart collaborative robots has the following outstanding features.

2.1 Core Technology is Independently Controllable

Among domestic collaborative robots, AUBO i5 is the first national brand to achieve complete localization. It is currently the only "pure domestic" collaborative robot in China with all intellectual property rights. From operating systems to application software, from controllers to core components, they are all independently developed by AUBO Company. The biggest advantage of localization lies in its autonomy and controllability, which has enabled China to successfully overcome the long-standing dilemma of being constrained by the "four major families" of international robots in the core technology of collaborative robots. Due to its autonomy and controllability, AUBO collaborative robots are particularly suitable for automated production of equipment in safety critical departments such as national defense and military industry, as well as auxiliary automated maintenance of military weapons and equipment.

2.2 Cost-Effective

In terms of key performance, the AUBO smart collaborative robots has fully reached the level of similar international products. For example, the AUBO i5 collaborative robot has a repeat positioning accuracy of ± 0.02 mm, a continuous operating life of 30000 h, a maximum angular velocity of 223°/s, a joint rotation range of $\pm 360°$, and a working radius of 886.5 mm. The above performance indicators are at the same level as international collaborative robot brands such as UR, Rethink, and Franka. However, due to being completely domestically produced, its price is only half or even one-third of similar international products. Therefore, from a global perspective, AUBO collaborative robots have a high cost-effectiveness advantage among similar products.

2.3 Security

Based on the visual perception and multi axis force perception technology independently developed by AUBO Company, the AUBO smart collaborative robots have achieved the requirements of adjustable stiffness and compliant control. When obstacles unexpectedly appear in their operating environment, the AUBO smart collaborative robot can automatically reduce its running speed; when unexpected collisions occur with obstacles, it can stop in a timely manner to maximize collision protection. At present, the AUBO i-series robots have passed the EU CE, North American NRTL, South Korean KCs, and China CR technology safety certification, fully meeting the safety protection requirements for human-machine cooperation.

2.4 Plug and Play

AUBO Smart Collaborative Robot provides drag teaching function and a visual graphical user operation interface. Users only need to perform simple operations on the interface, and can use the online automatic programming function provided by the system to manually drag and teach to set the robot's running trajectory. Therefore, compared with equipment such as CNC machine tools, the AUBO Collaborative Robot meets the requirements of plug and play. Users do not need to master specialized knowledge and technology in the field of robotics, nor do they need to perform complex manual programming. They can proficiently use the AUBO Collaborative Robot to carry out various operations by simply dragging and teaching.

2.5 Easier to Be Extended

Perhaps no robot can fully meet the constantly changing practical needs of users. Therefore, in robot applications, users sometimes need to expand the functionality or even adjust the structure of the robot. For users with a foundation in robotics technology, this is especially true. However, the vast majority of existing industrial robots have not provided users with secondary development interfaces, making it difficult or even impossible to expand their functions. To solve this problem, the AUBO Smart Collaborative Robot provides users with free open bus protocols, IO extension interfaces, and software secondary development API interfaces, and its control system supports ROS development. In addition to functional expansion, users can also make structural adjustments. The AUBO Collaborative Robot adopts a modular design in structure. By using joint modules and linkage modules, users can change the number of degrees of freedom of the robotic arm according to their own needs. Therefore, compared to other industrial robots, the AUBO Collaborative Robot has the advantages of easy expansion in both function and structure.

2.6 Artificial Intelligence

Although artificial intelligence has been widely applied in the fields of service and educational robotics, its application in industrial robotics is still very limited. In recent years, AUBO Company has comprehensively improved the artificial intelligence level of its collaborative robots from the following aspects:

The AUBO collaborative robot is equipped with an intelligent vision system, which improves obstacle recognition function, enhances assembly flexibility and accuracy. The AUBO collaborative robot and intelligent vision system have been deeply integrated. Traditional robotic arm programming usually uses a teaching method with multiple arm motion points, which is less flexible and requires a large number of fixtures to fix the workpiece or surrounding processing machinery. Once an earthquake or external force changes the relative position of the arm and objects in the work area, all action points need to be re taught. In contrast, the combination of AUBO collaborative robots and machine vision can bring significant advantages. Through visual recognition and compensation capabilities, we can flexibly adjust the position of the arm, effectively reducing the use of fixtures. Common hand eye relationships include two types: eye to hand and eye in hand. In eye to hand mode, the camera is hung on the end axis of the arm; In eye in hand mode, the camera and arm are fixed in two separate positions. The hand eye relationship is the key to the coordination between the robotic arm and the camera. AUBO adopts an eye to hand mode for fast and efficient camera calibration, (see Fig. 1).

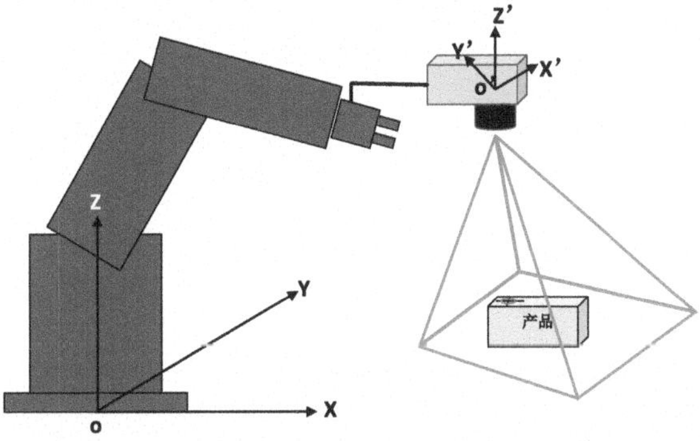

Fig. 1. Eye to Hand

OXY is the coordinate system of the robotic arm (see Fig. 2), and O'X'Y' is the camera coordinate system. Theta is the angle between two coordinate systems. Assuming the position of point P on the image is as shown in the figure, then P has a coordinate in the robot coordinate system and a coordinate in the image coordinate system [9].

Any point on the image can be converted into a coordinate point on the robotic arm's coordinate system through the built-in vision of the AUBO robotic arm: P (Machine) = f (P (Image)). OXY is the coordinate system of the robotic arm, and O'X'Y' is the transformation relationship of the camera coordinate system.

$$x = x' * r * \cos(\text{theta}) - y' * r * \sin(\text{theta}) + x0 \quad (1)$$

$$y = x' * r * \sin(\text{theta}) + y' * r * \cos(\text{theta}) + y0 \quad (2)$$

where r is the millimeter to pixel ratio, (mm/pixel) is the number of pixels corresponding to one millimeter, theta is the angle between two coordinate systems, and (x0, y0) is the

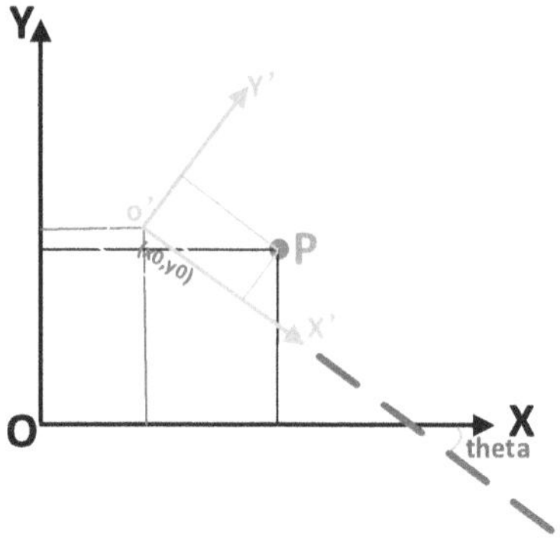

Fig. 2. Hand-eye Calibration

distance from the image coordinate origin to the mechanical coordinate origin. Simplify abstract formulas, assuming:

$$a = r * \cos(\text{theta}) \tag{3}$$

$$b = r * \sin(\text{theta}) \tag{4}$$

$$c = x0 \tag{5}$$

$$d = y0 \tag{6}$$

Obtain a simplified formula:

$$x = x' * a - y' * b + c \tag{7}$$

$$y = x' * b + y' * a + d \tag{8}$$

To solve for the unknowns a, b, c, and d in this equation, at least two sets of corresponding coordinate points are required, namely the corresponding points in the robot coordinate system and the image coordinate system. For example, the following two sets of corresponding relationships can be set:

Group 1: Image coordinate points (xImage1, yImage1) correspond to mechanical coordinate points (xMachine1, yMachine1).

Group 2: Image coordinate points (xImage2, yImage2) correspond to mechanical coordinate points (xMachine2, yMachine2).

By using these two sets of corresponding relationships, the values of a, b, c, and d can be solved, thus quickly establishing a transformation relationship between the image coordinate system and the robot coordinate system.

Added voice control function in some scenarios to improve user experience.

We have built a remote operation and maintenance platform for robots, based on big data analysis technology, to achieve remote monitoring of robot health status, analysis and evaluation of typical faults, and online upgrading of software systems.

3 Collaborative Robots Break Through Production Pain Points

The demand for collaborative robots in the automotive parts industry is mainly reflected in the following aspects:

Flexible production and lightweight design of collaborative robots (with a load capacity of less than 10 kg) meet the production needs of small accessories and support rapid tool replacement. For example, AUBO customized an automated production line for a certain car lighting enterprise, which achieved flexible adjustment of the production line through modular expansion, and quickly adapted to the needs of multiple varieties and small batches.

Cost optimization, collaborative robots are priced at only 1/2 to 1/3 of traditional robots, significantly reducing enterprise equipment investment. The localization advantage of AUBO i5 further reduces supply chain costs.

Safe human-machine collaboration, based on visual perception and multi axis force perception technology, collaborative robots can detect the position of workers in real time and adjust their operating speed to avoid collision accidents. The AUBO i5 has passed international safety certifications such as EU CE and North American NRTL, fully meeting the requirements for human-machine cooperation safety.

Intelligent upgrade, AUBO collaborative robot open bus protocol and software secondary development API interface, supporting users to customize functions, realizing full process logistics automation, improving production efficiency and quality, and increasing production line feeding response speed to minute level.

After a certain car lighting company introduced the AUBO i5 collaborative robot, labor costs were reduced by 30%, and the production line switching time was shortened from 2 h to 15 min, quickly achieving the transformation from labor-intensive to intelligent manufacturing. This practice has verified the irreplaceability of collaborative robots in the automotive parts industry.

4 Application Case of AUBO Intelligent Collaborative Robot

To address these industrial challenges, collaborative robots offer flexible automation solutions. Their adaptive capabilities are particularly suited for complex manufacturing environments.

4.1 The Current Situation and Problems

The types and models of car headlights are diverse, with small and numerous components and complex shapes and structures (see Fig. 3). The production and manufacturing process involves many processes such as injection molding, spray painting, gluing, coating curing, welding, coating, ultrasonic cleaning, component surface treatment, electronic device installation, and assembly, with strong professionalism. Due to the reasons above, the production of most car lights currently relies on manual operation of specialized equipment by workers, resulting in low automation levels and high labor intensity in the production process. The manual assembly line of a certain car lamp enterprise (see Fig. 4). It can be seen that the assembly line of existing car lights basically does not have

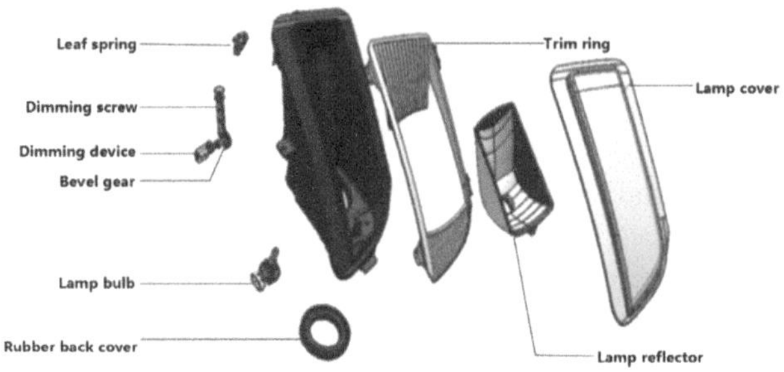

Fig. 3. Construction of front fog lights for automobiles

Fig. 4. Handmade assembly line of a certain car lamp enterprise

automated equipment, and all assembly process operations are manually completed by workers.

Research shows that there are five typical problems with existing manual production lines:

Low production efficiency and insufficient production capacity make it difficult to meet the explosive growth in automotive sales in recent years.

The product consistency is not good enough, the production quality is not stable enough, and the scrap rate is high.

The flexibility of the production line is low, and every update of the car light design requires a lot of time for personnel training.

The demand for workers is high, and the problem of difficult recruitment is becoming increasingly prominent in the current situation where the demographic dividend is gradually disappearing and labor force is becoming increasingly scarce.

The production process is monotonous, repetitive, boring, and fast-paced, causing extreme visual fatigue and physical and mental overload damage to workers.

4.2 Automation Upgrade of Car-Lamp Production Lines

Among all the car lamp production lines of the enterprise, the assembly lines have the most complex production process, the highest labor intensity, and the most urgent need for transformation and upgrading. Therefore, we first start with the assembly line to automate the car light production line of the enterprise [10, 11].

Taking the automation transformation plan of the front fog lamp assembly line of automobiles as an example, the following will be introduced. This assembly line is a fully automated system built from the main body of the AUBO smart collaborative robot, intelligent vision system, intelligent detection system, intelligent control system, automatic feeding equipment, thread locking equipment, gluing equipment, anti-static conveyor belt, etc. The entire assembly line adopts modular design to meet the needs of flexible production. The AUBO smart collaborative robot is equipped with an intelligent vision system, making assembly operations more flexible, efficient, and accurate. The assembly line has added an intelligent detection system to strengthen the inspection of assembly quality (see Fig. 5). The entire assembly line is controlled by an intelligent control system, achieving one click start and one click shutdown operations, and the assembly process does not require manual intervention.

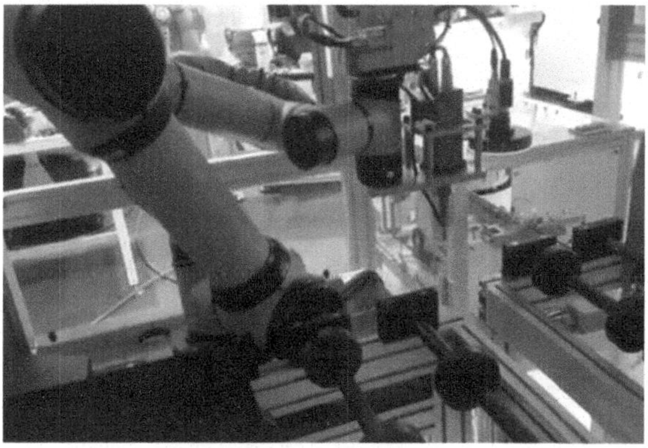

Fig. 5. Collaborative Robot Vision System

The front fog light of a car is composed of a shell (see Fig. 3), a dimming device, a dimming screw, an umbrella gear, a spring plate, a reflector, a decorative ring, a lampshade, a light bulb, a rear cover, a breathable cover, and other parts. The automated production line is divided into two parts (see Fig. 6). The first part (see Fig. 3) mainly completes the assembly of the shell, dimming device, dimming screw, umbrella gear, spring plate, reflector, decorative ring, and lampshade, including 12 processes such as loading, dust removal and collection, shell installation, dimming device installation, dimming screw installation, umbrella gear pressing, spring plate locking, dimming device, reflector and decorative ring installation, gluing, lampshade pressing, and unloading, Using a total of 12 AUBO smart collaborative robots; The second part (see Fig. 7) mainly completes the assembly of the light bulb, back cover, and breathable cover, including 9 processes such as loading, rolling of the rubber back cover, light bulb assembly, dimming, CCD electrical inspection, airtightness detection, installation of the breathable cover, marking, and cutting, using a total of 9 AUBO smart collaborative robots [12, 13].

Fig. 6. Automatic assembly line for front fog car-lamp Part 1

Fig. 7. Automatic assembly line for front fog car-lamp Part 1

Fig. 8. Automatic assembly line for front fog car-lamp Part 2

4.3 The Operation Effect

As the first phase of the transformation and upgrading of the car lamp production line, the automation transformation of the car lamp assembly line started in the first half of 2024, was completed and put into use by the end of the year. For several months, the assembly line has been running steadily and has played a significant role as follows.

Eliminated outdated production capacity, significantly improved production efficiency, and better met the demand for large-scale shipments.

The product consistency has been fully guaranteed (see Fig. 8), the production quality has been significantly improved, the occurrence of waste has been completely avoided, consumption has been reduced, and costs have been saved (Fig. 9).

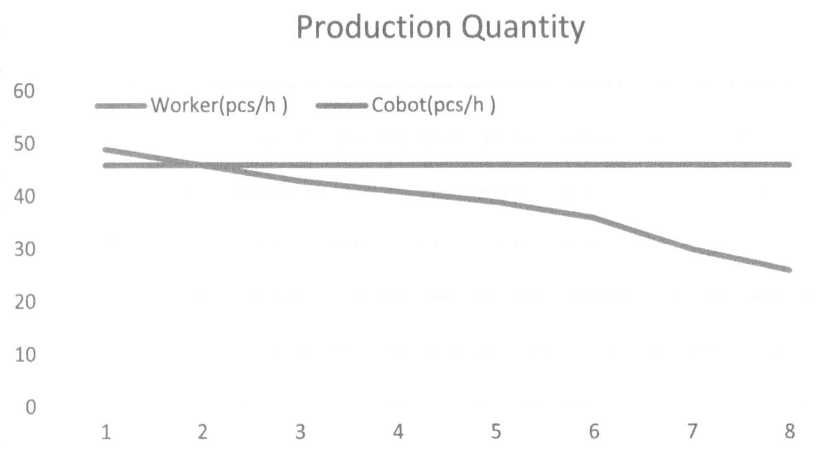

Fig. 9. Comparison of Production Efficiency

The level of flexible manufacturing has greatly improved, and the time for each update of the production line has been shortened from half a year to half a month, accelerating the transformation and upgrading of enterprises.

Implemented unmanned and intelligent production, alleviating the problem of difficult recruitment for enterprises; liberating productivity, reducing work-related accidents, and achieving the goal of putting people first.

5 Conclusion

As a new generation of industrial robots, collaborative robots play a crucial role in the transformation and upgrading of the manufacturing industry. Due to its outstanding cost performance advantage and independent and controllable technology, the AUBO Smart Collaborative Robot has become an international brand in the same field and a popular automation equipment in the industrial manufacturing industry. Automobiles are a pillar industry in the manufacturing industry, and the level of automation in domestic vehicle manufacturing is basically on par with developed countries. However, domestic automobile parts production still generally relies on manual operation of specialized machines. Aiming at the transformation and upgrading development opportunities of the automotive parts manufacturing industry, relying on the AUBO i-series intelligent collaborative robot, AUBO Company has developed an automatic production line for car lights for a domestic enterprise. The production line has been running smoothly for several months, bringing significant economic and social benefits to the enterprise. As long as the construction plan and operation mode of the production line are slightly modified, it can be promoted and applied to different car lighting enterprises, thereby promoting the overall automation level of the domestic car lighting industry to achieve rapid improvement.

References

1. Hou, X., Yang, Z.: On the application of industrial robots in the field of automotive manufacturing. Times Autom. **2**, 69–70 (2018)
2. Gu, Z.: On the application of robots in automotive intelligent manufacturing. China High Tech Zone **3**, 32 (2018)
3. Ren, Y.: Human machine collaboration to create a new pattern of intelligent manufacturing. Shanghai Inf. Technol. **3**, 46–49 (2018)
4. Hull, T., Minarcin, M.A.: Considerations in collaborative robot system designs and safeguarding. SAE Int. J. Mater. Manuf. **9**(3), 545–551 (2016)
5. Brown, A.S.: Rethink launches precise collaborative robot. Mech. Eng. **137**(12), 10 (2015)
6. Zhao, C.: Research on the Safety Control System of Human Machine Collaborative Robots. Northeastern University, Shenyang, pp. 2–4 (2019)
7. Si, N.: Making manufacturing processes more beautiful - universal robots UR5 and UR10 robots entering the Chinese market. Eng. Mach. Maint. **10**, 100 (2012)
8. Pransky, J.: The Pransky interview: Dr Esben Ostergaard, inventor, co-founder and CTO of Universal Robots. Ind. Robot-Int. J. **42**(2), 93–97 (2015)
9. Yellow base Research on Soft Control of Collaborative Robots with Visual servoing. University of Electronic Science and Technology of China, Chengdu, pp. 29–30 (2024)
10. Song, Y.: Research on the assembly process of automotive lights based on mechanical manufacturing. Taxation (01), 248 (2018)
11. Ke, Y., Zhou, B.: Improvement of automotive lamp assembly line based on industrial engineering technology. Mech. Manuf. (5) (2017)
12. Yu, Z., Pan, M., Liu, Q.: Design of an automated system for screw locking of car lights based on dual robot collaboration. China Light. Appl. (09), 65–66 (2025)
13. Zhang, H.: Design and implementation of a machine vision based screw automatic assembly system. Inf. Commun. (11), 78–79 (2020)

Deformation Flexible Gear in Self-sensing Harmonic Reducers

Ya Chen, Wenxiu Zhang, Shujing Liu(✉), Yinxiao Chen, Wei Guo, Dianjun Wang, and Liang Wang

College of Mechanical Engineering, Beijing Institute of Petrochemical Technology, Beijing 102617, China

Abstract. The self-sensing harmonic reducer integrating a torque sensor on the flexible gear is one of the methods to solve the technical problems of integrated joint torque measurement for collaborative robots. Due to the extremely complex deformation and strain conditions of the flexible gear during movement, the measurement of the output torque is difficult. Studying the deformation and strain laws of flexible wheels during motion is particularly important for determining a reasonable torque measurement scheme and improving the accuracy of torque measurement. Starting from the transmission principle of the harmonic reducer, the mechanical characteristics of the flexible gear under no-load and load conditions were systematically analyzed, and the theoretical equations of the omnidirectional displacement and stress distribution of the flexible gear at any cross-section were derived. The static simulation of the flexible wheel was carried out through ANSYS-Workbench software. Meanwhile, by setting four key observation paths, the distribution law of the maximum deformation at different cross-sections of the flexible wheel was clarified, and the bonding position of the strain gauge was optimized to obtain the high-precision torque signal. Then, the transient dynamics simulation analysis was carried out to obtain the stress distribution and deformation of the flexible gear under different working conditions and different rotational speeds. The transient dynamics simulation results are closer to the actual working conditions than the static analysis. The research results provide a theoretical basis for the torque measurement scheme of the self-sensing harmonic reducer.

Keywords: Integrated joint torque measurement · Self sensing harmonic reducer · Stress and deformation analysis of flexible wheels · Finite element simulation

1 Introduction

Traditional collaborative robot joints require the installation of torque sensors at the joint output end to measure output torque. This, however, leads to reduced joint stiffness, increased volume, weight, and structural complexity. To address these issues, this paper studies a self-sensing harmonic reducer that integrates a torque sensor on the flexspline for torque measurement. This self-sensing harmonic reducer uses strain gauges attached to the flexspline to obtain deformation signals for torque calculation.

H. Wei and J. Niu (Eds.): COBOT 2025, CCIS 2893, pp. 28–50, 2027.
https://doi.org/10.1007/978-981-92-2074-8_3

To establish the relationship between strain and torque, numerous studies have been conducted by researchers domestically and internationally. Wang Peisheng [1] established a model of a hat-shaped harmonic reducer including a complete flexible bearing and used the finite element method to analyze the stress distribution and deformation patterns of both components under different loads. Wu Can [2] took the CSG-32-80 harmonic reducer as the research object and conducted transient analysis on the flexspline using Ansys Workbench to obtain its stress and strain distribution. Wang Xiaodong [3] proposed a design scheme for a torque sensor integrated inside the harmonic reducer. Through finite element analysis, the relationship between the strain at the flexspline bottom and the applied torque was clarified, and a response surface model between the structural parameters of the sensitive grid and the Pearson correlation coefficient of strain-torque was established to determine the optimal parameters. The Soviet scholar Ivanov [4] equivalent the flexspline to a smooth ring model to calculate deformation and deformation forces, and built a simplified calculation model for the force situation of the gear teeth. Furthermore, he emphasized the impact of bending stress on flexspline strength, focusing on the stress distribution at the junction of the gear ring and the cylinder. However, this method results in stress concentration at the arc junction, leading to insufficient accuracy. Shen Yunwen [5], consistent with Ivanov's approach in studying the geometric deformation and stress distribution of harmonic reducers, simplified the flexspline model through extensive experiments, abandoned the complex flexspline tooth profile, and used a smooth cylindrical shell considering friction to explore the stress and deformation state of the flexspline cylinder. This yielded a simplified calculation model that revealed the intrinsic relationship between the flexspline's structural parameters, applied loads, and its cylindrical deformation. Due to the complexity of the flexspline structure and the miniature nature of its tooth features, varying degrees of simplification are generally required to facilitate subsequent analysis and calculation. Fu Junfeng [6, 7] initially did not consider the circular spline, equivalent the flexspline gear ring to an equal-thickness cylinder, and established a rigid-flexible contact model between it and the wave generator before proceeding with circular spline contact analysis. However, this approximate method does not align with reality, and the analysis results may contain significant errors.

W. Shen [8] developed a torque sensor using a strain gauge rosette ring as the sensing element to solve the problem of ripple signal interference in torque measurement caused by the periodic extrusion of the wave generator. M. A. A [9] proposed a simplified sensorless torque estimation method to address the high cost and system complexity of traditional sensor-based solutions. Based on the structural damping friction characteristics of the harmonic reducer, a simplified linear dynamic model was constructed to estimate transmitted torque by measuring the torsional angular rate. Ostapski [10, 11] improved the Newton-Raphson method, providing a solution for the elastic deformation problem of thin-walled shell structures with complex shapes. This method has been applied to the study of harmonic drives, and a finite element 3D model more suitable for ANSYS calculation was designed and simulated, making the simulation results closer to the actual situation. Gao Haibo et al. [10] used the APDL language to build a parameterized equivalent contact model for the cup-shaped harmonic reducer flexspline. Through finite element simulation of a simplified contact model of the thermally coupled flexspline and circular spline, they obtained the variation curve of the maximum stress in the flexspline with temperature. Kikuchi [11] equivalent the rolling ball bearing to an elastic element in the finite element analysis but did not consider the influence of friction factors. H Dong [12, 13] equivalent the flexspline to an equal-thickness shell and used the APDL module to establish and optimize a parametric finite element analysis model for the harmonic reducer, solving the spatial elastic deformation problem of cup-shaped flexsplines with arbitrary parameters. Yang Jian [14] studied the strength and failure mechanism of harmonic reducers, considered the influence of the flexible bearing, established a simplified 3D model, and performed finite element simulation calculations for the reducer assembly and dynamic operation process, focusing on analyzing the deformation response and stress distribution regularity of the flexspline. Jeon H S et al. [15] obtained the relationship between the vibration of the harmonic reducer and the stress and deformation of the flexspline through simulation analysis of the flexspline.

In view of this, this study aims to establish a three-dimensional model of the harmonic reducer to replicate the state of the flexspline under actual working conditions. Considering the transmission characteristics of the flexspline under different working conditions, theoretical models for flexspline deformation and stress will be constructed. Finite element simulation calculations will be performed to verify the distribution patterns of flexspline deformation and stress, with the goal of obtaining more accurate and reliable analysis results.

2 Analysis of Mechanical Characteristics of the Flexspline in Self-sensing Harmonic Reducers

After the harmonic reducer is assembled, the rotation of the wave generator causes the flexspline to deform. Variations in the type of wave generator lead to different deformation characteristics and stress distribution patterns in the flexspline. Among the various types of wave generators, this study focuses on the elliptical cam wave generator as the research object. Since many factors influence the deformation of the flexspline during actual operation, the following analysis examines its transmission deformation and stress behavior under both no-load and load conditions.

After assembly, the flexspline is compressed and deformed by the wave generator. A microelement (ABCD) is selected on the circular ring of the flexspline as the object of study, and a force analysis diagram under no-load conditions is established, as shown in Fig. 1.

Through analysis, the stress distribution and displacement of the flexspline under no-load conditions can be obtained as follows [18]:

$$\begin{cases} \omega = \omega_0 \cos 2\theta \\ v = \frac{3\rho\omega_0}{2r_m} \sin 2\theta - 2\omega_0 \sin 2\theta \end{cases} \tag{1}$$

$$\begin{cases} \varepsilon_\rho = 0,\ \tau_{\rho\theta} = 0,\ \gamma_{\rho\theta} = 0 \\ \varepsilon_\theta = 3\omega_0 \cos 2\theta \left(\frac{1}{\rho_m} - \frac{1}{\rho}\right) \end{cases} \tag{2}$$

$$\begin{cases} \sigma_\rho = \frac{3\omega_0 E\mu \cos 2\theta}{(1-2\mu)(1+\mu)} \left(\frac{1}{\rho_m} - \frac{1}{\rho}\right) \\ \sigma_\theta = \frac{3\omega_0 E(1-\mu) \cos 2\theta}{(1-2\mu)(1+\mu)} \left(\frac{1}{\rho_m} - \frac{1}{\rho}\right) \end{cases} \tag{3}$$

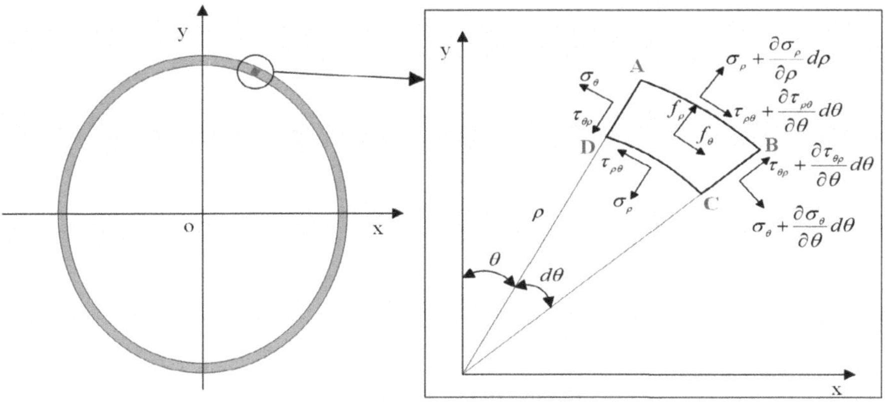

Fig. 1. Microelement force analysis diagram

In the equation: ω represents the radial displacement, v the circumferential displacement, ε_ρ the radial normal strain, $\tau_{\rho\theta}$ the shear stress, ε_θ the circumferential normal strain, $\gamma_{\rho\theta}$ the shear strain, σ_ρ the circumferential normal stress, and σ_θ the radial normal stress.

Under load conditions, the flexspline cup is simultaneously influenced by the wave generator and the circular spline, resulting in complex and irregular deformation characteristics. The flexspline tooth ring is affected by the applied load torque and can no longer maintain a stable elliptical shape as in the no-load state. Portions of the flexspline lose contact with the wave generator, causing significant distortion in these regions [15], as shown in Fig. 2.

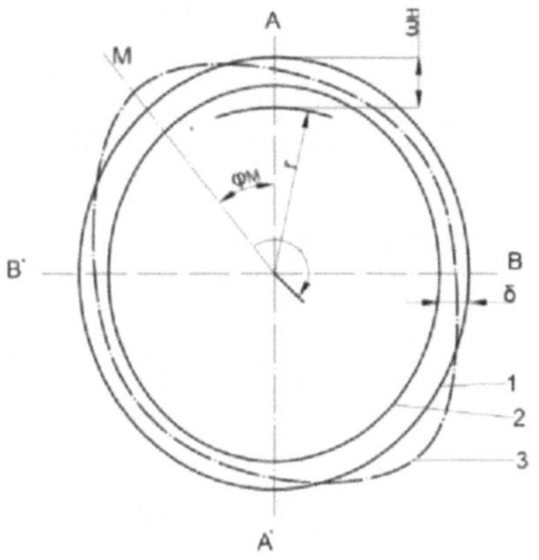

Fig. 2. Deformation diagram of flexible wheel under load

Under load conditions, the displacement and deformation pattern of the arc segment AB can be expressed as [19]: For the interval $0 \leq \varphi \leq \varphi_M$:

$$\begin{cases} \omega = \omega_{HO} - \delta - \delta_{KO} \\ v = -(\omega_{HO} - \delta - \delta_{KO})\varphi \\ \theta = -\frac{1}{r}(\omega_{HO} - \delta - \delta_{KO})\varphi \end{cases} \tag{4}$$

For the interval $\varphi_M \leq \varphi \leq \pi/2$:

$$\begin{cases} \omega = (\omega_{HO} - \delta - \delta_{KO})\begin{bmatrix} B_1\left(\frac{\pi}{2} - \phi\right)\cos\phi \\ +B_2\sin\phi + C \end{bmatrix} \\ v = -(\omega_{HO} - \delta - \delta_{KO})\begin{bmatrix} B_1\left(\frac{\pi}{2} - \phi\right)\sin\phi - \\ (B_1 + B_2)\cos\phi \\ +C\phi + C_1 \end{bmatrix} \\ \theta = \frac{1}{r}(\omega_{HO} - \delta - \delta_{KO})(2B_1\cos\phi - C\phi - C_1) \end{cases} \tag{5}$$

In the equation: ω represents the radial displacement, v the circumferential displacement, θ the normal rotation angle, ω_{HO} the initial radial displacement, δ the maximum radial displacement, δ_{KO} the total radial clearance of a standard bearing, B_1, B_2, C_1, C, and φ_M are constants.

During the harmonic gear transmission process, the flexspline undergoes changes in deformation, number of engaged teeth, and meshing area. The boundary effects within the contact region are difficult to estimate accurately, so the stress calculation is usually simplified. When the calculation length of the flexspline is taken as (L), the stress expression of the flexspline under load conditions can be written as follows [20]:

$$\sigma_Z = K_{rt}K_MK_dC_\sigma\frac{v\omega_0E\delta}{r_m^2} \tag{6}$$

$$\sigma_\varphi = K_{rt}K_MK_dC_\sigma\frac{\omega_0E\delta}{r_m^2} \tag{7}$$

$$\tau_{\varphi t} = K_{rt}K_MK_dC_\tau\frac{\omega_0E\delta}{r_mL} \tag{8}$$

The shear stress generated by the applied torque is expressed as shown in Eq. (9):

$$\tau_T = K_{pr}K_d\frac{T}{2\pi r_m^2\delta} \tag{9}$$

In the equation: K_M is the stress amplification factor caused by flexspline distortion; K_d is the dynamic load factor; K_{pr} is the nonuniform shear stress distribution coefficient; K_{rt} is the stress amplification factor related to the influence of the tooth ring; C_σ is the normal stress coefficient; C_τ is the shear stress coefficient; δ is the wall thickness at the flexspline tooth ring position; v is the Poisson's ratio of the material; and E is the elastic modulus of the flexspline's material.

To visualize the deformation behavior of the flexspline during the operation of the harmonic reducer, simulations were conducted using MATLAB. The resulting circumferential, radial, and axial displacements are shown in Figs. 3, 4, and 5, respectively.

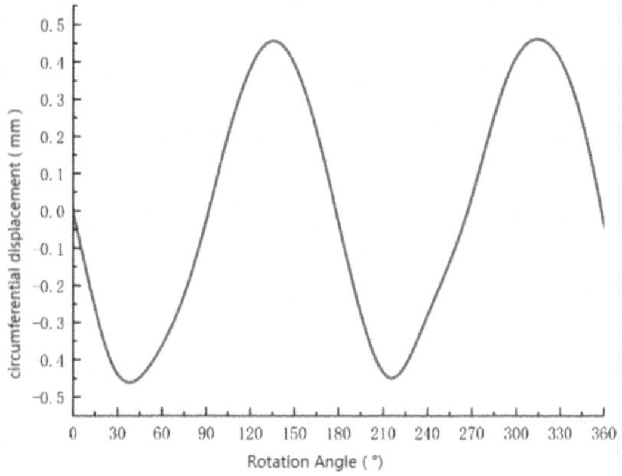

Fig. 3. Circumferential displacement trend chart

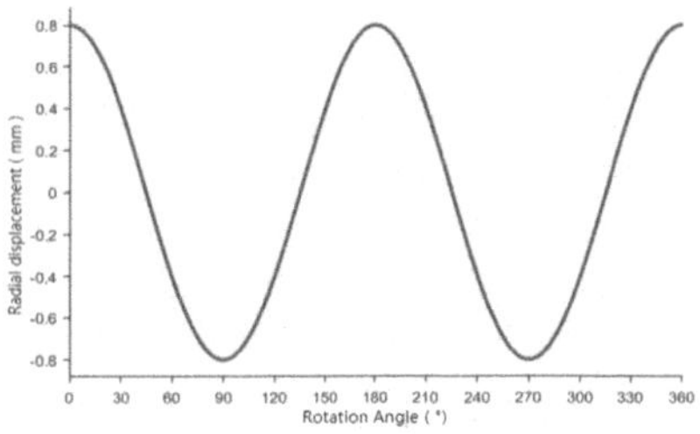

Fig. 4. Radial displacement trend chart

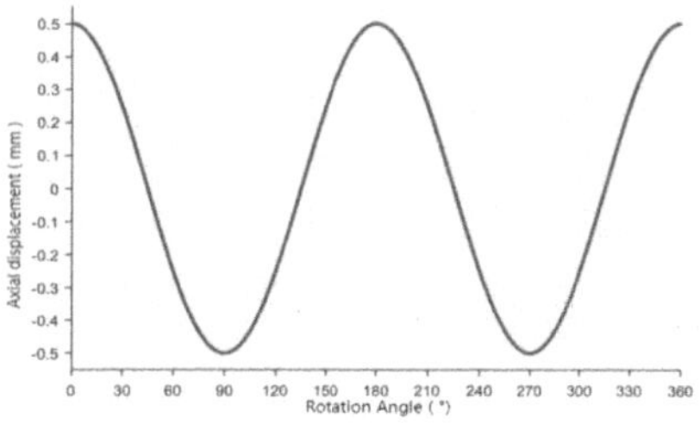

Fig. 5. Axial displacement trend chart

3 Static Simulation of the Self-sensing Harmonic Reducer

This article selects the LHSG-25-120-C-I-S11D model harmonic reducer from Suzhou Lead Harmonic Drive Co., Ltd. The flexspline is made of 40CrNiMoA material, the circular spline is made of 45 steel, and the simplified wave generator has lower material requirements, which can be made of 40Cr. The specific parameters are shown in Table 1 below.

Table 1. Harmonic reducer material parameters

Material model	Elastic Modulus (GPa)	Poisson's Ratio	Density (kg/m^3)	Yield Strength (MPa)
40CrNiMoA	209	0.295	7.85×10^3	850
45	209	0.269	7.89×10^3	360
40Cr	1500210	0.29	7.85×10^3	1000

The Johnson-Cook [16] constitutive equation for material 40Cr based on quasi-static and dynamic experiments [17]

$$\sigma_{eq} = \left(500 + 197.39\varepsilon_{eq}^{0.9184}\right)\left(1 + 0.03\ln\dot{\varepsilon}^{*}\right)\left(1 - T^{*0.891}\right) \quad (10)$$

where, $\dot{\varepsilon}^{*} = \dot{\varepsilon}/\dot{\varepsilon}_0$is reference strain rate, $\dot{\varepsilon}_0 = 10^{-3}\text{s}^{-1}$.

$T^{*} = \frac{T-T_r}{T_m-T_r}$is reference temperature, T_r =20 °C, Melting point $T_m = 1500$ °C.

Under constant stress amplitude fatigue conditions, the strain output of 40Cr material is simultaneously influenced by the coupling effects of its fatigue performance and temperature rise. From the perspective of fatigue performance, the tempered sorbite (metastable microstructure) of this quenched and tempered steel exhibits cyclic softening during the initial stages of cyclic loading, with the strengthening effects of dislocations and carbides weakening, leading to a gradual increase in strain amplitude under the same stress amplitude as the number of cycles increases. As the fatigue process progresses, local strain surges during the crack initiation phase, but macroscopic strain fluctuations remain relatively small. After crack propagation, the effective load-bearing area decreases, and to maintain a constant stress amplitude, the strain output accelerates until fracture, characterized by sharp fluctuations. Overall, the strain output follows an evolutionary trend of "slow rise → stabilization → accelerated rise to fracture."

From the temperature rise perspective, the heat generated by hysteresis energy dissipation during fatigue initially reduces the flow stress of 40Cr through thermal softening, further increasing the strain amplitude under constant stress amplitude and compounding with cyclic softening to accelerate deformation. When the temperature exceeds 300 °C, creep mechanisms intervene and interact with fatigue, introducing a time-dependent creep component to the strain output, manifested as continuous strain accumulation over time. Additionally, uneven temperature fields caused by significant temperature rises generate thermal stresses, disrupting the stability and uniformity of strain signals.

In summary, the strain output of 40Cr is primarily governed by fatigue performance, dictating its staged evolution trend, while temperature rise not only directly amplifies strain amplitude through thermal softening but also introduces creep behavior at high temperatures, adding a time-dependent characteristic to the strain output.

Convert the constitutive model into the corresponding input format and import it into Workbench software. In actual motion, friction contact exists between the wave generator and the inner surface of the flexible cylinder (friction coefficient 0.15), as well as between the flexible gear ring and the steel gear ring. The normal contact stiffness is set to 0.1 to balance penetration issues and convergence, employing symmetric contact behavior and the generalized Lagrangian method to enhance the reliability of the finite element model simulation.

3.1 Total Deformation Analysis of the Flexspline Under Different Working Conditions

A ring of strain gauges, each with a thickness of 0.03 mm, was attached around the middle section of the flexspline cup for finite element simulation. The layout is shown in Fig. 6.

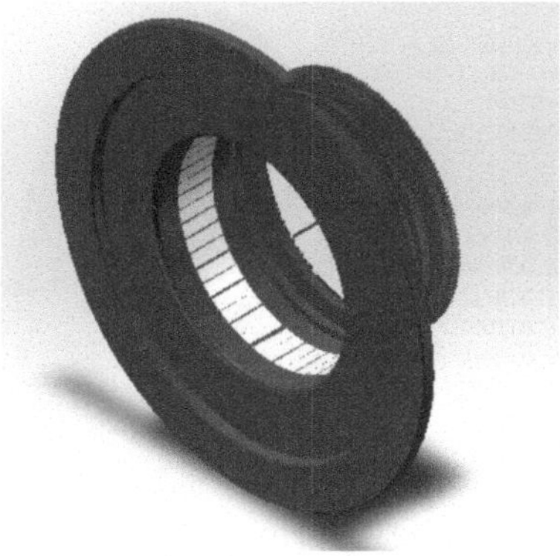

Fig. 6. Layout of strain gauges in 3D models

To investigate the overall displacement of the flexspline in the self-sensing harmonic reducer under different operating conditions, simulations were conducted to analyze the stress and deformation of the flexspline under no-load and load conditions. The results include the total displacement cloud map, strain cloud map, and the distributions of displacement and stress along the radial, circumferential, and axial directions. After assembling the 3D simulation model of the self-sensing harmonic reducer, the major axis of the wave generator was aligned with the X-axis (at 0° and 180°), the minor axis was aligned with the Y-axis (at 90° and 270°), and the axial direction was aligned with the Z-axis in the simulation software.

The total deformation cloud map of the flexspline under no-load conditions in the self-sensing harmonic reducer was obtained through finite element simulation. As shown in Fig. 7, under the action of the wave generator, the maximum deformation of the flexspline is 0.50575 mm, occurring in the contact region between the flexspline and the major axis of the wave generator. The total deformation gradually decreases along the axial direction of the flexspline cup, from the open end—where it contacts both the major and minor axes of the wave generator—toward the flange. This deformation exhibits a smooth gradient, and near the outer edge of the flange, the displacement approaches zero. The overall deformation distribution of the flexspline is perfectly symmetric with respect to the major and minor axes, which is consistent with the actual operating behavior.

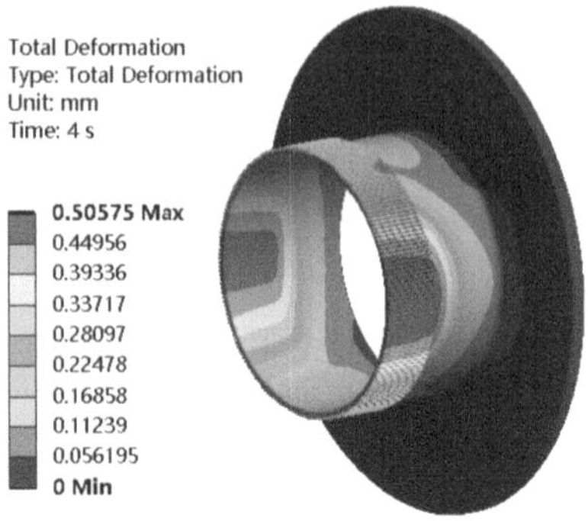

Fig. 7. Total deformation of flexible wheel under no-load

The total deformation cloud map of the flexspline under load conditions in the self-sensing harmonic reducer is shown in Fig. 8. According to the results, the deformation of the flexspline gradually decreases from the contact regions with the major and minor axes of the wave generator toward the flange. The maximum deformation occurs in the contact area between the flexspline and the major axis of the wave generator. The overall deformation distribution of the flexspline under load conditions is consistent with that under no-load conditions, exhibiting symmetry with respect to the major and minor axes, which aligns well with the actual operating behavior.

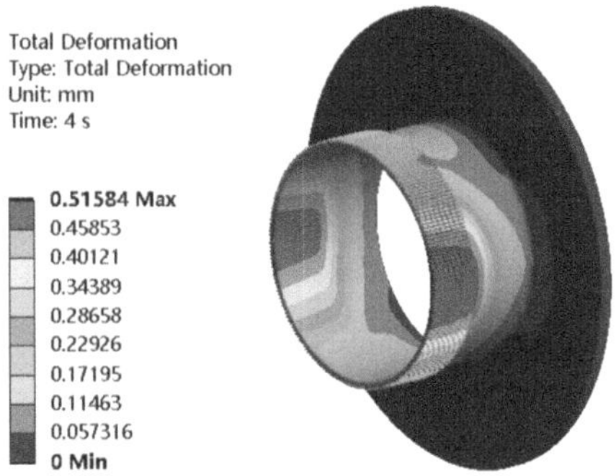

Fig. 8. Total deformation of flexible wheel under load

3.2 Analysis of Displacement Components of the Flexspline Under Different Working Conditions

The directional displacement cloud maps of the flexspline under no-load conditions in the self-sensing harmonic reducer, obtained through finite element simulation, are shown in Fig. 9. Figure 9(a) presents the radial displacement cloud map, where the maximum deformation value is 0.49172 mm, occurring at the position of the major axis of the wave generator. The deformation gradually decreases from the contact regions of the major and minor axes toward the flange, reaching its minimum along the minor axis. The distribution is symmetric overall, and the radial displacement of the flexspline varies periodically along the rotation direction of the wave generator—decreasing from the major axis to the minor axis and increasing again from the minor axis to the major axis. The results indicate that at the major axis, the displacement is positive and large, showing that the flexspline cup is stretched outward; while at the minor axis, the displacement is negative and small, indicating inward compression of the flexspline cup. This behavior is consistent with the actual working conditions of the harmonic reducer. Figure 9(b) shows the circumferential displacement cloud map, with a maximum deformation of 0.19165 mm, approximately 0.3 mm smaller than the radial displacement. The distribution is also symmetric, though the symmetry plane differs from that of the radial displacement. Figure 9(c) shows the axial displacement cloud map, whose overall distribution pattern is similar to that of the radial displacement. The deformation is symmetric with respect to both the major and minor axes of the wave generator, but its magnitude is the smallest among the three directions, with a maximum value of only 0.13803 mm. This indicates that during operation, the axial force acting on the flexspline is relatively small, resulting in minimal axial displacement. As shown in the figure, the axial displacement near the major axis of the wave generator is positive, meaning that both the cup wall and flange of the flexspline experience axial tension. Conversely, near the minor axis, the axial displacement is negative, indicating that the cup wall and flange undergo axial compression under the action of the wave generator.

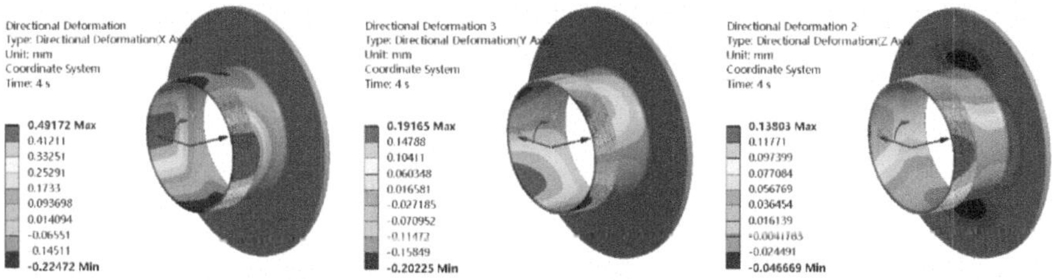

(a) Radial displacement (b) Circumferential displacement (c) Axial displacement

Fig. 9. Displacement cloud maps of the flexspline in all directions under no-load conditions

By analyzing the total deformation and directional displacement cloud maps of the flexspline in the self-sensing harmonic reducer, it can be observed that the maximum deformation occurs near the tooth ring of the flexspline under the action of the wave generator. To further study this behavior, four equally spaced cross-sections were selected along the flexspline tooth ring, and four neutral layer paths were created on these sections. The starting and ending points of each path were defined at the neutral layer node located on the major axis (0°) position, ensuring uniform node distribution along each path. The total deformation and directional displacements along these paths were obtained through simulation. The positions of the four paths are shown in Fig. 10.

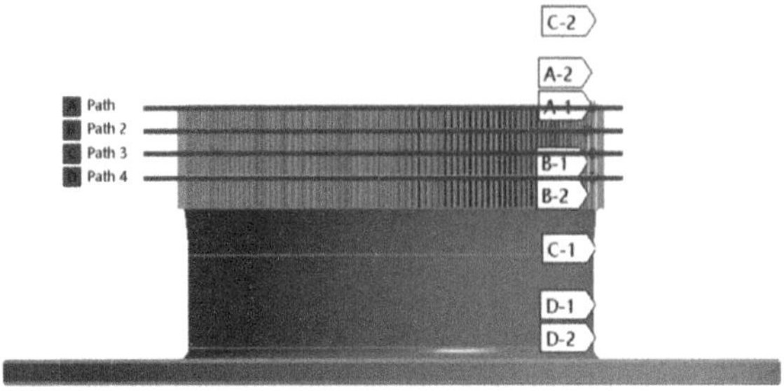

Fig. 10. The four path positions

The radial displacement curves along each path of the flexspline under no-load conditions, obtained through finite element simulation, are shown in Fig. 11. The radial displacement of the central layer at the tooth ring cross-section exhibits a periodic variation that is symmetric about the rotation angle of 180°. Within the rotation range of 0°–180°, the central layer's radial displacement curve is symmetric about 90°, and when the wave generator completes a full rotation of 360°, the radial displacement of the nodes along each path shows a sinusoidal-like periodic pattern. As the rotation angle increases, the radial displacement of the nodes varies cyclically. Among the four neutral layer paths, Path 1 exhibits the largest radial displacement variation, reaching 0.49155 mm. Additionally, as the cross-section moves farther from the open end of the flexspline, the magnitude of the radial displacement gradually decreases.

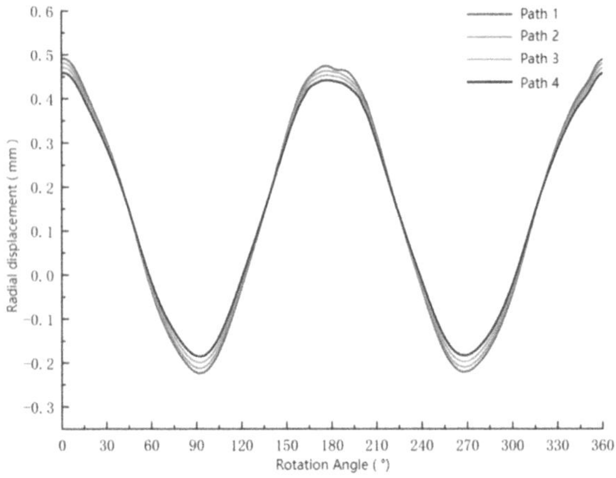

Fig. 11. Radial deformation curve under no-load

The circumferential displacement curves along each path of the flexspline under no-load conditions, obtained through finite element simulation, are shown in Fig. 12. Compared with the radial displacement curves, the circumferential displacement curves exhibit distinct phase characteristics. The maximum circumferential displacement is 0.17777 mm, and the maximum and minimum deformation values do not occur at rotation angles of 0°, 90°, 180°, or 270°, but rather at 45°, 135°, 235°, and 315° along each path. Moreover, the displacement magnitude decreases uniformly from Path 1 to Path 4. At the positions corresponding to the major and minor axes of the wave generator, the circumferential displacement of the flexspline tooth ring is approximately zero, showing little variation with position, and the curves for all paths nearly coincide.

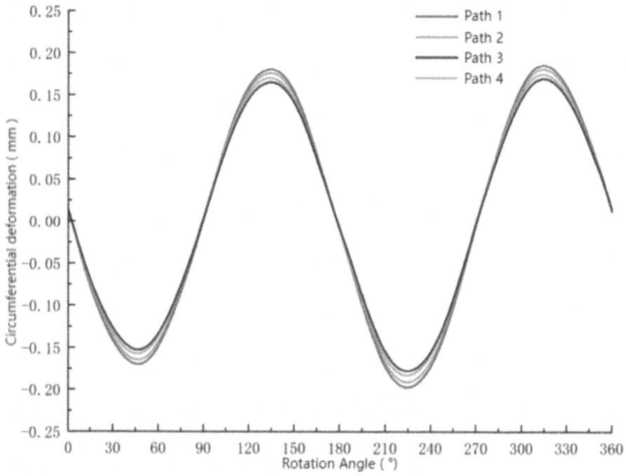

Fig. 12. Circumferential deformation curve during no-load

The axial displacement curves along each path of the flexspline under no-load conditions, obtained through finite element simulation, are shown in Fig. 13. Compared with the radial and circumferential displacement curves, the axial deformation values are relatively small, with a maximum displacement of 0.13088 mm. When the wave generator rotates through 360°, the nodes along each path exhibit a sinusoidal-like variation in axial displacement. As shown in the figure, the axial displacement is positive at rotation angles of 0° and 180°, and negative at 90° and 270°. Combined with the previous analysis of the axial displacement cloud map, it can be concluded that at the major axis of the wave generator, the flexspline deforms in the positive Z-axis direction, while at the minor axis, it deforms in the negative Z-axis direction.

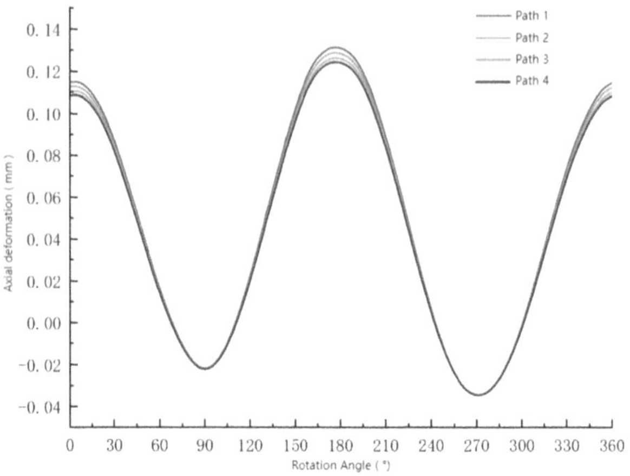

Fig. 13. Axial deformation curve under no-load

By comparing the total displacement cloud maps of the flexspline under no-load and load conditions, noticeable differences can be observed in the deformation behavior of the flexspline in the self-sensing harmonic reducer. Therefore, a rated load of 9.5 Nm was applied to the circular spline, and finite element simulations were performed to obtain the radial, circumferential, and axial displacement cloud maps of the flexspline under load conditions, as shown in Fig. 14. According to the results, the overall deformation distribution of the flexspline under load is similar to that under no-load conditions, both exhibiting symmetric deformation patterns. The maximum and minimum radial displacements occur at the open end of the tooth ring. However, when comparing the displacement magnitudes in different directions, the maximum deformation under load is significantly greater than that observed under no-load conditions.

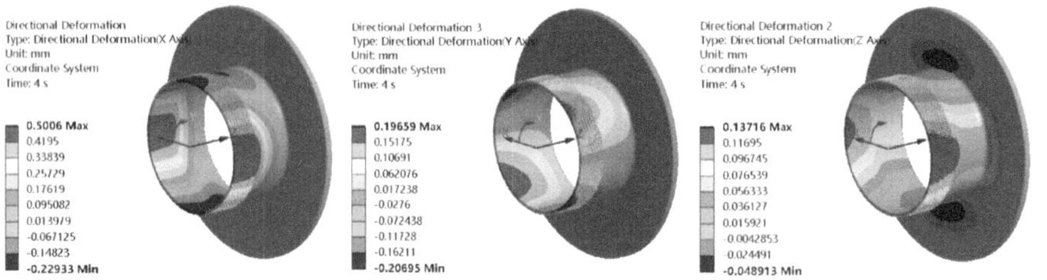

(a) Radial displacement (b) Circumferential displacement (c) Axial displacement

Fig. 14. Displacement cloud maps of the flexspline in all directions under load conditions

The same sectional paths used in the no-load simulation were selected for comparison, and their total deformation and directional displacements were computed through finite element analysis. The radial, circumferential, and axial displacement curves of the flexspline under load conditions in the self-sensing harmonic reducer are obtained from the simulation results and are shown in Figs. 15, 16, and 17, respectively.

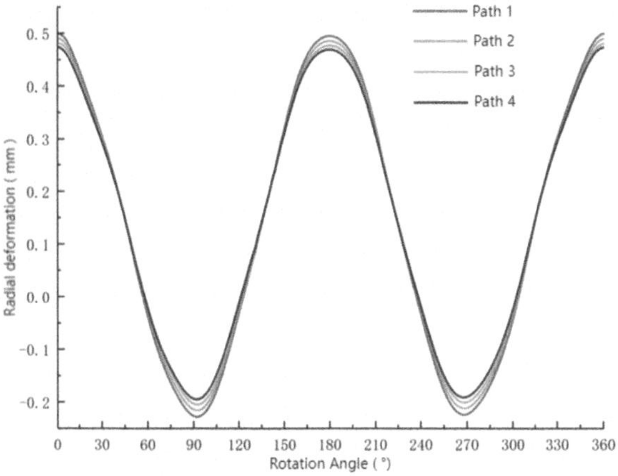

Fig. 15. Radial deformation curve under load

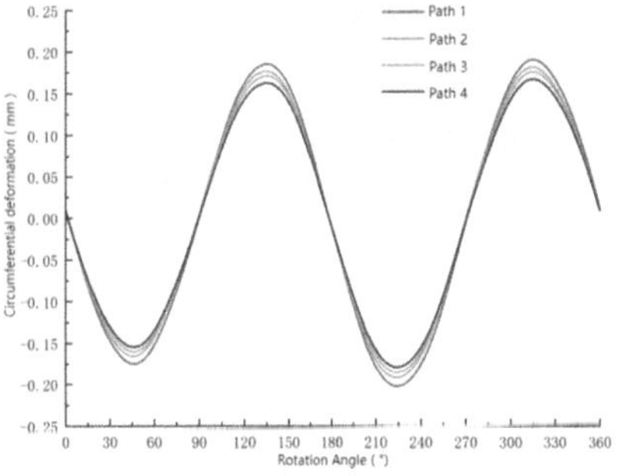

Fig. 16. Circumferential deformation curve under load

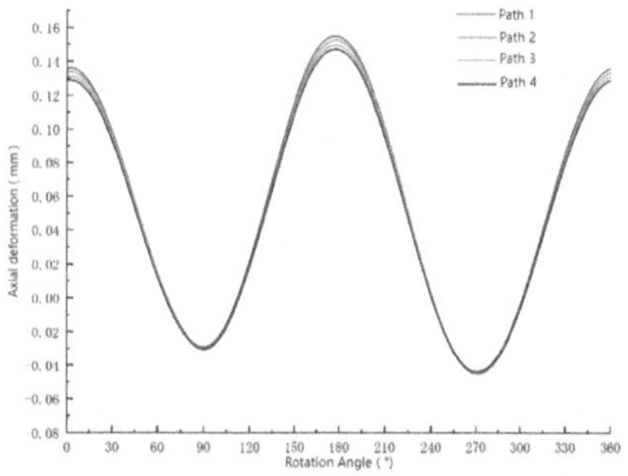

Fig. 17. Axial deformation curve under load

Under load conditions, the maximum radial displacement of the flexspline occurs slightly to the left of the wave generator's rotation angles of 0° and 180°, forming an approximate 5° offset. The maximum radial deformation reaches 0.50022 mm, which is greater than the 0.49155 mm observed under no-load conditions. This indicates that the applied load not only increases the radial displacement deformation of the flexspline gear rim but also shifts the location of the maximum deformation. The circumferential displacement is nonzero along the major axis (0° and 180°), showing a certain degree of directional deviation under the rated load. The maximum circumferential deformation reaches 0.1908 mm, an increase of 0.02 mm compared to the no-load condition, demonstrating that the load alters both the direction and magnitude of the circumferential displacement of the flexspline gear rim. The axial displacement no longer exhibits symmetry about the 180° rotation angle but instead shows a slight deviation with the change in rotation angle. The maximum axial deformation reaches 0.13683 mm, which is 0.006 mm higher than that under no-load conditions.

3.3 Stress Analysis of the Flexspline Under Different Working Conditions

A fixed constraint was applied to the flexspline, while a displacement load was applied to the wave generator. The resulting stress distribution of the flexspline under no-load conditions is shown in Fig. 18. As illustrated, the region of maximum stress in the self-sensing harmonic reducer's flexspline occurs at the tooth root on the rear end of the gear rim in contact with the major axis of the wave generator, exhibiting a pronounced stress concentration. The stress distribution is essentially centrally symmetric, and the stress magnitude gradually decreases from the rear end of the gear rim toward the bottom of the cup.

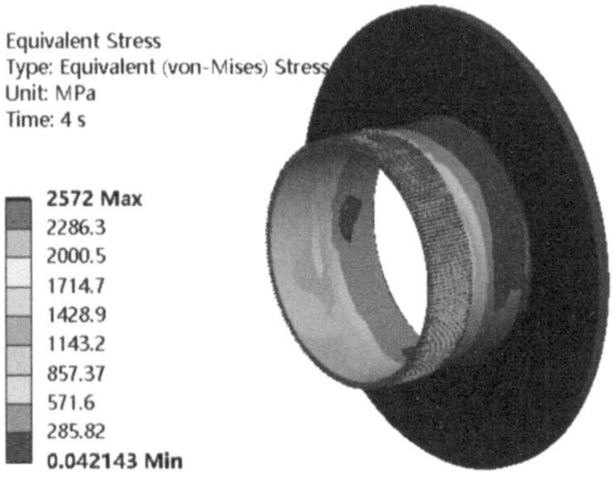

Fig. 18. Stress cloud map of the flexspline under no-load conditions

Under load conditions, a torque about the Z-axis—corresponding to the rated torque of the LHSG-25-120-C-I-S11D harmonic reducer—was applied to the output end face of the rigid wheel, in the direction opposite to the rotation of the wave generator. The resulting stress distribution of the flexspline under load is shown in Fig. 19. As depicted, the maximum stress region of the self-sensing harmonic reducer's flexspline under load remains concentrated at the tooth root on the rear end of the gear rim in contact with the major axis of the wave generator. The overall distribution pattern is similar to that under no-load conditions, indicating that when subjected to the rated torque, the global equivalent stress distribution of the flexspline does not change significantly. This suggests that the effect of the rated load on the overall stress distribution of the flexspline is relatively minor.

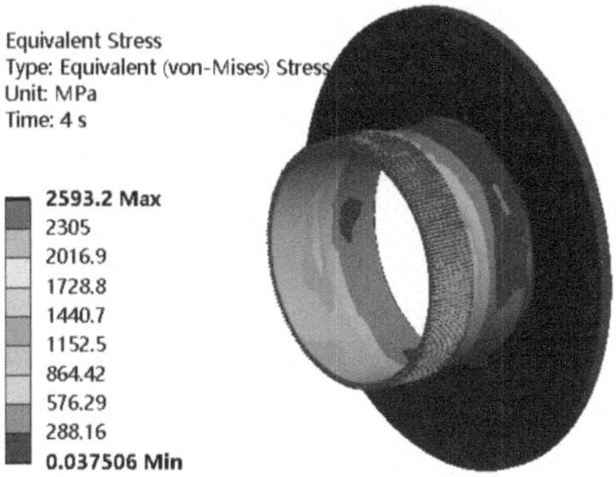

Fig. 19. Stress cloud map of the flexspline under load conditions

In summary, under different operating conditions, the variation trends of radial, circumferential, and axial displacements are generally consistent; however, the addition of load causes a slight angular deviation in each directional displacement. The equivalent stress distribution patterns under both no-load and load conditions are also similar, with stresses primarily concentrated at the tooth root on the rear side of the gear rim in contact with the major axis of the wave generator, exhibiting an approximately centrally symmetric distribution. Furthermore, compared with the no-load condition, the overall deformation and stress magnitude of the flexspline increase under load.

4 Transient Dynamics Simulation of the Self-sensing Harmonic Reducer

Through the static simulation analysis of the flexspline, its stress state can be obtained. However, during actual operation, the flexspline undergoes periodic deformation under the action of the wave generator, and the stress, displacement, and velocity of each point on the flexspline continuously change throughout the transmission process. Therefore, to more accurately reflect its performance, a transient dynamic simulation analysis of the self-sensing harmonic drive is conducted in ANSYS Workbench. Considering that, in real motion, each node on the flexspline experiences deformation following an approximately sinusoidal pattern, a sinusoidal excitation is applied to the inner surface of the flexspline cylinder, defined as $f(t) = \sin(3.14 * t)$. The corresponding sinusoidal excitation waveform is shown in Fig. 20.

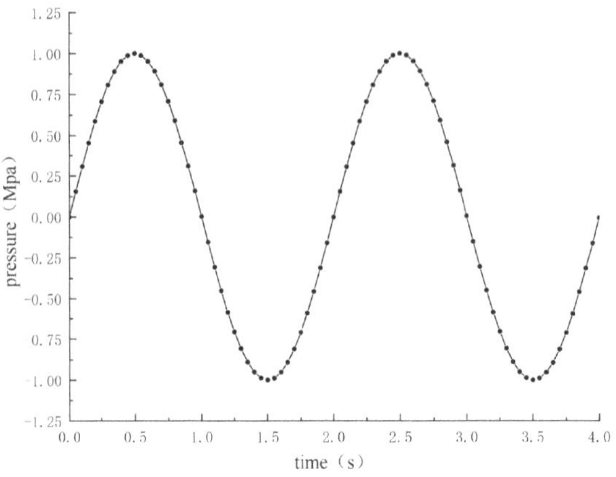

Fig. 20. Sinusoidal excitation

Based on the specifications of the LHSG-25-120-C-I-S11D harmonic reducer, a rated load of 9.5 Nm was applied to the rigid wheel, and a rotational speed of 2000 r/min was imposed on the wave generator. All other boundary conditions and contact settings were kept consistent with those used in the static simulation. A transient dynamic simulation of the self-sensing harmonic reducer was then performed, and the resulting total deformation and stress contour plots of the flexspline are shown in Figs. 21 and 22, respectively.

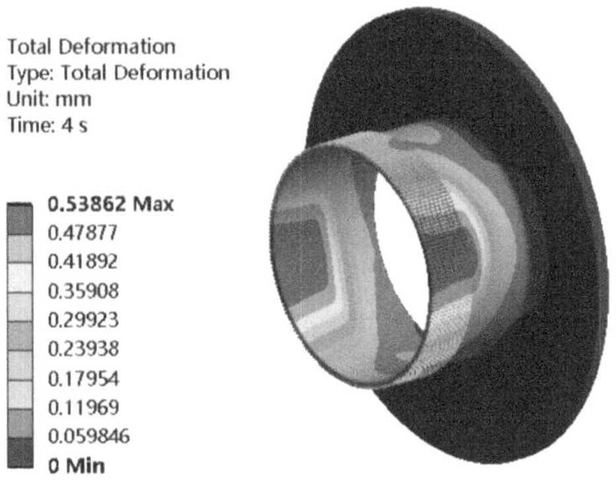

Fig. 21. Deformation cloud map of the flexspline under rated operating conditions

Fig. 22. Stress cloud map of the flexspline under rated operating conditions

As shown in Fig. 21, the maximum total deformation of the flexspline reaches 0.53862 mm, occurring in the region where the flexspline is in contact with the major axis of the wave generator. From the contact regions of the major and minor axes toward the flange, the deformation gradually decreases along the axial direction of the flexspline, exhibiting a smooth gradient. Near the outer edge of the flange, the displacement approaches zero. Overall, the total deformation of the flexspline displays a perfectly symmetrical distribution about the major and minor axes. As illustrated in Fig. 22, under the rated load, the stress distribution of the flexspline does not exhibit central symmetry but shows a certain angular deviation, indicating that the axial load on the gear teeth of the flexspline is non-uniform. The stress along the flexspline cylinder gradually decreases along the Z-axis, while the maximum equivalent stress, measured at 2924.1 MPa, is concentrated at the tooth root on the rear side of the flexspline gear rim. Moreover, since the transient dynamic simulation more closely reflects the actual operating conditions of the harmonic drive, it enables a more accurate analysis of the deformation and stress distribution at the locations where the strain gauges are attached on the flexspline.

Rotational speeds of 500 r/min, 1000 r/min, 1500 r/min, and 2000 r/min were applied to the end face of the wave generator, and the system was simulated for one full rotation (360°). The resulting total deformation of the flexspline under each rotational speed condition is shown in Fig. 23.

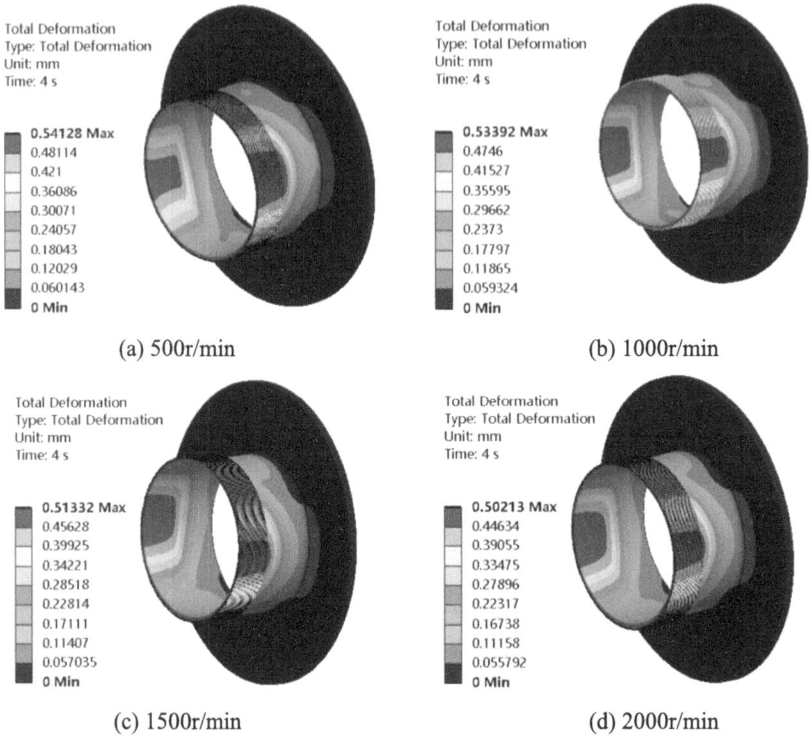

Fig. 23. Total deformation cloud maps of the flexspline at different rotational speeds

The simulation results were plotted as a box diagram, as shown in Fig. 24. As illustrated, the total deformation of the flexspline gradually decreases with the increase in the rotational speed of the wave generator, while the overall distribution pattern of deformation remains essentially consistent. This indicates that the rotational speed of the wave generator has a minimal effect on the total deformation of the flexspline.

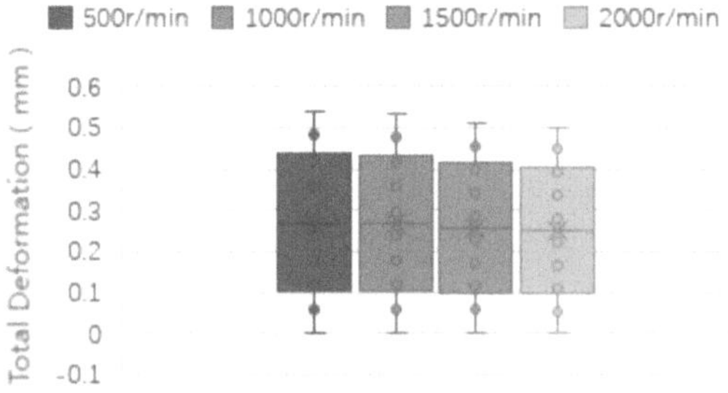

Fig. 24. Box plot of total deformation at different rotational speeds

The stress distributions of the flexspline under different rotational speeds are shown in Fig. 25. Although the stress distribution patterns vary slightly among the four speed conditions, in all cases the stress gradually decreases along the Z-axis direction, with the maximum equivalent stress concentrated at the rear tooth root of the flexspline gear ring.

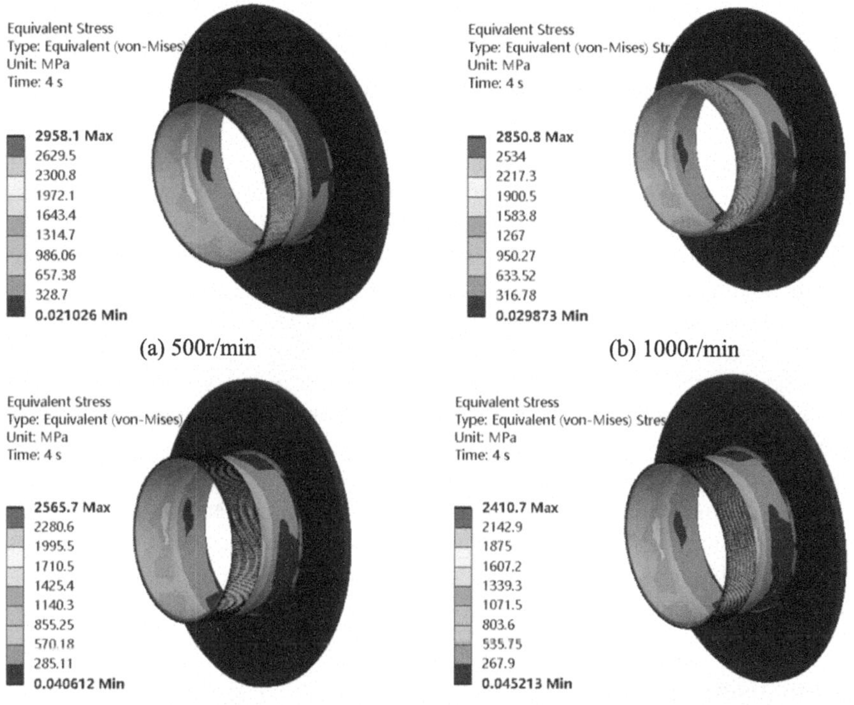

(a) 500r/min (b) 1000r/min

(c) 1500r/min (d) 2000r/min

Fig. 25. Stress cloud maps of the flexspline at different rotational speeds

The stress simulation data were plotted as a box diagram, as shown in Fig. 26. It can be observed that as the rotational speed of the wave generator increases, the stress within the flexspline gradually decreases.

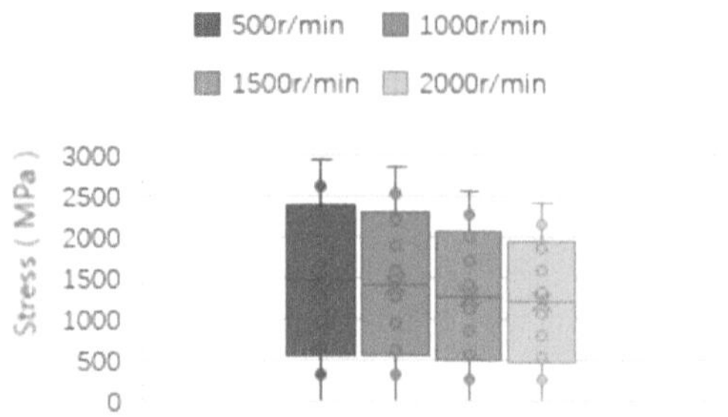

Fig. 26. Box plot of stress at different rotational speeds

It is noteworthy that the simulation results demonstrate a decreasing trend in both total deformation and equivalent stress of the flexspline as the wave generator's rotational speed increases. This phenomenon primarily stems from the system's dynamic inertial effects: at high rotational speeds, the centrifugal force generated by the flexspline tends to restore its original circular shape, partially offsetting the elliptical deformation induced by the wave generator. Simultaneously, inertial forces resist rapid deformation, leading to reduced time-averaged stress and deformation. Additionally, shortened contact duration and enhanced system damping at higher speeds further suppress dynamic responses. Consequently, increased rotational speed exerts a certain 'dynamic unloading' effect on the flexspline.

In summary, under rated operating conditions, the transient dynamic simulation of the self-sensing harmonic drive yields results that are greater than those obtained from the static analysis. This indicates that the transient dynamic simulation provides a representation that is closer to the actual motion behavior of the self-sensing harmonic drive.

5 Conclusion

To address the technical challenge of integrated joint torque measurement in collaborative robots, one effective approach is the development of a self-sensing harmonic drive that integrates a torque sensor into the flexspline. However, due to the complex deformation and strain behavior of the flexspline during operation, accurately measuring the output torque remains difficult. Therefore, it is necessary to conduct further research on torque measurement methods for such systems.

Based on the transmission principle and kinematic analysis of the harmonic drive, the mechanical characteristics of the flexspline were theoretically analyzed. The radial displacement, circumferential displacement, normal angular deformation equations, and stress distribution at any point on an arbitrary cross-section of the flexspline were derived under both no-load and load conditions.

Static simulations of the self-sensing harmonic drive were performed for both conditions. The results indicate that the deformation and stress distribution patterns of the flexspline are generally consistent under no-load and load conditions. However, the application of load causes a slight angular deviation in the total deformation and stress distribution of the radial, circumferential, and axial displacements. By defining four observation paths to monitor the maximum deformation of the flexspline at different cross-sections, the optimal placement positions for strain gauges were determined to obtain the most accurate torque signals.

Furthermore, by applying a sinusoidal excitation consistent with the actual transmission characteristics of the harmonic drive, a transient dynamic simulation was conducted under rated operating conditions. The results show that the transient dynamic response is greater than that of the static simulation, indicating that transient analysis provides a representation closer to real operating behavior.

Finally, dynamic simulations at different rotational speeds were performed to analyze the stress and deformation behavior of the flexspline. The obtained results improve the accuracy of torque measurement and provide a theoretical foundation for developing an integrated joint torque sensing scheme for collaborative robots.

References

1. Wang, P., Zhao, C., Li, J., Zhou, Z., Yan, H., Hu, J.: Analysis of meshing characteristics and stress deformation of harmonic reducer. Aeronaut. Sci. Technol. **34**(7), 73–80 (2023)
2. Wu, C., et al.: Analysis of fatigue life and mechanical properties of flexspline in harmonic reducer. J. Mech. Transm. **46**(1), 115–119 (2022)
3. Wang, X., Du, Z., Shen, W., et al.: Design and test of integrated torque sensor for harmonic reducer. Opt. Precis. Eng. **32**(19), 2899–2910 (2024)
4. Ivanov, M.H.: Harmonic Gear Transmission, 1st edn., pp. 13–29. National Defense Industry Press, Beijing (1987). Trans. by Shen Yunwen, Li Kemei
5. Shen, Y., Ye, Q.: Theory and Design of Harmonic Gear Drive, 1st edn., pp. 180–235. Machinery Industry Press, Beijing (1985)
6. Fu, J., Dong, H., Shen, Y.: Finite element analysis of flexspline stress in harmonic gear transmission. China Mech. Eng. **18**, 2210–2214 (2007)
7. Fu, J., Dong, H.: Finite element analysis of flexspline stress and reasonable selection of structural parameters. Mech. Sci. Technol. Aerosp. Eng. **9**, 1101–1104 (2007)
8. Shen, W., Lou, Z., Du, Z., Gao, H., Wang, X.: Torque sensor developed with a strain rosette ring for in situ integration in harmonic reducer. IEEE Sensors J. **24**(16), 25802–25813 (2024). https://doi.org/10.1109/JSEN.2024.3420754
9. Ismail, M.A.A., Windelberg, J., Liu, G.: Simplified sensorless torque estimation method for harmonic drive based electro-mechanical actuator. IEEE Robot. Autom. Lett. **6**(2), 835–840 (2021). https://doi.org/10.1109/LRA.2021.3052392
10. Ostapski, W.: Analysis of the stress state in the harmonic drive generator-flexspline system in relation to selected structural parameters and manufacturing deviations. Bull. Polish Acad. Sci. Tech. Sci., 683–698 (2010)
11. Kikuchi, N., Chung, K.Y., Torigaki, T., et al.: Adaptive finite element methods for shape optimization of linearly elastic structures. Optim. Shape: Autom. Struct. Des., 139–169 (1986)
12. Dong, H., Wang, D.: Elastic deformation characteristic of the flexspline in harmonic drive. In: International Conference on Reconfigurable Mechanisms and Robots, pp. 363–369 (2009)
13. Dong, H., Zhu, Z., Zhou, W., et al.: Dynamic simulation of harmonic gear drives considering tooth profiles parameters optimization. J. Comput. **7**(6), 1429–1436 (2012)
14. Yang, J.: Assembly of Harmonic Reducer and Strength Calculation in Dynamic Operation Process. Xiamen University, Xiamen (2017)
15. Jeon, H.S., Oh, S.H.: A study on stress and vibration analysis of a steel and hybrid flexspline for harmonic drive. Compos. Struct. **47**(1–4), 827–833 (1999)

16. Sirigiri, V.K.R., Gudiga, V.Y., Gattu, U.S., et al.: A review on Johnson Cook material model. Mater. Today: Proc. **62**(6), 3450–3456 (2022). https://doi.org/10.1016/j.matpr.2022.04.279
17. Wang, L.: Study on Constitutive Equation and Heat Treatment Performance of 40Cr Steel. Henan Polytechnic University, Henan (2022)
18. Yang, J.: Strength Calculation of Harmonic Gear Reducer in Assembly and Dynamic Operation Process. Xiamen University, Xiamen (2017)
19. Wang, Z.: Research on Mechanical Characteristics and Transmission Accuracy of Harmonic Drive. South China University of Technology (2021)
20. Jiang, S., Chen, C., Nie, S., et al.: Analysis and research on the load distribution law of flexible gear in harmonic gear transmission. Mech. Des. (6), 11–13 (2001)
21. Yi, W.: Finite Element Analysis and Structural Optimization of Harmonic Reducer. Kunming University of Science and Technology (2023)

Design of Humanoid Robot Arm System

Shujing Liu[1], Dong Wang[1], Ya Chen[1(✉)], Dianjun Wang[2], and Peng Wang[3]

[1] School of Mechanical Engineering, Beijing Institute of Petrochemical Technology, No.19 Qingyuan North Street, Huangcun Town, Daxing District, Beijing 102617, China
chenya@bipt.edu.cn

[2] School of Mechanical Engineering and Hydrogen Energy Research Centre, Beijing Institute of Petrochemical Technology, No.19 Qingyuan North Street, Huangcun Town, Daxing District, Beijing 102617, China

[3] School of Information Engineering, Beijing Institute of Petrochemical Technology, No.19 Qingyuan North Street, Huangcun Town, Daxing District, Beijing 102617, China

Abstract. With the growing research and application of robotic arms, they are widely used in medicine, industry, aerospace and other fields. Humanoid robotic arm research aims to replace humans in tedious and dangerous work by simulating human arm movements. This paper designs a 7-degrees-of-freedom (7-DOF) humanoid robotic arm system. To enhance flexibility and simulate human arm movement, it adopts a 7-DOF SRS configuration, divided into the wrist, the elbow and the shoulder. The wrist uses parallel linear joints (frameless motor rotation converted to linear motion via reverse planetary roller screw) for higher load. Kinematic analysis is done via the improved D-H method; joint limit angle measurement and SolidWorks trajectory planning experiments verify flexibility; manipulator dynamic simulation and part static mechanical simulation verify load compliance. Simulation shows the robotic arm meets set indicators. The wrist's parallel linear joint design enables more accurate control, higher energy efficiency and load than general humanoid robotic arms, with reference value for related research.

Keywords: Humanoid robotic arm · Seven degrees of freedom · Wrist core joint mechanism · Parallel linear mechanism

1 Introduction

1.1 Research Background

Humanoid robots are tasked with undertaking repetitive, tedious, and dangerous work to replace human labor [1, 2]. Currently, in domestic research on humanoid robotic arms, six-degrees-of-freedom (6-DOF) robotic arms are predominantly utilized. While such robotic arms are easy to implement and can fully replicate the degrees of freedom of human arms, they suffer from notably low flexibility [3, 4]. Conversely, the seven-degrees-of-freedom (7-DOF) solution achieved through series joints leads to excessive arm weight. Thus, as productivity advances and robot manufacturing capabilities improve, enhancing robot flexibility and pursuing lightweight design will emerge as a key trend in future robot research [5–7].

H. Wei and J. Niu (Eds.): COBOT 2025, CCIS 2893, pp. 51–71, 2027.
https://doi.org/10.1007/978-981-92-2074-8_4

Humanoid robots originated in the early 20th century, with a key milestone in 1969 when Japan's Waseda University developed the first bipedal humanoid robot. In recent years, firms like Boston Dynamics, Agile Robots and Tesla have launched humanoid robot products, moving the tech from lab research to practical use.Per QYResearch, the global bipedal robot market reached \$244 million in 2023 and is projected to hit \$697 million by 2029, with a 17.02% CAGR, showing great industry potential. In October 2023, China's Ministry of Industry and Information Technology, in its Guiding Opinions on the Innovative Development of Humanoid Robots, stressed that humanoid robots—integrating AI, high-end manufacturing and new materials—will be a disruptive product reshaping production and daily life.As a core component of humanoid robots, the humanoid robotic arm directly dictates functionality and flexibility, key to replacing humans in repetitive, tedious and dangerous tasks. However, current domestic research mainly focuses on 6-degree-of-freedom (6-DOF) robotic arms: easy to implement but low in flexibility (end-effectors often can't reach some workspace points due to joint limits). Also, traditional series rotational joints add weight, increasing costs and lowering load capacity.

Currently, the core design trends of 7-DOF anthropomorphic arms focus on three aspects: First, lightweight design – through the application of cable-driven technology and new lightweight materials, the self-weight of the arm is reduced while ensuring structural strength. Second, enhanced high flexibility – by virtue of redundant degree-of-freedom optimization algorithms, the motion posture limitations of 6-DOF arms are broken through, and the coverage of the working space is expanded. Third, innovation in wrist structure – replacing traditional series rotational joints with parallel linear mechanisms to balance control precision, load capacity, and energy efficiency. The SRS configuration + parallel linear wrist design adopted in this study is precisely a targeted response to these three trends. It not only enhances the anthropomorphic motion flexibility through the SRS structure but also addresses the pain point of excessive self-weight in traditional series 7-DOF arms by means of parallel linear joints.

Therefore, developing a more flexible and lightweight robotic arm is an urgent need, which motivates this study on a 7-DOF humanoid robotic arm.

1.2 Domestic and International Research Status

International Research Status. Foreign research on robotic arms has formed a relatively mature technological and structural system, allowing different structural designs to be applied to diverse scenarios [8–11]. In the field of flexible drive, representative cases include the cable-driven anthropomorphic arm developed by the Massachusetts Institute of Technology (MIT). This configuration exhibits obvious advantages: its transmission components are lightweight (excluding heavy rigid transmission shafts), the flexible cable ensures high safety during human-robot interaction, and the structural layout is highly adaptable to narrow working spaces. However, it also has inherent limitations: the elasticity of the cable easily causes end positioning errors, requiring complex tension coordination control algorithms to compensate; meanwhile, its rated load capacity is significantly lower than that of rigid-driven configurations, making it unsuitable for heavy-load operation scenarios [15].

In recent years, well-known enterprises have successively launched humanoid robot products, such as those from Boston Dynamics, Agile Robotics, and Tesla [12, 13]. Among them, Tesla's Optimus robot adopts a planetary roller screw in its wrist to realize a parallel linear joint design. Compared with general rotating joints, this design achieves higher control precision and energy efficiency [14]. In terms of structural configuration, the SRS configuration (a typical parallel-series hybrid structure) has more concentrated degrees of freedom compared with ordinary series configurations. It can better simulate the movement trajectory of the human arm and has initially achieved the goal of lightweight design, but its lightweight level still needs to be improved— there is a trade-off between structural compactness and mass reduction, and further optimization is required to balance load capacity and arm weight.

Against this background, the lightweight design of this study presents obvious differentiation. Aiming at the shortcomings of existing configurations (e.g., low load capacity of cable-driven arms, insufficient lightweight optimization of SRS configurations, and relatively complex structure of Tesla' s linear joints), this research takes "synergetic optimization of lightweight and performance" as the core. By selecting high-strength lightweight composite materials and optimizing the topological structure of multi-degree-of-freedom coupling components, it realizes further mass reduction of the arm while ensuring end positioning precision and rated load capacity, thus making up for the defect that existing lightweight designs often sacrifice partial performance.

Domestic Research Status. Domestic development is progressing rapidly, with products like UBtech ' s Walker X (featuring upgraded gait control) and Xiaomi's CyberOne (with emotion perception and 300 NM torque). In terms of actuation, electric drive is dominant for precision, while tendon drive is being explored for lightweight design—for example, the cable-driven 7-DOF arm developed by Harbin Institute of Technology. This cable-driven structure exhibits prominent technical advantages: its adoption of flexible cables instead of rigid transmission components greatly reduces self-weight, and it achieves a high load-to-weight ratio, enabling it to carry heavier payloads under the same self-weight condition, which addresses the weight bottleneck of traditional rigid-driven robotic arms. In application scenarios, it has broad prospects: in collaborative robots, its lightweight and flexible characteristics lower collision risks during human-robot collaboration, making it applicable to precision tasks such as electronic product assembly and medical assistance; in wearable devices, it provides a technical solution for rehabilitation exoskeletons and military assistive exoskeletons — its low self-weight reduces the physical burden on users while maintaining stable load-bearing performance, effectively improving the usability of wearable robotic systems.

Structurally, most domestic robotic arms adopt 6-DOF designs, which have limited solutions for specific postures. 7-DOF arms, such as UBtech's Panda Robot, offer redundant degrees of freedom, wider workspaces, and more flexible postures. However, challenges persist: series rotational joints result in heavy weight; parallel linear joints (used in the bipedal robot from Huazhong University of Science and Technology) improve energy efficiency but have a narrow wrist angle range. Urgent issues include optimizing 7-DOF structures, replacing series rotational mechanisms with parallel linear ones, and expanding wrist movement ranges.

1.3 Research Contents

This study focuses on designing a 7-DOF humanoid robotic arm with a parallel linear wrist joint, encompassing three core tasks:

Configuration Design. Drawing on human bionics, this study analyzes the degrees of freedom of the human arm to determine the 7-DOF structure. It (This study) investigates kinematics and workspace, and optimizes dimensions to maximize arm manipulability—addressing the flexibility limitations of 6-DOF arms.

Core Component Design. This study innovates the wrist structure by replacing traditional series rotational joints with a linear motion mechanism. Design key components like planetary roller screws to reduce self-weight, while ensuring load capacity and expanding wrist pitching and yawing angles.

Simulation and Verification. Construct a 3D model of the robotic arm, and use kinematics to calculate the workspace and joint relationships. Conduct dynamic simulations (via SolidWorks Motion) to verify load capacity, and perform static mechanical analysis on core components to ensure structural reliability.

2 Overall Scheme Design

2.1 Analysis of Design Requirements and Indexes

At present, the domestic robot arm has been widely used in aviation, aerospace, medical, industrial and other fields. With the development of science and technology and the advancement of research on humanoid robot arm, the research on humanoid robot arm aims to replace people in some repetitive and high-risk work. Therefore, in the design of the humanoid robot arm, different from the high load and high dead weight required by the industrial robot arm, the following factors should be considered in the design of the humanoid robot arm: (1) The humanoid robot arm itself needs to be closer to the structure of the human arm; (2) Humanoid robotic arms need to have a relatively high degree of flexibility, and sometimes even need to approach the degree of freedom of the human arm; (3) Humanoid robotic arms need to be more accurate in control; (4) The humanoid robot arm needs to be different from the industrial robot arm in terms of arm weight and arm length, and should have the weight and arm length close to the human arm. The requirements and indexes are shown in Table 1.

Table 1. Design Requirements and Indexes

Working space	Radius 650 mm
Load	Rated load 3 kg
	Maximum load 5 kg
Degrees of freedom	7
Total weight of robotic arm	$\leq$15 kg

(continued)

Table 1. (*continued*)

Joint velocity	30°/
Linear joint velocity	2 mm/s

2.2 Motion Analysis of Humanoid Robotic Arm

There are three operating states of shoulder: shoulder flexion/extension, shoulder abduction/adduction, shoulder abduction/adduction; There is a state of motion in the elbow: abduction/adduction of the elbow; There are three motion states of the wrist: wrist flexion/extension, wrist external/internal rotation, and wrist abduction/adduction. Seven motion states were analyzed for the whole robot arm. Through the disassembly of the human arm movement, it is found that the shoulder and wrist of the human body are actually a spherical joint, and the rotation axes of the first three joints of the SRS configuration are perpendicular to each other, which can simulate the spherical joints of the human shoulder, and the rotation axes of the last three joints are also perpendicular to each other, which is equivalent to the spherical joints of the wrist (see Fig. 1).

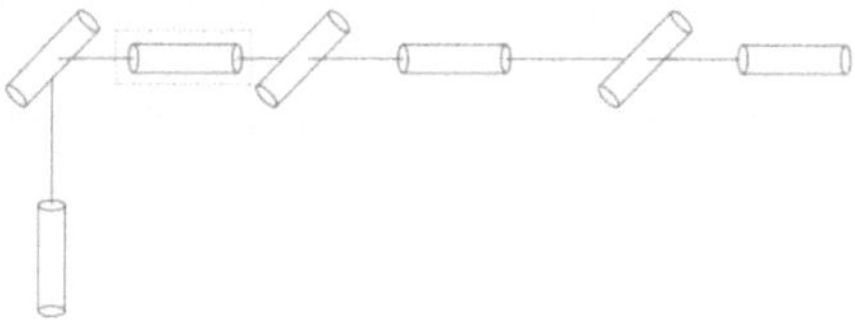

Fig. 1. SRS structure

2.3 Design of Wrist Core Joint Structure Scheme

In the structure design of the humanoid robot arm, the pitching joint of the three wrist joints is realized by a parallel linear mechanism, and the pitching motion of the degree of freedom is realized by the reciprocating motion of the parallel linear mechanism (see Fig. 2)

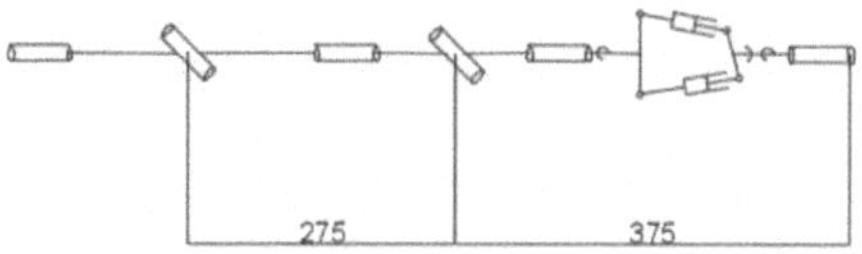

Fig. 2. Overall structure of humanoid robotic arm

3 Kinematics Modeling of Humanoid Robotic Arm

3.1 Table D-H

The kinematic analysis of the robot arm refers to the representation of each joint in a coordinate system, the joint Angles, deviation distance, connecting rod length and torsion Angles between each joint are listed in a D-H table, and the pose transformation relationship between each joint is described in the form of a matrix. The joint space is also calculated. The D-H table is established by adding coordinate system to each joint of the humanoid robot arm. Its Mechanical arm motion diagram, (see Fig. 3) (Table 2).

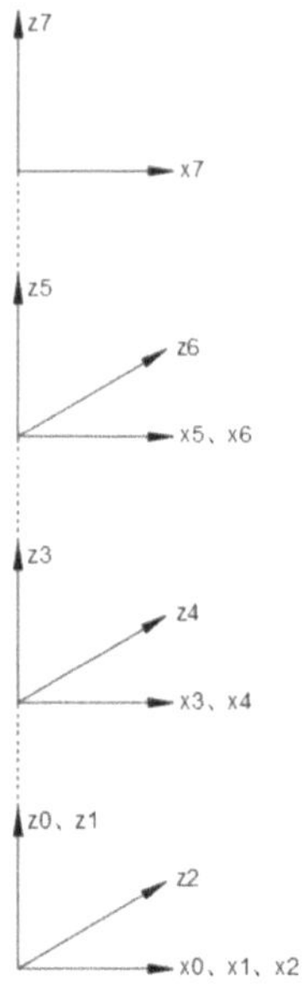

Fig. 3. Mechanical arm motion diagram

Table 2. Table D-H

	− 1	− 1		
1	0	0	0	1
2	–90	0	0	2
3	90	0	275	3
4	–90	0	0	4
5	90	0	295	5
6	–90	0	0	6
7	90	0	80	7

3.2 Kinematic Analysis

Based on the above DH table, it can be known that the transformation matrix between the two connecting rods is shown in Eq. (1):

$$\begin{bmatrix} c\theta_i & -s\theta_i & 0 & a_{i-1} \\ s\theta_i c\alpha_{i-1} & c\theta_i c\alpha_{i-1} & -s\alpha_{i-1} & -s\alpha_{i-1}d_i \\ s\theta_i s\alpha_{i-1} & c\theta_i s\alpha_{i-1} & c\alpha_{i-1} & c\alpha_{i-1}d_i \\ 0 & 0 & 0 & 1 \end{bmatrix} \quad (1)$$

Let the motion relationship matrix between link 1 and link 7 be as shown in Eq. (2):

$${}_7^0T = \begin{bmatrix} n_x & o_x & a_x & p_x \\ n_y & o_y & a_y & p_y \\ n_z & o_z & a_z & p_z \\ 0 & 0 & 0 & 1 \end{bmatrix} \quad (2)$$

The respective elements in the matrix can be obtained through calculations, as shown in Eq. (3) to (14).

$${}_1^0T = \begin{bmatrix} c\theta_1 & -s\theta_1 & 0 & 0 \\ s\theta_1 & c\theta_1 & 0 & 0 \\ 0 & 0 & 1 & 0 \\ 0 & 0 & 0 & 1 \end{bmatrix} \quad (3)$$

$${}_2^1T = \begin{bmatrix} c\theta_2 & -s\theta_2 & 0 & 0 \\ 0 & 0 & 1 & 0 \\ -s\theta_2 & -c\theta_2 & 0 & 0 \\ 0 & 0 & 0 & 1 \end{bmatrix} \quad (4)$$

$${}_3^2T = \begin{bmatrix} c\theta_3 & -s\theta_3 & 0 & 0 \\ 0 & 0 & -1 & 0 \\ s\theta_3 & c\theta_3 & 0 & 275 \\ 0 & 0 & 0 & 1 \end{bmatrix} \quad (5)$$

$${}_4^3T = \begin{bmatrix} c\theta_4 & -s\theta_4 & 0 & 0 \\ 0 & 0 & 1 & 0 \\ -s\theta_4 & -c\theta_4 & 0 & 0 \\ 0 & 0 & 0 & 1 \end{bmatrix} \quad (6)$$

$${}_5^4T = \begin{bmatrix} c\theta_5 & -s\theta_5 & 0 & 0 \\ 0 & 0 & -1 & 295 \\ s\theta_5 & c\theta_5 & 0 & 0 \\ 0 & 0 & 0 & 1 \end{bmatrix} \quad (7)$$

$${}_6^5T = \begin{bmatrix} c\theta_6 & -s\theta_6 & 0 & 0 \\ 0 & 0 & 1 & 0 \\ -s\theta_6 & -c\theta_6 & 0 & 0 \\ 0 & 0 & 0 & 1 \end{bmatrix} \quad (8)$$

According to the principle of forward kinematics, it can be known that ${}^{0}_{7}\mathrm{T} = {}^{0}_{1}\mathrm{T} * {}^{1}_{2}\mathrm{T} * {}^{2}_{3}\mathrm{T} * {}^{3}_{4}\mathrm{T} * {}^{4}_{5}\mathrm{T} * {}^{5}_{6}\mathrm{T} * {}^{6}_{7}\mathrm{T}$. Substituting the calculation results into different θ values enables the acquisition of pose transformation matrices between joints 1 to 7 of the robotic arm. The pose transformation matrices under different θ values obtained using the Robotics Toolbox in MATLAB are consistent with the calculated results. When $\theta_1 \sim \theta_7 = \mathrm{pi}/3 = 0$, the pose matrix of the end effector of the robotic arm can be calculated as:

$$ {}^{6}_{7}T = \begin{bmatrix} c\theta_7 & -s\theta_7 & 0 & 0 \\ 0 & 0 & -1 & 80 \\ s\theta_7 & c\theta_7 & 0 & 0 \\ 0 & 0 & 0 & 1 \end{bmatrix} \tag{9} $$

3.3 Inverse Kinematics Analysis

Inverse kinematics analysis refers to a process where, given the end - effector pose of a robotic arm, the joint angles of the robotic arm are obtained through mathematical or geometric analysis between the links of the robotic arm. Since the robotic arm in this study adopts an SRS configuration, the inverse kinematics solution of this robotic arm will be carried out using the arm angle parameter method.

As the shoulder joint and wrist joint of the SRS configuration can be regarded as two spherical joints, the plane formed by the upper arm and the forearm will not affect the end - effector pose of the robotic arm when performing self - motion via the spherical joints. The angle between the initial plane of the arm plane and the plane after self - motion is called the arm angle. The process of solving the analytical result of the robotic arm, which involves introducing the arm angle parameter, is known as arm angle parameterization (see Fig. 4).

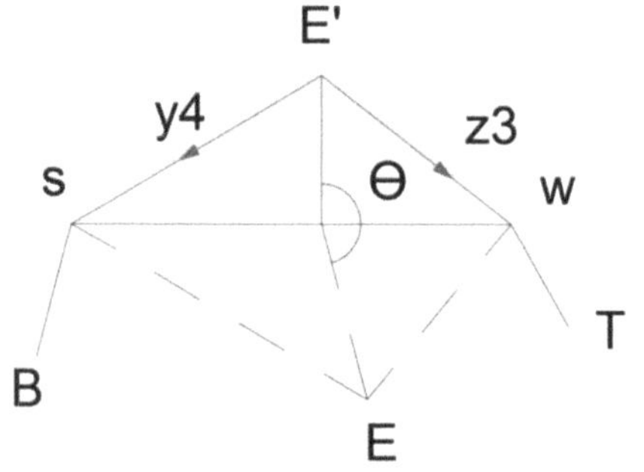

Fig. 4. Schematic Diagram of Arm Angle for SRS-Structure Manipulator

Based on the geometric relationships shown in the figure, joint angle 4 can be derived.

$$ \cos\angle SEW = \frac{d_{SE}^2 + d_{EW}^2 - \|P_{SW}\|^2}{2 \cdot d_{SE} \cdot d_{EW}} \tag{10} $$

$$\theta_4 = \pi - \angle SEW \tag{11}$$

For the solution of the shoulder joint angle, according to the coordinate system establishment rules of the DH parameter method and the geometric relationships, it can be known that:

$$p_{SW} + d_{SE} \cdot y_4 = d_{EW} \cdot z_3 \tag{12}$$

According to the kinematic model, the motion relationship between joint 3 and joint 4 is as follows:

$$ {}^3_4R = \begin{bmatrix} c_4 & -s_4 & \\ 0 & 0 & 1 \\ -s_4 & -c_4 & \end{bmatrix} \tag{13}$$

$${}^3_4y = \begin{bmatrix} -s_4 \; 0 \; -c_4 \end{bmatrix}^T \tag{14}$$

It can be derived from the above formula that:

$$y_4 = {}^0_4 R^{\psi=0} \cdot {}^3_4 y = \begin{bmatrix} x_3 \; y_3 \; z_3 \end{bmatrix} \cdot {}^3_4 y = -s_4x_3 - c_4z_3 \tag{15}$$

Substituting Eq. (20) into Eq. (17) gives x3, and Eq. (21) is as follows:

$$x_3 = \frac{d_{EW}z_3 + c_4z_3 - p_{SW}}{-s_4d_{SE}} \tag{16}$$

By introducing Rodrigues formula, we can derive z3.

$$z_3 - \mathrm{Rot}(u_{sw} \times \begin{bmatrix} 0 \; 0 \; 1 \end{bmatrix}, \angle EWS) \cdot u_{sw} \tag{17}$$

Y3 can be obtained by taking the cross product of x3 and z3. In summary, the rotation matrix of the first three joints can be derived as follows:

$${}^0_3R^{\psi=0} = \begin{bmatrix} x_3 \; y_3 \; z_3 \end{bmatrix} \tag{18}$$

$${}^0_3R = \mathrm{Rot}(u_{sw}, \psi) \cdot {}^0_3 R^{\psi=0} = A_s \sin\psi + B_s \cos\psi + C_s \tag{19}$$

According to the kinematic model, the following matrix can be obtained. By observing this matrix, the angles of joint 1, joint 2, and joint 3 can be calculated:

$${}^0_3R = \begin{bmatrix} * & * & c_1s_2 \\ * & * & s_1s_2 \\ -c_3s_2 & s_3s_2 & c_2 \end{bmatrix} \tag{20}$$

$$\cos(\theta_2) = {}^0_3 R(3, 3) \tag{21}$$

$$\tan(\theta_3) = -\frac{{}^0_3R(3, 2)}{{}^0_3R(3, 1)} \tag{22}$$

$$\tan(\theta_1) = -\frac{{}^0_3R(2,3)}{{}^0_3R(1,3)} \tag{23}$$

Given the angles of joints 1, 2, 3, and 4, the following derivation can be obtained through the kinematic model and the relationships between the joint angles:

$${}^0_7R = {}^0_3R \cdot {}^3_4R \cdot {}^4_7R \tag{24}$$

$${}^4_7R = \begin{bmatrix} * & * & c_5s_6 \\ c_7s_6 & -s_6s_7 & -c_6 \\ * & * & s_5s_6 \end{bmatrix} \tag{25}$$

$${}^4_7R = {}^3_4R^T \cdot {}^0_3R^T \cdot {}^0_7R = A_w \sin(\psi) + B_w \cos(\psi) + C_w \tag{26}$$

Among these.

$$A_w = {}^3_4R^T A_s^T \, {}^0_7R \tag{27}$$

$$B_w = {}^3_4R^T B_s^T \, {}^0_7R \tag{28}$$

$$C_w = {}^3_4R^T C_s^T \, {}^0_7R \tag{29}$$

By examining the wrist rotation matrix, the angles of joints 5, 6, and 7 can be derived as follows:

$$\cos(\theta_6) = -{}^4_7R(2,3) \tag{30}$$

$$\tan(\theta_7) = -\frac{{}^4_7R(2,1)}{{}^4_7R(2,2)} \tag{31}$$

$$\tan(\theta_5) = \frac{{}^4_7R(3,3)}{{}^4_7R(1,3)} \tag{32}$$

Substituting the pose transformation matrix obtained from Eq. (21–23), (30–32) into the aforementioned inverse kinematics formula for calculation, multiple sets of joint angles under this pose can be obtained. It can be found that θ_1~θ_7 = pi/3 is one of the solutions, confirming that the inverse kinematics solution is correct3.4 Kinematics modeling analysis based on MATLAB/RoboticsToolbox.

3.4 Kinematics Modeling Analysis Based on MATLAB/RoboticsToolbox

Based on the D-H table, the robot was built by programming the following program, and the working space of the robot arm was obtained by limiting each joint Angle and separating the time interval. The result is a spherical workspace with a radius of 570 mm. When establishing the D-H table, the improved D-H method adopted did not calculate the length of the end joint and the length from the first joint to the base in it. Here, the length of the seventh joint was 80 mm, and the radius of the working space was 650 mm, which met the requirements of the work design index, its Robot arm workspace (see Fig. 5).

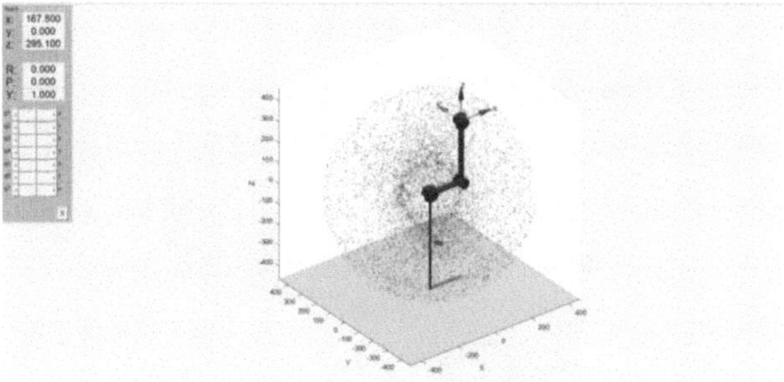

Fig. 5. Robot arm workspace

4 Joint Structure Design

4.1 Overall Structure Design of Mechanical Arm

The SRS configuration is selected. The spherical joint of the human shoulder is equivalent through the first three degrees of freedom pairwise and the axis is intersected at one point, and the spherical joint of the wrist is equivalent through the second three degrees of freedom pairwise and perpendicular. The parallel linear mechanism of the wrist joint is one of the key points of the design, and the rotating motion of the motor needs to be transformed into linear motion by the planetary roller screw. According to GB/T 10000-2023, the distance from the elbow to the wrist of an adult is about 375 mm, and the elbow to the shoulder is about 275 mm [15] (see Fig. 6).

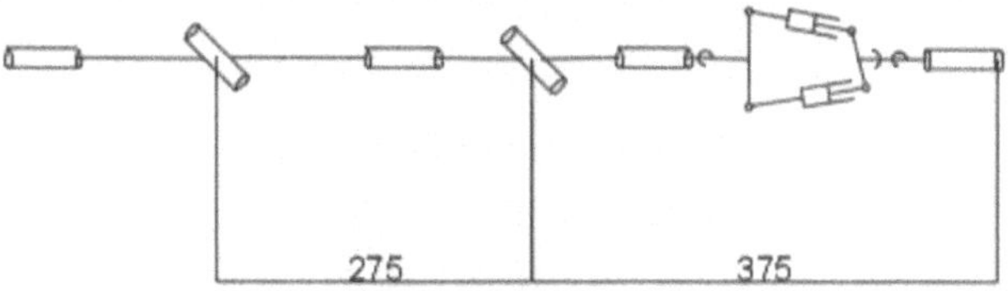

Fig. 6. Overall configuration of robotic arm

4.2 Mechanical Arm Shoulder Joint Structure Design

The shoulder joint of the robot arm is one of the most important parts in the robot arm, and its design and performance directly affect the range of motion, accuracy and bearing capacity of the robot arm. The shoulder has three degrees of freedom, in order to better simulate the human shoulder, the structure of three-axes intersection at one point simulated ball joint is adopted. The shoulder joint needs to be designed to withstand the overall weight of the robot arm and other external loads. The shoulder shell is made of

BSB plastic, while the main force-bearing structures is made of aluminum alloy 1060. The shoulder joint is a rotary joint, and the integrated joint eRob110H is used inside; The second shoulder joint is the pitching joint of the shoulder, and the integrated joint eRob90H is used inside; The three shoulder joints are rotary joints, and the integrated joint eRob80I is used inside (see Fig. 7)

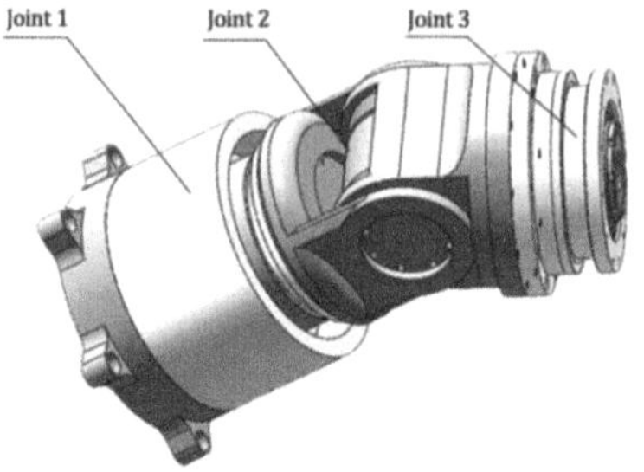

Fig. 7. Shoulder assembly

4.3 Mechanical Elbow Joint Structure Design

The elbow joint is another important part of the robotic arm. The elbow has only one degree of freedom, allowing the arm to bend and extend horizontally. Although the elbow joint usually bears less load than the shoulder joint, it still needs to be designed to withstand the weight of the mechanical arm and the external load of the structure, the elbow shell is made of BSB plastic, its internal flange is made of aluminum alloy, and the integrated joint eRob80H is used inside the elbow joint, (see Fig. 8).

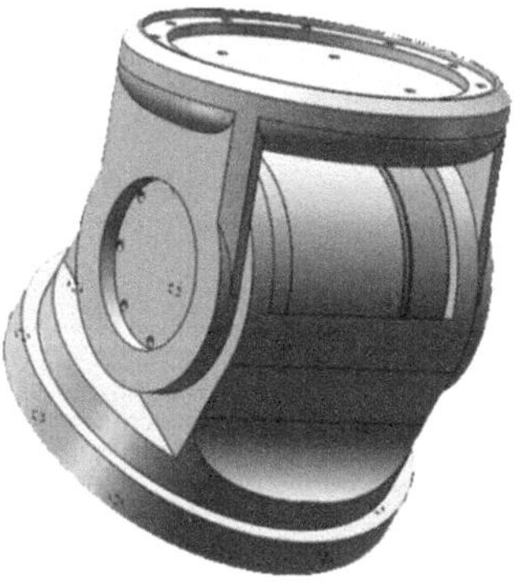

Fig. 8. Elbow assembly

4.4 Wrist Joint Design

The design of the wrist joint determines the attitude relationship of the robot arm and is a crucial part of the design of the robot arm. The wrist as a whole includes three degrees of freedom. The schematic diagram of the structure (see Fig. 9). Different from the front and back rotating joints, the wrist pitching joints will adopt a parallel linear mechanism, in which the parallel linear mechanism is realized through the directional planetary roller screw to transform the rotating motion into a linear motion. When designing the wrist structure, the whole structure design is focused on a straight joint in the wrist.

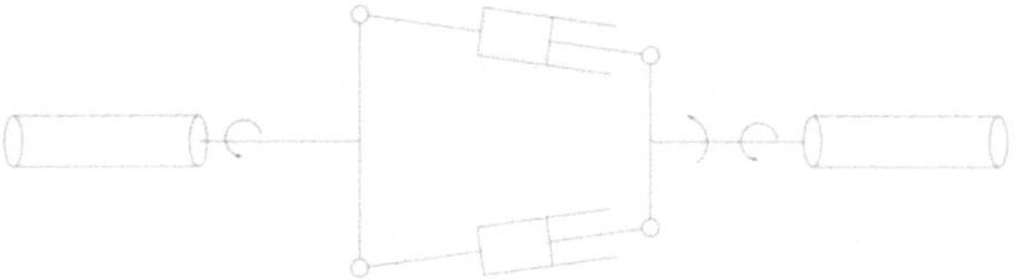

Fig. 9. Wrist structure diagram

The design of linear mechanism mainly includes the modeling assembly of the reducer and the design of the planetary roller screw. The design of planetary roller screw includes: hollow screw, roller, nut three parts design. The following table data is obtained through calculation (Table 3).

Table 3. Design Parameter Table of Roller Screw

	Hollow lead screw	Long nut	roller
Number of heads	$n_s = 3$	$n_n = n_s = 3$	$n = 1$
Thread profile	Triangle	Triangle	Triangle
Profile	90°	90°	90°
Angle lead	$L_s = 2$	$L_n = L_s = 2$	$L_R = 0.67$

Modeling through SolidWorks: hollow screw: middle diameter $d_s = 10$, lead $L_s = 2$ mm, number of thread heads $n_s = 3$, pitch $P_s = \frac{L_s}{n_s} = \frac{2}{3} = 0.67$ mm, number of teeth of spur gear at the end $z_s = 40$ rollers: Pitch $P_r = 0.67$ mm pitch diameter $d_r = \frac{d_s}{n_s} = \frac{10}{3} = 3.33$ mm, take the end helical gear teeth $z_r = 13$ long nut: Pitch diameter $d_n = d_s + 2d_r = 10 + 6.66 = 16.66$ mm, here take 17 mm, lead $L_n = 2$ mm, thread $n_n = n_s = 2$ mm, pitch $P_n = 0.67$ mm, the helical gears at the roller ends and the straight gear at the hollow screw end have a gear ratio of 1:3(see Fig. 10).

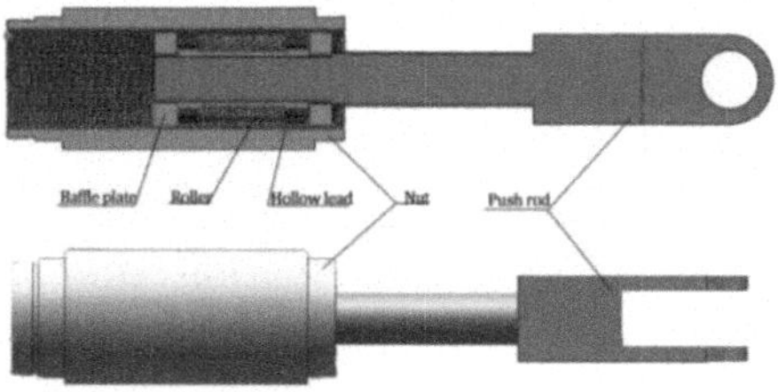

Fig. 10. Assembly diagram of planetary roller screw

In addition to the flexwheel and cross roller bearings, the harmonic reducer also includes steel wheel and wave generator, in which the steel wheel is matched with the bearing inner diameter, the flexwheel is matched with the bearing outer diameter, and the steel wheel and flexwheel size are obtained through the matching relationship, and the wave generator is installed with the flexwheel inside. The harmonic reducer is modeled with SolidWorks (see Fig. 11).

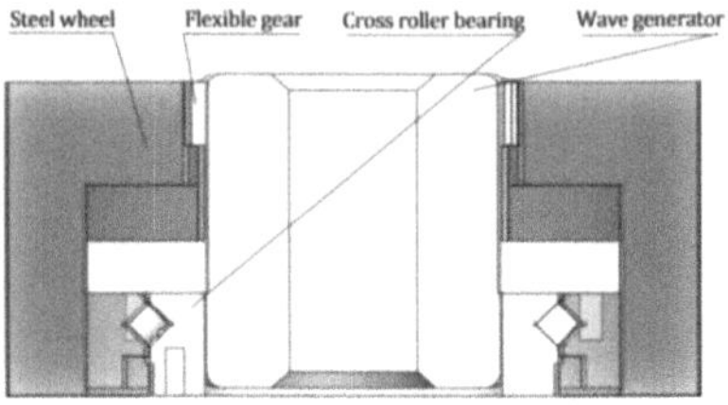

Fig. 11. Harmonic drive

After the design of the directional planetary roller screw and the modeling of the reducer are completed, the following assemblies are assembled into the designed housing in the order of planetary roller screw, roller screw - reducer connection flange, harmonic reducer, shaft and frameless motor, (see Fig. 12).

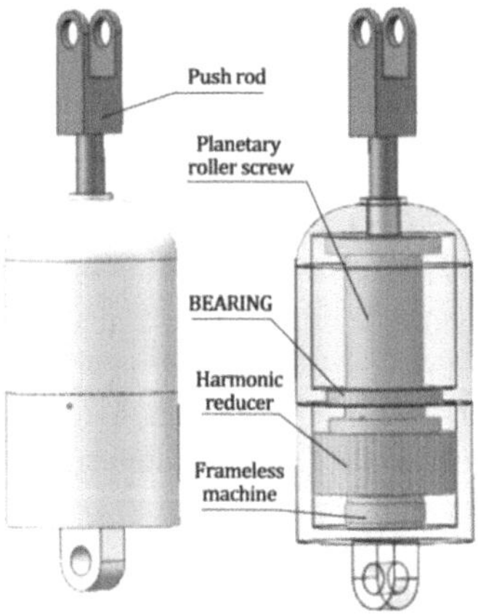

Fig. 12. Harmonic reducer assembly

Wrist joint 5 and joint 7 are rotary joints, wrist joint 5 adopts integrated joint eRob70H, and joint 7 adopts integrated joint eRob70F. Most of the shell of the wrist is made of BSB plastic material, of which the H-shaped plate with larger load is made of aluminum alloy material, and the planetary roller screw push-rod part is made of ordinary carbon steel. The two linear joints are placed in parallel to get a parallel linear mechanism, and the parallel linear mechanism designed previously is connected with two rotating joints in series to get the entire wrist model, (see Fig. 13).

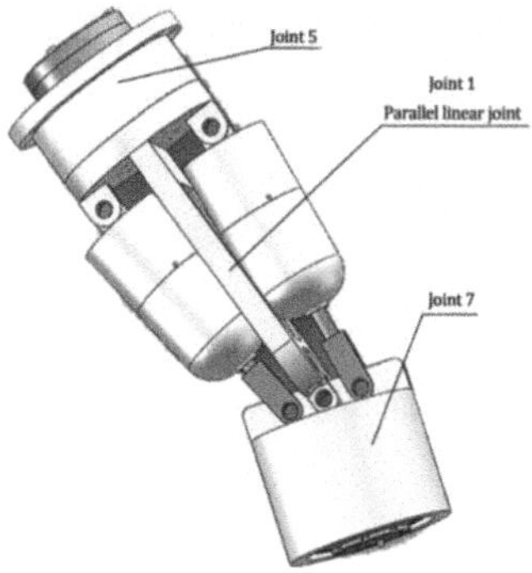

Fig. 13. Wrist assembly

4.5 Overall Assembly of Humanoid Robotic Arm

After designing the three joints, the shoulder, elbow and wrist joints are assembled in series to obtain a complete seven-degrees-of-freedom humanoid robot arm assembly model,as shown in Fig. 14.

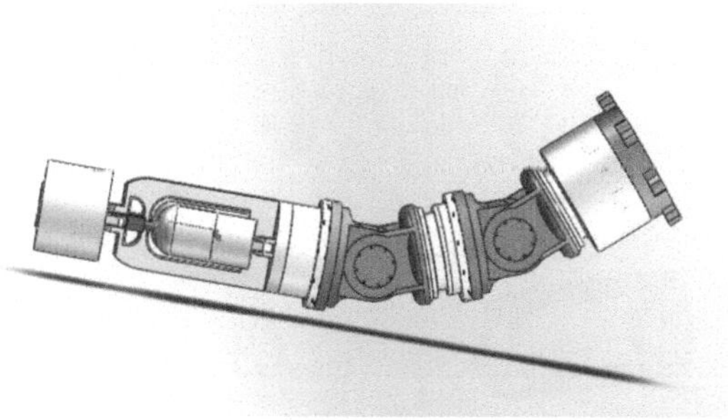

Fig. 14. Humanoid robotic arm assembly

The weight simulation was carried out by SolidWorks. The weight of the shell measured with SolidWorks was 4.417 kg and the mass of the linear joint was 0.226 kg, (see Fig. 15). It can be seen that the mass of joint eRob70F is 0.77 kg, that of node eRob70F is 0.87 kg, that of joint eRob110H is 2.88 kg, that of eRob80H is 1.19 kg, and that of eRob80I is 1.19 kg. The mass of eRob90H is 1.87 kg, and the total mass calculated is 0.77+2.88+1.19*2+0.226*2+0.87+1.87+4.417=13.639 kg, which meets the requirements of lightweight robot arm.

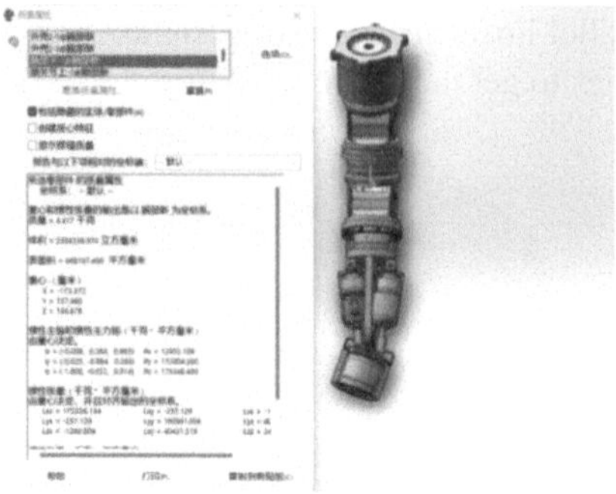

Fig. 15. Calculation of weight of robotic arm

5 Static Analysis

5.1 Static Mechanical Simulation of Core Components

In addition to dynamic analysis of each joint, static mechanical analysis of some core parts is also carried out to ensure that they can meet the design requirements, using static mechanical analysis of parts in SolidWorks: (1) First, the parts are fixed; (2) Apply load or torque to the part under load; (3) Build a grid; (4) Perform static mechanical analysis using SolidWorks to obtain the results, (see Fig. 16).

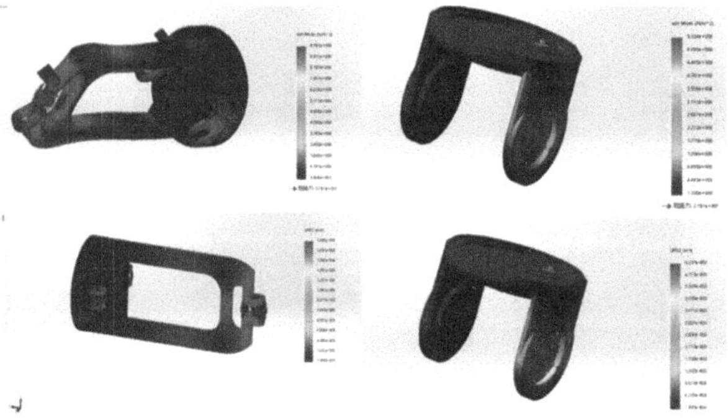

Fig. 16. Static mechanics analysis

Through the simulation structure, it can be seen that the maximum force required for the joint to produce obvious deformation is obviously less than the yield limit of the material, which proves that the selected material is reasonable.

5.2 Based on SolidWorks, the Dynamic Analysis and Verification of the Robot Arm are Carried Out

In order to verify whether each joint can meet the load requirements of the robot arm, it is necessary to verify and analyze it. Here, the motion plug-in in SolidWorks is used to verify it. By installing a 5 kg load on the end of the robot arm, the rotary motor is simulated for each joint part. The load required by each joint of the manipulator is judged by simulation analysis, and whether the selected joint can meet the requirements is judged by comparison. The steps of SolidWorks dynamic simulation are as follows: (1) A load of 5 kg is applied to the end joint (2) a rotary motor is added to each joint, and the movement is carried out according to the acceleration - uniform speed - deceleration curve (3) Gravity is added to the mechanical arm (4) The highest torque required by each joint is calculated, and the selected joint is compared for analysis and judgment to determine whether the selected joint can meet the requirements

Through the static mechanics analysis of the wrist linear joint and the pitching joint, it is judged that they meet the design requirements. The dynamic simulation of each joint was carried out, and the joint selection design met the dynamic requirements.

6 Simulation and Analysis of Humanoid Robotic Arm Motion

6.1 Robot Arm Joint Motion Range Verification

In the kinematic analysis, it is not necessary to consider the joint limit Angle of the robot arm, but in fact, due to the overall shape of the robot arm, the movement of many joints is not a 360-degree rotation. Here, the rotation Angle of each joint is measured to determine the Angle of each joint.

Measurement steps: (1) Open the joint model of the robot arm in SolidWorks and add a rotating motor or a linear motor to each joint. (2) Add solid contact to the parts that may contact. (3) Perform motion simulation. (4) Measure the motion Angle of each joint when the motion stops. (5) In order to avoid parts contact during actual motion, The joint limit Angle should be slightly less than the measured Angle (Table 4).

Table 4. Joint angle limit

Joint	Joint angle limit
Joint 1	$\pm 180^{\circ}$
Joint 2	$\pm 70^{\circ}$
Joint 3	$\pm 180^{\circ}$
Joint 4	$\pm 70^{\circ}$
Joint 5	$\pm 180^{\circ}$
Joint 6	$\pm 50^{\circ}$
Joint 7	$\pm 180^{\circ}$

By comparing the joint motion angles of the human arm shown in Summary of Normal Motion Range of Human Joints on Orthopedic Basis, it can be seen that the joint motion angles of the designed robotic arm joints are close to those of the human arm.

6.2 Experiment on Trajectory Planning of Robotic Arm

In order to understand the motion ability of the robot arm, the trajectory planning experiment is carried out in the motion space of the robot arm. The principle of trajectory planning experiment is as follows: set the trajectory required by the robot arm in advance, discretize the trajectory to a number of points, invert each point one by one, obtain the joint Angle of the robot arm, and obtain the motion process of the robot arm through the joint Angle to complete the inverse solution, (see Fig. 17).

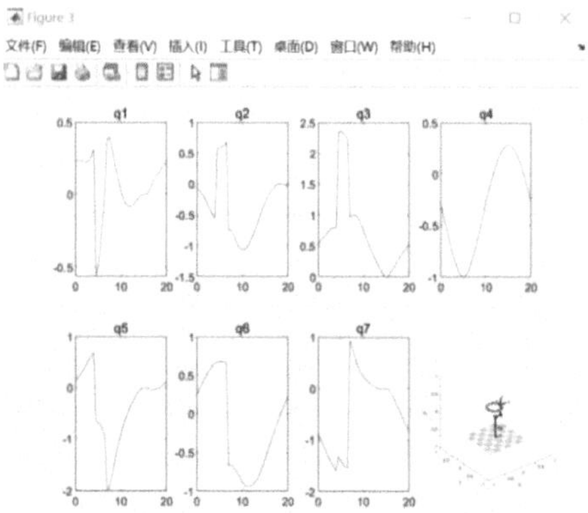

Fig. 17. Joint angle variation diagram

By discretizing the trajectory directly, the joint Angle changes too sharply with time, which is easy to impact the joint movement of the robot arm. In this paper, seventh-degrees polynomial is used to optimize the trajectory, so that the joint Angle changes smoothly with the trajectory (see Fig. 18).

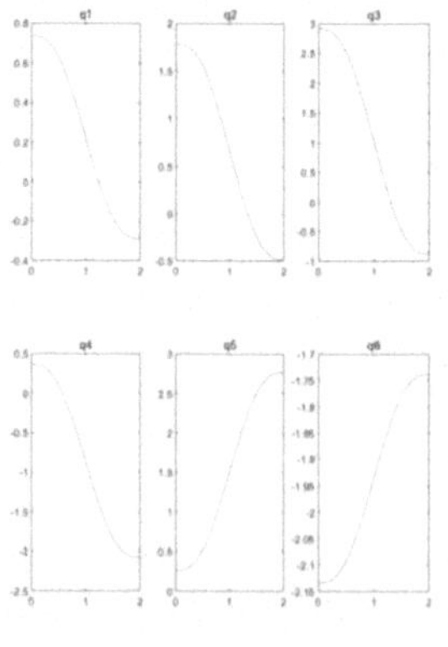

Fig. 18. Joint angle variation diagrams

After the improvement, the joint Angle of the manipulator moves smoothly with time, and the impact on the joint is small, which has certain reference significance for the actual trajectory movement of the manipulator.

7 Conclusions

7.1 Main Outcomes of the Study

This study focused on addressing the core demands of flexibility, anthropomorphism, and lightweight design in humanoid robotic arms, and achieved three key outcomes:

Configuration Innovation: A 7-degrees-of-freedom (7-DOF) humanoid robotic arm with an SRS (Shoulder-Rotator-Spherical) configuration was developed. The SRS structure mimics the spherical motion characteristics of the human shoulder and wrist, while the wrist module was innovatively designed with parallel linear joints—forming a integrated structure that balances anthropomorphic motion and load stability.

Key Parameter Optimization: Through material selection and structural thickness optimization of arm links and joint casings, the total mass of the robotic arm was controlled within 3.5 kg (15% lighter than conventional 7-DOF arms with similar load capacity). High-torque servo motors with harmonic drives were adopted for shoulder and elbow joints to balance torque output and weight, further enhancing energy efficiency.

Performance Index Achievement: The optimized robotic arm met all predefined design targets, including a workspace coverage rate of over 90% in typical front-of-body manipulation scenarios, a maximum stable load capacity of 2 kg, and joint motion ranges exceeding minimum requirements.

7.2 What Experiments/Simulations Demonstrate

Multiple simulation tools and performance tests were employed to verify the robotic arm's performance, with clear validation results as follows:SolidWorks Simulations: Verified the rationality of lightweight design by analyzing overall weight distribution; confirmed that redundant mass was reduced while structural rigidity was maintained (no excessive deformation under rated loads).MATLAB Kinematic Modeling: Simulated workspace and motion trajectories, demonstrating that the 7-DOF configuration achieves high flexibility—supporting continuous motion in complex spatial paths and precise replication of human arm movements. Finite Element Analysis (FEA): Conducted static mechanical simulations, proving that the arm can stably bear a 2 kg load without structural deformation (stress concentration remained below the yield strength of the selected aluminum alloy). Adams Dynamic Simulations: Validated dynamic response stability; under a 1 kg payload, the arm maintained smooth motion with minimal vibration, confirming its adaptability to high-load dynamic operations. Performance Tests: Confirmed joint motion ranges and lightweight advantages, providing direct evidence for design feasibility.

7.3 Limitations of the Current Design

Despite meeting core performance targets, the current design has three notable limitations:

(1) Lack of Physical Prototype Validation: All performance data were derived from simulations; no physical prototype has been fabricated to verify real-world operational performance.
(2) Limited Environmental Adaptability: The design did not consider the impact of extreme environments on joint motion stability and structural durability, restricting its application in complex scenarios.
(3) Basic Control and Interaction Capabilities: The study focused on structural design and performance simulation, but did not develop advanced control algorithms or conduct human-robot interaction tests, limiting practical usability.

7.4 Directions for Future Work

To address the above limitations and expand the robotic arm's application value, future research will focus on four directions:

(1) Physical Prototype Development and Validation: Fabricate a full-scale prototype based on the current design; conduct real-world tests to verify motion accuracy, load capacity, and long-term reliability, and optimize structural details based on test feedback.
(2) Advanced Control Algorithm Optimization: Develop adaptive control algorithms to improve the arm's resistance to external disturbances; integrate collision avoidance logic to enhance safety during human-robot interaction.
(3) Environmental Adaptability Enhancement: Modify structural materials (e.g., corrosion-resistant coatings) and joint designs to test performance under extreme temperatures, humidity, and dust conditions, expanding applicable scenarios.
(4) Application-Oriented Function Expansion: Combine the robotic arm with sensors to develop application-specific solutions for fields such as medical auxiliary operations, industrial assembly, and home service robotics.

Acknowledgments. I thank my supervisor for guidance on 7-DOF SRS configuration and kinematic modeling. Gratitude to Beijing Institute of Petrochemical Technology for providing simulation software and lab resources. Thanks to colleagues for discussions on component design, and my family for constant support.

Disclosure of Interests. The authors declare no known competing financial interests or personal relationships that could have appeared to influence the work reported in this paper on the design of the 7-DOF humanoid robot arm system. No funding from commercial entities related to robotics manufacturing was received for this research, ensuring impartiality in the study's design, data analysis, and conclusions.

References

1. Howard, B.: Check out this scary robotic-arm charger for the Tesla Model S. ExtremeTech.com **2015**(7), 1–15 (2015)
2. Tesla Motors, Inc.: Patent issued for robotic processing system and method. J. Eng. **2015**(7), 5134–5137 (2015)
3. Shabnom, M., et al.: Mathematical model and evaluation of dynamic stability of industrial robot manipulator: universal robot. Syst. Soft Comput. **6**, 62–71 (2024)
4. Shabnom, M., et al.: Mathematical model and evaluation of dynamic stability of industrial robot manipulator: universal robot. Syst. Soft Comput. **6**, 200071–200074 (2024)
5. Ye, W., Zhang, P., Chang, H.: A data-driven optimal time-delayed control approach and its application to aerial manipulators. Control Eng. Pract. **142**, 105754–105757 (2024)
6. Yamokoski, J., Radford, N.: Robonaut, Valkyrie, and NASA Robots, pp. 201–214. Springer, Amsterdam (2018)
7. Anonymous: R&D Effort to Focus on Robotic Dexterity. Assembly **65**(10), 10–20 (2022)
8. Gheorghe, A.C.: Robotic humanoid arm. Sci. Bull. Electr. Eng. Fac. **20**(1), 37–39 (2020)
9. Roschi, S.: The Meca500: small and ready to conquer the robot world. Mach. Des. **89**(7), 15–26 (2017)
10. Xiao, X.: Study on Dynamic Characteristics and Safety of Pneumatic Muscle Humanoid Robotic Arm. Southwest Jiaotong University, Chengdu (2021). https://doi.org/10.27414/dc.nki.Gxnju.2020.000521
11. Keller, J.: DARPA asks Boston Dynamics to build enhanced version of legged infantry-support robot. Military Aeros. Electron. **24**(12), 15–30 (2013)
12. Jackson, D.: Tesla CEO Elon Musk plans 'futuristic' robotaxis, optimus bot to start production. Urgent Commun. **2022**(11), 26–56 (2022)
13. Liu, L., et al.: Robots for sustainability: evaluating ecological footprints in leading AI-driven industrial nations. Technol. Soc. **76**, 102460–102463 (2024)
14. Lu, Y.: Arm Design of Biped Robot. Huazhong University of Science and Technology, Wuhan (2020)
15. Zhihai, L., Yongfeng, Y., Shaolu, T., et al.: Statics analysis of 6-DOF manipulator based on ANSYS. Coal Mine Mach. **41**(12), 75–77 (2020)
16. National Ergonomics Standardization Technical Committee: Chinese adult body size. China National Standardization Administration, Beijing (2024)

Gait Planning and Simulation of a Bipedal Humanoid Robot Based on ROS/MuJoCo

Dianjun Wang[1], Xiaofan Yang[1], Ya Chen[1](✉), Chaoqiang Xu[1], and Peng Wang[2]

[1] College of Mechanical Engineering, Beijing Institute of Petrochemical Technology, Beijing 102617, China
chenya@bipt.edu.cn

[2] College of Information Engineering, Beijing Institute of Petrochemical Technology, Beijing 102617, China

Abstract. Gait planning is a core technology for bipedal humanoid robots, as stable and smooth walking remains a key challenge in their development. To address the limitations of existing methods and improve walking stability, a multi-stage ZMP-based gait planning method integrated with the linear inverted pendulum (LIP) model and Levenberg-Marquardt (L-M) algorithm is proposed. The method involves designing trajectory generators for reference ZMP, CoM (based on LIP), and ankle joints, followed by optimizing joint trajectories via the L-M algorithm. A simulation platform based on ROS, MuJoCo, and Qt is constructed to verify the method through flat-ground straight-line walking experiments. Simulation results demonstrate that the ZMP trajectory stably stays within the support polygon, while the CoM and joint trajectories maintain good smoothness. This work provides a practical and reliable solution for bipedal robot gait planning, offering valuable reference for related research.

Keywords: biped humanoid robots · ZMP criterion · Gait planning

1 Introduction

In recent years, robotics has advanced rapidly, emerging as one of the hottest fields in the high-tech industry. Among various types of robots, bipedal humanoid robots stand out due to their uniqueness and potential value [1]. In the early development stage of humanoid robots, Waseda University in Japan developed the walking robot WABOT-I in 1973, based on the Zero Moment Point (ZMP) concept [2–4]. In recent years, humanoid robots have become a major research focus: for instance, Chinese company Xiaomi unveiled its humanoid robot CyberOne in August 2022 [5], while another Chinese firm Unitree Technology launched the full-size general-purpose humanoid robot Unitree H1 in August 2023. In 2024, American company Tesla released its humanoid robot "Tesla Optimus Gen 2" [6].

However, despite the significant progress achieved in the mechanical structure design and perception system integration of biped humanoid robots, they still face numerous urgent technical challenges in the research and development process, such as adaptation

H. Wei and J. Niu (Eds.): COBOT 2025, CCIS 2893, pp. 72–89, 2027.
https://doi.org/10.1007/978-981-92-2074-8_5

to complex dynamic environments and coordination of multi-objective motions. As the core foundation for humanoid robots to achieve human-like interaction and autonomous operations, stable, fast, and flexible walking capability is not only a key functional indicator but also a research focus and cutting-edge hot topic in the current robotics field. To break through the existing technical bottlenecks and realize the stable, fast, and flexible walking performance of biped humanoid robots in dynamic scenarios, the design and optimization of efficient and robust control algorithms are undoubtedly the core key and technical support.

Gait planning is a core and key technology for humanoid robots to achieve stable and human-like walking. Researchers have conducted extensive studies on biped robot gait planning and achieved numerous results in the past. In early research, the capturability theory [7, 8], central pattern generator (CPG)-based trajectory generation method [9], gait planning method based on the zero-moment point (ZMP) stability criterion [10, 11], and gait planning methods based on inverted pendulum and flywheel models [12–15], etc., have been successively proposed. These classic methods each have their own characteristics and applicable scenarios, but in practical applications, there is still a significant deviation between the robot's actual operating state and the expected target, and its comprehensive motion performance such as flexibility and stability still has a considerable gap compared with human walking. In recent years, biped robot gait control methods based on model predictive control (MPC) and Whole-Body Control (WBC) have become common technical solutions in this field due to their strong constraint handling capabilities and dynamic optimization characteristics. Nevertheless, MPC and WBC often suffer from high computational complexity, which may limit their real-time performance in embedded systems for humanoid robots. Additionally, these methods require accurate modeling of complex dynamics and strict parameter tuning, increasing the implementation difficulty for practical applications. To quickly verify the correctness, effectiveness, and reliability of control algorithms, simulation testing is an efficient and low-cost verification method.

To address the limitations of existing methods, this paper proposes a ZMP-based gait planning method with distinct advantages in computational efficiency and implementation simplicity. Compared with classic ZMP/LIP methods, the proposed approach introduces the Levenberg-Marquardt (L-M) algorithm for joint trajectory optimization, which enhances trajectory smoothness and tracking accuracy by effectively solving nonlinear least squares problems with strong robustness to initial values. Compared with MPC and WBC, our method avoids complex dynamic modeling and online rolling optimization, reducing computational overhead while ensuring walking stability for pre-planned motions. The core innovations of this work are threefold: 1) Integrating ZMP stability criterion with L-M optimization to generate high-precision and smooth joint trajectories; 2) Balancing control performance and computational efficiency, making it more suitable for efficient offline trajectory generation in robot systems; 3) Building a ROS/Qt/MuJoCo co-simulation platform that enables systematic verification of the gait planning workflow. Finally, the effectiveness and reliability of the proposed gait planning method are verified through simulation experiments.

2 Bipedal Humanoid Robot Model

This paper focuses on a bipedal humanoid robot, with the DOF configuration shown in Fig. 1.

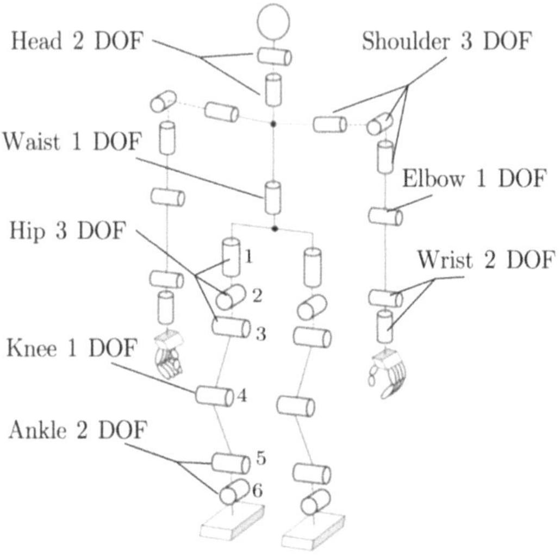

Fig. 1. DOF configuration of a humanoid robot

As shown in Fig. 1, the bipedal humanoid robot studied in this paper has a total of 27 degrees of freedom, with each leg possessing 6 degrees of freedom. The hip joint, knee joint, and ankle joint have 3, 1, and 2 degrees of freedom, respectively. The 3D model of the humanoid robot was created using SolidWorks software, as illustrated in Fig. 2.

Fig. 2. 3D model of a biped humanoid robot

3 Gait Parameters and Planning Process

3.1 Gait Parameter Settings

In this paper, the scenario of straight-line walking on flat ground is studied. A walking cycle of the humanoid robot is divided into a right-leg stance-swing cycle and a left-leg stance-swing cycle [16], as shown in Fig. 3.

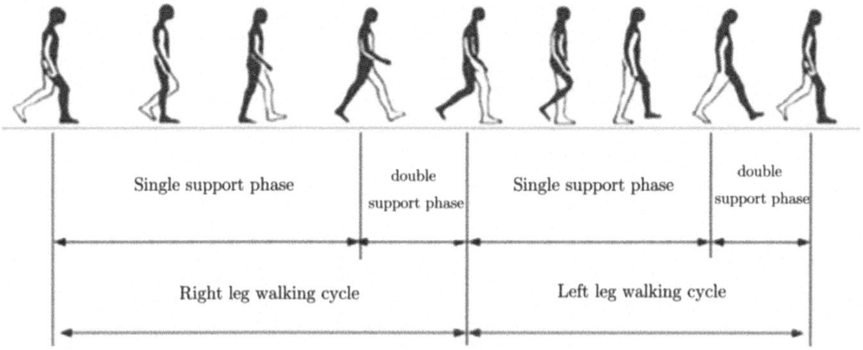

Fig. 3. Walking cycle diagram

Gait parameters during walking are configured based on human gait characteristics and the mechanical constraints of the robot's joint motion range, with specific values listed in Table 1.

Table 1. Gait Parameters

Gait parameters	Parameter value
Step length	0.125 m
Single-leg lift height	0.025 m
Single support phase duration	0.7 s
Double support phase duration	0.7 s
Distance between feet	0.233 m
Step count	4

Among them, the step length is the distance the swinging leg advances within one walking cycle; the single-leg lift height is the maximum height during ankle joint movement; the distance between feet is the distance between the ankle joints of both legs when the humanoid robot is fully upright; and the step count refers to the number of walking cycles the humanoid robot undergoes during the walking process.

3.2 Process of Gait Planning

The gait planning process for humanoid robots is illustrated in Fig. 4. Firstly, gait parameters are configured. Then, a reference joint trajectory is generated via the sequential operation of a ZMP trajectory generator, a centroid trajectory generator, and an ankle joint trajectory generator.

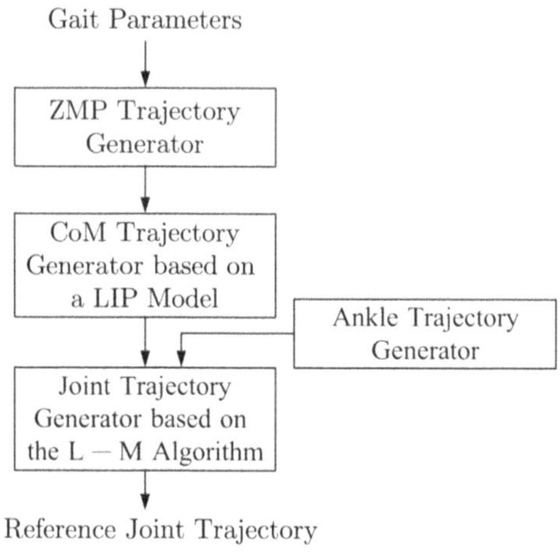

Fig. 4. Process of gait planning

4 Gait Trajectory Generation

4.1 Reference ZMP Trajectory Generation

The ZMP criterion for the walking stability of humanoid robots [17] states that the robot can walk stably when the ZMP trajectory always lies within the support polygon. To obtain a stable walking gait, based on the gait parameters, a reference ZMP trajectory is generated, as shown in Fig. 5.

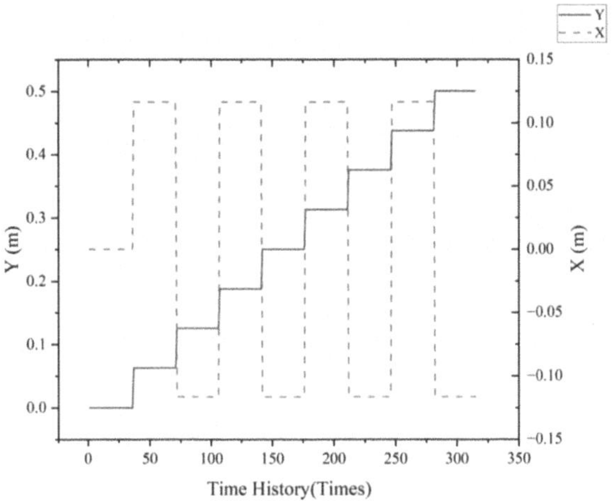

Fig. 5. Reference ZMP trajectory

4.2 Reference CoM Trajectory Generation Based on a LIP Model

To effectively describe the dynamic characteristics of a humanoid robot walking in a straight line on flat ground, a three-dimensional linear inverted pendulum (LIP) model was established for the robot's legs [18, 19]. The LIP can be decomposed into two planar linear inverted pendulums in the forward and lateral directions, which move in the X-Z and Y-Z planes, respectively. It is assumed that the height of the CoM of the LIP (i.e., the Z-coordinate of CoM) remains constant during walking.

Taking the planar LIP in the Y-Z plane as an example, the contact point between the LIP and the ground is the ZMP, as shown in Fig. 6.

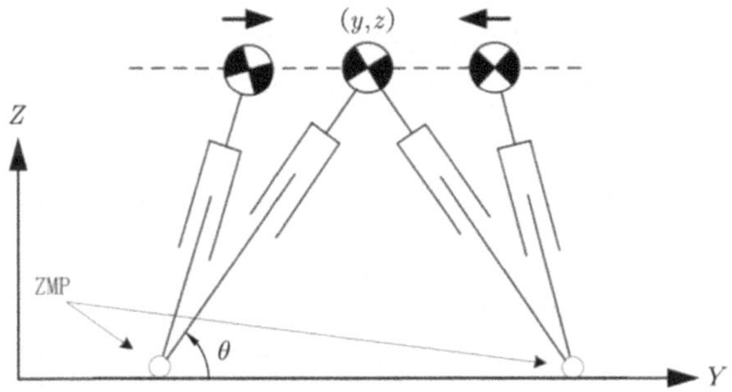

Fig. 6. Planar LIP in the Y-Z plane

Assuming the CoM coordinates of the LIP are (y, z), where the height z is constant, and the angle between the LIP and the positive Z-axis is θ, then the equation of motion for the CoM of the LIP is

$$\ddot{y} = g \tan\theta = g\frac{y}{z} \tag{1}$$

The Y-coordinate of ZMP is

$$p_{ZMPy} = y - \frac{z}{\tan\theta} = y - \frac{z}{g}\ddot{y} \tag{2}$$

The position, velocity, and acceleration of the CoM are taken as state variables, while the derivative of the CoM acceleration u_y is designated as the system input

$$u_y = \frac{\mathrm{d}}{\mathrm{d}t}\ddot{y} \tag{3}$$

Then the state-space equation is

$$\begin{cases} \frac{\mathrm{d}}{\mathrm{d}t}\begin{bmatrix} y \\ \dot{y} \\ \ddot{y} \end{bmatrix} = \begin{bmatrix} 0 & 1 & 0 \\ 0 & 0 & 1 \\ 0 & 0 & 0 \end{bmatrix}\begin{bmatrix} y \\ \dot{y} \\ \ddot{y} \end{bmatrix} + \begin{bmatrix} 0 \\ 0 \\ 1 \end{bmatrix} u_y \\ p_{ZMPy} = \begin{bmatrix} 1 & 0 & -\frac{z}{g} \end{bmatrix}\begin{bmatrix} y \\ \dot{y} \\ \ddot{y} \end{bmatrix} \end{cases} \tag{4}$$

Set the sampling time as T, discretize Eq. (4), then

$$\begin{cases} \mathbf{y}(k+1) = \mathbf{A}\mathbf{y}(k) + \mathbf{b}u_y(k) \\ p_y(k+1) = \mathbf{c}\mathbf{y}(k) \end{cases} \tag{5}$$

where

$$\begin{cases} \mathbf{y}(k) = \left[y(kT), \dot{y}(kT), \ddot{y}(kT)\right]^{\mathrm{T}} \\ u_y(k) = u_y(kT) \\ p_y(k) = p_y(kT) \end{cases}$$

$$\mathbf{A} = \begin{bmatrix} 1 & T & T^2/2 \\ 0 & 1 & T \\ 0 & 0 & 1 \end{bmatrix} \quad \mathbf{b} = \begin{bmatrix} T^3/6 \\ T^2/2 \\ T \end{bmatrix} \quad \mathbf{c} = \begin{bmatrix} 1 & 0 & -\frac{z}{g} \end{bmatrix}$$

According to the preview control theory [20], the reference ZMP trajectory is treated as future reference information, and the current CoM position is adjusted according to changes in the ZMP trajectory, thereby the CoM trajectory is generated.

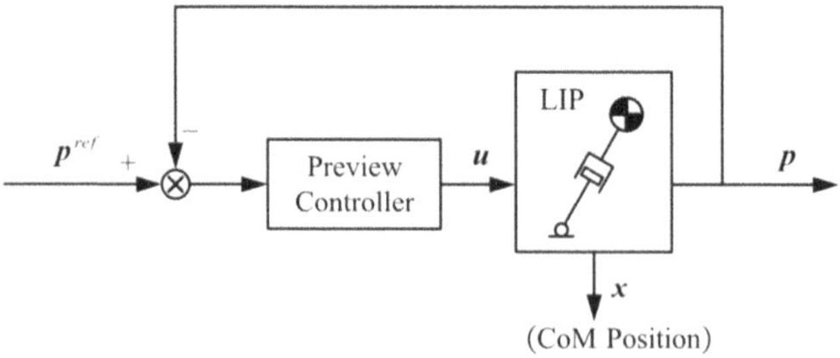

Fig. 7. Block diagram of preview control

The structure diagram of the preview control system is shown in Fig. 7, where $\mathbf{p}^{ref}$ is the reference ZMP trajectory, u is the input of the LIP, $\boldsymbol{p}$ is the output of the LIP, i.e., the output ZMP trajectory. The cost function for ZMP tracking is

$$J = \sum_{i=k}^{\infty}[Q_e \boldsymbol{e}(i)^T \boldsymbol{e}(i)+ \Delta\boldsymbol{x}^T(i)\mathbf{Q}_x\Delta\mathbf{x}(i) + R\Delta u^2(i)] \tag{6}$$

where

$$\begin{cases} \mathbf{e}(i) = \mathbf{p}_x(i) - \mathbf{p}_x^{ref}(i) \\ \Delta\mathbf{x}(i) = \mathbf{x}(i) - \mathbf{x}(i-1) \\ \Delta u(i) = u(i) - u(i-1) \end{cases}$$

where $\mathbf{e}(i)$ is the ZMP tracking error, $\mathbf{p}_x^{ref}(i)$ is the X-coordinate of reference ZMP at time iT, $\Delta\boldsymbol{x}(i)$ is the state vector increment, $\Delta u(i)$ is the input increment, Q_e and R are positive gains, and $\mathbf{Q}_x$ is a third-order positive definite matrix.

Assuming the preview step is N_L, then the optimal input is

$$\underset{u_x(k)}{\arg\min} J = -\mathbf{G}_i \sum_{i=0}^{k} \mathbf{e}(k) -\mathbf{G}_x\mathbf{x}(k) - \sum_{j=1}^{N_L} \mathbf{G}_p(j)\mathbf{p}_x^{\mathrm{ref}}(k+j) \tag{7}$$

Among them, $\mathbf{G}_i$, $\mathbf{G}_x$, $\mathbf{G}_p$ are gains, where

$$\begin{cases} \mathbf{G}_i = \left[R + \tilde{\mathbf{b}}^{\mathbf{T}}\tilde{\mathbf{K}}\tilde{\mathbf{b}}\right]^{-1}\tilde{\mathbf{b}}^{\mathbf{T}}\tilde{\mathbf{K}}\tilde{\mathbf{I}} \\ \mathbf{G}_x = \left[R + \tilde{\mathbf{b}}^{\mathbf{T}}\tilde{\mathbf{K}}\tilde{\mathbf{b}}\right]^{-1}\tilde{\mathbf{b}}^{\mathbf{T}}\tilde{\mathbf{K}}\tilde{\mathbf{F}} \\ \mathbf{G}_p = \left[R + \tilde{\mathbf{b}}^{\mathbf{T}}\tilde{\mathbf{K}}\tilde{\mathbf{b}}\right]^{-1}\tilde{\mathbf{b}}^{\mathbf{T}}\tilde{\mathbf{X}}(l-1) \end{cases} \tag{8}$$

where

$$\tilde{\mathbf{b}} = \begin{bmatrix} \mathbf{cb} \\ \mathbf{b} \end{bmatrix}, \tilde{\mathbf{I}} = \begin{bmatrix} \mathbf{I}_p \\ 0 \end{bmatrix}, \tilde{\mathbf{F}} = \begin{bmatrix} \mathbf{cA} \\ \mathbf{A} \end{bmatrix}, \tilde{\mathbf{Q}} = \begin{bmatrix} Q_e & 0 \\ 0 & \mathbf{Q}_x \end{bmatrix}, \tilde{\mathbf{A}} = [\tilde{\mathbf{I}}\ \tilde{\mathbf{F}}]$$

$$\begin{cases} \tilde{\mathbf{K}} = \tilde{\mathbf{A}}^{\mathrm{T}}\tilde{\mathbf{K}}\tilde{\mathbf{A}} - \tilde{\mathbf{A}}^{\mathrm{T}}\tilde{\mathbf{K}}\tilde{\mathbf{b}}[R + \tilde{\mathbf{b}}^{\mathrm{T}}\tilde{\mathbf{K}}\tilde{\mathbf{b}}]^{-1}\tilde{\mathbf{b}}^{\mathrm{T}}\tilde{\mathbf{K}}\tilde{\mathbf{A}} + \tilde{\mathbf{Q}} \\ \tilde{\mathbf{X}}(1) = -\tilde{\mathbf{A}}_c^{\mathrm{T}}\tilde{\mathbf{K}}\tilde{\mathbf{I}} \\ \tilde{\mathbf{X}}(l) = \tilde{\mathbf{A}}_c^{\mathrm{T}}\tilde{\mathbf{X}}(l-1), l = 2, \cdots, N \\ \tilde{\mathbf{A}}_{\mathbf{c}} = \tilde{\mathbf{A}} - \tilde{\mathbf{b}}\left[R + \tilde{\mathbf{b}}^{\mathrm{T}}\tilde{\mathbf{K}}\tilde{\mathbf{b}}\right]^{-1}\tilde{\mathbf{b}}^{\mathrm{T}}\tilde{\mathbf{K}}\tilde{\mathbf{A}} \end{cases}$$

4.3 Reference Ankle Trajectory Generation

To generate a reference ankle trajectory, key points during the walking process are specified, and a smooth ankle joint trajectory is generated based on the cubic spline interpolation method [21].

Assume the robot's single-step motion time is T_c, and the single-support phase duration is T_d. Then the execution time of the k-th step is from kT_c to $(k+1)T_c$, where $k = 0, 1, \cdots, n$. For the initial or terminal step of the motion, several key points on the starting or ending foot are selected, and then the trajectory $S_s(t) = (S_{sy}(t), S_{sz}(t))$ is planned via cubic spline interpolation. For intermediate steps during the motion, several key points on the swing foot during one step are selected, and the ankle trajectory $S_l(t) = (S_{ly}(t), S_{lz}(t))$ is planned via cubic spline interpolation, then the right ankle trajectory is described by

$$y_r(t) = \begin{cases} 0 & t \in [0, T_c] \\ \frac{k-1}{2} S_{ly}(T_c) + S_{ly}(t) & t \in [kT_c, kT_c + T_d] \\ \frac{k+1}{2} S_{ly}(T_c) & t \in [kT_c + T_d, (k+1)T_c], k = 1, 3, \cdots, n \\ \frac{n-1}{2} S_{ly}(T_c) + S_{sy}(t) & t \in [nT_c, nT_c + T_d] \\ \frac{n-1}{2} S_{ly}(T_c) + S_{sy}(T_c) & t \in [nT_c + T_d, (n+1)T_c] \end{cases} \tag{9}$$

$$z_r(t) = \begin{cases} 0 & t \in [0, T_c] \\ S_{lz}(t) & t \in [kT_c, kT_c + T_d] \\ 0 & t \in [kT_c + T_d, (k+1)T_c], k = 1, 3, \cdots, n \\ S_{sz}(t) & t \in [nT_c, nT_c + T_d] \\ 0 & t \in [nT_c + T_d, (n+1)T_c] \end{cases} \tag{10}$$

4.4 Reference Joint Trajectory Generation Based on L-M Algorithm

The deviation between the actual centroid position of the robot and the reference centroid position is

$$\mathbf{r}(\boldsymbol{\Theta}) = \mathbf{p}_c(\boldsymbol{\Theta}) - \mathbf{p}_c^{ref}(\boldsymbol{\Theta}) \tag{11}$$

The following optimization problem can be constructed to solve for the joint variables $\boldsymbol{\Theta}$ that minimize the norm of the deviation.

$$\begin{cases} \min S = \mathbf{r}(\boldsymbol{\Theta})^{\mathrm{T}} \mathbf{r}(\boldsymbol{\Theta}) \\ s.t. \quad lb_i \le \theta_i \le ub_i \quad i = 1, 2, \cdots, 6 \end{cases} \tag{12}$$

where lb_i and ub_i are the lower and upper bounds of the i-th joint variable, respectively. Taking the right leg as an example, for the k-th time point in a walking cycle, there is

$$\begin{aligned} \mathbf{r}(\boldsymbol{\Theta}_k) &= \left({}^{W}\mathbf{p}_{aR} + {}^{R}\mathbf{f}_c(\boldsymbol{\Theta}_k)\right) - {}^{W}\mathbf{p}_{ck} \\ &= {}^{0}_{6}\mathbf{R}^{-1}\left({}^{0}_{6}\mathbf{p}(\boldsymbol{\Theta}_k) - {}^{0}_{6}\mathbf{p}^{ref}(\boldsymbol{\Theta}_k)\right) \end{aligned} \tag{13}$$

where ${}^{0}_{6}\boldsymbol{R}$ is the rotation matrix of the right leg end-effector frame relative to the right leg base frame, ${}^{0}p(\boldsymbol{\Theta}_k)$ and ${}^{0}p^{ref}(\boldsymbol{\Theta}_k)$ are the actual and reference CoM positions in the world frame respectively, and $\mathbf{f}_c^R(\mathbf{\Theta}_k) = [px_c, py_c, pz_c]^T$ represents the coordinates of the actual CoM position in the right leg ankle joint frame.

$$\begin{cases} px_c = \sum_{i=0}^{n} m_i p_{xci} / \sum_{i=0}^{n} m_i \\ py_c = \sum_{i=0}^{n} m_i p_{yci} / \sum_{i=0}^{n} m_i \\ pz_c = \sum_{i=0}^{n} m_i p_{zci} / \sum_{i=0}^{n} m_i \end{cases} \tag{14}$$

where (p_{xi}, p_{yi}, p_{zi}) represents the positional components of the homogeneous transformation matrix ${}_{ci}^{6}\mathbf{T}$ from the CoM of the i-th link of the right leg to the end-effector frame of the right leg, and m_i is the mass of the i-th link of the right leg.

Assuming that the robot's feet only undergo translation without rotation during walking, then the Jacobian matrix of the deviation r with respect to $\mathbf{\Theta}_k$ is:

$$\mathbf{J}(\mathbf{\Theta}_k) = {}_{6}^{0}\mathbf{R}^{-1} \frac{\partial\left({}^{0}\mathbf{p}(\mathbf{\Theta}_k) - {}^{0}\mathbf{p}^{ref}(\mathbf{\Theta}_k)\right)}{\partial \mathbf{\Theta}_k} \tag{15}$$

Therefore, the submatrix corresponding to the right leg in the Jacobian matrix is

$$\mathbf{J}_R(\mathbf{\Theta}_k) = \begin{bmatrix} {}_{6}^{0}\mathbf{R}^{-1} & \mathbf{0}_{3\times3} \\ \mathbf{0}_{3\times3} & \mathbf{0}_{3\times3} \end{bmatrix} \mathbf{J}(\mathbf{\Theta}_k) \tag{16}$$

The optimization problem can be solved via the Levenberg-Marquardt (L-M) algorithm [22], with its iterative formula is given as

$${}^{m+1}\mathbf{\Theta}_k = {}^{m}\mathbf{\Theta}_k - \left(\mathbf{J}_R^{\mathrm{T}}\mathbf{J}_R + \lambda\mathbf{I}\right)^{-1} \mathbf{J}_R^{\mathrm{T}}\mathbf{r} \tag{17}$$

where m is the number of iterations and λ is a non-negative constant. Given the initial value of the joint variables, the joint variables at the next moment can be calculated iteratively by Eq. (17), thereby obtaining a reference joint trajectory.

The reliability of the L-M algorithm solution is verified by a simulation. Given an initial right leg joint variables $\mathbf{\Theta}_1 = [0.2461, 0.3054, 1.1135, 0.4398, 0, 0.7645]^T$, the corresponding CoM position is $\boldsymbol{R}_{fc1} = [0.032421{,}0.032421{,}0.148695]^T$, the target CoM position is set to $\boldsymbol{R}_{fc2} = [0.034424{,}\text{-}0.606823{,}0.146817]^T$, by the L-M algorithm, the right leg joint variable $\mathbf{\Theta}_2$ corresponding to $\mathbf{R}_{fc2}$ can be determined. By setting the allowable error for each joint to 0.0001 and the maximum number of iterations to 100, the joint variables reach the solution accuracy at the 13th iteration, as shown in Fig. 8.

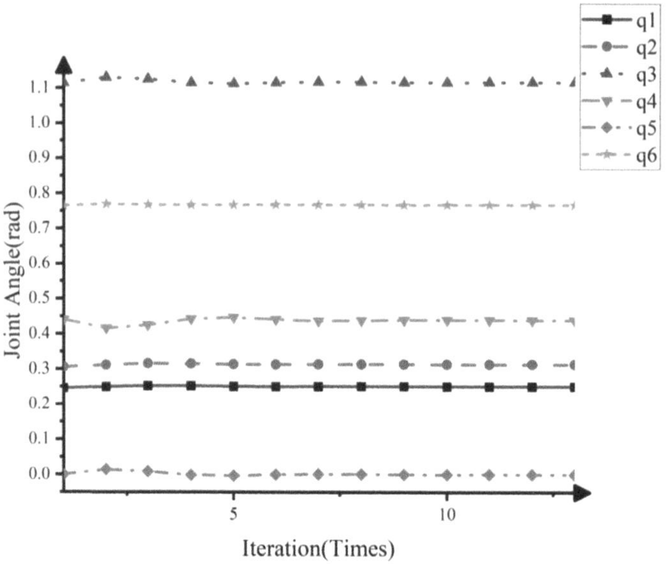

Fig. 8. Iterative process for joint angles

Substituting $\mathbf{\Theta}_2$ into the Eq. (14), the obtained CoM position relative to the right ankle joint is $[0.034432, -0.606831, 0.146813]^T$, compared with the desired CoM position $\mathbf{R}_{fc2}$, the difference is $[0.000008, 0.000008, 0.000004]^T(rad)$, thus the maximum relative error is 0.02%, which verified the correctness and reliability of the L-M algorithm.

5 Gait Planning Simulation Based on ROS/MuJoCo

5.1 Simulation Platform Construction

Considering the simulation requirements, a simulation platform was constructed on Ubuntu 20.04 (64-bit), powered by a 13th Gen Intel Core i9-13900H CPU, an NVIDIA GeForce RTX 4060 Laptop GPU (8 GB), and equipped with 32 GB RAM (5200 MT/s), based on ROS, Qt, and MuJoCo [23]. The architecture of this platform is illustrated in Fig. 9.

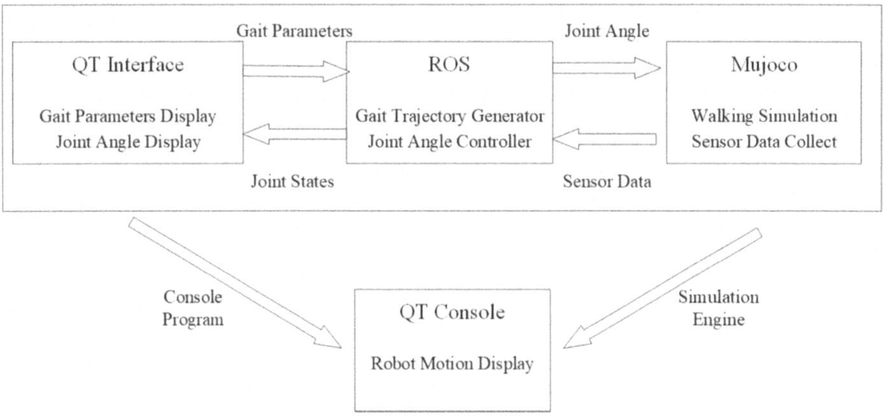

Fig. 9. Simulation platform of gait planning

The robot description file (URDF file) is exported from SolidWorks software, and imported into the Rviz simulation interface, as shown in Fig. 10.

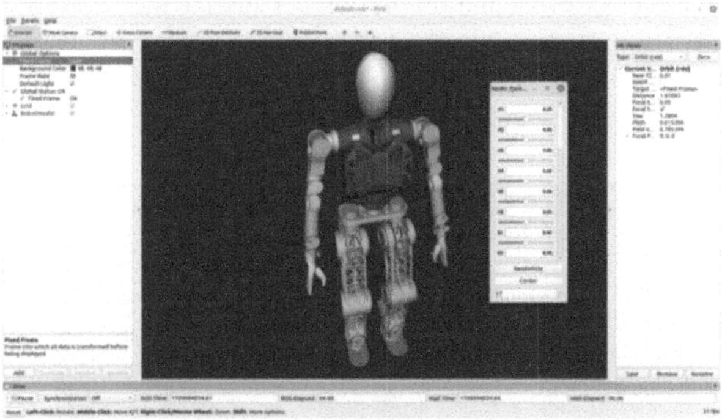

Fig. 10. Robot model in the Rviz simulation interface

Develop the host computer interface using QT and achieve communication between QT and ROS through UI controls, as shown in Fig. 11.

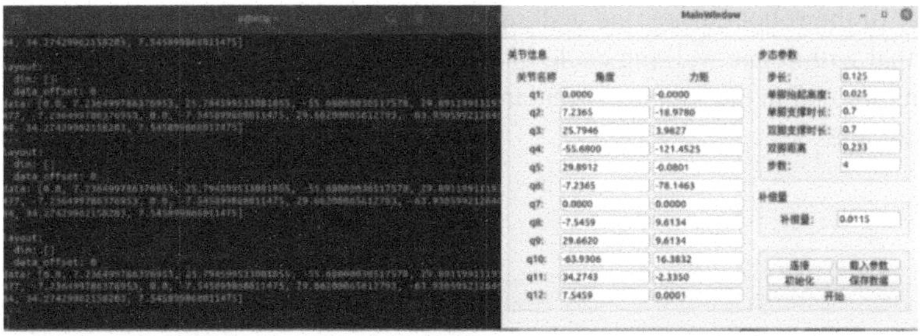

Fig. 11. UI controls for QT and ROS communication

Each trajectory generator for gait planning is realized as a ROS node, the node subscription diagram is shown in Fig. 12.

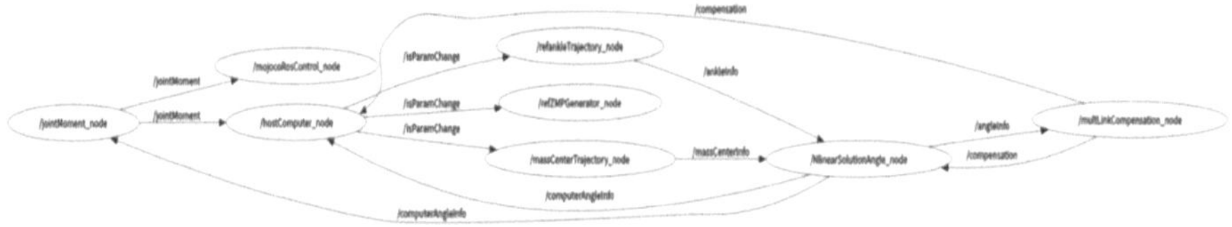

Fig. 12. Node subscription diagram

Input the gait parameters to the QT control interface, and the reference ZMP trajectory, reference centroid trajectory, and joint trajectory are generated by corresponding ROS nodes, which are then imported into the MuJoCo simulation engine.

5.2 Gait Planning Simulation

Set the simulation step size to 0.0001s, design the integrator as Runge-Kutta integration, set the gravitational acceleration, and set the contact label to the allowed state, then input the gait parameters from Table 1 into the QT control interface, the QT control platform is shown in Fig. 13.

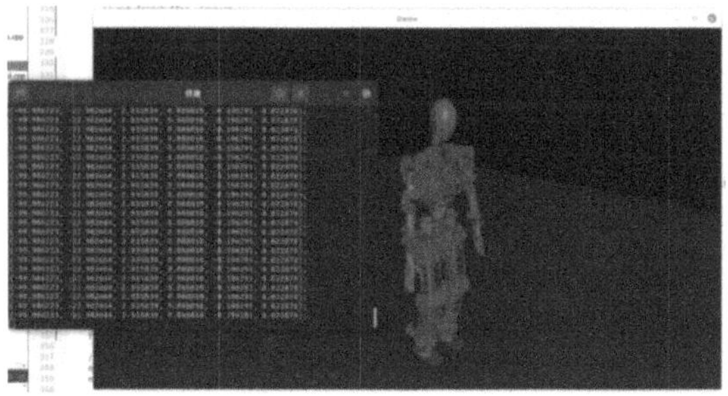

Fig. 13. QT control platform

The ZMP trajectory, centroid trajectory, and joint trajectory are recorded. The ZMP trajectory of the robot during one walking cycle is shown in Fig. 14 and Fig. 15, it can be seen that within one walking cycle of the robot, the ZMP trajectories in the X and Y directions are within the support boundaries with a stability margin of 32.74%, ensuring stable walking.

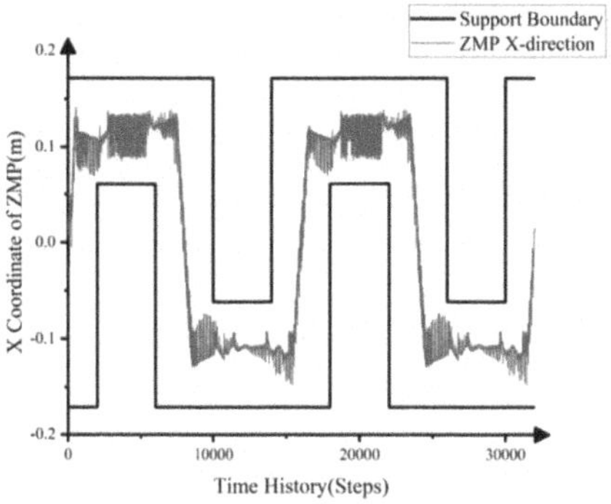

Fig. 14. ZMP trajectory in X direction

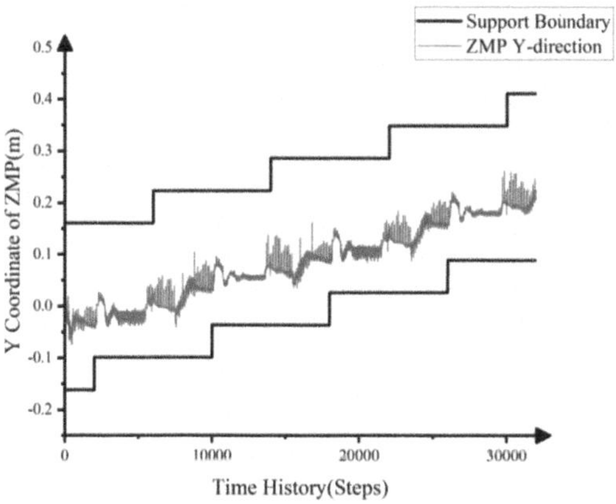

Fig. 15. ZMP trajectory in Y direction

The CoM position of the robot during one walking cycle is shown in Fig. 16 and Fig. 17.

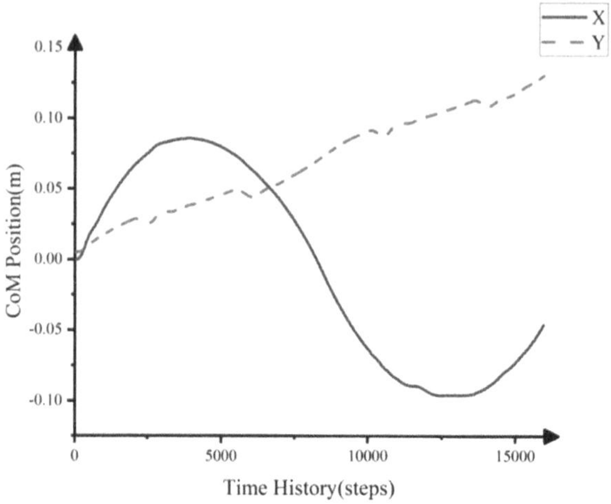

Fig. 16. CoM position in X and Y direction

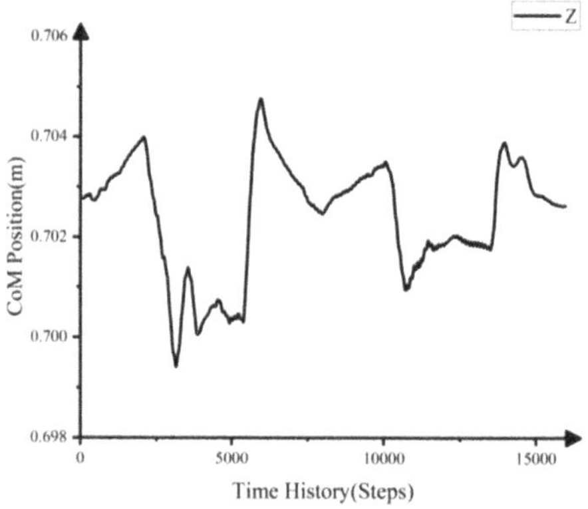

Fig. 17. CoM position in Z direction

As shown in Fig. 16 and Fig. 17, the CoM position in the X and Y directions are relatively smooth, indicating stable motion. The position in the Z direction remains around 0.703 m with a maximum amplitude of 0.00534 m, demonstrating that the centroid height is basically constant and stable, thus verifying the effectiveness of the centroid trajectory generator.

The joint variables of the robot's right leg during one walking cycle are shown in Fig. 18.

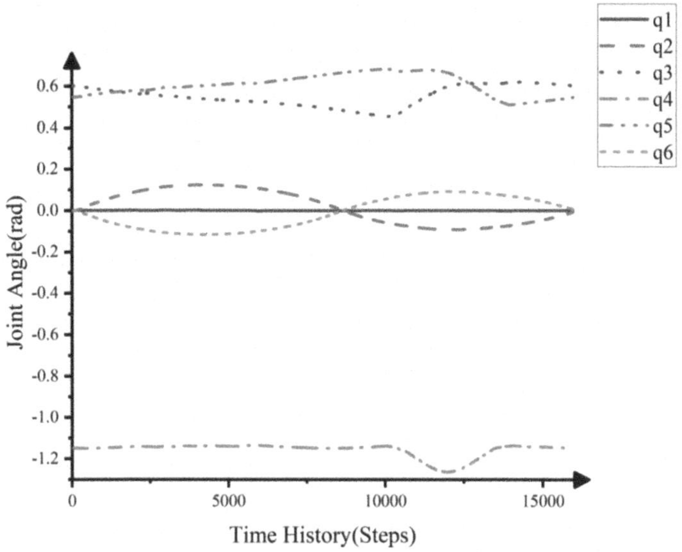

Fig. 18. Right leg joint trajectory during walking process

As shown in Fig. 18, the joint variable curves of the right leg are relatively smooth, indicating stable joint motion, which verifies the effectiveness of the joint trajectory generator.

The simulation results demonstrate that each stage of the gait planning process designed in this paper can effectively generate motion trajectories. The ZMP trajectory remains relatively stable within the support region, the centroid height is maintained at approximately 0.703m, and the final generated joint motion trajectories for the legs are smooth. The simulation results validate the feasibility and effectiveness of the gait planning method.

6 Conclusion

To address the gait planning problem for bipedal humanoid robots, a multi-stage, full-process gait planning method is proposed. The DOF configuration of the bipedal humanoid robot is completed, the basic framework of gait planning and gait parameters are configured. The design of the gait trajectory generator is accomplished, including the reference ZMP trajectory generator, reference CoM trajectory generator based on a LIP model, reference ankle joint trajectory generator, and joint trajectory generator based on the L-M algorithm.

To verify the effectiveness of the proposed method, a simulation platform based ROS, Qt, and MuJoCo is built to simulate the straight-line walking of a humanoid robot on flat ground. The ZMP trajectory, CoM trajectory, and joint trajectories during walking are recorded. Simulation results show that the ZMP trajectory consistently remains stable within the support polygon with a stability margin of 32.74%, while the CoM and joint trajectories maintain good smoothness, demonstrating that the robot can achieve stable and smooth walking motion.

The multi-stage gait planning method proposed in this paper can effectively generate the bipedal robot's motion trajectory, ensuring stable and smooth walking. It provides a practical solution with certain reference significance for the gait planning of bipedal humanoid robots.

References

1. Naksuk, N.: Whole-body human-to-humanoid motion transfer with balance manipulability ellipsoid. Doctoral Dissertation, Purdue University, West Lafayette (2006)
2. Hashimoto, K., Yoshimura, Y., Kondo, H., et al.: Realization of quick turn of biped humanoid robot by using slipping motion with both feet. In: Proceedings of 2011 IEEE International Conference on Robotics and Automation, pp. 2041–2046. IEEE (2011)
3. Hashimoto, K., Kang, H., Nakamura, M., et al.: Realization of biped walking on soft ground with stabilization control based on gait analysis. In: Proceedings of 2012 IEEE/RSJ International Conference on Intelligent Robots and Systems, pp. 2064–2069. IEEE (2012)
4. Kato, I.: Development of WABOT 1. Biomechanism **2**, 173–214 (1973)
5. Zhao: Xiaomi releases full-size humanoid bionic robot "CyberOne". Robot Technol. Appl. (05), 8 (2022)
6. Anonymous: Tesla releases the new generation humanoid robot Tesla Optimus Gen 2. Logist. Technol. Appl. **29**(01), 64 (2024)
7. Koolen, T., De Boer, T., Rebula, J., et al.: Capturability-based analysis and control of legged locomotion, part 1: theory and application to three simple gait models. Int. J. Robot. Res. **31**(9), 1094–1113 (2012)
8. Liu, G.L., Habib, M.K., Watanabe, K., et al.: CPG based control for generating stable bipedal trajectories under external perturbation. In: Proceedings of IEEE Annual Conference, pp. 1019–1022. IEEE (2007)
9. Huang, Q., Yokoi, K., Kajita, S., et al.: Planning walking patterns for a biped robot. IEEE Trans. Robot. Autom. **17**(3), 280–289 (2001)
10. Kajita, S., Kanehiro, F., Kaneko, K., et al.: Biped walking pattern generation by using preview control of zero-moment point. In: Proceedings of IEEE International Conference on Robotics and Automation, pp. 1620–1626. IEEE (2003)
11. Kamogawa, Y., Yamada, K., Masuta, H., et al.: Stability control and pattern generation for biped humanoid robot. In: Proceedings of International Conference on Control, Automation and Systems, pp. 910–915. IEEE (2013)
12. Ishiwaki, K., Kagami, S.: Strategies for adjusting the ZMP reference trajectory for maintaining balance in humanoid walking. In: Proceedings of IEEE International Conference on Robotics and Automation, pp. 4230–4236. IEEE (2010)
13. Mahani, M.A.N., Jafari, S., Rahmatkhah, R.: Novel humanoid push recovery using knee joint. In: Proceedings of IEEE International Conference on Humanoid Robots, pp. 446–451. IEEE (2011)
14. Yoshida, Y., Takeuchi, K., Sato, D., et al.: Balance control of humanoid robots in response to disturbances in the frontal plane. In: Proceedings of IEEE International Conference on Robotics and Biomimetics, pp. 2241–2247. IEEE (2011)
15. Urata, J., Ishiwaki, K.: Online decision of foot placement using singular LQ preview regulation. In: Proceedings of IEEE International Conference on Humanoid Robots, pp. 13–18. IEEE (2011)
16. Han, Y.H., Cho, B.K.: Slope walking of humanoid robot without IMU sensor on an unknown slope. Rob. Auton. Syst. (2022)

17. Kim, J., Bertram, J.E.A.: Compliant walking appears metabolically advantageous at extreme step lengths. Gait Posture **64**, 84–89 (2018)
18. Mishra, P., Jain, U., Choudhury, S., et al.: Footstep planning of humanoid robot in ROS environment using Generative Adversarial Networks (GANs) deep learning. Rob. Auton. Syst. (2022)
19. He, X.Y., Xu, J.C., Wang, Z.W., et al.: Research on preview control of humanoid robots based on inverted pendulum model. Inf. Technol. **43**(12), 154–158+164 (2019)
20. Kajita, S.: Humanoid Robotics (Translated by Guan, Y.S.). Tsinghua University Press, Beijing (2007)
21. Zhang, X.H., Wang, C.Q.: Research on manipulator trajectory planning based on LM algorithm. Software Guide **19**(11), 174–177 (2020)
22. Zamparelli, A., Scianca, N., Lanari, L., et al.: Humanoid Gait generation on uneven ground using intrinsically stable MPC. In: Proceedings of IFAC Symposium on Robot Control. IFAC (2019)

Path Planning for 6R Collaborative Robots Based on RRT Algorithm

Ya Chen[1], Linhao Yuan[1], Dianjun Wang[1(✉)], Peng Wang[2], and Shujing Liu[1]

[1] School of Mechanical Engineering, Beijing Institute of Petrochemical Technology, No.19 Qingyuan North Street, Huangcun Town, Daxing District, Beijing 102617, China
yuan13582273125@163.com

[2] School of Information Engineering, Beijing Institute of Petrochemical Technology, No.19 Qingyuan North Street, Huangcun Town, Daxing District, Beijing 102617, China

Abstract. With the ongoing transformation of the manufacturing industry, the demand for collaborative robots capable of operating in complex and constrained environments continues to rise. To maximize the efficiency of 6R collaborative robots in such settings, a key challenge lies in evaluating and selecting an appropriate path-planning algorithm from numerous available methods, which requires a comprehensive analysis of robot kinematics and planning performance. In this study, a high-fidelity kinematics model of a six-axis collaborative robot is established using the modified Denavit–Hartenberg (D-H) method, enabling the derivation of both forward and inverse kinematic solutions. The robot's workspace is further determined through the Monte Carlo method to quantitatively characterize its reachable domain. Focusing on motion planning in obstacle-rich environments, this paper compares the Rapidly-exploring Random Tree (RRT) algorithm and its improved variants, RRT* and RRTconnect. Simulation experiments conducted in MATLAB and ROS-MoveIt! demonstrate that RRTconnect achieves the fastest planning speed due to its bidirectional search mechanism, whereas RRT* generates smoother and more optimal trajectories at the expense of higher computation cost. Physical experiments performed on the Aubo-E5 collaborative robot further validate the simulation results and confirm the efficiency, adaptability, and practical feasibility of sampling-based path-planning algorithms in real-world constrained environments. Overall, the findings of this study provide a structured reference for selection of suitable path-planning algorithms for collaborative robots and offer practical guidance for enhancing operational safety,reliability, and efficiency in intelligent manufacturing.

Keywords: Collaborative robot · robotic arm · Kinematics · Path planning · RRT · ROS

1 Introduction

With the significant growth of the global manufacturing industry, particularly the rapid expansion of pharmaceuticals, food processing,and 3C electronics, production methods are shifting from large-scale and standardized processes to more flexible and customized

H. Wei and J. Niu (Eds.): COBOT 2025, CCIS 2893, pp. 90–108, 2027.
https://doi.org/10.1007/978-981-92-2074-8_6

modes [1]. This transformation has placed higher demands on manufacturing automation, while traditional industrial robots, due to their fixed operational patterns and limited adaptability, are increasingly unable to meet these evolving requirements. Therefore, collaborative robots, as a new type of robotic technology, have become widely adopted in modern manufacturing environments [2, 3].

With the continuous development of robotic technology, both domestic and international scholars have conducted extensive research on kinematic and path-planning algorithms.Wang Meng et al. analyzed the forward and inverse kinematics of a 7-DOF serial robot [4], while Ma Guoqing and colleagues used MATLAB to perform the kinematics modeling and simulation of the UR10 robot [5]. Zhu Qinghao et al. established the kinematic model of the IRB120 robot and verified it through RobotStudio and MATLAB simulations [6]. In the field of mobile robots, Wang Dianjun proposed an improved A* algorithm to enhance path smoothness by removing redundant nodes [7], and Xu Kaibo et al. introduced a two-layer ant colony algorithm capable of global and local optimization [8].

For robotic manipulators, sampling-based planning algorithms such as the Rapidly-exploring Random Tree (RRT) family have been widely studied due to their scalability and probabilistic completeness. The RRT algorithm incrementally explores the configuration space through random sampling without requiring an explicit environmental model [9]. Its advantages—including simplicity, computational efficiency, and suitability for high-dimensional spaces—make it a strong candidate for manipulator path planning. However, challenges such as algorithm selection, planning efficiency, and the generation of smooth and feasible trajectories still remain in practical robotic applications.

This paper investigates the kinematic modeling and path-planning performance of the Aubo-E5 collaborative robot. A high-fidelity kinematic model was developed using the modified Denavit–Hartenberg method, and the robot's reachable workspace was obtained through the Monte Carlo approach. Additionally, a systematic comparison of the RRT, RRT*, and RRTconnect algorithms was conducted. By integrating simulation and physical experiments, this study evaluates the planning time and trajectory characteristics of the three algorithms, providing practical guidance for improving the operational efficiency of 6-DOF collaborative robots in complex environments.

2 Overall Scheme Design

2.1 Scheme Design of the Motion Simulation

This study aims to investigate the motion simulation and path-planning performance of a six-axis collaborative robot. First, a modified Denavit-Hartenberg (D-H) model of the Aubo-E5 robot is establish, and its kinematic characteristics are analyzed using Matlab, laying a solid foundation for subsequent deployment in the Robot Operating System (ROS) simulation. Next, several representative path-planning algorithms suitable for robotic manipulators are examined, and through a series of comparative experiments, the most appropriate algorithm for the Aubo-E5 robot is identified. Finally, a ROS-based simulation platform is constructed, and the selected algorithm is integrated into this platform to enable complete motion-planning execution within ROS.

2.2 Robot Selection

The Aubo-E5 collaborative robot was selected as the experimental platform for this research. The Aubo-E5 features a lightweight construction, high versatility, and flexible configuration, making it well-suited for the simulation and experiment tasks conducted in this work. The physical model of the robot is shown in Fig. 1.

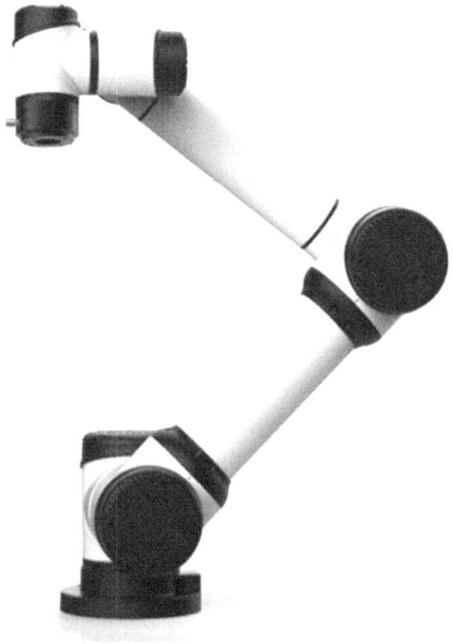

Fig. 1. Aubo-E5 Collaborative Robot

2.3 Selection of Robotic Arm Path Planning Algorithms

Currently, there are three main categories of path planning algorithms, including search-based path planning algorithms, bionic intelligent path planning algorithms, and sampling-based path planning algorithms [10].

Path planning algorithms based on search have advantages in finding optimal solutions and completeness, but they have drawbacks in search efficiency, real-time performance, and computational resource consumption. When selecting an algorithm, it is necessary to consider specific application scenarios and requirements. The initial goal of path planning algorithms based on bionics is to achieve global optimization. However, in practical implementations, they often fail to obtain truly optimal paths, and their computation burden makes them unsuitable for high-dimensional manipulator planning [11].

For a 6-DOF robotic manipulator, factors such as system degrees of freedom, joint constraints, link interactions, and workspace characteristics must be comprehensively considered. Considering the characteristics of the Aubo-E5 robot and the complexity of the environment, this study selects the Rapidly-exploring Random Tree (RRT) algorithm and its variants (RRT* and RRTconnect) for experimental verification, due to their high scalability, probabilistic completeness, and suitability for high-dimensional configuration spaces.

3 Kinematic Simulation Analysis

3.1 Kinematic Model

The Denavit-Hartenberg (D-H) parameter method is a widely used for the kinematic modeling of open-chain robotic mechanisms because it provides a unified and standardized framework for describing the spatial relationships between adjacent joints. Based on the structural characteristics of the Aubo-E5 robot, the coordinate frames for each joint were assigned following the modified D-H convention, as illustrated in Fig. 2. This formulation enables a clear and systematic representation of link transformations, allowing the forward and inverse kinematics of the manipulator to be derived analytically.

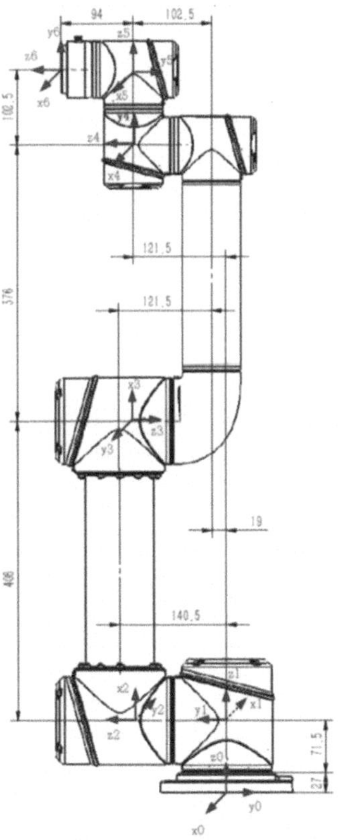

Fig. 2. Robot Link Coordinate System

According to the established link coordinate system, the complete D-H parameter table of the Aubo-E5 robot is presented in Table 1. Utilizing these parameters, a kinematic model was constructed using the Link function from the MATLAB Robotics Toolbox, which provides the foundation for subsequent workspace computation and algorithmic simulation.

Table 1. D-H Parameter Table for Aubo-E5 Robot

Joint i	Twist Angle α_{i-1}	Link Length a_{i-1} (mm)	Link Offset d_i (mm)	Joint Angle θ_i
1	0	0	98.5	θ_1
2	$-\pi/2$	0	121.5	θ_2
3	π	408	0	θ_3
4	π	376	0	θ_4
5	$-\pi/2$	0	102.5	θ_5
6	$\pi/2$	0	194	θ_6

3.2 Forward Kinematics Solution

Forward kinematics determines the pose (position and orientation) of the robot's end effector given a set of joint variables [14]. Using the D-H parameters, the homogeneous transformation matrix of each link relative to its predecessor is derived, and the end-effector pose can be obtained by multiplying all link transformations sequentially. In order to simplify the representation, the following conventions are adopted: $\sin\theta_1 = s\theta_1$, $\cos\theta_2 = c\theta_2$, $\sin(\theta_1 + \theta_2) = s\theta_{12}$, $\cos(\theta_1 + \theta_2) = c\theta_{12}$.

The transformation matrix of link i relative to link i-1 can be written as:

$$
{}_{i}^{i-1}T = \begin{bmatrix} c\theta_i & s\theta_i & 0 & \alpha_{i-1} \\ s\theta_i c\alpha_{i-1} & c\theta_i c\alpha_{i-1} & -s\alpha_{i-1} & -ds\alpha_{i-1} \\ s\theta_i s\alpha_{i-1} & c\theta_i s\alpha_{i-1} & c\alpha_{i-1} & d_1 c\alpha_{i-1} \\ 0 & 0 & 0 & 1 \end{bmatrix} \tag{1}
$$

By substituting the D-H parameters from Table 1 into Eq. (1) and performing successive matrix multiplications, the final homogeneous transformation matrix describing the pose of link 6 relative to the base frame is obtained, as shown in Eq. (2):

$$
{}_{6}^{0}T = {}_{1}^{0}T\,{}_{2}^{1}T\,{}_{3}^{2}T\,{}_{4}^{3}T\,{}_{5}^{4}T\,{}_{6}^{5}T \tag{2}
$$

This expression represents the forward kinematics solution. To verify the correctness of the model, the joint angles were assigned as [0, π/3, π/4, π/6, π/2, 0], and the resulting transformation matrix was compared with the output of the MATLAB fkine function. The two results were identical, confirming the accuracy and reliability of the forward kinematic model.

3.3 Inverse Kinematics Solution

Inverse kinematics refers to the process of determining the necessary joint angles to achieve a target position and orientation of the robot's end effector, given the target position and orientation as known inputs [15]. The mathematical formulation of the Aubo-E5 inverse kinematic can be expressed as:

$$\begin{bmatrix} n_x\ o_x\ a_x\ p_x \\ n_y\ o_y\ a_y\ p_y \\ n_z\ o_z\ a_z\ p_z \\ 0\ \ 0\ \ 0\ \ 1 \end{bmatrix} = {}^0_1T\,{}^1_2T\,{}^2_3T\,{}^3_4T\,{}^4_5T\,{}^5_6T \tag{3}$$

To solve for the joint variables, the inverse transformation of each preceding link is applied successively, allowing the target pose to be decomposed into solvable components. This process yields the analytical solution of the joint angles corresponding to a desired end-effector position and orientation.

To validate the derived inverse kinematics, the end-effector pose obtained through forward kinematics was used as the target input, and MATLAB's ikine function was employed for comparison. The resulting joint configuration T = [0, 1.0471, 0.7853, 0.5235, 1.5707, 0] erfectly matched the original joint angles, demonstrating that the inverse kinematics model is mathematically correct and numerically consistent with the forward calculations.

3.4 Workspace Analysis

The workspace of a robot refers to the set of all reachable points that the robot's end effector can reach under given conditions. It is an important indicator for measuring the robot's working capability [16]. Among the methods for solving the robot's workspace, the Monte Carlo method is primarily used.

In this study, a large number of joint configurations were randomly generated within their allowable ranges, and forward kinematics was used to compute the corresponding end-effector positions. All feasible points were recorded to obtain the three-dimensional workspace and its projections onto the XOY, YOZ, and XOZ planes, as shown in Fig. 3(a)–(d).

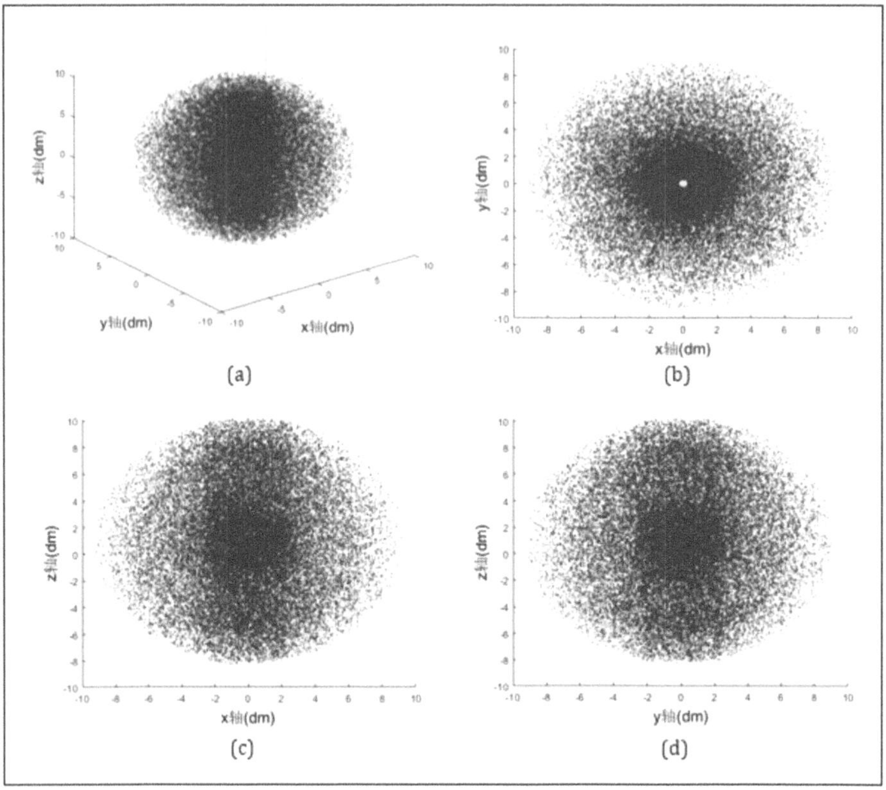

Fig. 3. Workspace and its projections:(a)Aubo-E5 Workspace (b)projection on the XOY plane (c)projection on the XOZ plane (d)projection on the YOZ plane.

The resulting workspace exhibits a hollow ellipsoidal shape, with its boundaries primarily determined by the robot's joint limits and link lengths. This visualization reflects the inherent motion capability of the Aubo-E5 robot and provides essential constraints for subsequent path-planning simulations.

4 Path Planning Algorithms

4.1 RRT

The Rapidly-exploring Random Tree (RRT) algorithm is a classical sampling-based method designed to efficiently search high-dimensional configuration spaces [17]. It incrementally expands a tree structure by randomly sampling the configuration space and steering the tree toward unexplored regions, enabling fast exploration without requiring an explicit environmental model. Owing to its probabilistic completeness, computational efficiency, and suitability for complex and high-dimensional environments, RRT has been widely applied in robotic path planning. The pseudo-code for the RRT is presented in Table 2.

Table 2. The pseudo-code for the RRT algorithm

RRT Algorithm
T.add(xinit);i = 0
While i < N
Xrand = UniformSampling();i + +
Xnearest ← nearest_neighbor(T,Xrand)
Xnew = Steer(Xnearest.Xrand);
if(!collision_free(Xnearest,Xnew))then
CONTINUE
T.add(Xnew);
if ReachGoal(Xgoal,T)
return Path(T);
else
Return Failure;

4.2 RRT*

Due to the relatively high cost and suboptimal paths generated by the traditional RRT algorithm in path planning, an improved version called RRT* algorithm has been proposed [18]. The core idea of RRT* is to iteratively optimize the tree structure during expansion by selecting the parent node that results in the minimum cumulative cost and performing a rewiring process to update neighboring nodes. This enables RRT* to asymptotically converge toward an optimal solution. Compared with the original RRT, RRT* is more suitable for complex, non-convex, and optimal-planning tasks, such as robotic manipulation, autonomous driving, and motion planning in cluttered environments. The pseudo-code for the RRT* algorithm is presented in Table 3.

Table 3. The pseudo-code for the RRT* algorithm

RRT* Algorithm
T.add(xinit);
While i < N
Xrand = GoalBiasSampling();i + +
Xnearest ← nearset_neighbor(T,Xrand)
Xnew = Steer(Xnearest.Xrand);
if(!collision_free(Xnearest,Xnew))then
Ynear = Near(Xnew)
for all Xnear in Ynear
c = Xnear.c + Cost(Xnear,Xnew)
if c < Xnew.c
Xnew.parent = Xnear
T.add(Xnew);
for all Xnear in Ynear\Xnear.parent
if Xnear.c > Cost(Xnear,Xnew) + Xnear.c then
Xnear.parent = Xnew;
if ReachGoal(Xgoal,T)
return Path(T);
else
Return Failure;

4.3 RRTconnect

The RRTconnect (Bidirectional Rapidly-exploring Random Tree) algorithm is a bidirectional extension of the RRT framework and was proposed to overcome the slow convergence and exploration inefficiency of single-tree approaches [19]. Two trees are simultaneously grown from the start and goal configurations, and the algorithm aggressively attempts to connect them through the Extend and Connect operations. RRTconnect leverages bidirectional rapid exploration to improve planning efficiency and reduce redundant sampling, making it well suited for high-dimensional manipulators operating in constrained environments; however, the absence of a rewiring procedure often results in feasible but non-optimal paths. The pseudo-code for the RRTconnect algorithm is presented in Table 4.

Table 4. The pseudo-code for the RRTconnect algorithm

RRTconnect (Xinit,Xgoal)
V1.add(xinit);V2.add(Xgoal);i = 0
While i < N do
Xrand = GoalBiasSampling();i + +
if not(Extend(V1.Xrand) = Trapped) then
if(Connect(V2,Xnew) = Reached) then
Return Path(V1,V2)
Swap(V1,V2)
Return Failure

4.4 Simulation Experiment Analysis

In MATLAB, a uniform 3D simulation environment was constructed to ensure fair comparisons among RRT, RRT*, and RRTconnect. The simulation environment included identical workspace dimensions, consistent obstacle configurations, and fixed start and goal points. This setup allowed for controlled and repeatable testing of each algorithm's performance. The visualization of the three algorithms is presented in Fig. 4(a) to (c).

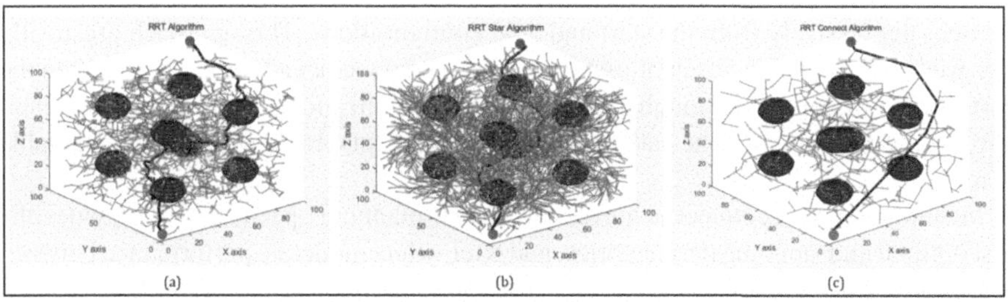

Fig. 4. Comparison of path planning results for the three algorithms in the MATLAB simulation environment. (a) RRT Algorithm, (b) RRT Algorithm, (c) RRTconnect Algorithm.*

In this simulation: the red dot represents the start point, the blue dot represents the goal point, and the brown spheres represent obstacles. The solid black line is the final path generated by the algorithm; the red and green nodes/edges represent the search tree explored by the algorithm. (The simulation environment was set to a 100 $\times$ 100 $\times$ 100 3D space containing 7 obstacles).

Table 5 summarizes the planning times required by each of these three algorithms.

Table 5. Time consumption for path planning of the three algorithms

Algorithm	Time Consumption
RRT	0.141
RRT*	1.308
RRTconnect	0.066

Based on the above charts and data, it can be concluded that the three algorithms—RRT, RRT*, and RRTconnect—are all capable of finding a feasible solution when a valid path exists, thereby verifying their probabilistic completeness. However, clear performance differences are observed in terms of path quality, computational efficiency, and search behavior.

For the RRT algorithm, the generated paths are feasible but generally longer and less smooth due to its purely random sampling strategy. The lack of optimization mechanisms results in redundant nodes and unnecessary detours, which not only reduce path quality but also increase planning time.

In contrast, the RRT* algorithm introduces two optimization steps—Choose Parent and Rewiring—to continually refine the tree structure. These operations enable the algorithm to asymptotically approach the optimal solution, resulting in the shortest and smoothest paths among the three methods. However, the additional optimization computation significantly increases the time cost, leading to the longest average planning time (1.308 s).

The RRTconnect algorithm adopts a bidirectional tree-growing strategy, extending two trees aggressively from the start and goal configurations. This approach drastically improves search efficiency, enabling the algorithm to generate a feasible path within the shortest time (0.066 s). Although it lacks a global optimization process and thus cannot guarantee path optimality, the path quality remains acceptable, making it highly suitable for real-time applications.

In summary, RRTconnect achieves the fastest planning speed, RRT* provides the most optimal and smoothest trajectories, and RRT offers moderate performance between the two algorithm. These complementary characteristics indicate that algorithm selection should be based on the trade-off between computation time and path optimality required by specific robotic tasks.

5 Motion Simulation

5.1 ROS Simulation Environment Setup

To validate the effectiveness of the selected path-planning algorithms in a high-fidelity environment, a ROS–MoveIt! simulation platform was constructed for the Aubo-E5 collaborative robot. MoveIt! provides integrated support for motion planning, collision detection, kinematic computation, and trajectory execution, enabling a simulation process that closely approximates real-world robot behavior.

In this study, the robot model was configured using the Unified Robot Description Format (URDF), which includes accurate link geometries, joint limits, inertial parameters, and mesh-based collision models [20]. The MoveIt! Setup Assistant was employed to generate the Semantic Robot Description Format (SRDF) file, define the planning groups, configure the end-effector, and establish the default planning pipeline.

These configurations ensure that the simulation environment accurately reflects the physical characteristics and constraints of the Aubo-E5 robot. The final simulation model, as loaded in Rviz, is depicted in Fig. 5.

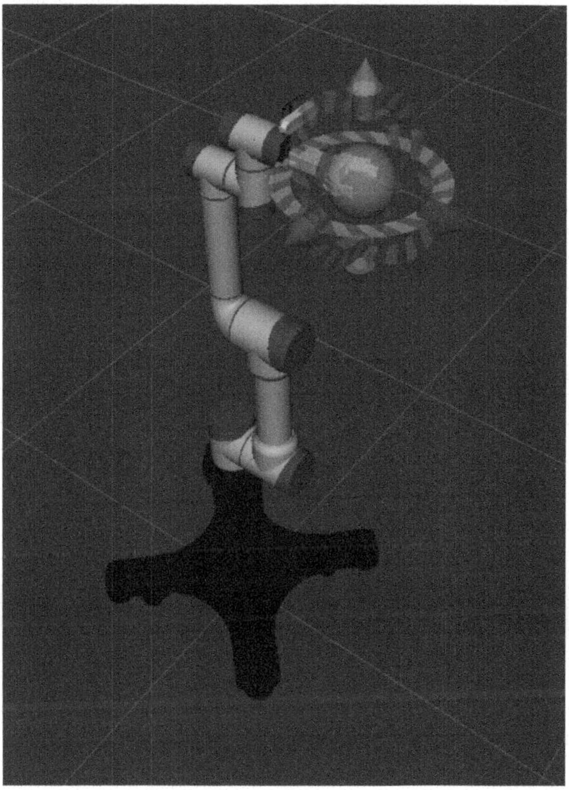

Fig. 5. Aubo-E5 Robot

5.2 Motion Planning

A corresponding robotic arm controller node was developed to initialize the position of the robotic arm. The robot poses were represented in the form of joint angle vectors. The initial and target configurations were set to [0, 0, 0, 0, 0, 0] and $[-125°, -89°, -44°, 45°, 125°, 0°]$. The position setting are illustrated in Fig. 6(a) and (b).

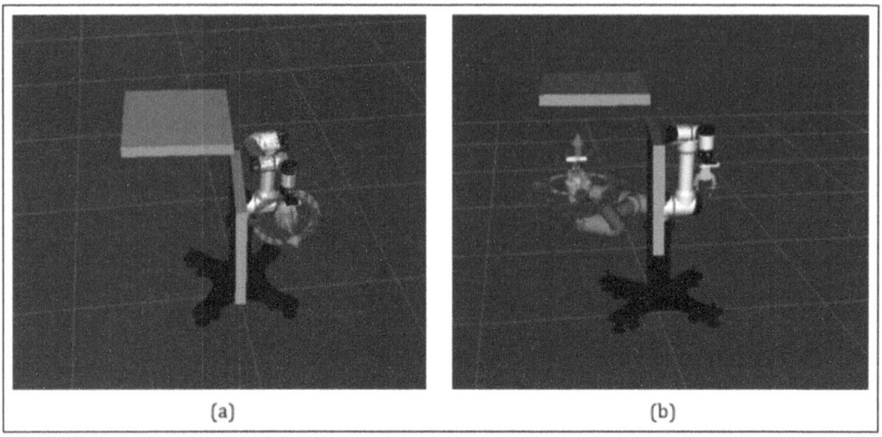

Fig. 6. The position setting (a) The starting point (b) The target point

Obstacle avoidance simulation experiments were performed using the RRT, RRT*, and RRTconnect algorithms. The trajectories generated by the three algorithms are illustrated in Fig. 7, and the planning times are summarized in Table 6.

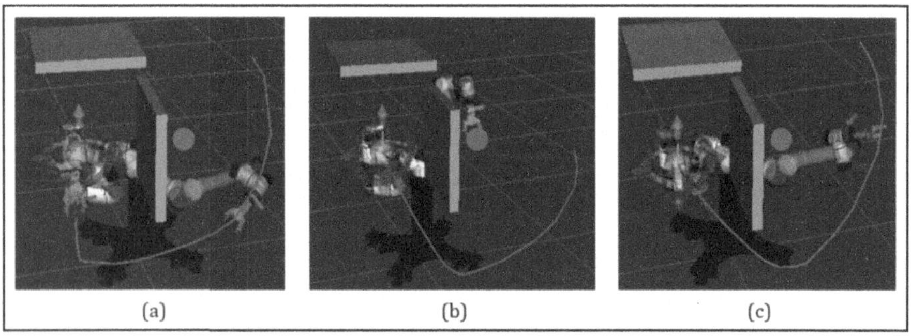

Fig. 7. Trajectory diagram of three algorithms in the ROS-MoveIt! simulation: (a) RRT Algorithm (b) RRT* Algorithm (c) RRTconnect Algorithm

In the simulation scene: the red dot represents the start position, and the interactive marker represents the goal pose. The green objects are the obstacles added to the planning scene. The cyan curve shows the end-effector trajectory planned by each algorithm.

Table 6. Time Required for Each Algorithm to Plan the Path

Algorithm	Planning Time
RRT	5.161
RRT*	5.947
RRTconnect	0.853

The ROS simulation results (Table 6) highlight significant performance differences in this high-fidelity environment. The RRT and RRT* algorithms require considerable planning time (5.161s and 5.947s, respectively). This is primarily due to two inherent factors in the ROS-MoveIt! platform: the exponential complexity of searching the six-dimensional configuration space (C-space) of the 6-DOF robot, and the high computational cost of mesh-based collision checking associated with the complete URDF model.

In this high-dimensional and computationally intensive environment, the RRTconnect algorithm (0.853s) demonstrates markedly superior performance. Its bidirectional search strategy rapidly establishes a connection between the two trees, thereby reducing the number of invalid samples and minimizing expensive collision-checking iterations.

This simulation analysis confirms the feasibility and effectiveness of the three algorithms in a high-fidelity environment, providing a solid foundation for the subsequent physical experiments.

6 Path Planning Experiment

6.1 Experimental Environment Setup

To validate the simulation results and test the algorithms' performance in a real-world context, a physical experiment was conducted using the Aubo-E5 collaborative robot. The experimental scenario was established by utilizing the control cabinet as the supporting platform and placing cardboard boxes as physical obstacles.The objective was for the robotic arm to successfully plan and execute a trajectory from a defined starting pose to a target pose while safely navigating around the obstacles, as shown in Fig. 8.This setup simulates a typical manufacturing environment in which a robot must operate within constrained space containing static objects. To account for the randomness inherent in sampling-based algorithms, the path planning experiment for each of the three algorithms (RRT, RRT*, and RRTconnect) was executed 30 times to ensure and representative results.

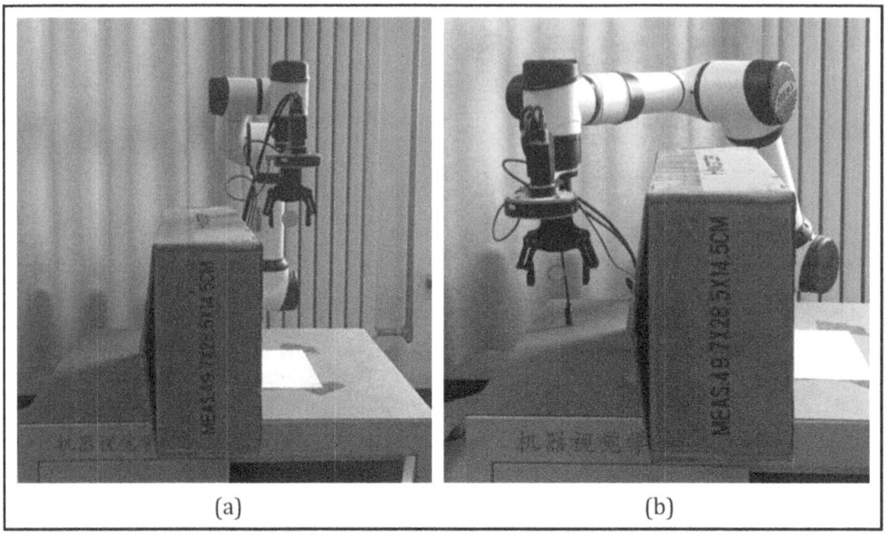

Fig. 8. Experimental Scene:(a) The starting point (b) The target point

In the real-world scene: the red dot represents the start position as set in Fig. 8(a), and the blue dot represents the goal pose as set in Fig. 8(b).

6.2 Algorithm Experiment

Physical experiments were conducted using the three algorithms to validate the simulation findings and assess real-world performance. The motion processes for the RRT, RRT*, and RRTconnect algorithms are shown in Fig. 9. These figures demonstrate that the Aubo-E5 robotic arm was able to successfully navigate around the obstacle under all three planning strateg.

Fig. 9. Arm Motion Process: (a) RRT Algorithm-Robotic Arm Motion Process (b) RRT* Algorithm-Robotic Arm Motion Process (c) RRTconnect Algorithm-Robotic Arm Motion Process

The corresponding end-effector motion trajectories, as visualized in Rviz during the execution, are illustrated in Fig. 10.

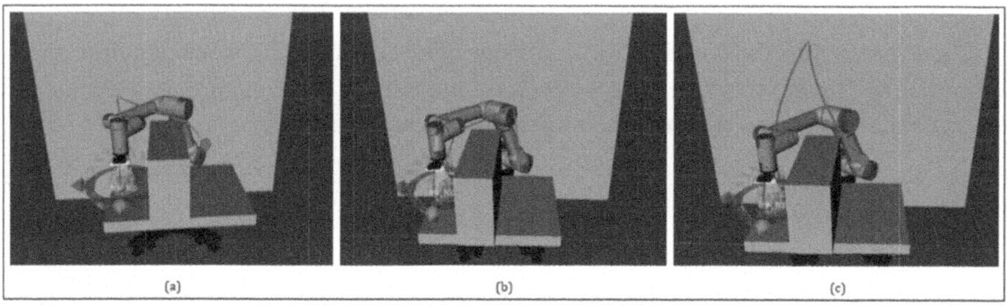

Fig. 10. End-Effector Trajectories of the Three Algorithms: (a) RRT Algorithm (b) RRT* Algorithm (c) RRTconnect Algorithm

In the simulation scene: the red dot represents the start position, and the interactive marker represents the goal pose. The green objects are the obstacles added to the planning scene. The cyan curve shows the end-effector trajectory planned by each algorithm.

The average planning time for the three algorithms is presented in Table 7, ensuring that the reported results are consistent and statistically reliable.

Table 7. Time required for path planning by each algorithm

Algorithm	Planning Time
RRT	5.060
RRT*	5.012
RRTconnect	0.057

6.3 Results Analysis

Based on the results of real robotic arm path planning experiments, all three algorithms successfully planned and executed obstacle-avoidance motions on the physical robot, thereby verifying the robustness and practical feasibility of these sampling-based path-planning methods under real-world physical constraints.

Among the three methods, the RRTconnect algorithm, which adopts a bidirectional search strategy, demonstrated the fastest planning speed; however, the resulting paths are relatively more complex and longer, making it more suitable for scenarios that prioritize search efficiency over path optimality.

The RRT* algorithm excels at generating near-optimal and smoother paths, making it preferable for applications with higher path quality requirements, though it may occasionally fail to find a feasible solution due to its stricter optimization process.

In contrast, the RRT algorithm produces paths of lower smoothness and quality compared with RRT*, but it remains highly reliable in consistently finding feasible paths.

Therefore, in practical applications, a balance must be struck between planning speed and path quality depending on the specific requirements of tasks, such as palletizing and material handling, so as to select the most appropriate algorithm.

7 Conclusions

To address the challenge of selecting an optimal path-planning algorithm for 6R collaborative robots operating in complex environments, this study first established a high-fidelity kinematic model of the Aubo-E5 robot. The modified Denavit-Hartenberg (D-H) method was adopted to derive the forward and inverse kinematics, while the Monte Carlo method was used to determine the robot's workspace, thereby providing a solid theoretical foundation for subsequent algorithm evaluation. Building on this foundation, a systematic comparative analysis of sampling-based path planning algorithms (RRT,

RRT*, and RRTconnect) was conducted. The results confirmed that these methods are highly suitable for manipulator path planning due to their computational efficiency in high-dimensional spaces and their independence from explicit environmental modeling.

Extensive experiments were performed using MATLAB, ROS, and an actual Aubo-E5 robotic arm to further validate the performance of the RRT family of algorithms. The findings demonstrate that the RRTconnect algorithm achieves the fastest planning speed, although the resulting paths are typically more complex and longer, making it applicable to scenarios that prioritize rapid planning. The RRT* algorithm yields the smoothest and most optimal trajectories but incurs significantly higher computational costs. In contrast, the RRT algorithm consistently identifies feasible paths, although with lower path quality. Therefore, the selection of a suitable algorithm should depend on the requirements of specific tasks—such as material handling or palletizing—balancing planning speed against path optimality.

Through the integration of theoretical modeling, simulation, and real-world experimentation, this study provides valuable insights and practical references for industrial applications. The results contributes to enhancing the safety, efficiency and operational stability of collaborative robots, thereby reducing unnecessary waste and improving overall economic performance in industrial production.

References

1. Xiaoling, H.: Research on Trajectory Planning of Six-Degree-of-Freedom Cooperative Robot. Guilin University of Electronic Technology (2021)
2. Probst, E.: Job shop automates using cobot arm. Mod. Mach. Shop **93**(9), 78–81 (2021)
3. Bader, K., Hendley, S.A., Bollen, V.: Assessment of Collaborative Robot (Cobot)-Assisted histotripsy for venous clot ablation. IEEE Trans. Biomed. Eng. **68**(4), 1220–1228 (2021)
4. Wang Meng, F., Tie, D.H., et al.: Kinematic analysis of a 7-DOF serial robot. Mech. Des. Manuf. **08**, 8–11 (2016)
5. Guoqing, M., et al.: Kinematic analysis and simulation of the UR10 Robot Based on MATLAB. Manuf. Autom. **41**(10), 87–90 (2019)
6. Qinghao, Z., et al.: Kinematic analysis and simulation of a 6R industrial robot. China Sci. Technol. Pap. **15**(08), 953–958 (2020)
7. Dianjun, W.: Indoor mobile robot path planning based on an improved A* algorithm. J. Tsinghua Univ. (Nat. Sci. Ed.) **52**(08), 1085–1089 (2012)
8. Kaibo, X., et al.: Robot path planning method based on double-layer ant colony algorithm and dynamic environment. Acta Electron. Sin. **47**(10), 2166–2176 (2019)
9. Lozano, P., Tomas, W.: An algorithm for planning collision-free paths among polyhedral obstacles. Commun. ACM **22**(10), 560–570 (1979)
10. Yongfeng, B.: Research on Path Planning for Substation Inspection Robot Based on Improved Ant Colony Fusion Algorithm. Tianjin University of Technology (2023)
11. Jie, H.: Research on Obstacle Avoidance Path Planning of Six-Degree-of-Freedom Manipulator. Jiangnan University (2023)
12. Da, X., Yang, L., Ye, L.: Forward kinematics analysis of ammunition loading robot. Mach. Des. Manuf. **11**, 160–162 (2015)
13. Singh, A., Singla, A., Soni, S.: Extension of D-H parameter method to hybrid manipulators used in robot–assisted surgery. Proc. Inst. Mech. Eng. H **229**(10), 703–712 (2015)
14. Fei, W., et al.: Obstacle avoidance path planning for robotic arm based on improved APF-Informed-RRT*. Mod. Mach. Tool Autom. Manuf. Techn. **08**, 60–65 (2024)

15. Ke, W., Xuecheng, P.: Kinematic analysis and trajectory planning based on six-axis robot. Mod. Mach. Tool Autom. Manuf. Tech. **07**, 17–22 (2024)
16. Qin, W., et al.: Kinematic analysis and trajectory planning simulation of 6DOF manipulator. Manuf. Autom. **46**(04), 42–47 (2024)
17. Qin, Z., et al.: Motion planning method and experiments of tomato bunch harvestingmanipulator. Trans. CSAE **37**(9), 149–156 (2021)
18. Karaman, S., Frazzoli, E.: Sampling-based algorithms for optimal motion planning. Int. J. Robot. Res. **30**(7), 846–894 (2011)
19. Cheng, X., et al.: An improved RRT-Connect path planning algorithm of robotic arm for automatic sampling of exhaust emission detection in Industry 4.0. J. Ind. Inf. Integr. **33**(01), 100436 (2023)
20. Huiting, J.: Research on Motion Planning of Kiwifruit Picking Manipulator Based on ROS. Guizhou Normal University (2024)

Research on Detection Method of Taraxacum Kok-Saghyzs and Weeds Based on Improved YOLOv11n

Taohong Wang[1(✉)], Xiaodong Zhang[1,2], Guowei Tang[1], and Zhipeng Yin[2]

[1] School of Mechanical Engineering, Xinjiang University, Urumqi 830047, China
19166901091@163.com

[2] School of Mechanical Engineering, Xi'an Jiaotong University, Xi'an 710049, China

Abstract. To address issues of low detection accuracy, frequent missed detections, and false detections caused by phenotypic similarities between Taraxacum kok-saghyzs (TKS) and weeds, an improved weed recognition and detection algorithm (CGEL-YOLOv11n) for TKS based on YOLOv11n is proposed in this paper. Firstly, a Content-Aware ReAssembly of Features (CARAFE) upsampling module is incorporated to capture sufficient semantic information, and the original GhostSPPCSPC convolutional module is replaced to reduce the model's parameter count. Secondly, an Efficient Channel Attention (ECA) mechanism is introduced to enhance the capability of extracting features such as edges, textures, and shapes of TKS and weeds. Finally, a lightweight detection head module (LSD) is designed to significantly improve the model's ability to capture target features while controlling the parameter count, thereby enhancing the overall detection performance of the network. Experimental results demonstrate that the improved CGEL-YOLOv11n model achieves a 7.31-percentage-point increase in mean Average Precision (mAP@50), with the parameter count and computational complexity reduced by 10.5% and 1.5%, respectively. The model not only enhances detection performance but also realizes a lightweight design, striking a favorable balance between detection accuracy and model complexity.

Keywords: YOLOv11n · Taraxacum kok-saghyzs · Weed detection · CARAFE · Attention mechanism

1 Introduction

Taraxacum kok-saghyzs(TKS), a perennial herbaceous plant, is regarded as a critical alternative resource for natural rubber. Its roots possess the capacity for rubber synthesis, demonstrating remarkable advantages in ecological adaptability and economic value [1]. However, in major TKS cultivation areas such as Xinjiang, planting fields are plagued by diverse weed species with rapid growth rates. Traditional manual weeding is plagued by low efficiency and high costs, while the overuse of broad-spectrum herbicides not only increases agricultural inputs but also may have adverse impacts on TKS root development and rubber synthesis. In severe cases, it can even result in direct TKS mortality and soil ecological pollution [2]. Thus, it is particularly crucial to develop a targeted algorithm for weeding robots to achieve accurate identification of weeds in TKS planting fields.

H. Wei and J. Niu (Eds.): COBOT 2025, CCIS 2893, pp. 109–119, 2027.
https://doi.org/10.1007/978-981-92-2074-8_7

In recent years, significant progress has been achieved in farmland weed recognition using deep learning-based object detection technologies. YOLO series algorithms have become mainstream due to their real-time performance advantages: for instance, Wang Huizheng et al. [3] improved YOLOv8n to achieve a 91.2% recognition accuracy for lawn weeds; Luo Liuming et al. [4] utilized YOLOv8n to enhance the small-target detection capability for sugarcane field weeds; Zhu Shipinget al. [5] adopted DeepLabv3 + to complete pixel-level segmentation of potato seedlings and weeds; Jian Cui et al. [6] optimized soybean field weed recognition based on Faster R-CNN. To adapt to field equipment, researchers have realized model lightweighting through techniques such as model pruning and quantization: for example, Chen Xue et al. [7] improved YOLOv5 for soybean fields, and Shu Aijing et al. [8] reduced the parameter count by 40% based on YOLOv8. These studies cover various crops, including corn, sugar beet, and rice. Nevertheless, as an alternative crop for natural rubber, TKS is mainly cultivated in arid regions with intense sunlight, such as Xinjiang, where weeds are dominated by lamb's-quarters (Chenopodium album) [9]. Additionally, the leaves of TKS at the seedling stage are highly similar in morphology to weeds of the genus Taraxacum [10]. Existing algorithms lack sufficient robustness under complex lighting conditions, while Gao Qiang et al. [9] only studied the herbicide control effect in TKS fields without integrating intelligent recognition technologies.

In summary, three major limitations remain in existing research: it lacks consideration for the growth cycle characteristics of perennial plants such as TKS. Second, unique environmental factors in Xinjiang's production areas result in insufficient robustness of existing models under complex backgrounds. Third, TKS is highly similar in morphology to weeds of the genus Taraxacum, with a misjudgment rate as high as 38% for traditional recognition methods based on color or texture [11].

Based on the above limitations, the difficulty for dual-arm robots to distinguish weeds in TKS planting fields will be addressed in this paper, and an improved, accurate recognition algorithm for TKS and weeds will be proposed, enabling weeding robots to accurately distinguish between TKS and weeds with highly similar morphologies under harsh conditions such as intense sunlight. This work lays a certain technical foundation for the research on dual-arm robot collaboration technology characterized by "one arm protecting TKS and the other removing weeds".

2 Baseline Model YOLOv11n

YOLO has been widely applied in agricultural detection due to its advantages of fast detection speed and high accuracy, enabling target detection under diverse environmental conditions. YOLOv11n is a fast and accurate object detection network, suitable for large-scale image data processing and analysis, with good real-time performance and low computational complexity. However, TKS and weeds exhibit characteristics such as similar morphology, complex backgrounds, and small-sized weed targets, which affect the detection performance of conventional models. Therefore, this study proposes an improved YOLOv11n network model to further enhance detection performance, aiming to realize the recognition and localization of TKS and weeds for unmanned vehicles. The overall structure of the YOLOv11n network is illustrated in Fig. 1.

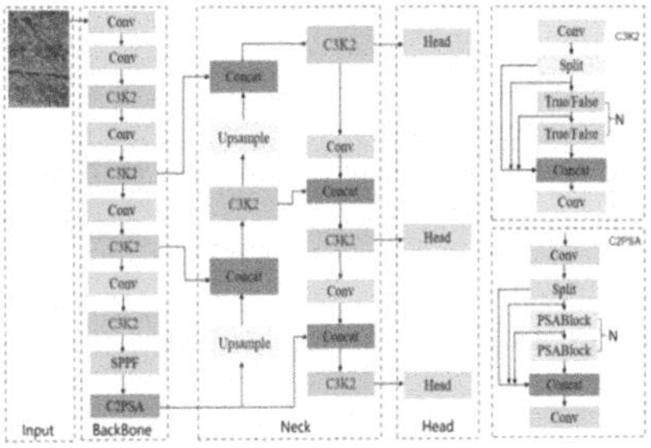

Fig. 1. Overall Architecture of the YOLOv11n Network

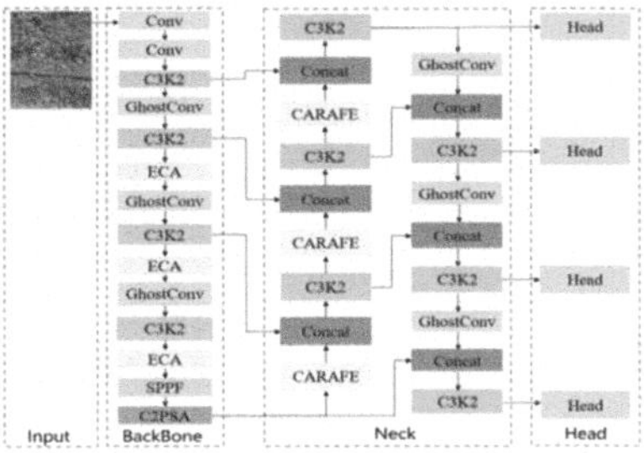

Fig. 2. Overall Architecture of CGEL-YOLOv11n

3 CGEL-YOLOv11n Network

The performance of general object detection algorithms in crop and weed detection is influenced by multiple factors, including target size, background complexity, and the detected objects themselves. By incorporating the Content-Aware ReAssembly of Features (CARAFE) upsampling operator, replacing the GhostSPPCSPC module, introducing the Efficient Channel Attention (ECA) mechanism, and designing the lightweight small-target detection head (LSD), the model's multi-scale feature fusion capability is further enhanced. This not only enhances the model's generalization ability, reduces the number of parameters, and boosts the localization accuracy of target detection. The overall network architecture of CGEL-YOLOv11n is illustrated in Fig. 2. Therefore, for the detection of TKS and weeds in this paper, an improved high-recognition model (CGEL-YOLOv11n) based on YOLOv11n is proposed.

3.1 Upsampling Module: Content-Aware ReAssembly of Features (CARAFE)

Feature upsampling is a key operation in the neck network, and the YOLOv11n network adopts conventional upsampling for this process. However, conventional upsampling can only restore the image spatial dimension and fails to capture the rich semantic information required for dense prediction tasks. Therefore, this study replaces the original upsampling in the network with the CARAFE upsampling module to extract more abundant semantic information, thereby improving model performance. The structure diagram of the CARAFE upsampling module is illustrated in Fig. 3.

CARAFE is mainly divided into two modules: upsampling kernel prediction and feature reassembly. The efficiency and performance of CARAFE depend on its core parameters: upsampling ratio σ, reassembly kernel size k_{up}, and content encoding kernel size $k_{encoder}$. Among these parameters, $k_{encoder}$ (content encoding kernel size) determines the size of the contextual receptive field captured when generating adaptive reassembly weights which sum to 1. To achieve the optimal balance between improving performance (e.g., enhancing small target detection capability) and maintaining lightweight computational overhead, we adopt the empirically validated optimal settings: reassembly kernel size $k_{up} = 5$ and content encoding kernel size $k_{encoder} = 3$. To clarify the specific calculation process of feature reassembly, the logical flow of the CARAFE module is summarized in Algorithm 1.

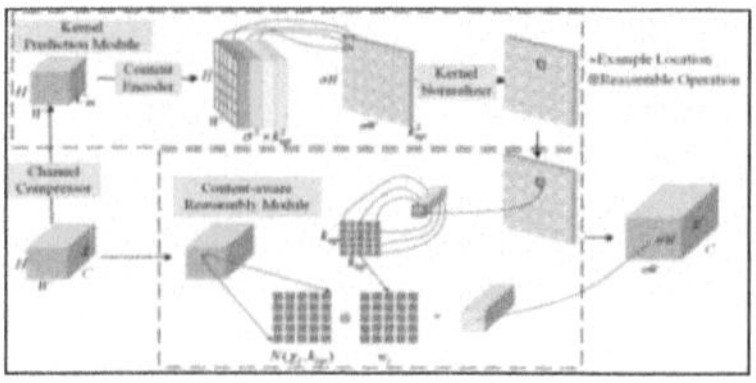

Fig. 3. CARAFE Sampling Module

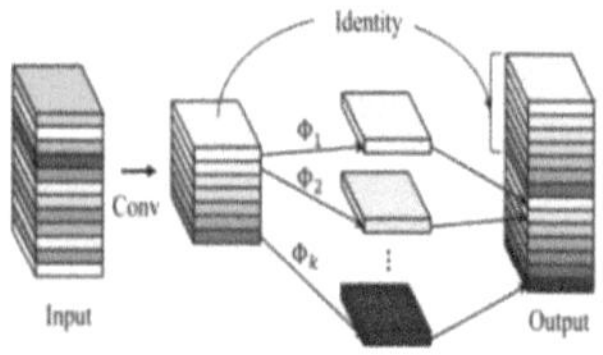

Fig. 4. GhostConv Convolution Structure

3.2 Convolutional Module: GhostSPPCSPC

GhostConv (GC) is an innovative convolutional neural network structure designed to optimize network efficiency by reducing computational complexity while preserving recognition performance, and its structure is illustrated in Fig. 4. In the YOLOv11n model, the SPPCSPC module is utilized to enhance multi-scale target detection capability, but it suffers from high memory consumption and computational overhead. Therefore, to reduce parameter count while ensuring accuracy, GC is introduced into the SPPCSPC module to replace the original conventional convolution, which is defined as the GhostSPPCSPC convolutional module. To demonstrate the reduction in computational redundancy, Algorithm 2 outlines the integration of GhostConv into the SPPCSPC structure.

Algorithm 1: CARAFE Upsampling Flow

CARAFE Upsampling Flow

Input: Feature Map X of size (C, H, W); Upsample ratio σ

Output: Upsampled Feature Map Y of size $(C, \sigma H, \sigma W)$

Parameters: k_{up} (Reassembly kernel); k_{enc} (Encoder kernel)

1. **Kernel Prediction Module**:
   - Compress channels of X to C_m via 1×1 Conv
   - Encode Content: Apply Conv layer (kernel k_{enc}) to generate reassembly kernels
   - Expand: Reshape output to $(\sigma^2 \times k_{up}^2, H, W)$
   - Pixel Shuffle: Rearrange to $(\sigma H, \sigma W, k_{up}^2)$
   - Normalize: Apply Softmax spatially to ensure kernel weights sum to 1
2. **Feature Reassembly Module**:
   - Map each location in Y back to X
   - Extract $k_{up} \times k_{up}$ region centered at input location
   - Perform dot product between predicted kernel and local region
3. **Return Y**

Algorithm 2: GhostSPPCSPC Construction

GhostSPPCSPC Construction

Input: Tensor X

Step 1: Split Processing

Branch A: $X \xrightarrow{GhostConv} Y_1$

Branch B:

1. $X \xrightarrow{GhostConv} Y_{tmp}$
2. $Y_{tmp} \xrightarrow{GhostConv} Y_{tmp1}$
3. $Y_{tmp1} \xrightarrow{SPP} \{P_1, P_2, P_3\}$ (MaxPolling with sizes 5, 9, 13)
4. $Y_{SPP} = Concat(Y_{tmp1}, P_1, P_2, P_3)$
5. $Y_{SPP} \longrightarrow Y_2$

Step 2: Fusion

$Y_{cat} = Concat(Y_1, Y_2)$

$Y_{final} = GhostConv(Y_{cat})$

Output: Y_{final}

3.3 Attention Mechanism: ECA

To enable the network to focus more on important feature channels and enhance the discriminative ability of feature information, this study introduces the ECA (Efficient Channel Attention) mechanism. This mechanism weights and adjusts the channels of feature maps to enhance the model's detection performance. The schematic diagram of the ECA module is shown in Fig. 5. The efficient computation of channel weights without dimensionality reduction is detailed in Algorithm 3.

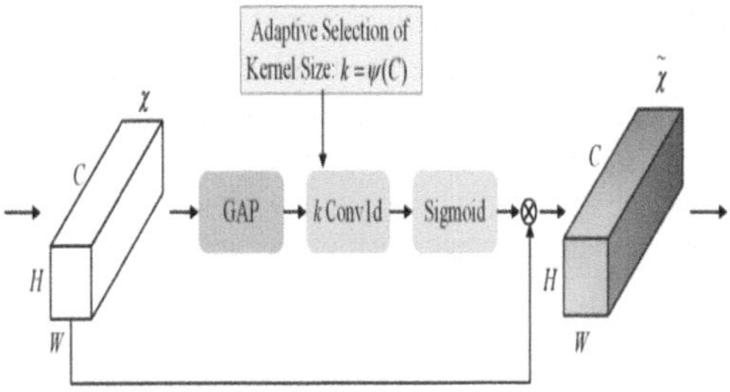

Fig. 5. ECA Module Structural Diagram

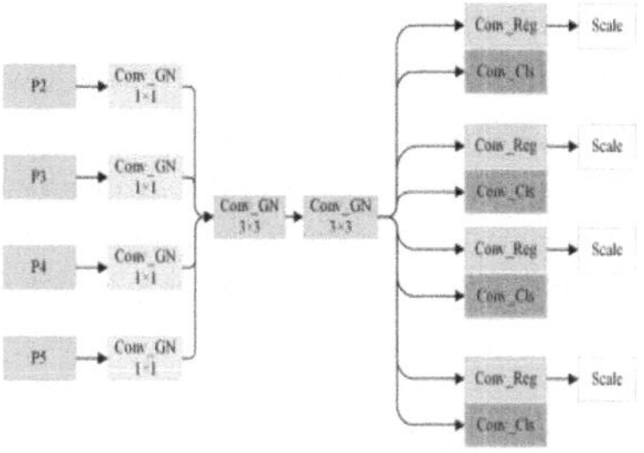

Fig. 6. LSD Detection Head Module

3.4 LSD Detection Head

To further lightweight the network model and improve recognition accuracy, a lightweight detection head, LSD, is designed, as illustrated in Fig. 6. By receiving input feature maps of different scales, the LSD head captures object information of various sizes, enhancing the model's capability to detect objects of varying scales. To avoid the instability of Batch Normalization (BN) layers, weak fine-grained feature discrimination, and redundant parameter computation under small-batch training after lightweighting, the convolutional normalization layer is replaced with a Group Normalization (GN) layer—achieving stable small-batch training, improved fine-grained feature extraction capability, reduced parameter count and computational complexity, and alleviated overfitting. The structural difference—specifically the replacement of normalization layers within the LSD head—is formally specified in Table 1.

Algorithm 3: ECA Mechanism

ECA Mechanism

Input: Feature Map of size (C, H, W)

Output: Refined Feature Map $\tilde{X}$

1. **Global Information Aggregation:** $y = GlobalAvgPool(X) \rightarrow Shape(C,1,1)$
2. **Adaptive Kernel Size Selection:** Calculate kernel size k based on channel dimension C : $k = \Psi(C) = \left|\frac{\log_2(C)}{\gamma} + \frac{b}{\gamma}\right|$

   (where γ and b are hyperparameters)
3. **Channel Interaction:** $\omega = Conv1D_k(y) \rightarrow$ Perform 1D convolution across channels
4. **Weight Activation:** $scale = \sigma(\omega) \rightarrow$ Apply Sigmoid function
5. **Feature Refinement:** $\tilde{X} = X \otimes scale \rightarrow$ Channel-wise multiplication
6. Return $\tilde{Y}$

4 Experiments and Analysis

4.1 Dataset and Preprocessing

The experimental dataset of this study was collected from the Anningqu Rubber Grass Breeding Base of Xinjiang Academy of Agricultural Sciences in Urumqi, covering two target categories: TKS and weeds, with 280 original images. The images were annotated using the VOTT tool, with label categories of "weeds" and "rubber". To prevent overfitting during model training, data augmentation techniques were applied to the collected images, including random rotation, Gaussian noise injection, vertical flipping, brightness adjustment, and random shadow addition—to improve the model's generalization ability. Finally, the dataset was expanded to 1980 images, which were divided into a training set, a validation set, and a test set at a ratio of 7:2:1 (Fig. 7).

Algorithm 4: The pseudocode for the key implementation points of the model

```
Training Procedure of the CGEL-YOLOv11n
Training Procedure of the CGEL-YOLOv11n
PROCEDURE TRAIN_MODEL(Model, DataLoader, Epochs, lr0)
    Initialize_Weights(Model)
    Optimizer = Initialize_Optimizer(Model, lr0)
    Scheduler = Initialize_Scheduler(Optimizer, Epochs)
    FOR epoch = 1 TO Epochs
        Model.Train()
        FOR batch IN DataLoader
            Imgs, Targets = batch
            Preds, Losses = Model(Imgs, Targets)
            Loss = SUM(Losses)
            Optimizer.Zero_Grad()
            Loss.Backward()
            Optimizer.Step()
        Scheduler.Step()
        IF epoch % Save_Interval == 0
            Model.Evaluate()
            Model.Save_Checkpoint()
    END FOR
END PROCEDURE
```

Fig. 7.Dataset Samples

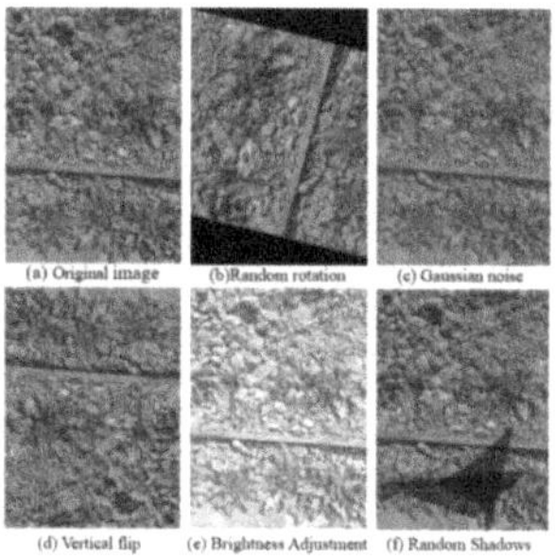

Fig. 7. Dataset Samples

4.2 Configurations and Parameters

The software and hardware environment for the model construction experiment is the Ubuntu 20.04 operating system. The server is equipped with 40 GB of memory, an RTX 3080 graphics card (with the corresponding driver installed), and an Intel Xeon(R) Platinum 8255C processor. The parallel computing framework is CUDA 11.3, and the model is implemented based on the PyTorch deep learning framework. The initial learning rate of the model is set to 0.01, the momentum is 0.937, and the weight decay is 0.0005. The number of training epochs is 300, the batch size is 16, and the IoU threshold is set to 0.5. The pseudocode is shown in Algorithm 4.

4.3 Ablation Experiments

To address the issues of compromised recognition accuracy, false positives, and missed detections that are prone to occur due to the relatively similar phenotypic characteristics between TKS and weeds, this paper proposes an improved YOLOv11n model for the detection of these two target categories (TKS and weeds). Ablation experiments are designed to verify the performance improvement brought by each proposed module to the network. Specifically, the experiments involve sequentially adding the CARAFE upsampling module to the baseline YOLOv11n network, replacing the original SPPCSPC convolutional module with the GhostSPPCSPC module, incorporating the ECA attention mechanism, and designing the lightweight LSD detection head with these operations denoted as Operations A, B, C, and D, respectively. The results are presented in Table 2.

Table 1. Comparison of Normalization layers in Detection Head

Structure Type	Mathematical Formulation
Standard Head	Output = Act(BN(Conv(F_{in})))
	Note: BN depends on batch size N.
LSD Head	Output = Act(GN(Conv(F_{in})))
	Note: GN is independent of batch size.

Table 2. Comparison of Ablation Experiments for CGEL-YOLOv11n

A	B	C	D	mAP (%)	P (%)	R (%)	Parameter/10^6	FLOPs/G	FPS
				87.27	91.44	85.33	2.58	6.4	57.6
√				89.74	92.41	85.68	2.66	6.4	57.0
	√			88.53	91.86	86.92	2.35	6.3	131.7
		√		88.91	92.15	87.43	2.59	6.4	56.8
			√	89.12	92.37	87.85	2.14	6.2	68.5
√	√			90.69	93.57	89.87	2.21	6.3	122.0
√		√		91.05	93.62	90.11	2.67	6.4	56.5

(*continued*)

Table 2. (*continued*)

A	B	C	D	mAP (%)	P (%)	R (%)	Parameter/10^6	FLOPs/G	FPS
√			√	91.28	93.74	90.36	2.23	6.2	66.3
	√	√		90.83	93.49	89.95	2.36	6.3	128.5
	√		√	91.45	93.86	90.52	2.02	6.1	116.3
		√	√	91.67	93.98	90.78	2.15	6.2	65.1
√	√	√		92.36	93.96	91.05	2.28	6.3	119.3
√	√		√	92.89	94.53	91.21	2.09	6.1	112.7
	√	√	√	93.14	94.76	91.38	2.03	6.1	110.5
√	√	√	√	94.58	95.28	91.64	2.31	6.3	108.2

Note: √ *indicates that the module is adopted*

Ablation experiment results show that compared with the baseline model YOLOv11n, CGEL-YOLOv11n model's mAP@50 increases by 7.31 percentage points, Precision by 3.84 percentage points, and Recall by 6.31 percentage points, with the parameter count reduced by 10.5%. The proposed model exhibits superior detection accuracy and overall performance for TKS and weeds in rubber grass fields. As shown in Fig. 8, with the addition of the improved modules, all three core metrics of the model have been significantly improved, verifying the effectiveness of the proposed CGEL-YOLOv11n model in the detection of TKS and weeds in rubber grass fields.

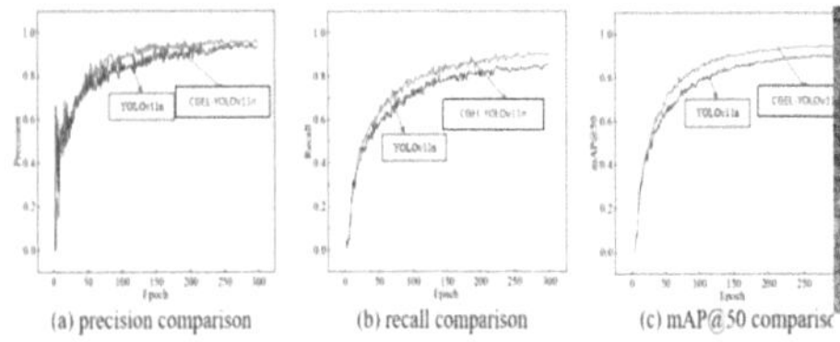

Fig. 8. Comparison of Precision, Recall, and mAP@50

Fig. 9. Heatmap effect comparison

4.4 Heatmap Visualization Comparison

Currently, in the field of object detection, the decision-making process of models often lacks transparency. To enhance model interpretability, this study adopts Grad-CAM technology to generate heatmaps, which are fused with original images to intuitively reveal whether the network effectively captures key target features. A visualization comparison of heatmaps between CGEL-YOLOv11n and the baseline YOLOv11n is presented in Fig. 9. Red regions represent the model's key recognition areas, followed by yellow regions, while blue regions correspond to non-critical features. Experimental results demonstrate that CGEL-YOLOv11n can more accurately focus on the key morphological features of TKS and weeds, significantly reducing attention to soil backgrounds and non-target plants, and effectively suppressing environmental interference.

4.5 Performance Comparison of Different Object Detection Algorithms

To verify the effectiveness of the proposed CGEL-YOLOv11n model, comparative experiments are conducted between CGEL-YOLOv11n and mainstream lightweight object detection models under the same experimental conditions using the TKS and weed dataset. The experimental results are presented in Table 3.

Table 3. Comparative Experimental Results of Different Detection Models

Model	mAP@50 (%)	P (%)	R (%)	Parameter/10^6	FLOPs/G
YOLOv5n	83.94	80.87	79.54	2.51	7.1
YOLOv6n	79.69	82.62	74.56	4.24	11.8
YOLOv7-tiny	83.27	85.44	82.33	6.01	13.9
YOLOv8n	86.42	89.68	84.38	3.01	8.1
YOLOv9t	85.44	88.32	80.49	1.97	6.7
YOLOv10n	83.62	85.68	80.06	2.70	6.5
YOLOv11n	87.27	91.44	85.33	2.58	6.4
YOLOv12	84.13	89.79	80.55	2.56	6.4
CGEL-YOLOv11n	94.58	95.28	91.64	2.31	6.3

As can be seen from the comparative results in Table 3, the Precision (P), Recall (R), and mean Average Precision (mAP@50) of the proposed CGEL-YOLOv11n model in this paper are all higher than those of the other mainstream lightweight comparative models. The mAP@50 value of CGEL-YOLOv11n reaches 94.58%, which is 10.64, 17.89, 11.31, 8.16, 9.14, 10.96, 7.31, and 10.45 percentage points higher than that of YOLOv5n, YOLOv6n, YOLOv7-tiny, YOLOv8n, YOLOv9t, YOLOv10n, YOLOv11n, and YOLOv12 models, respectively. For the detection of TKS and weeds, the parameter count of CGEL-YOLOv11n is reduced by 10.5% compared with the baseline YOLOv11n, and the computational complexity is reduced by 1.5%compared with the baseline YOLOv11n. Overall, CGEL-YOLOv11n achieves significantly higher Precision, Recall, and mAP@50 for both TKS and weeds compared with all comparative models, while reducing both the parameter count and computational complexity.

4.6 Visualization Comparison of Recognition Performance

To intuitively demonstrate the recognition performance of the CGEL-YOLOv11n model, a comparative analysis is conducted between the baseline YOLOv11n model and the proposed CGEL-YOLOv11n model.

As presented in Fig. 10, YOLOv11n suffers severe missed detections and frequent background false detections in crop-weed overlapping and occlusion scenarios. In contrast, CGEL-YOLOv11n resolves missed detections via an edge enhancement mechanism and effectively avoids background misclassification through multi-scale feature fusion and classification network optimization, ensuring high detection accuracy.

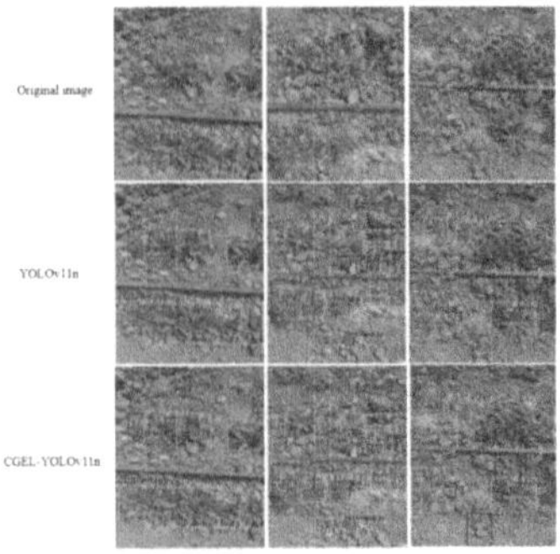

Fig. 10. Recognition Performance Under Different Scenarios

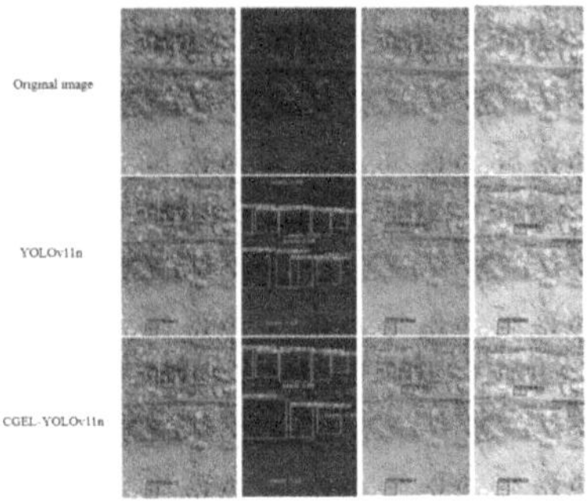

Fig. 11. Recognition Performance Under Different Lighting

As presented in Fig. 11, YOLOv11n is significantly affected by lighting variations (missed detections and false positives under low light, and missed detections of small weeds under strong light). In contrast, CGEL-YOLOv11n, via feature enhancement and multi-scale fusion, achieves accurate target recognition across all lighting conditions, adapting well to complex field lighting with more stable detection performance.

5 Conclusions

To address the issues of high missed detection and false positive rates in TKS and weed detection, the baseline YOLOv11n model is improved through four key optimizations: (1) integrating the Content-Aware ReAssembly of Features (CARAFE) upsampling module to enhance semantic information capture; (2) replacing the original SPPCSPC module with the GhostSPPCSPC module to reduce the parameter count and accelerate detection speed; (3) introducing the Efficient Channel Attention (ECA) mechanism to improve feature extraction capability and detection accuracy; (4) designing the lightweight detection head (LSD) to strengthen the model's ability to capture target features. Ablation experiment results demonstrate that the improved CGEL-YOLOv11n model achieves a mean Average Precision (mAP@50) of 94.58%, Precision of 95.28%, and Recall of 91.64%, with the parameter count and computational complexity reduced by 10.3% and 1.5%, respectively, compared with the baseline. The proposed model outperforms existing mainstream lightweight object detection models in terms of detection accuracy, can effectively identify TKS and weeds, and exhibits strong adaptability to complex field environments. This research provides technical support for weed detection in TKS fields and the collaborative application of dual-arm weeding and protection robots.

References

1. Shuyang, C., et al.: Research status and future prospect of Taraxacum kok-saghyz (TKS) Rubber Industry. Chin. Agric. Sci. Bull. **36**(10), 33–38 (2020)

2. Qiang, G., et al.: Control effects of five pre-sowing herbicides on weeds and their selectivity evaluation in Taraxacum koksaghyz fields. Weed Sci. **42**(03), 60–68 (2024)
3. Huizheng, W., et al.: Detecting lawn heterogeneous weed using lightweight improved YOLOv8n. Trans. Chin. Soc. Agric. Eng. 1–9 (2025)
4. Liuming, L., Yanzhou, L., Meiqi, S., Xin, H., Xi, C.: A identification method for sugarcane weed based on YOLOv8n. J. Chin. Agric. Mechanization **46**(02), 237–244 (2025)
5. Chao, Z., Bin, L., Kun, L.: Research on lightweight weed recognition algorithm based on improved YOLOv8. Appl. Electron. Tech. **51**(01), 80–85 (2025)
6. Jian, C., et al.: Weed identification in soybean seedling stage based on UAV images and faster R-CNN. Comput. Electron. Agric. **227**(P1), 109533 (2024)
7. Xue, C.H.E.N., Feng, T.A.N.: Research on lightweight soybean field weed identification method based on improved YOLOv5. Agric. Mach. Maintenance **02**, 1–7 (2025)
8. Aijing, S., Yanchao, Z.: Lightweight weed detection model based on improved YOLOv8. Softw. Eng. **27**(10), 18–22 (2024)
9. Yan, Z., et al.: Evaluation of excellent germplasm resources of Taraxacum kok-saghyz in Xinjiang. ChinaSeed Ind. **06**, 197–198 (2024)
10. Xinle, Z., et al.: Weed identification in soybean seedling stage based on optimized faster R-CNN algorithm. Agriculture **13**(1), 175 (2023)
11. Jian, Q., et al.: Research advances and perspectives on Rubber-producing Taraxacum. Chin. Bull. Botany **50**(01), 133–141 (2015)

An Intelligent Robotic Welding Platform Integrating Machine Vision and Offline Programming

Zhang Qian-kun[1,4], Jin Rui[2(✉)], Cheng Qi-bin[1], and Qi Rui-bin[3]

[1] Zhejiang Construction Investment Innovation Technology Co., Ltd, Hangzhou 310013, China
[2] Zhejiang Construction Investment Group Co., Ltd., Hangzhou 310013, China
Construction_Robots@163.com
[3] Zhejiang Sci-Tech University, Hangzhou 310018, China
[4] College of Civil Engineering and Architecture, Zhejiang University, Hangzhou 310058, China

Abstract. The assembly, welding, and processing of non-standard structural steel components require highly precise operational systems. However, current steel structure production equipment lacks effective visual feedback mechanisms, resulting in reduced flexibility and efficiency in online teaching and programming. This paper presents an intelligent welding and processing system for steel components that integrates machine vision and offline programming technologies. Through algorithms such as hand–eye calibration, point cloud filtering, and boundary fitting, the system achieves three-dimensional spatial positioning of steel components and supports multi-layer and multi-pass weld planning. Combined with robotic offline programming technology, the system can automatically generate and interpret welding motion trajectories for industrial robotic arms. In addition, a multi-source cooperative communication control system has been developed for command dispatching and data transmission, allowing the system to achieve fully automated and intelligent machining of steel components. Experimental results from a long-seam welding workstation demonstrated stable operation and high weld quality, with an overall weld qualification rate reaching 97%, verifying the reliability and industrial applicability of the proposed system.

Keywords: Machine vision · Offline programming · Intelligent welding system · trajectories

1 Introduction

In recent years, rapid infrastructure development and urbanization have greatly increased the demand for industrial robots and steel structure buildings [1–4]. In response, national policies have actively encouraged the integration of industrial robotics with intelligent construction technologies. The application of industrial robots in steel structure fabrication and welding not only improves productivity and reduces labor costs but also ensures stable and consistent welding quality. In the manufacturing of steel structures, industrial robotics mainly focus on two aspects: offline programming and automated welding

H. Wei and J. Niu (Eds.): COBOT 2025, CCIS 2893, pp. 120–131, 2027.
https://doi.org/10.1007/978-981-92-2074-8_8

systems. Offline programming technology enables the design and optimization of robot trajectories in a virtual environment [5–7], thereby reducing on-site teaching operations and equipment debugging time [8–11]. The implementation of intelligent and automated welding systems further enhances production flexibility and efficiency [12–15].

Significant progress has been made in the integration of vision systems with robotic welding. For example, Qinghua Wu et al. [16] proposed a vision-guided robotic welding approach using a linear structured-light sensor to extract and plan complex three-dimensional weld seams on large steel components. Wei Zhang et al. [17] reviewed the implementation of robotic technologies in steel structure fabrication, covering hardware architecture, parameter optimization, and programming strategies. Lei Yang et al. [18] systematically analyzed the use of structured-light sensors in intelligent welding robots, discussing weld seam localization, tracking, molten pool monitoring, and quality assessment. Although these studies demonstrate significant progress, the integration of multiple technologies in complex welding environments remains insufficient.

Current offline programming platforms such as ABB and FANUC support high-fidelity virtual simulation, but their performance is often inadequate when dealing with small-batch, non-standard, or complex-shaped steel components [19–22]. The main challenges include: (1) incomplete functionality and low efficiency in offline programming; (2) limited penetration and lack of adaptive control in medium-thick plate welding; and (3) weak data integration between offline programming and welding systems, leading to poor automation and low flexibility.

In conclusion, the existing steel component welding equipment integrating machine vision and offline programming has problems such as low penetration at important nodes of medium and thick plate steel box girder components, lack of automatic matching and correction functions for welding process parameters, and low efficiency in multi-device collaboration.To overcome these challenges, this study proposes an intelligent robotic welding system that integrates machine vision with offline programming. The system performs weld seam localization and geometric feature extraction through automated image acquisition and three-dimensional point cloud processing. A curve-fitting algorithm generates accurate welding trajectories and automatically assigns the corresponding welding process parameters to each welding node, while the PQArt offline programming platform provides closed-loop control from weld path planning to robotic execution. This approach supports digital, intelligent, and flexible manufacturing of large-scale steel components, offering an advanced solution for constructional steel welding.

2 Architectural Design of Intelligent Robotic Welding Systems

The proposed system consists of three main modules: a data preprocessing module, a data parsing and post-processing module, and an automatic control and execution module. This integrated framework establishes an end-to-end workflow that covers three-dimensional data acquisition, weld seam detection, curve fitting, trajectory planning, program generation, and robotic execution. It allows data to flow seamlessly from perception to operation, ensuring that each step in the process is traceable and automatically updated. Based on the PQArt platform, the system bridges virtual simulation and physical equipment through Python-based data parsing and logic conversion interfaces, enabling flexible interaction between digital models and real-world welding operations. The system architecture is shown in Fig. 1.

Following a data-driven automatic programming principle, the process begins with a structured-light depth camera capturing high-resolution point cloud data of weld seams on steel components. The raw point cloud is then filtered, segmented, and registered to eliminate noise and enhance the geometric features of weld regions. Using clustering and region-growing algorithms, the system identifies weld edges and feature points with high spatial accuracy. Subsequently, cross-sectional profiles and corner coordinates of components are calculated through iterative merging and surface fitting methods.

Based on the recognized geometry and welding specifications, the system automatically determines the number of welding layers and passes, generates corresponding trajectory and orientation data, and constructs a complete weld path plan. The parsed data is then post-processed into robot-executable commands that conform to the PQArt format, supporting flexible program generation for different robot brands and controller types.

Finally, the post-motion operation data is read through the multi-communication collaborative control system to achieve the task issuance and scheduling of robot photo scanning and multi-layer and multi-pass welding. The multi-communication collaborative control system is developed based on the C#. It establishes communication mechanisms with robot equipment and camera hardware through network IP. The data transmission rate is 50-100M/s, and the response delay for instruction issuance is 5-10ms. This system is equipped with functions such as rapid parsing of post-motion data instructions for robots, real-time monitoring of equipment operation status, and visualization of robot motion status. As the "dispatching brain", the multi-communication collaborative control system can ensure the stable operation of the welding workstation, high operation accuracy and consistent weld quality of steel components.

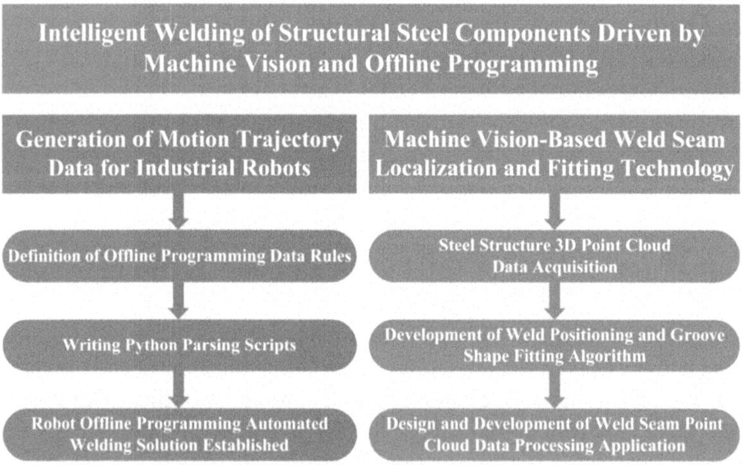

Fig. 1. System Architecture of Intelligent Robotic Welding

3 Machine Vision-Based Weld Seam Detection and Path Planning

In the manufacturing of steel structures, spatial localization of weld seams and extraction of geometric features are essential prerequisites for automated programming and welding path planning. This paper presents a weld seam detection and groove fitting system that uses three-dimensional structured light vision, enabling fully automated processing across the entire workflow, as shown in Fig. 2.

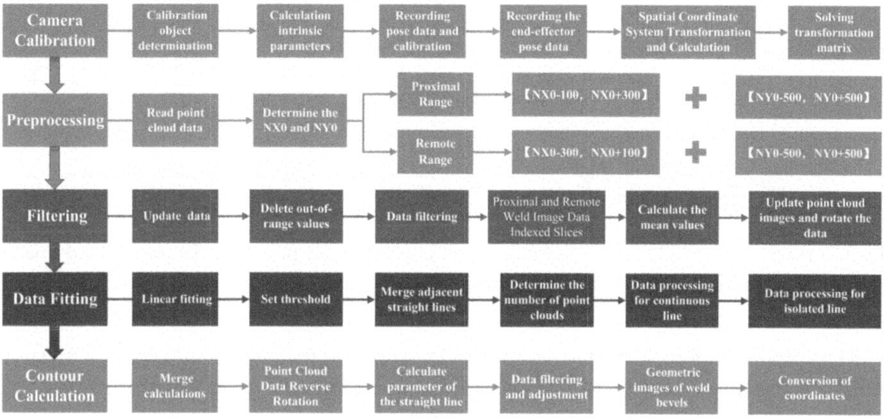

Fig. 2. Weld seam detection and groove fitting algorithm

The Photoneo M-series monocular structured-light depth camera is used to capture high-precision point clouds of the workpiece, as shown in Fig. 3(a) and (b). Hand–eye calibration between the robotic arm and the vision sensor ensures spatial correspondence between coordinate systems. The definition of component parameters is shown in Fig. 3(c).

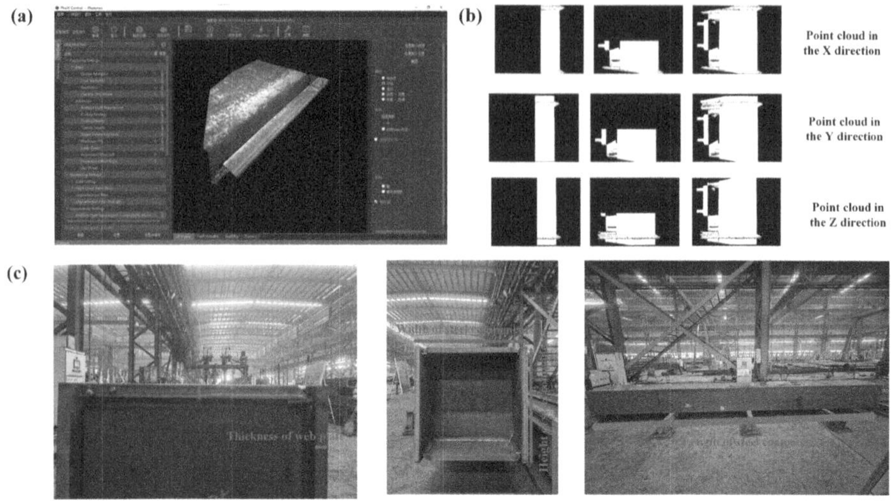

Fig. 3. (a) Structured light scanning camera and control driver program; (b) point cloud image data; (c) Measurement of geometric dimensions of steel components

The captured point cloud data contain three-dimensional coordinates (X, Y, Z). After preprocessing, the system performs the following operations, as shown in Fig. 4(a):

(1) **Coordinate decomposition:** splitting the 3D data into X, Y, and Z channels;
(2) **Two-stage filtering:** removing points below 500 mm and eliminating continuous empty rows;
(3) **Central region slicing:** reducing the dataset size to improve computational efficiency;
(4) **Rotation and segmented fitting:** aligning inclined or vertical boundaries to optimal computational positions, and restoring the fitted section via inverse rotation.

The scanning accuracy of the camera used in the workstation is 0.5mm. At each weld seam scanning point, the component point cloud data in the X, Y, and Z directions of the camera coordinate system can be obtained. The size of the point cloud data matrix in each direction is 16000*16000. If the operation is based on the entire matrix, it will impose a very high demand on the computing load capacity of the upper computer of the welding workstation. Due to the presence of discrete point clouds in the original data, filtering operations are carried out by setting a height threshold in the Z direction (500 mm $< Z_h <$ 900 mm) and reducing the slicing range of the matrix (with a size of 1000*1000) to eliminate other invalid discrete point clouds and those that affect the component positioning and fitting. Only a portion of high-quality point cloud data is saved for processing. Through a series of calculations such as spatial rotation, linear fitting, line segment merging, and bevel corner coordinate conversion of these high-quality point cloud data, the geometric dimensions and physical spatial coordinates of the weld bevel can be quickly obtained. The overall calculation time of the weld point cloud image is shortened from the original 5 min to 1 min, and the fitting error of the weld size is about 0.1–0.2 mm. In addition, the stability and accuracy of the vision system of the welding workstation are based on hand-eye calibration between the camera and the robotic arm. By scanning the calibration program that comes with the camera, the repeatability of the calibration can be ensured.

Through these processes, the geometric parameters of the weld groove and the spatial position of the weld centerline are accurately extracted from the reconstructed surface data. By applying polynomial curve fitting and spatial smoothing algorithms, the groove profile is refined to eliminate local noise and ensure continuity of the weld path. The extracted parameters—including groove angle, depth, and root gap—serve as critical inputs for multi-layer and multi-pass welding trajectory planning. These parameters guide the determination of welding sequences, layer thicknesses, and torch orientations, ensuring that the planned path conforms precisely to the actual groove geometry. The results of groove fitting and the corresponding welding spot layout are presented in Fig. 4(b) and (c), respectively, demonstrating the accuracy of the feature extraction and the rationality of the planned welding distribution.

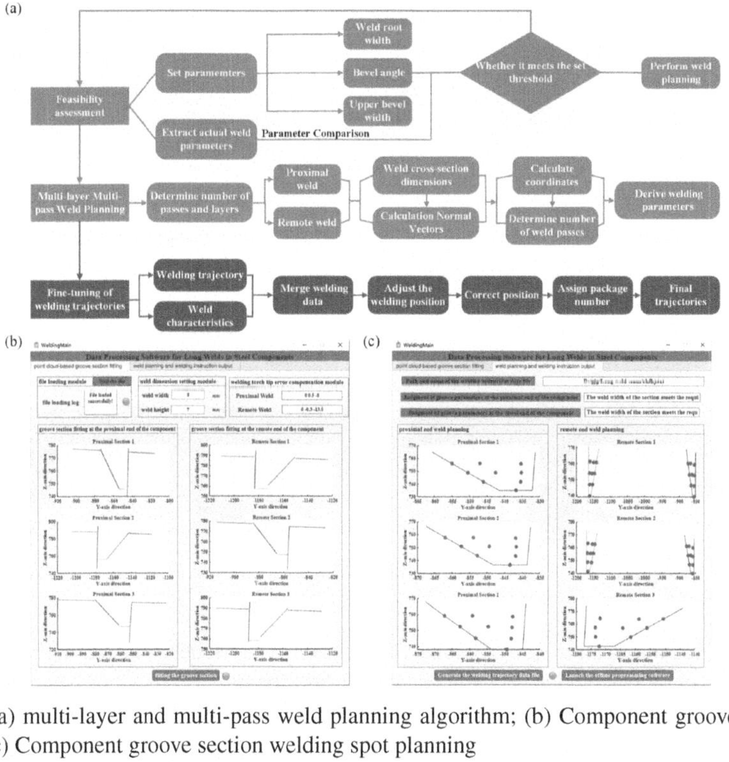

Fig. 4. (a) multi-layer and multi-pass weld planning algorithm; (b) Component groove section fitting; (c) Component groove section welding spot planning

4 Automated Offline Programming Scheme

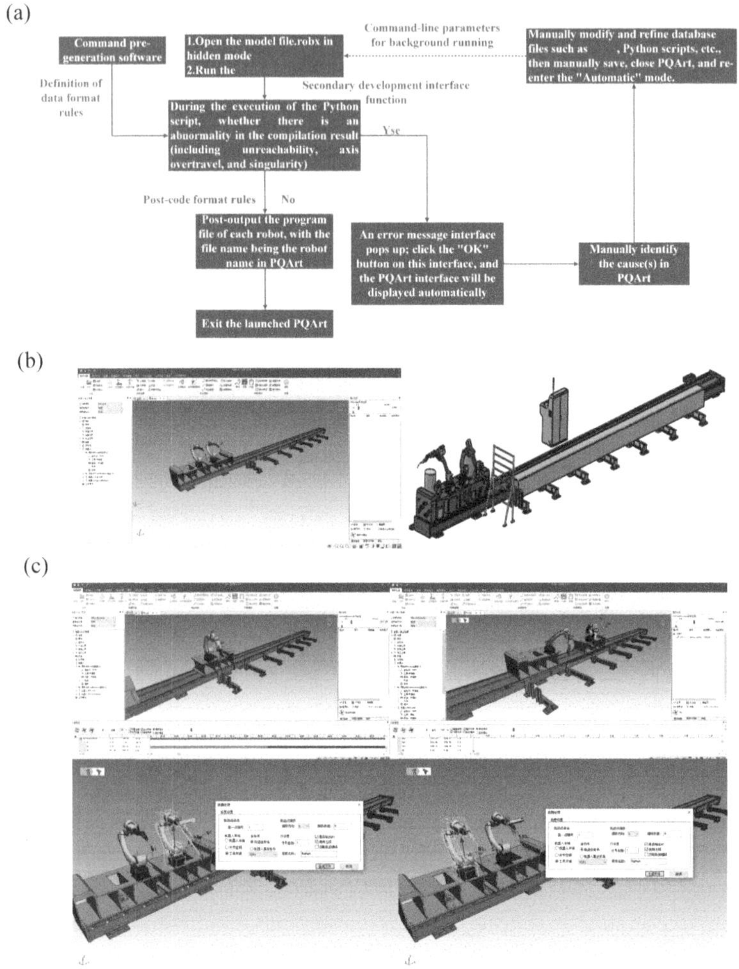

Fig. 5. (a) Automated offline programming framework; (b) 3D Model of the welding workstation; (c) Analysis of robotic motion data utilizing the PQArt platform

Three-dimensional models of the welding environment are first created using CAD software according to real operating conditions, and then imported into the PQArt offline programming platform for assembly, device binding, and coordinate calibration. The process framework and simulation system platform are shown in Fig. 5(a) and (b), respectively (Table 1).

Table 1. Partial data format rules for motion trajectories and control instructions

Serial Number	Symbol	Meaning	Note
1	A	Arc	Arc initiation
2	B	/	Parameters of the pendulum arc, denoted by a numeral directly succeeding the character "B"
4	D	Die	Extinguish the arc
5	E	/	Execute the photo-taking command

Based on this framework, an automated offline programming model for structural steel welding is constructed. The model integrates data acquisition, preprocessing, and trajectory planning within a unified workflow. In this process, the Python-based preprocessing module removes comment symbols and non-ASCII characters from the original program file and converts the filtered data into a standardized input format for subsequent trajectory parsing. The simplified logic is shown in Table 2.

The automated execution module is responsible for distributing robotic control instructions after post-processing. Operating through a command-line interface, the system invokes PQArt's hidden mode for compilation. If unreachable trajectories, axis limit violations, or collisions are detected, it prompts manual intervention; otherwise, post-processed program files are automatically generated and transmitted through the host computer communication system, achieving smooth transition from virtual programming to actual robotic execution.

Table 2. Python-based parsing workflow for welding trajectory data

```
import re, io
# Remove comments and non-ASCII characters from source code
clean_data = []
with io.open('input.txt', 'r', encoding='gbk') as f:
    for line in f:
        if "#" not in line:
            line = re.sub('[\u4e00-\u9fa5]', '', line).strip()
            clean_data.append(line)
# Write processed results for offline parsing
with open('output.txt', 'w') as g:
    g.write('\n'.join(clean_data))
```

5 Integration Testing of Welding Systems

The integrated testing and demonstration of the long-seam welding workstation were conducted in a steel structure fabrication workshop, with the component list summarized in Table 3. The tests aimed to verify data processing accuracy, program parsing correctness, software–hardware coordination, and overall welding stability. Using the proposed offline programming platform, the system automatically identified weld seam point clouds, planned trajectories, and generated executable programs (Fig. 6(a)), while the corresponding welding parameters and inspection results are shown in Fig. 6(b) and (c).

Table 3. The component list (uint: mm)

ID	Length	Thickness	Height	Width	Weld segment 1	Weld segment 2	Weld segment 3	Weld segment 4	Weld segment 5
1	5370	30	500	500	1195	2370	1805	/	/
2	5565	30	500	500	500	2560	2505	/	/
3	5680	30	600	600	485	2770	2425	/	/
4	7250	30	600	600	1530	1950	2130	14802	160
5	6826	26	500	500	150	550	1900	4056	170
6	7660	30	500	500	190	1730	1800	2310	1630

Although the steel structure welding workstation has carried out stable welding processing operations on the test components and the components of the actual engineering project, there are still some problems in the processing. The highly reflective surface of steel components will have a significant impact on the scanning of the camera, resulting in poor imaging effect of point cloud data. If the box-shaped components formed through the pre-processing procedures have significant processing errors, the vision system of the welding workstation will not fit the bevel well, thereby affecting the welding accuracy of the steel components.

The long-seam welding workstation has been demonstrated in a steel structure manufacturing enterprise. As shown in Fig. 6(c), the welds produced under this system exhibited stable formation. Ultrasonic testing confirmed an overall qualification rate of about 97%, meeting workshop production standards and verifying the system's reliability in engineering applications.

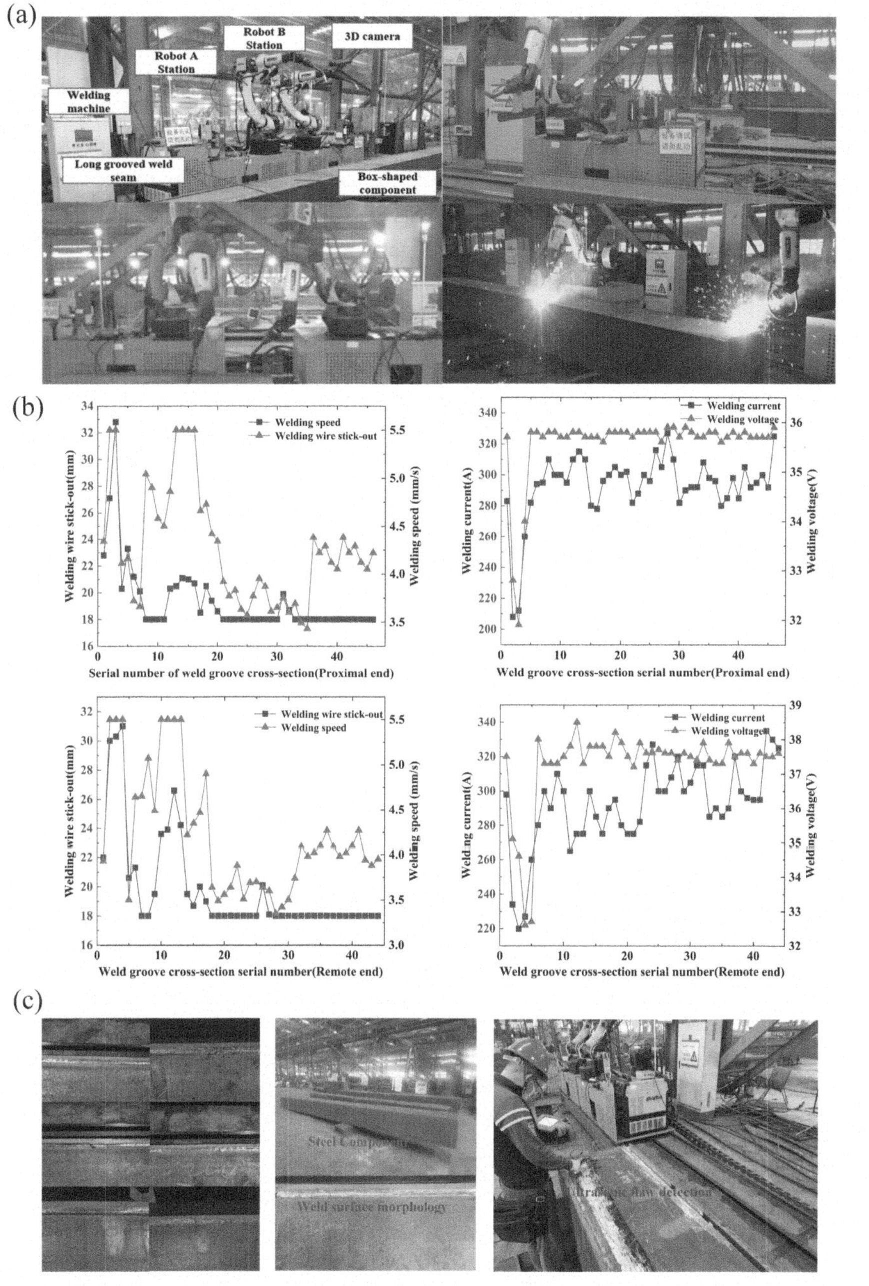

Fig. 6. (a) The long-seam welding workstation performs fabrication tasks; (b) Welding parameters of steel components; (c) Quality inspection of weld seams in steel components

6 Conclusions

By integrating weld seam identification and localization algorithms with offline programming technology, the proposed system enables automated welding of complex medium- and thick-plate steel components with improved accuracy and operational efficiency. The structured-light vision–based weld recognition and groove-fitting method enhances the precision and automation of three-dimensional information acquisition, which forms a reliable data basis for subsequent trajectory planning. Meanwhile, the PQArt-based workflow establishes an effective connection between robotic motion information and executable processing instructions, thereby improving weld planning efficiency and reducing manual intervention as well as operator-induced variability.

The long-seam welding workstation further demonstrated stable weld formation and full penetration in box-beam applications, and all tested components met workshop acceptance requirements. Although the system has shown dependable performance, its recognition accuracy remains sensitive to dimensional deviations in fabricated components, which may introduce errors in visual identification. Future efforts will therefore focus on refining the fitting algorithm to improve its robustness and adaptability to the geometric variations commonly encountered in steel structure fabrication.

Acknowledgments. The work in this paper was supported by the National Key R & D Program under Grant No. 2022YFB4700400, and Zhejiang Provincial Science and Technology Plan Project under Grant No.2025C01062.

Disclosure of Interests. The authors have no competing interests to declare.

References

1. Kou, K., Yang, G., Zhang, W., Wang, C., Yao, Y., Zhou, X.: uav autonomous navigation based on multi-modal perception: a deep hierarchical reinforcement learning method. In: Yu, Z., Hei, X., Li, D., Song, X., Lu, Z. (eds.), Intelligent Robotics. CIRAC 2022. Communications in Computer and Information Science, vol. 1770. Springer, Singapore (2023). https://doi.org/10.1007/978-981-99-0301-6_4
2. Peng, Y.: The application and research progress of steel structures in construction industrialization. In: IOP Conference Series: Earth and Environmental Science, vol. 330, p. 022003. IOP Publishing, Bristol (2019). https://doi.org/10.1088/1755-1315/330/2/022003
3. Mengesha, G.: Revolutionary steel structures: a comprehensive review of current trends and future directions. Int. J. Emerging Sci. Eng. (IJESE) (2024)
4. Zentay, P., Kutrovacz, L., Ottlakan, M., Szalay, T.: Aspects of industrial applications of collaborative robots. In: Yuschenko, A. (ed.) Modern Problems of Robotics. MPoR 2020. Communications in Computer and Information Science, vol. 1426, Springer, Cham (2021). https://doi.org/10.1007/978-3-030-88458-1_1
5. Liu, J.: Simulation study of industrial robots based on offline programming. In: Proceedings of the 1st International Conference on Data Science and Engineering (ICDSE 2024), pp. 537–541 (2024)
6. Marjan, D., Elisaveta, D., Damjan, P.: Overcoming industrial robotics challenges and the role of offline programming. Mech. Eng.-Sci. J. **42**(1), 33–41 (2024)

7. Zheng, C., et al.: Hybrid offline programming method for robotic welding systems. Robot. Comput.-Integr. Manuf. **73**, 102238 (2022)
8. Abebe, R., Gopal, M.: Optimization and simulation of arc welding robot parameters based on offline programming. Mater. Today: Proc. (2023)
9. Babcinschi, M., Cruz, F., Duarte, N., Santos, S., Alves, S., Neto, P.: Offline robot programming assisted by task demonstration: an AutomationML interoperable solution for glass adhesive application and welding. Int. J. Comput. Integr. Manuf. 1–12 (2024)
10. Sheng, Y.: Research on offline programming technology for industrial robot laser cutting. MW Metal Forming 50–53 (2025)
11. Li, Y., Xu, Y., Wang, X., Ma, X., Wang, Q., Zhang, H: A feature-extraction localization algorithm research for teaching-free automated robotic welding based on 3D point cloud. Int. J. Adv. Manuf. Technol. 1–16 (2025)
12. Zakharova, I., Shchetynin, S., Shchetynina, V., Zusin, A., Volenko, I.: Use of robotic and automated systems in welding and restoration of parts. Mach. Energetics **16**(1) (2025)
13. Lee, D., Han, K.: Vision-based construction robot for real-time automated welding with human-robot interaction. Autom. Constr. **168**, 105782 (2024)
14. Lee, D., Han, K.: Autonomous navigation and positioning of a real-time and automated mobile robotic welding system. J. Constr. Eng. Manag. **151**(5), 04025033 (2025)
15. Curiel, D., Suárez, A., Veiga, F., Aldalur, E., Villanueva, P.: Advanced welding automation: Intelligent systems for multipass welding in Butt Double V-Groove and Tee Double Bevel configurations. MethodsX **13**, 103027 (2024)
16. Wu, Q., Li, Z., Gao, C., Biao, W., Shen, G.: Research on welding guidance system of intelligent perception for steel weldment. IEEE Sens. J. **23**(5), 5220–5231 (2023)
17. Zhang, W., Li, J.: The application of intelligent welding robots and visual detection algorithms in building steel structures. Scalable Comput.: Pract. Experience **25**(5), 4265–4273 (2024)
18. Yang, L., Liu, Y., Peng, J.: Advances techniques of the structured light sensing in intelligent welding robots: a review. Int. J. Adv. Manuf. Technol. **110**(3), 1027–1046 (2020)
19. Pasang, P., Kerdsomya, P., Kunanoppadol, J., Mettanant, V., Katejanekarn, T.: Simulation of a welding station with RobotStudio. Sci. Eng. Health Stud. 24040010 (2024)
20. Gao, M., Li, R., Kong, B., Zhang, H., Xiao, L., Bu, K.: Simulation design and optimization of multi-robot production line based on RobotStudio. In: Proceedings of the Fourth International Conference on Mechanical Engineering, Intelligent Manufacturing, and Automation Technology (MEMAT 2023), vol. 13082, pp. 87–93. SPIE, Bellingham (2024)
21. Vladimirov, B., Nikolov, S., Tsolov, S.: Programming industrial robots in the Fanuc ROBOGUIDE environment. Eng. Proc. **70**(1), 20 (2024)
22. Di, G., Xu, W.: Simulation design of industrial robot handling workstation based on ROBOGUIDE. In: International Conference on Innovative Computing, pp. 242–249. Springer Nature, Singapore (2025)

An Integrated Joint Assembly Workstation System Based on Collaborative Robots

Dianjun Wang[1], Zheng Wang[2], Peng Wang[3], Ya Chen[2](✉), Zhiguo Xie[2], and Zhihu Song[2]

[1] School of Mechanical Engineering and Hydrogen Energy Research Centre, Beijing Institute of Petrochemical Technology, No. 19 Qingyuan North Street, Huangcun Town, Daxing District, Beijing 102617, China

[2] School of Mechanical Engineering, Beijing Institute of Petrochemical Technology, No. 19 Qingyuan North Street, Huangcun Town, Daxing District, Beijing 102617, China
chenya@bipt.edu.cn

[3] School of Information Engineering, Beijing Institute of Petrochemical Technology, No. 19 Qingyuan North Street, Huangcun Town, Daxing District, Beijing 102617, China

Abstract. To improve the assembly quality of integrated joints for collaborative robots, an efficient and intelligent integrated joint assembly workstation was designed based on collaborative robots. This workstation is implemented based on a hierarchical PLC control architecture and a dual-arm robot system, integrating a set of vision inspection systems, and employs Lua scripts to perform real-time error compensation for hole position deviations. Experimental results demonstrate that the positioning accuracy of the robot end-effector is significantly enhanced after compensation: the average error is reduced from 1.535 mm to 0.640 mm, representing an improvement of 58.3%. This system effectively improves the assembly quality of integrated joints and provides a valuable design reference for high-precision robot assembly.

Keywords: Integrated joint assembly · Workstation design · Offline simulation · Vision system · Error compensation

1 Introduction

1.1 Research Background

Collaborative robots are increasingly being applied in flexible manufacturing, precision assembly, and small-batch customized production. Among these applications, collaborative integrated joint assembly poses higher demands on robot trajectory accuracy, visual recognition, and multi-arm collaborative control due to small part sizes, tight fit clearances, and high positioning accuracy requirements. Despite the good repeatability of collaborative robots, they are still affected by factors such as workpiece placement errors, camera calibration errors, fixture positioning offsets, and changes in ambient light in complex assembly scenarios. This results in deviations of up to millimeters between

H. Wei and J. Niu (Eds.): COBOT 2025, CCIS 2893, pp. 132–151, 2027.
https://doi.org/10.1007/978-981-92-2074-8_9

the actual landing point and the planned trajectory, leading to assembly failures, increased insertion forces, and even damage to parts. Therefore, achieving online error compensation for collaborative robots in precision assembly holds significant engineering value.

1.2 Domestic and International Research Status

In recent years, scholars both domestically and internationally have conducted extensive research on visual positioning, error modeling, and compensation methods for collaborative robots. Some literature focuses on the pose optimization of collaborative robots in offline path planning, improving trajectory accuracy through kinematic analysis, error propagation modeling, or simulation-based compensation methods [1, 2, 3, 4]. Other studies utilize 2D/3D vision to achieve workpiece recognition, positioning, or pose estimation, and incorporate interpolation algorithms for simple deviation correction [11, 13, 15]. Still, other works propose enhancing visual robustness through hand-eye calibration, template matching, or deep learning methods [5, 7, 11]. However, based on existing research, there are still three main deficiencies: (1) many methods only compensate errors in the offline stage, lacking the real-time closed-loop compensation capability embedded in the robot controller, making it difficult to adapt to time-varying deviations caused by random placement of workpieces; (2) a significant amount of research remains at the laboratory-level verification stage, lacking engineering implementation in practical assembly tasks (such as bolt hole positioning, fine tolerance fitting, etc.); (3) existing research often lacks detailed replication of the entire process from "visual positioning → coordinate transformation → robot compensation execution," resulting in insufficient reproducibility, as well as a lack of systematic error statistics and significance analysis.

Addressing the aforementioned issues, this paper designs and implements a dual-arm collaborative assembly station that integrates joint assembly tasks for collaborative robots, incorporating visual inspection and real-time error compensation. The system consists of a "vision recognition arm + execution assembly arm + PLC central control + offline simulation (PQArt)". It captures hole image data through an industrial camera, calculates deviations using template matching, and corrects trajectories in real-time through Lua scripts in the robot controller, thereby forming a closed-loop control process of "perception-decision-execution-correction". Compared to existing methods, the main technological innovations and contributions of this study are as follows:

An embedded real-time compensation method is proposed: directly transmitting visual detection results to the collaborative robot controller, and updating TCP coordinates in real-time during the underlying trajectory execution phase through Lua scripts, achieving online compensation without the need for an external PC and enhancing adaptability under multi-source deviations.

Constructing a precision assembly station for dual-arm collaboration: The vision arm is responsible for positioning and transmitting compensation values, while the assembly arm performs tightening/inserting actions. Through PLC coordination of various modules, a stable industrial-grade collaborative assembly process is achieved.

Provide reproducible and complete implementation details and system verification, including camera calibration errors, visual detection parameters, coordinate transformation methods, compensation pseudo-codes, and measurement processes; and systematically evaluate the compensation effect through paired experiments, error statistics, and significance analysis.

At the engineering level, the workstation structure constructed in this study is compact and has a clear process, compatible with various collaborative robot systems. It can provide reference for precision assembly, repetitive tasks, and flexible production line transformation. Experimental results show that the proposed visual compensation strategy can significantly reduce assembly positioning errors. In typical experiments, the error decreased from 1.535 ± 0.244 mm to 0.640 ± 0.250 mm, a reduction of about 58.3%. The paired t-test results indicate a highly statistically significant difference ($p \ll 0.001$).

In summary, this paper aims to address the positioning deviation issue of collaborative robots in precision assembly scenarios, establish a real-time compensation mechanism that integrates vision and control, and comprehensively evaluate system performance through a verification process combining simulation and real-machine testing. This provides an engineering practical basis for the further application of collaborative robots in factory automation.

2 Overall Scheme Design of Assembly Workstation

2.1 Assembly Process and Requirements for Integrated Joint Module

As the core components of collaborative robots, the characteristics of the integrated joint play a decisive role in the performance of collaborative robots. The assembly process quality of the integrated joint is one of the key factors to ensure its performance. The joint model is shown in Fig. 1. This integrated design not only enhances the overall performance of the robot but also imposes higher requirements on the assembly process. A correct assembly process can ensure high precision and reliability of the modules, and it is also a key factor in improving production efficiency and reducing costs [5–7].

Fig. 1. Integrated joint model

The integrated joint module is a complex system, primarily composed of motors, harmonic reducers, bearings, torque sensors, encoders, drivers, controllers, and machined housing components, as shown in Figs. 2 and 3.

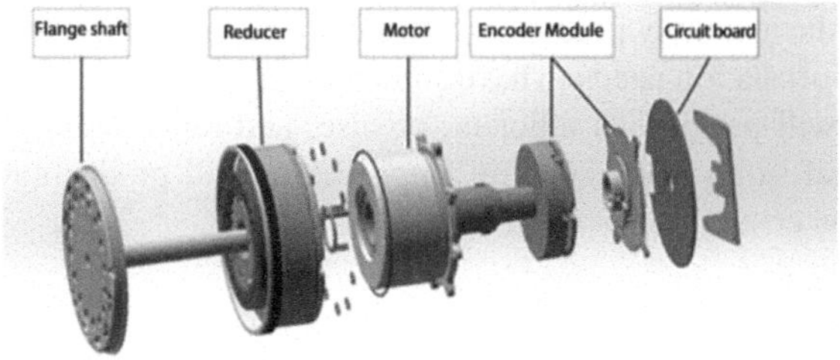

Fig. 2. Exploded view of the integrated joint model

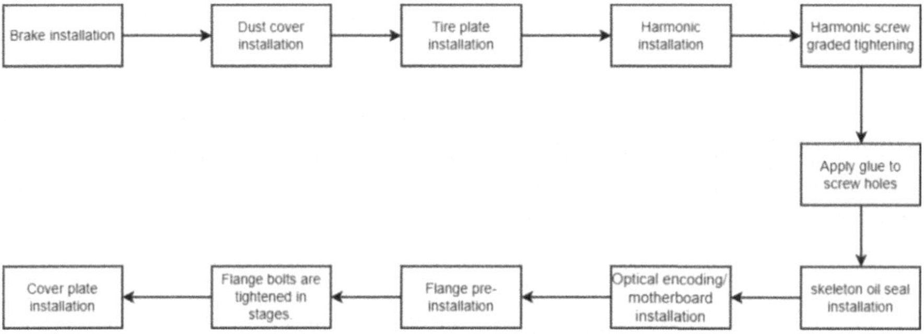

Fig. 3. Assembly process of integrated joint module

During the screw attachment process, the positioning accuracy of the screw holes should generally be within 1.05mm. Exceeding this deviation may lead to assembly failure or product damage. Meanwhile, torque control of the screws is crucial for ensuring the assembly quality of the integrated joint module. The screws in the joint module need to be locked with different torques, graded from small to large. Both being too loose and too tight may affect product performance. The harmonic reducer is tightened in three steps, and the flange shaft is tightened in two steps, with different torques applied each time, increasing gradually. The specific parameters are shown in Table 1.

Table 1. Assembly Process Parameter Table

Assembly steps	Data (N·m)	Measuring instrument
Harmonic primary fastening	1	Smart electric screwdriver
Harmonic secondary fastening	2.5	Smart electric screwdriver
Harmonic third-level fastening	4	Smart electric screwdriver
Flange Class I fastening	2	Smart electric screwdriver
Flange Grade 2 tightening	5	Smart electric screwdriver

According to the assembly process and process parameter requirements of the integrated joint, the workstation needs to have automatic positioning and locking functions. The workstation will provide an efficient, precise, and reliable assembly solution. It ensures operational safety and economy, improves overall production efficiency, and reduces production costs [8–10].

2.2 Overall Scheme Design

The overall design of the workstation primarily encompasses the design of the robot system, end tools, machine platform, conveyor belt, lifting mechanism, and control system. The workstation was designed with a height of 140 cm and width of 60 cm for an optimal ergonomic footprint. To ensure mobility and reconfigurability, universal wheels were selected for the base.

For operational convenience and safety, a control button panel was designed and integrated on the machine, which includes operation buttons such as the power switch, start button, and emergency stop button. Additionally, cable connectors and pneumatic interfaces were designed on both sides of the machine to facilitate wiring, debugging, maintenance, and repair work. The overall model is shown in Fig. 4.

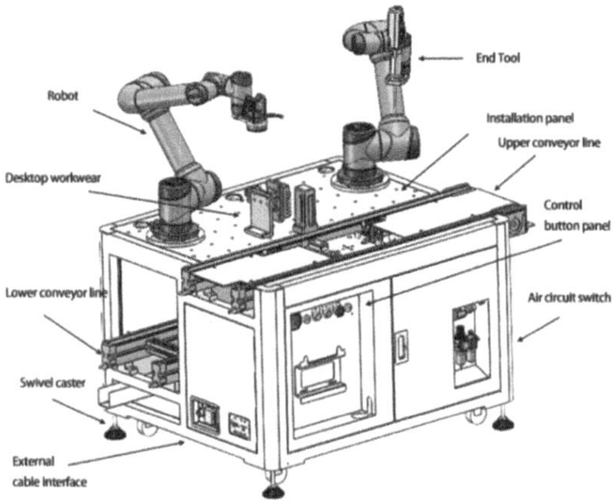

Fig. 4. Overall design of the integrated joint assembly workstation

A dual-level conveyor system was designed to separate the workflows: the upper belt was designated for transporting the integrated joint module trays, while the lower belt was implemented for recycling the trays. The upper and lower conveyor lines can efficiently transport the integrated joint module trays to the designated assembly position for assembly tasks. Aluminum profiles were selected for the main framework due to their high strength-to-weight ratio and modularity. This framework supports a belt-driven system designed to transport the integrated joint module trays. The structure is shown in Fig. 5.

To achieve secondary precise positioning of the joint module, a lifting mechanism was designed and integrated into the workstation. This mechanism ensures high precision and quality in assembly work through its lifting action. The structure is shown in Fig. 6.

Fig. 5. Conveyor belt

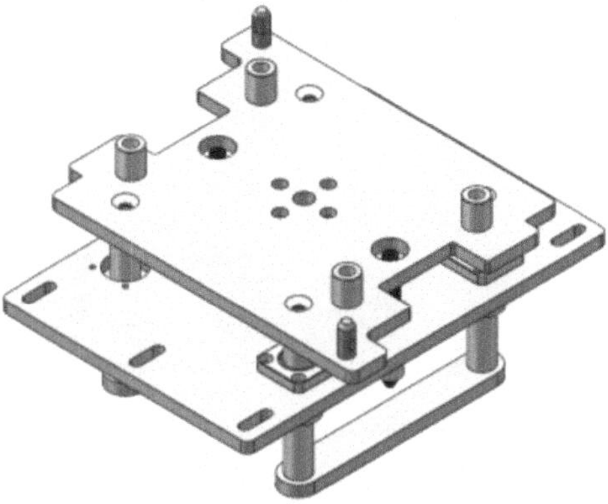

Fig. 6. Lifting mechanism

A four-level hierarchical control system was designed to manage and control various components of the entire system, as shown in Fig. 7.

Information layer: including devices such as robot teaching pendant, three-color lights, and displays. These devices are responsible for receiving operator commands and displaying the system's operating status.

Control layer: It includes devices such as robot controllers, Programmable Logic Controller (PLC), vision controllers, and electric screwdriver controllers. These devices communicate through networks and buses to achieve precise control and scheduling of various execution units in the system.

Drive layer: This layer includes devices such as relays, solenoid valves, and speed control drivers. It is responsible for electrically driving the execution units through control signal triggering, and achieving operations such as switching and speed control of the execution units.

Execution layer: It includes various execution units, such as cameras, robots, cylinders, electric screwdrivers, etc. The execution units perform corresponding actions according to control signals to complete various tasks of the assembly workstation. Such as the movement of robots in the workstation, the movement of electric screwdrivers, the action of cylinders, etc.

A central PLC was selected to manage the core control logic, as it provides the robust I/O management and deterministic timing required for synchronous operation. This design choice enabled the implementation of detailed I/O allocation and electrical

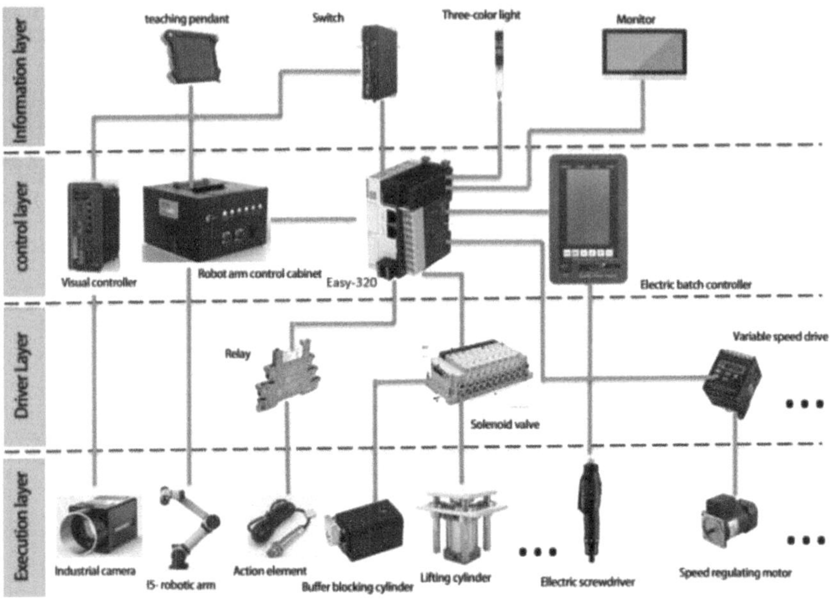

Fig. 7. Architecture diagram of electronic control system

circuit designs (such as the main circuit, switching power supply, and safety buttons) to ensure stable operation (Table 2).

Table 2. Key components and technical parameters of the workstation

Category	Component	Model
Robotics System	Collaborative Robot	AUBO-i5
	PLC Controller	Easy-320
	Smart Electric Screwdriver	ESTIC Handy 2000 Touch
Vision System	Industrial Camera	Gemini 335
	Camera Resolution	1920 × 1080 pixels
	Vision Software	VisionMaster
Calibration & Measurement	Calibration Method	Zhang's Method[16]
	Calibration Error (R1.2)	Average re-projection error: < 0.1 pixels
	Repeatability Error (R1.2)	Calibration repeatability error: < 0.1 mm
	Measurement Tool	Vernier Caliper
	Tool Precision (R1.3)	± 0.02 mm
Software Platform	Offline Simulation	PQArt

(continued)

Table 2. (*continued*)

Category	Component	Model
	Compensation Scripting	Lua (Embedded)

2.3 Integrated Joint Locking

To meet the locking requirement, a dual-arm robotic system was designed using two AUBO-i5 robots. Distinct roles were assigned to each arm: one was equipped with an industrial camera for detecting and calibrating the locking hole position, while the other utilizes an intelligent electric screwdriver for locking work. The camera and screw locking tooling are shown in Fig. 8. The electric screwdriver and controller are shown in Fig. 9.

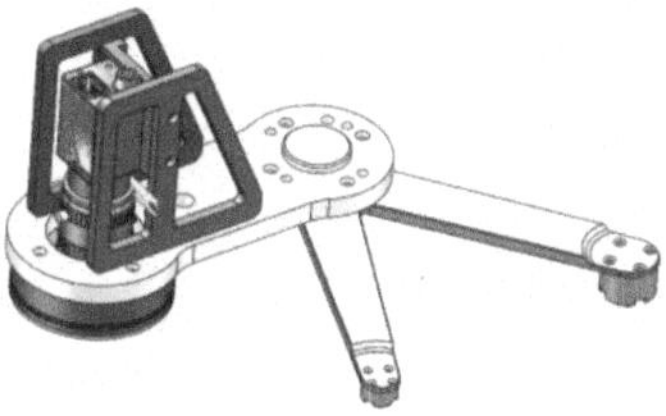

Fig. 8. Camera and screw locking fixture

Fig. 9. Smart electric screwdriver and controller

The overall steps for screw hole position recognition are as follows: The robot is positioned at the initial assembly position, and the red switch knob is rotated clockwise by 90 degrees to power on the workstation and initiate the operation. The manual valve air source switch is opened to supply air to the workstation. When the sensor of the lifting mechanism detects that the joint module tray is in place, it lifts the tray for secondary positioning. The robot moves to the imaging position to capture images of the joint module for screw hole position recognition, and then tightens the screws one by one.

During the screw locking process, multiple transition points are added to avoid accidental collisions during robot movement. Figure 10 is a detailed flowchart of the screw locking process.

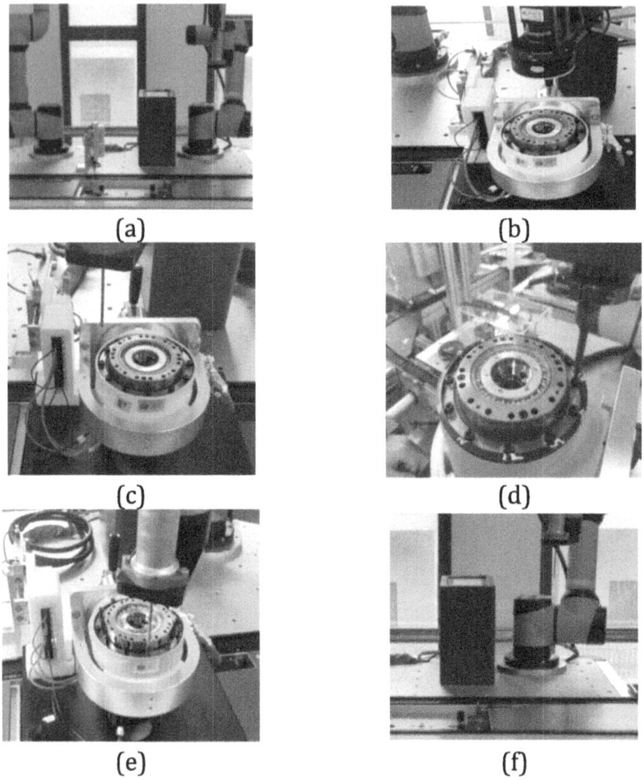

Fig. 10. Screw locking process: (a) Initial position, (b)Photographing position, (c) Preparation position, (d) First locking position, (e) Second locking position, (f)End position

Figure 10(a) shows the initial position of the robot, waiting for the joint module to be in place before starting the screw locking operation; Fig. 10(b) shows the robot's camera position, where the industrial camera takes photos of the joint module and identifies the position of the screw holes; Fig. 10(c) shows the robot's preparation position for starting the screw locking operation, where the robot moves to above the screw hole and prepares for the tightening action; Fig. 10(d) and (e) show the robot's screw locking positions, where the robot will tighten the screws of the joint module one by one according to the programmed path; Fig. 10(f) shows the robot's assembly completion position, where the robot returns to the initial position at the beginning of the assembly work, indicating the completion of the locking task.

3 Offline Simulation of Assembly Workstation

Offline simulations were conducted using PQArt software to verify the correctness of the overall scheme and to check whether the robot works in the correct posture. Offline simulation can vividly display the robot's movement path and state, facilitating a comprehensive understanding of the workstation's motion prior to real-machine operation and reducing errors during operation [11, 12]. When using offline simulation software for verification, the entire process can be divided into three main stages: model building, custom mechanism setting, and trajectory generation and simulation [13] (Fig. 11).

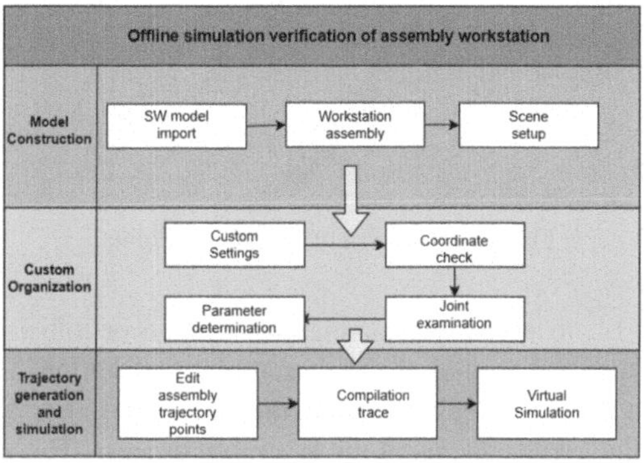

Fig. 11. Offline programming simulation verification steps

After completing the model building and defining the mechanism operation, it is necessary to edit the TCP point trajectory of the robot and compile and perform virtual simulation verification on the obtained trajectory. The TCP point is the working point of the electric screwdriver at the end of the robotic arm. According to the assembly process requirements, the tightening method is diagonal locking and cross-tightening, which can reduce the deformation or distortion of the assembly parts caused by applying excessive pressure in one area, thus ensuring the uniformity and stability of the overall assembly (Fig. 12).

(a) (b)

Fig. 12. Screw locking and trajectory sequence: (a) Screw locking sequence, (b) simulated locking trajectory

Compilation was compiled to assess its feasibility, including checking for potential issues such as unreachable positions, singular points, axis overruns, or collisions. At the same time, multiple intermediate transition points were added to the software to prevent collisions during the robot's movement. Figure 13 shows the added transition points.

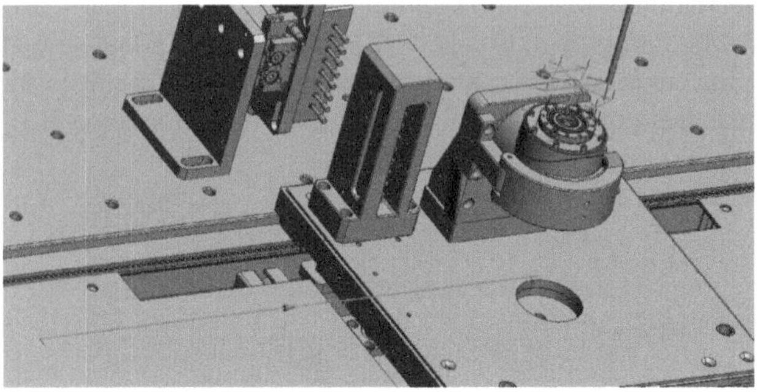

Fig. 13. Addition of Transition Points

As shown in Fig. 14, no issues such as singularity points or collisions occurred during the virtual simulation of the integrated joint assembly workstation. The total simulation time from the start of the screw locking task to the end of the entire process was 39.84 s, which meets the timing requirements of the assembly process. The screw locking work of the integrated joint module was successfully completed, verifying the correctness of the overall design of the assembly workstation.

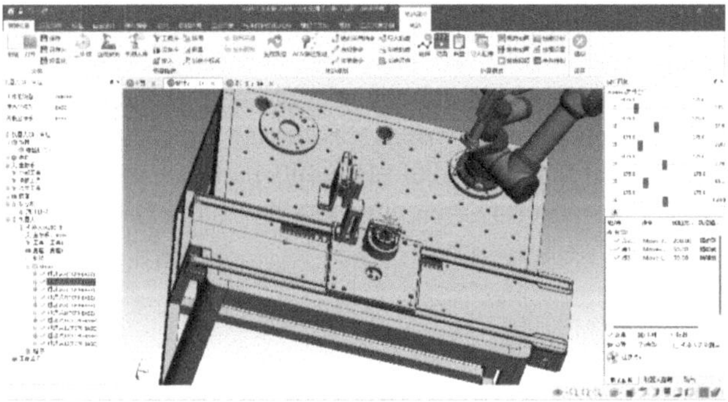

Fig. 14. Offline simulation demonstration of assembly workstation

4 Visual Error Compensation

4.1 Visual System

The precise positioning of screw holes is crucial for the assembly quality during the assembly process of integrated joint modules. However, hole offsets of workpiece screws often occur, necessitating the use of a vision system for error compensation.

A typical machine vision system comprises a camera, a light source, an image acquisition card/vision processor, firmware-related interface cables, and other accessories. In an integrated joint assembly workstation, the composition of the workstation's vision system is illustrated in Fig. 15.

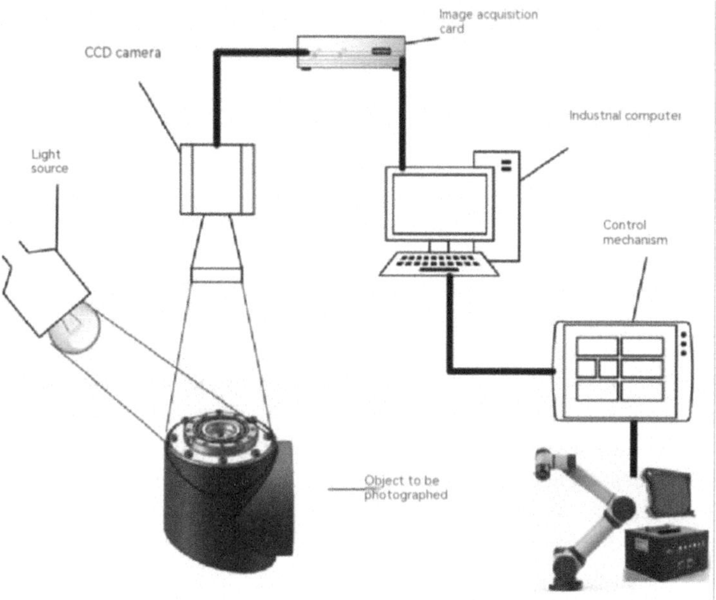

Fig. 15. Assembly workstation machine vision system

The composition of a vision system can generally be summarized into two main forms: one is a system based on personal computers, which includes front-end imaging devices and back-end data processing components, with the latter typically located inside an industrial computer; the other is a system centered around an intelligent camera, which incorporates lighting, imaging, and processing functions, enabling all vision tasks to be completed with just one intelligent camera [13, 14].

4.2 Camera Calibration Principle

By calibrating the camera, the actual coordinate system of the screw hole positions in the integrated joint module can be transformed to obtain the coordinate system of the screw hole positions in the camera's space. After obtaining this coordinate system, the robot can locate the screw hole positions of the integrated joint in real space.

Camera calibration mainly involves four coordinate systems: the pixel coordinate system, the image coordinate system, the industrial camera coordinate system, and the world coordinate system. According to the derivation, the transformation relationship between the imaging point (u, v) in the pixel coordinate system and the coordinate point

(x_w, y_w, z_w) in the world coordinate system in the physical real world is as follows:

$$z_c\begin{bmatrix} u \\ v \\ 1 \end{bmatrix} = \begin{bmatrix} f_x & 0 & u_0 & 0 \\ 0 & f_y & v_0 & 0 \\ 0 & 0 & 1 & 0 \end{bmatrix}\begin{bmatrix} R & T \\ 0 & 1 \end{bmatrix}\begin{bmatrix} x_w \\ y_w \\ z_w \\ 1 \end{bmatrix} \tag{1}$$

In Eq. (1), $f_x = \frac{f}{d_x}$ and $f_y = \frac{f}{d_y}$ represent the normalized focal lengths of the x-axis and y-axis, respectively. Under normal circumstances, the following is adopted:

$$M = \begin{bmatrix} f_x & 0 & u_0 & 0 \\ 0 & f_y & v_0 & 0 \\ 0 & 0 & 1 & 0 \end{bmatrix}, N = \begin{bmatrix} R & T \\ 0 & 1 \end{bmatrix} \tag{2}$$

From Eq. (2), it can be seen that the matrix is related to the parameters M, f_x, f_y, u_0 and v_0, hence it is referred to as the intrinsic parameter matrix of the industrial camera. The matrix N is determined by R and T, thus it is called the extrinsic parameter matrix of the industrial camera. Through a series of derivations and calculations, the internal and external parameter information of the industrial camera can be obtained, thereby determining the transformation relationship between the three-dimensional real space and the two-dimensional camera space.

During the camera shooting process, various factors may cause deviations in the image angle. Accurate calculation and handling of the camera's deviation angle are crucial for practical applications. Therefore, it is necessary to take corresponding calibration and compensation measures for errors to minimize their impact on the assembly process.

Horizontal offset compensation. Two shots were taken by the camera to compare and calculate the deviation of the No. 2 target screw hole in the X and Y directions. The camera calculated the central coordinates of the target screw hole in the two images according to the image processing algorithm, denoted as (x_1, y_1) and (x_2, y_2) respectively. By comparing the central coordinates of the target screw hole obtained from the two shots, the deviation of the screw hole in the X and Y directions is calculated:

$$\Delta x_1 = x_2 - x_1 \tag{3}$$

$$\Delta y_1 = y_2 - y_1 \tag{4}$$

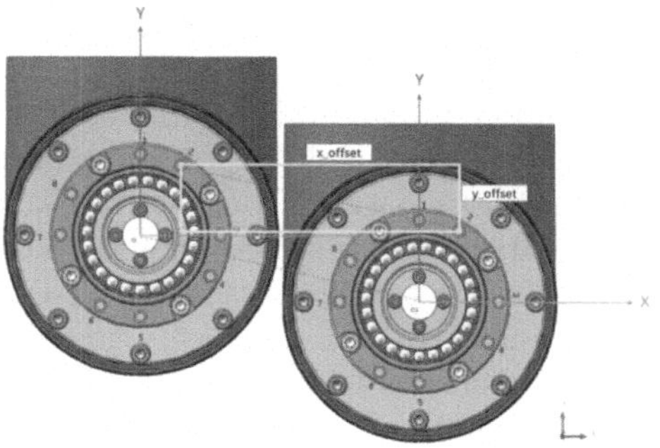

Fig. 16. Compensation for horizontal offset of hole positions

As shown in Fig. 16, x_offset is the deviation Δx_1, in the X direction, representing the displacement of the No. 2 screw hole in the X direction; y_offset is the deviation Δy_1 in the Y direction, representing the displacement of the No. 2 screw hole in the Y direction.

Rotational deviation Compensation. Figure 17 shows the angle change of the screw hole of the joint module.

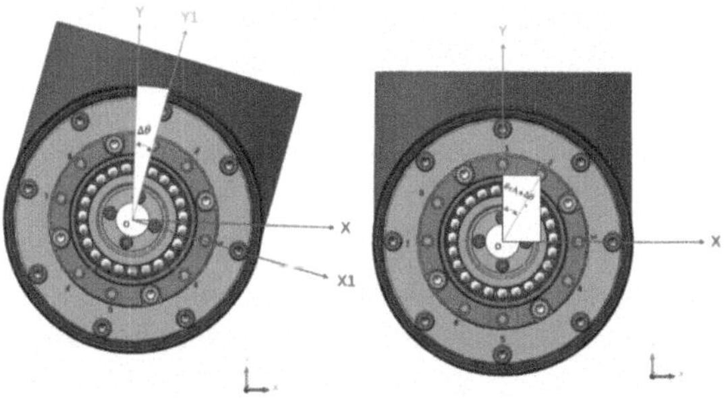

Fig. 17. Hole position rotation offset compensation

Two images of the screw hole taken by the camera were used to calculate the change in the horizontal rotation angle $\Delta\theta$ of the hole before and after shooting. Based on the angle of No. 2 screw hole and the distance from the screw hole to the rotation center O, the horizontal deviation Δx_2 and the vertical deviation Δy_2 were calculated.

$$\Delta x_2 = 38.5^{*}\cos(\theta \mathrm{rA} + \theta \mathrm{rT}) \tag{5}$$

$$\Delta y_2 = 38.5^{*}\sin(\theta \mathrm{rA} + \theta \mathrm{rT}) \tag{6}$$

4.3 Hole Position Recognition

Hole position recognition was achieved by employing the template matching method to complete the identification of the target image and the precise localization of workpiece information. Template matching is based on the comparison of similarity between a predefined template image and the target image. This method is mainly divided into two categories: matching methods based on grayscale values and shape-based matching methods. While this approach is intuitive, it is time-consuming and susceptible to environmental lighting conditions, which limits its practicality in industrial applications. Shape-based matching methods focus on comparing the shape contours of the template and target images. This method exhibits better robustness to lighting variations and partial occlusions, making it suitable for accurately identifying and locating workpieces with specific shape characteristics [15].

Given the specific shape characteristics of the screw hole positions in the joint module, this study adopted shape-based template matching technology. The process was implemented on the VisionMaster vision software platform for identifying and positioning workpieces. The template configuration diagram is shown in Fig. 18.

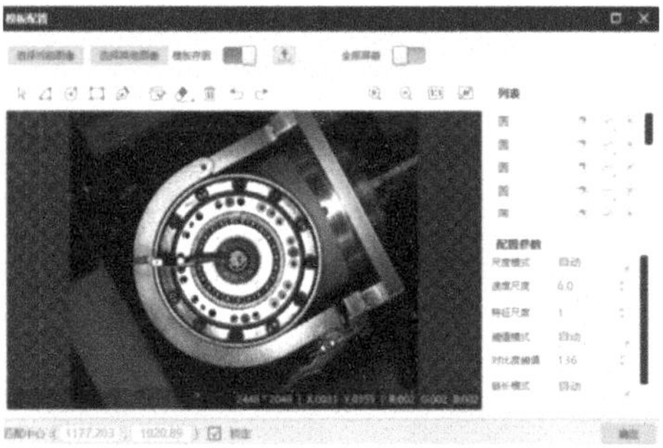

Fig. 18. Template Configuration

Based on the template configuration results, the positioning coordinates of the target workpiece image were obtained. The template matching score is 0.97, which reflects the similarity between the recognition image and the template. The value ranges from 0 to 1, with a score closer to 1 indicating a higher similarity between the image and the template, and a better matching effect.

During the camera shooting process, various factors may cause horizontal and rotational deviations in the images. To ensure the accuracy of screw hole recognition, Lua scripts were selected to implement the compensation logic, due to their ability to be embedded directly into the robot's control system for real-time calculation and compensation of error offsets.

In the hole position offset compensation process, by analyzing the sources of error, the calculation of horizontal and rotational offsets for screw hole positions was implemented using Lua scripts. Subsequently, the scripts were embedded into the robot's control system, and communication between the robot and the camera is established. The camera captured images and calculated the offsets, and the Lua scripts compensated these offset values to the robot's motion control system, enabling the robot to accurately position the target hole position.

The scripts calculated the horizontal and rotational offsets based on different scenarios, and then store the results in global variables for use by other components. In this way, it is possible to dynamically calculate the offsets of the screw hole positions based on the image data captured by the camera and the current position of the robot, and make corresponding compensations to ensure the accuracy and stability of the assembly.

5 Error Compensation Experiment

This experiment aims to verify the impact of offset compensation on the positioning accuracy of screw hole locations. In the experiment, a vernier caliper was used to measure and record the distance between the robot's end TCP point and the center point of the screw hole before and after each offset compensation. Figure 19 shows the measurement method for the positioning offset of screw hole locations. To facilitate measurement, a needle is fixed at the robot's end TCP (Tool Center Point) to simulate an Allen wrench head, and the distance from the needle to the farthest point of the screw hole is measured.

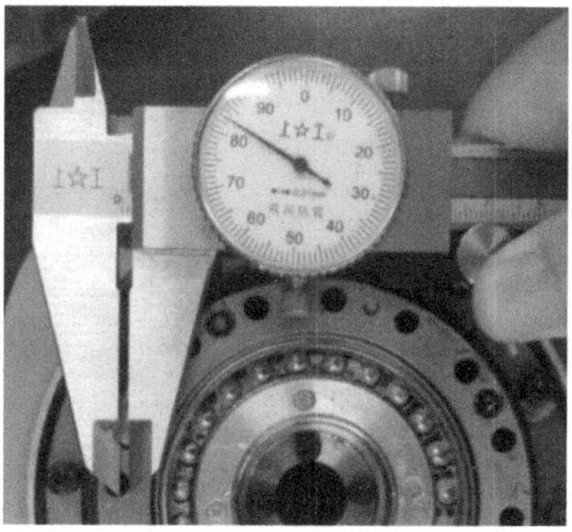

Fig. 19. Experimental data measurement method

Hole offsets before and after compensation were measured for the screw holes of the three integrated joint modules, respectively. The experimental data measurement results are presented in the following three tables and figures (Tables 3, 4 and 5) (Figs. 20, 21 and 22).

Table 3. First set of measurement data

Hole number	Error value before compensation (mm)	Error value after compensation (mm)
1	1.64	0.65
2	1.58	0.58
3	1.88	0.86
4	1.43	0.81
5	1.37	0.62
6	1.75	0.51
7	1.45	0.38

(*continued*)

Table 3. *(continued)*

Hole number	Error value before compensation (mm)	Error value after compensation (mm)
8	1.63	0.57

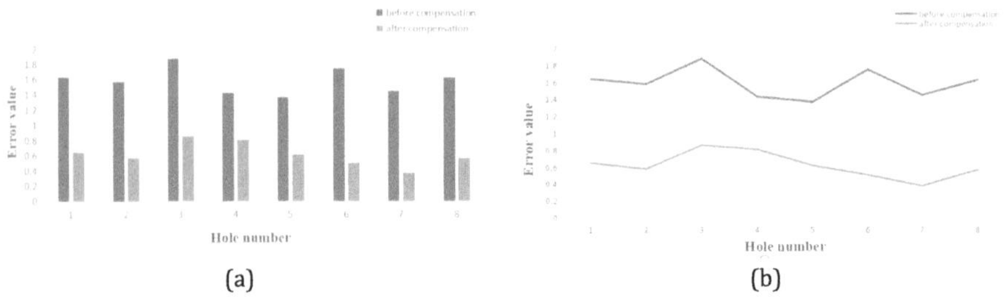

Fig. 20. Measurement data for error compensation of the offset of the first set of hole positions: (a)Bar graph of error before and after compensation, (b)Line chart of error before and after compensation

Table 4. Second set of measurement data

Hole number	Error value before compensation (mm)	Error value after compensation (mm)
1	1.85	0.44
2	1.11	1.05
3	1.59	0.58
4	1.64	0.82
5	1.36	0.37
6	1.92	0.79
7	1.47	1.19
8	1.24	0.29

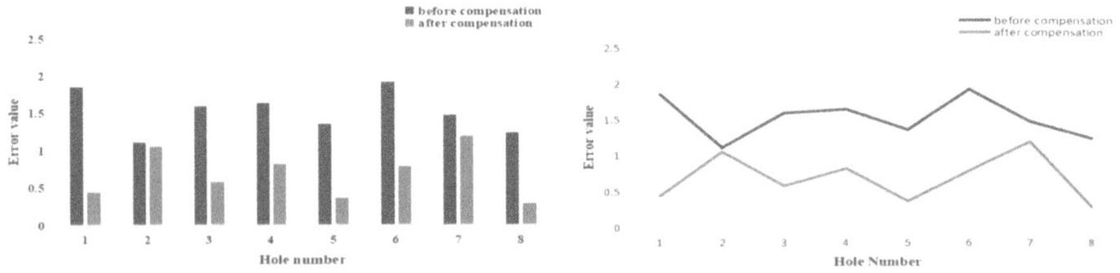

Fig. 21. Measurement data for error compensation of the offset of the second set of hole positions: (a)Bar graph of error before and after compensation, (b)Line chart of error before and after compensation

Table 5. Third set of measurement data

Hole number	Error value before compensation (mm)	Error value after compensation (mm)
1	1.42	0.34
2	1.83	0.88
3	1.19	0.62
4	1.7	0.97
5	1.3	0.71
6	1.08	0.2
7	1.77	0.43
8	1.63	0.69

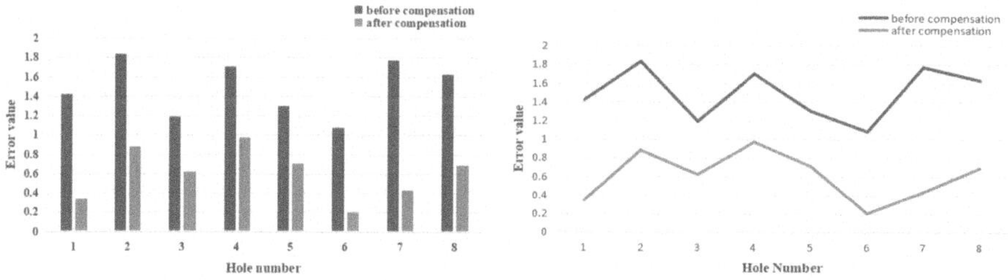

Fig. 22. Measurement data for error compensation of the offset of the third set of hole positions: (a)Bar graph of error before and after compensation, (b)Line chart of error before and after compensation

To verify the effectiveness of the proposed method, a comparative experiment was conducted. Firstly, the positioning error without visual compensation was measured and served as the Benchmark Method. This benchmark represents the accuracy achievable solely through offline programming and robotic repetitive positioning. Subsequently, the real-time visual compensation system proposed in this paper was activated.

Comparing the data before and after compensation, the impact of offset compensation on the positioning accuracy of the robot's end-effector electric screwdriver head can be observed. This experiment was repeated three times under the same conditions, yielding essentially consistent results, confirming the stability and effectiveness of the method. According to the experimental results, the average error before hole position offset compensation was 1.535 mm, and the average error after compensation was 0.640 mm, representing a 58.3% improvement in positioning accuracy compared to before compensation. This method shows promising practical application prospects.

6 Conclusion

This paper presents the design of an integrated joint assembly workstation system, which achieves hole position error compensation based on industrial cameras. A hole position offset compensation method based on industrial cameras is proposed and implemented. Through practical application in an experimental environment, it is verified that this hole position offset compensation method effectively improves the accuracy of screw hole positioning, with a positioning accuracy improvement of 58.3% compared to before compensation. The method demonstrates significant practical applicability. The accuracy improvement of this workstation is the result of multi-module collaboration. Future work will include conducting ablation studies to quantitatively separate the contributions of the secondary mechanical positioning of the lifting platform and visual servo compensation to the final assembly accuracy.

The designed integrated assembly workstation can meet various performance indicators, ensure the quality of joint assembly, and provide a reference for the design of assembly-type workstations.

Funding. This work was supported by the Ministry of Industry and Information Technology's 2022 Composite Robotics Project(Grant numbers TC220H05D) and R&D Program of Beijing Municipal Education Commission·(Grant numbers KM202410017007).

References

1. Papakostas, N., Alexopoulos, K., Kopanakis, A.: Integrating digital manufacturing and simulation tools in the assembly design process: a cooperating robots cell case. CIRP J. Manuf. Sci. Technol. **4**(1), 96–100 (2011)
2. Ore, F., Vemula, B.R., Hanson, L., et al.: Human–Industrial robot collaboration: application of simulation software for workstation optimisation. Procedia CIRP **44**, 181–186 (2016)
3. Liu, H.Y., Su, Y., Su, L.: Design and simulation of robot gluing and assembly workstation based on C# and SMART. J. Liuzhou Vocat. Techn. Coll. **21**(1), 99–104 (2021)
4. Zhou, Z.Y.: Design and Motion Simulation Analysis of Robotic Welding Workstation for Main Chord of Standard Section. Central South University of Forestry and Technology (2023)
5. Sun, Z.F.: Localization and Object Grasping Technology of Service Robots in Dynamic Environments. Southeast University (2017)
6. Vasiliev, A., Dalyaev, I.: Simulation method for the transport system of a small-sized reconfigurable mobile robot. Machines **9**(1), 52–63 (2021)
7. Yang, J.Z., Le, Y., Zhang, J.B., Zhou, Y.H., Zhao, C.X., Chen, Q.T.: Research on real-time compensation strategy for positioning accuracy of mobile robots. J. Mech. Eng. **58**(14), 44–53 (2022)
8. Ye, S.Q., Liu, P., Luo, J.M., Yu, L., Wang, X.N.: A self-standing mobile robot with image function for power inspection. Hubei Electr. Power **46**(2), 120–126 (2022)
9. Nagy, Á., Vajk, I.: Nonconvex time-optimal trajectory planning for robot manipulators. J. Dyn. Syst. Meas. Contr. **141**(11), 111007 (2019)
10. Liu, K.: High-beat intelligent assembly planning for final assembly automation. Automot. Manuf. **6**, 16–18 (2023)
11. Feng, L.Y., et al.: Simulation design of robotic assembly workstation based on quick-change fixtures and vision. Mod. Mach. Tool Autom. Manuf. Tech. **8**, 126–130+134 (2021)

12. Yong, Y.Y., et al.: Design and research on robotic assembly workstation system based on machine vision. Mach. Build. Autom. **47**(5), 184–186+209 (2018)
13. Zhao, W.B.: Design of collaborative robotic assembly workstation based on machine vision. Mod. Mach. Tool Autom. Manuf. Tech. **6**, 162–164+168 (2020)
14. Oruc, R., Baklacioglu, T.: Cruise range modeling of different flight strategies for transport aircraft using genetic algorithms and particle swarm optimization. Energy **294**, 130917 (2024)
15. Luo, L., Guo, T., Cui, K., et al.: Trajectory planning in robot joint space based on improved quantum particle swarm optimization algorithm. Appl. Sci. **13**(12), 7031 (2023)
16. Zhang, Z.: A flexible new technique for camera calibration. IEEE Trans. Pattern Anal. Mach. Intell. **22**(11), 1330–1334 (2000)

Light Swin: A Lightweight Swin Transformer Based on Sparse Attention InterWindow Blocks for Image Classification

Yiqi Xiong[1(✉)], Tao Zuo[2], Yulong Tu[2], Zhaowei Wu[2], Huasong Min[2], and Yunhan Lin[3]

[1] School of Artificial Intelligence and Automation, Wuhan University of Science and Technology, Wuhan 430081, China
zuotao@wust.edu.cn
[2] School of Electronic Information, Wuhan University of Science and Technology, Wuhan 430081, China
zuotao@wust.edu.cn
[3] School of Computer Science and Technology, Wuhan University of Science and Technology, Wuhan 430081, China

Abstract. In recent years, the integration of attention mechanisms and deformable convolutional networks has significantly advanced the field of computer vision, enhancing models' ability to capture complex spatial relationships and adapt to various geometric transformations. However, these improvements often come at the cost of increased computational demands and model complexity. At the same time, existing lightweight classification networks tend to perform poorly on fine-grained classification tasks. To address these challenges, this paper introduces two key components: InterWindow blocks and Sparse attention mechanisms with l_∞. InterWindow blocks facilitate efficient information exchange between non-adjacent regions while preserving information extraction from adjacent areas. This design effectively captures long-range dependencies without the computational overhead associated with traditional global attention methods. Sparse attention mechanisms with l_∞ focus on the most relevant parts of the input data, ensuring the comprehensive capture of critical information with limited parameters. By integrating InterWindow blocks and Sparse attention, this work aims to balance model performance and computational feasibility, making it particularly suitable efficiently processing large volumes of data. Experiments demonstrate that the proposed model exhibits superior robustness and accuracy compared to traditional CNN and Transformer models.

Keywords: Computer vision · Image classification · Lightweight modified Swin Transformer · Sparse attention

H. Wei and J. Niu (Eds.): COBOT 2025, CCIS 2893, pp. 152–175, 2027.
https://doi.org/10.1007/978-981-92-2074-8_10

1 Introduction

Currently, image classification and semantic segmentation are two of the most popular research directions in the field of computer vision. The key to image classification research is to design a model that can efficiently extract information from images. Convolutional neural networks (CNN) construct powerful feature representations by mapping features to image pixel units, and have become the main backbone network for various visual tasks. With the intensive study of convolutional neural networks, the architecture of CNN has gradually increased in complexity [21], with more extensive connection, and a more complex convolutional form [22–24].

Lightweight networks have gained significant attention due to their ability to deliver competitive performance while being computationally efficient. For instance, convolutional neural networks (CNNs) like RepVGG [5] and EfficientNet [6] have proven to be highly effective for image classification tasks with relatively fewer parameters compared to Transformers. These lightweight CNNs are designed to balance model accuracy and computational efficiency.

On the other hand, Transformer-based models, which face challenges related to their high parameter counts and computational complexity. [1] While architectures like Vision Transformer [2], Swin Transformer [3], and MobileViT [4] have sought to reduce these costs, they still often require careful tuning and trade-offs between performance and efficiency.

Developing more efficient Transformer variants or hybrid models that combine the strengths of both CNNs and Transformers has become an area of active research. These efforts aim to push the boundary of what is achievable in terms of both accuracy and efficiency in image classification and beyond.

Sparse attention is an important research direction to improve the computational efficiency of attention mechanisms in recent years. Some sparse attention uses fixed patterns to limit the number of tags involved, such as sliding Windows [35, 36], extended sliding Windows [37, 38], combinations of patterns [39], or domain-specific patterns [40]. Recent studies aim to achieve a constant memory cost during inference through a predefined heuristic cache reclaim strategy [41]. Anagnostidis et al. [42] proposed a learnable, irreversible key-value pair pruning method that improves the efficiency of inference time memory by relaxing pruning operations to the concept of cumulative gating. However, the problem of quadratic complexity still exists in the training process of this method, which affects its ability to speed up the training process.

In order to optimize the existing problem and improve the model representation, this paper proposes a novel Light Swin backbone structure. This structure uses Swin Transformer to extract low-level features and feeds them as input to the InterWindow block for processing high-level features. The main design idea of the InterWindow block is to window the image and perform segmentation. As illustrated in Fig. 1, in the l layer, each window performs one attention calculation. In the subsequent layer ($l+1$), the corresponding pixel elements of each window only pay attention to the pixels in the corresponding position in the patch. This allows the pixels of each window to encode the information of all the pixels in the image. In addition, this paper introduces a sparsity regularization term to control the weight distribution. This enables the model to dynamically maximize the largest attention weights while making the other attention weights close to zero [29, 30].

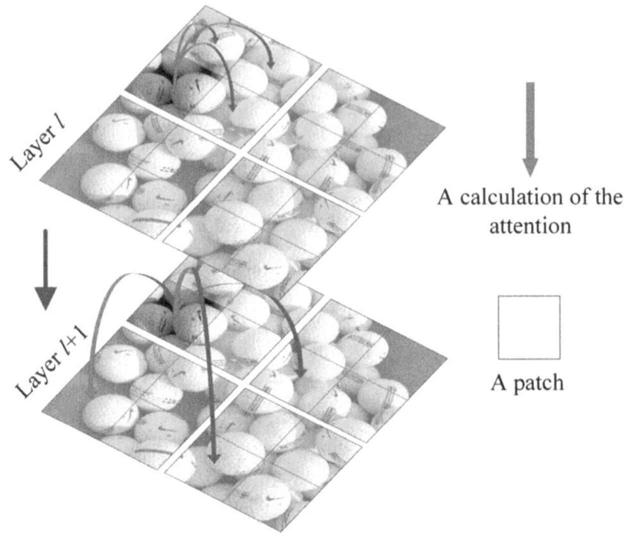

Fig. 1. Schematic of the method of calculating self-attention in the InterWindow block

The proposed Light Swin model demonstrate superior performance in image classification compared to several existing models. The Light Swin model has a significantly smaller number of parameters, 17.03M, compared to Swin-T [3] model with 29M parameters and VIT-B [2] model with 85M parameters. Furthermore, the reduction in the number of parameters in Light Swin is accompanied by an improvement in accuracy, when compared to studies with the same size of parameter counts. The accuracy of the Light Swin model on the ImageNet-100 [13], CIFAR-100 [14], CIFAR-10 [14] and iNaturalist 2018 (INAT-18) datasets is 88.28%, 85.8%, 97.46% and 76.8%, respectively. In summary, the contributions of this paper are as follows:

- This paper proposes a network architecture based on InterWindow blocks, which improves the performance and efficiency of the model for image feature extraction in the deeper layers of the network.
- The concept of Sparse Attention is introduced, which effectively constrains the attention allocation and makes the attention weights show sparsity.
- The prediction accuracy of the model proposed in this paper on ImageNet-100, CIFAR-100, CIFAR-10 and INAT-18 benchmark datasets is improved compared to the prediction accuracy of other models.
- In comparison to the state-of-the-art Transformer model, the model proposed architecture exhibits a more streamlined parameter count, thereby demonstrating a higher efficiency of parameter utilization.

2 Related Work

2.1 CNN and Variants

Convolutional neural networks (CNN) have been the standard network model in the field of computer vision for the past decade [25]. Since the introduction of AlexNet [7], CNNs have gradually become a mainstream research direction, and researchers have worked on developing deeper and more efficient convolutional neural networks, such as VGG [8], Googlenet [9], ResNet [10], DenseNet [11], and EfficientNet [6]. In order to capture image detail information more comprehensively, all of these convolutional networks must apply dense convolution at the pixel level while pooling layers are used to spatially reduce the feature maps, exhaustive convolution on dense cells is still required, which leads to a limitation in the speed of the network model. ECA-Net [48] improves channel attention by avoiding dimensionality reduction and using local cross-channel interaction via one-dimensional convolution. However, ECA-Net ignores spatial dependencies, which limits its ability to capture geometric structures inherent in point image data.

2.2 Swin Transformer

In the previous study on Vision Transformer [2], it was observed that the complexity of the attention computation grew quadratically with the input image size. To overcome this challenge, Swin Transformer [3] proposes a backbone algorithm that performs linear computational complexity with respect to the image size. Swin Transformer employs an innovative approach that begins with a small image. Patches are initially processed and then gradually merged with neighboring patches in the deeper layers of the Transformer. This strategy allows Swin Transformer to have a more flexible shifted window, which effectively reduces the computational complexity. Swin Transformer also makes significant progress in terms of accuracy, reaching 84%, a 5% improvement over the ViT model. In this paper, Swin Transformer is utilized to extract low-level image features, and an InterWindow block is introduced on the generated output to further enhance the performance of the model.

3 Method

3.1 Overall System Solution

Figure 2 presents an overview of the Light Swin architecture, which comprises the Swin Transformer Block and the InterWindow Block. The entire architecture is divided into four main stages, where in images are downsampled at each stage to generate multi-scale feature representations. In this study, this paper adheres to the Swin Transformer Block [3] design as previously studied.

In Stage 1, the input image is accurately divided into non-overlapping patches by the Patch Partition module, which adheres to the partitioning strategy of the Swin Transformer. This paper sets that every 4×4 neighboring pixels form a patch, and that the input is an RGB three-channel image. Consequently, the feature dimension of each patch is $4 \times 4 \times 3 = 48$. In the initial phase, the image features are mapped to a predetermined dimension (C) by a linear embedding layer. Subsequently, these tokens ($\frac{H}{4} \times \frac{W}{4}$) are passed through two Swin Transformer blocks to extract the base features of the image. In Stage 2, the Patch Merging layer is responsible for downsampling the tokens to ($\frac{H}{8} \times \frac{W}{8}$) size and adjusting the output dimension to $2C$. This process aims to reduce the size of the feature map. Subsequently, the Swin Transformer block is applied to perform the feature transformation. This process is repeated in Stage 3, where four Swin Transformer blocks are applied and the feature transformation is performed by an InterWindow Block. Finally, in Stage 4, the features are transformed directly using the InterWindow Block after being downsampled through the Patch Merging layer. In particular, the InterWindow block used in the text takes into account both local feature extraction of the window and cross-window spatial feature extraction capabilities, and finally connects this transformation to the final classification network. In order to control effective sparsity, $l\infty$ regularization function is introduced in the sparse self-attention layer of W-MSA and IW-MSA to impose constraints on the distribution of attention weights. This method avoids model overfitting and enhances the robustness of the model.

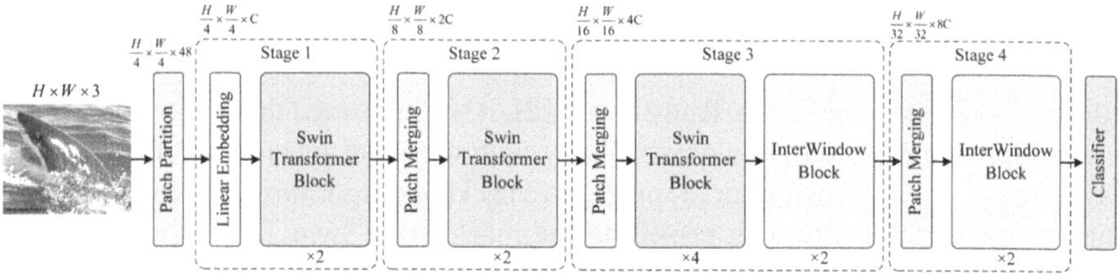

Fig. 2. Light Swin Model Architecture

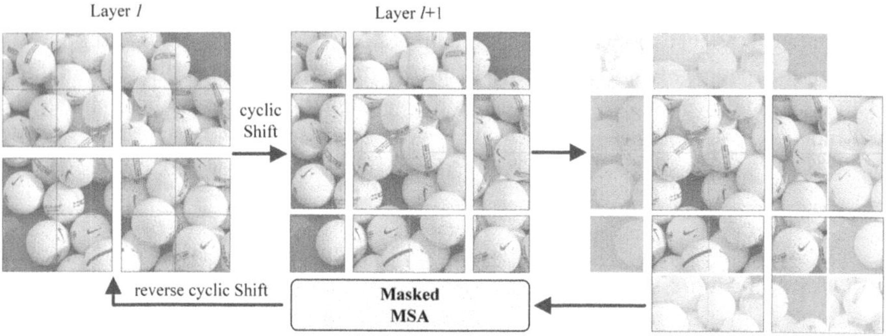

Fig. 3. Illustration of a batch computation approach for self-attention in shifted window partitioning

3.2 Swin Transformer Block

Taking the original image as input, the patch partition block divides the input image into small non-overlapping blocks. As can be seen in Fig. 3, the gray square-like partitioned region represents a token, the large red square represents the local window, and the token exchanges information with other tokens in small gray squares within the large red square. As the token moves to the next layer, the window is shifted, and the Swin Transformer block in the next layer uses the shifted W-MSA (SW-MSA) module in the second layer. Figure 4 shows that in addition to the W-MSA and SW-MS modules, tokens must pass through the LayerNorm (LN) module, the remaining link, and the Multilayer Perceptron (MLP) module. By traversing the Swin Transformer, feature representations can be obtained from patches. Then, the patch merging layer reduces the feature resolution and increases the feature dimension to obtain a hierarchical feature representation of the image.

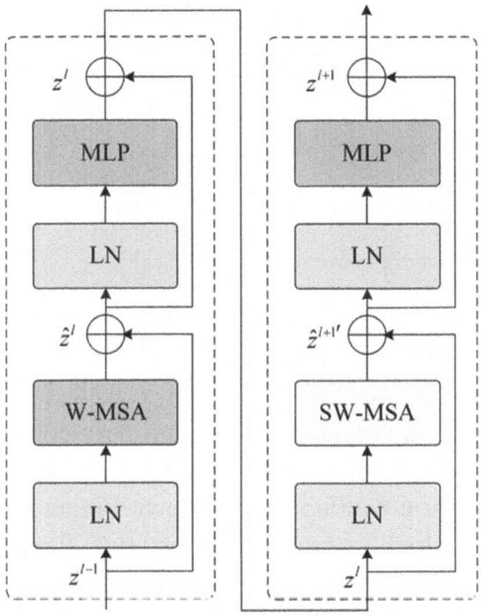

Fig. 4. Swin Transformer Block

The problem with moving window partitions is that it results in more windows and, in a mobile configuration, inconsistent window sizes after partitioning. Therefore, it is necessary to use a cyclic moving approach, as shown in Fig. 4. After such a move, there will be non-adjacent sub-windows in the batch window, so a relative position bias must be introduced into the self-attention calculation [20, 26–28].

3.3 InterWinodw Block

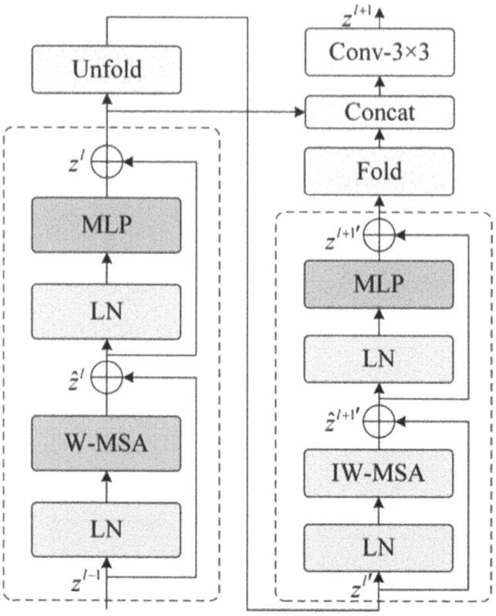

Fig. 5. InterWinodw Block

The InterWindow Block is composed of two primary components: the local attention feature extraction module for non-overlapping windows and the spatial attention extraction module between non-overlapping windows. Figure 5 provides a detailed illustration of the InterWindow Block's structural composition. First, the features extracted in the previous stage are passed through the W-MSA (Windows Multi-head Self-Attention) module for local information extraction. Next, the feature map is expanded by the Unfold operation, and then, the InterWindow Multi-head Self-Attention (IW-MSA) module is passed through the Unfold operation for spatial information extraction. Finally, the feature map is folded back to its original shape via the Fold operation, and then fused with the residual branches and mapped into the shape of W, H, and C via convolution. The result is then passed to the next stage. Throughout the process, the InterWindow Block fuses local and spatial information through the interaction of the W-MSA and IW-MSA modules. This enables the orderly extraction of local features and inter-window spatial relations within non-overlapping windows.

3.4 IW-MSA and W-MSA

The W-MSA reduces the amount of computation by dividing the image into several windows. The normal MSA (Muti-head Self-Attention) needs to be de-computed with all the pixels in the image during the Self-Attention computation. However, when using Windows Muti-head Self-Attention (W-MSA), the feature map is first divided into windows according to the size of $M \times M$,and then Self-Attention computation is performed for each window internally, and this approach restricts the Self- Attention computation is limited to each window, which effectively reduces the computational complexity.

The spatial modeling capability of the window-based Self-Attention module is limited due to the lack of cross-window connections. In this paper, we propose the IW-MSA module, inspired by MobileViT [4], to ensure the spatial modeling capability of the model while further reducing the computational complexity.

The IW-MSA module is used in the InterWindow Transformer as shown in Fig. 6. First, the partitioned feature maps extracted from the previous layer are processed by a relative positional spreading operation, and then the spread feature maps are inputted into the Transformer with the IW-MSA block for attention computation. In this process, the IW-MSA performs a subregional sparse attention calculation for pixels with the same relative position, that is, the attention calculation between the same color blocks in the diagram. Because the InterWindow calculation is carried out at high latitude, it can extract global features while ensuring efficiency. Subsequent to this, the results are refolded back to the shape of the original. The feature maps are then used for global feature modeling. These feature maps are subsequently stitched together with the original feature map inputs via the shortcut branching along the channel direction. Finally, a convolutional layer with a convolutional kernel size of 3 × 3 is used for feature fusion to obtain the output feature map.

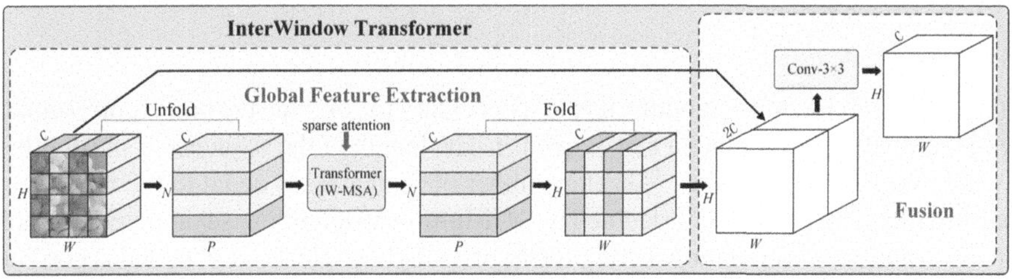

Fig. 6. InterWindow Transformer Block

In comparison to the displacement window partitioning method utilized in Swin Transformer, the IW-MSA module circumvents a series of issues associated with sliding windows, obviating the necessity for relative position encoding, thereby simplifying the model design. By establishing the InterWindow connection, the IW-MSA module facilitates cross-window information interaction, which enhances the global perception of the model and enables it to more effectively capture the global correlation of image features. Furthermore, given the prevalence of data redundancy in image data, the traditional pixel-level attention computation leads to a significant amount of wasted computational resources. The design of the IW-MSA module significantly improving the computational efficiency of the model.

3.5 Computational Complexity

3.5.1 MSA and W-MSA

The computational complexity of the multi-head self-attention module can be derived from the following equation:

$$\Omega(MSA) = 4hwC^2 + 2(hw)^2C \tag{1}$$

the computational complexity of W-MSA is calculated as follows:

$$\Omega(W-MSA) = 4hwC^2 + 2(M)^2hwC \tag{2}$$

(the feature map has a total of hw pixels, and the dimension of each pixel is C, M denotes the window size, the width and height of each local window.).

A comparison of the W-MSA computational complexity Eq. (2) with the MSA computational complexity Eq. (1) reveals that the computational complexity of W-MSA is smaller.Consider a feature map with dimensions h and w of 56, M of 7, and C of 96. In this case, the W-MSA module can be expected to save approximately 1.85 GFLOPs compared with the MSA module.

3.5.2 IW-MSA

In the case of the IW-MSA module, it is first necessary to combine the pixels corresponding to the small windows into a new sequence. The length of the sequence is the number of windows, which is equal to $\frac{hw}{M^2}$. The number of sequences is equal to the number of small window tokens, which is denoted by M. Multi-head attention is computed for each sequence, and the computational complexity of a single sequence is obtained according to Eq. (1).

$$4\frac{hw}{M^2}C^2 + 2\left(\frac{hw}{M^2}\right)^2C$$

Given that there are M^2 pairs of sequences, the overall computational complexity is:

$$M^2 \cdot \left(4\frac{hw}{M^2}C^2 + 2\left(\frac{hw}{M^2}\right)^2C\right) = 4hwC^2 + 2\left(\frac{hw}{M}\right)^2C$$

Consequently, the computational complexity of IW-MSA is calculated as follows:

$$\Omega(IW-MSA) = 4hwC^2 + 2\left(\frac{hw}{M}\right)^2C \tag{3}$$

A comparison of the IW-MSA computational complexity Eq. (3)with the MSA computational complexity Eq. (1)reveals that the two formulas differ only in the latter term. It can be demonstrated that the IW-MSA computational complexity is smaller than that of the MSA, due to the fact that $\frac{hw}{M}$ is always smaller than hw. The IW-MSA module more suitable for deeper image processing tasks.

In the context of deep network architecture design, by comparing the computational complexity formulas of IW-MSA and W-MSA, Eqs. (2) and (3)are subtracted to get Eq. (4)

$$2\left(M^2 - \frac{hw}{M^2}\right)hwC \tag{4}$$

Equation (4)demonstrate that the computational complexity of the IW-MSA and W-MSA modules is equivalent when the height and width of the feature map, h and w, are equal to the width and height of the small window, M^2. For example, when the M of the small window is 7, and the h and w of the feature map are M^2, that is, 49, the computational complexity of W-MSA and IW-MSA is the same. However, in the designed network structure, the height-width of the feature map decreases as the network level continues to deepen. Consequently, in the deeper stages of the network, the transformer layer with IW-MSA is employed instead of the SW-MSA layer in Swin Transformer, which not only effectively reduces the computational complexity, but also enhances the model's spatial modeling capability.

3.6 Sparse Attention

Sparse attention reduces the complexity by clustering strategy. It only calculates the attention in the cluster and ignores other global information. Consequently, a self-regulating attentional sparsification method is required to achieve effective sparsification of attention while maintaining model performance. This paper introduces the l_∞ paradigm to achieve attentional sparsification for W-MSA and IW-MSA.

The l_∞ paradigm is a metric that measures the maximum absolute value of elements in a vector, which can effectively capture the extreme values in the vector. In the context of the attention mechanism, the introduction of the l_∞ paradigm enables the regularisation of attention weights and the control of their distribution sparsity through the hyperparameter λ.

For W-MSA and IW-MSA, the calculation method of attention weight is:

$$w \in W - MSA = SoftMax\left(\frac{QK^T}{\sqrt{d}}\right) \tag{5}$$

$$w \in IW - MSA = softMax\left(\frac{Q^N K^T}{\sqrt{d}}\right) \tag{6}$$

$$Regularization = \lambda\left(\max_{w \in W-MSA-weights} |w| + \max_{w \in IW-MSA-weights} |w|\right) \tag{7}$$

In Eq. (7), we incorporate the maximum values of the attention scores of W-MSA and IW-MSA in the InterWindow block as part of the regularization. Specifically, This paper employs the attention weights under the l_∞ paradigm constraint as part of the loss function. The relationship between the sparsity of the attention weights and the performance of the model is balanced by adjusting the hyperparameter λ in the regularization term.

$$Toalloss = -\sum_{i=1}^{n} y_i \log(p_i) + Regularization \tag{8}$$

In Eq. (8), $-\sum_{i=1}^{n} y_i \log(p_i)$ represents the classification cross-entropy loss. Additionally, an attention weight regularization term is introduced to dynamically adjust the distribution of attention scores during the training process. The model proposed in this paper is capable of regulating the sparsity of the attention weights during the training process, thereby enabling the optimal attention distribution to be identified. Enabling it to adapt more effectively to different tasks and scenarios.

4 Experiments

In this paper, cross-dataset experiments are conducted using the CIFAR-100 [14], CIFAR-10 [14], Imagenet-100 [13] and INAT-18 image classification datasets to evaluate the performance of the Light Swin model. On the small-scale CIFAR-10[14] dataset, we expected to observe the performance of the model under image resource constraints, while on Imagnet-100, we only selected 8 high-level taxonomic species to verify the performance of the proposed method in fine-grained classification.

To verify the role of sparse attention on the generalization ability of the model, this paper conducts regularized ablation experiments on the attention weights in the Inter-Window Block. Three different regularization methods, l_1, l_2, and l_∞, are used in this paper. l_1 regularization tends to produce sparse weight matrices, i.e., many of the weights are close to zero, in contrast to l_2 regularization, which does not directly produce sparse weights, but instead encourages the elements of the weight matrix to be as small as possible. By using these different regularization methods, it is possible to better understand the impact of using l_∞ regularization for sparse attention on the generalization ability of the model. Furthermore, this paper compares the best experimental results obtained with other state-of-the-art studies.

4.1 Model Parameter Selection

In this study, the W-MSA and IW-MSA attention layers follow the design of the hyperparameters of the Swin-T model [3], setting their dimensions (dim) and number of heads (head) to 384, 12 and 768, 24, respectively.

This paper conducts a series of training experiments on the Imagenet-100 [13] image classification dataset by comparing the model performance under different learning rates and optimizers in order to find the most suitable hyperparameter configuration for this model.

(1) Learning rate

This experiment uses three learning rate values, 1e-3, 1e-4, 1e-5, are selected for the experiment and decayed using the lambda function, respectively. Through six experiments, this paper compares the recognition accuracy of the model classification under different learning rate adjustment methods.

Through the experiment, the highest accuracy is achieved when training with the lambda decay function at a learning rate of 1e-4.

(2) Optimizer

The choice of optimizer also directly affects the speed at which the model learns. To select a suitable optimizer, this experiment uses several popular optimizers: SGD, Momenutum, Adam, Nesterov, and AdaMax. Through five experiments, the optimizer selects Adam for training with the highest accuracy.

(3) Pretrained models

Since Light Swin low-level feature extraction uses the architecture of Swin-T, this paper adopts the pre-trained Swin-T model on Imagenet-1K to speed up the training process while maintaining the advantages of Swin Transformer in low-level image feature extraction. To verify the effectiveness of the pre-trained model, a series of comparison experiments are conducted in this paper. The experimental results are shown in Fig. 7.

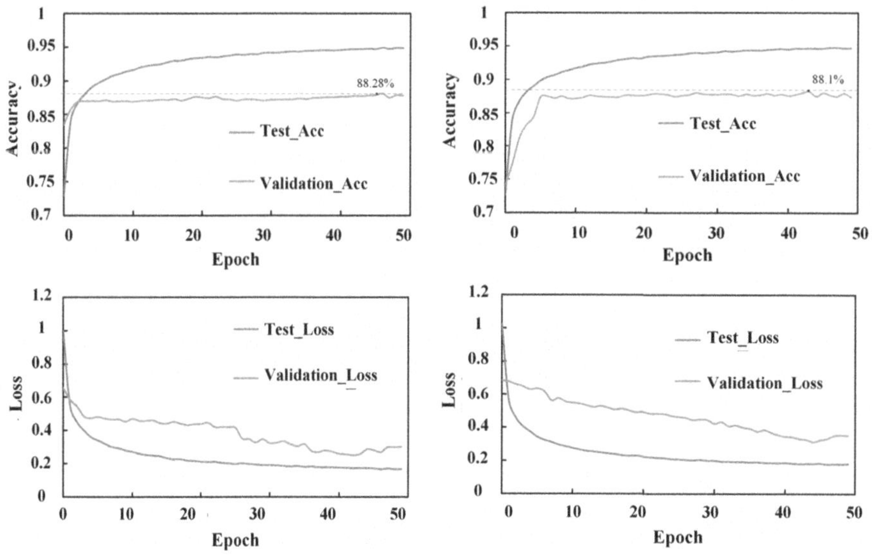

Fig. 7. Training accuracy and loss variation of Light Swin with and without freezing pre-trained models

Observing the experimental results, it can be found that in the case of freezing the pre-trained model, the accuracy and loss of the model show a stable trend after the model has been trained to 30 epochs, and the highest accuracy of the model reaches 88.28%.

In comparison, when the pre-trained model is not frozen, the Accuracy and Loss of the model gradually stabilize after 40 epochs of training, and the highest accuracy of the model is 88.1%. It can be seen that using the pre-trained Swin-T model not only significantly accelerates the convergence speed of the training process, but also helps to improve the feature extraction ability of Light Swin.

4.2 Experiment Result and Analysis

This paper compares Light Swin with other state-of-the-art models on the CIFAR-100 [14] dataset. The experimental results are shown in Table 1 under the input image test with a resolution of 224^2. Although the models are similar in the number of parameters, Light Swin shows a significant advantage in prediction accuracy. Specifically, compared to Swin-T [3], Nest-S [19], ResNeXt-50 [16], DenseNet-BC-250 [17] and DenseNet-BC-190 [17], Light Swin model's prediction accuracy improves by 7%, 3.38%, 0.66%, 4.72% and 2.26%, respectively.

Furthermore, this paper also conducted tests on the CIFAR-10 [14] dataset to evaluate the performance of Light Swin on smaller datasets. The experimental results are shown in Table 2. The results show that the prediction accuracy of Light Swin is improved by 3%, 0.49%,0.03%, and 2.65%, compared to Swin-T [3], NesT-S [19], SparseSwin [15], and DenseNet-BC-250 [17], respectively. It is even 0.16% higher than DenseNet-BC-190 [17] with Mixup [18] image enhancement technique. This result fully demonstrates that the Light Swin model has consistent efficacy on datasets of different dimensions.

This paper presents the results of comparison experiments conducted with various state-of-the-art models on the CIFAR dataset. The findings demonstrate that Light Swin exhibits excellent performance in image prediction tasks. Light Swin demonstrates superior feature extraction capabilities and prediction accuracy on both large and small datasets. Additionally, Light Swin exhibits clear advantages in the utilization efficiency of the number of parameters spent, as illustrated in Figs. 8 and 9. The Light Swin model using sparse attention with l_∞ demonstrates enhanced accuracy in translating the number of parameters into model predictions on multiple datasets. This is evidenced by a reduction in the number of parameters by 8.57M and an improvement in accuracy by 2.26% over the Convolution architecture of DenseNet-BC-190 [18] on CIFAR-100 [13] dataset. The model achieved a reduction in the number of parameters by 6.37M and an improvement in accuracy by 0.49% over the Transformer architecture of NesT-S [19] on the CIFAR-10 [14] dataset.

Table 1. CIFAR-100 Training Model

Model	#Params	Model Type	Top-1 acc
Swin-T [3]	29M	Transformer	78.08
NesT-S [19]	23.4M	Transformer	81.7
ResNeXt-50 [16]	25.03M	Transformer	84.42
DenseNet-BC-250 [17]	15.3M	Convolution	80.36
DenseNet-BC-190 [17]	25.26M	Convolution	82.82

(*continued*)

Table 1. (*continued*)

Model	#Params	Model Type	Top-1 acc
Light Swin (ours)	17.03M	Transformer	84.15
Light Swin with l_∞ (ours)	17.03M	Transformer	85.08

Table 2. CIFAR-10 Training Model

Model	#Params	Model Type	Top-1 acc
Swin-T [3]	29M	Transformer	94.46
NesT-S [19]	23.4M	Transformer	96.97
SparseSwin [15]	17.58M	Transformer	97.43
DenseNet-BC-250 [17]	15.3M	Convolution	94.81
DenseNet-BC-190 + Mixup [18]	25.26M	Convolution	97.3
Light Swin (ours)	17.03M	Transformer	97.03
Light Swin with l_∞ (ours)	17.03M	Transformer	97.46

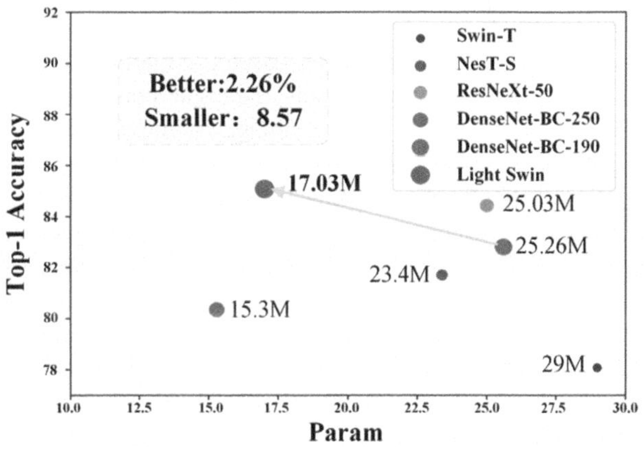

Fig. 8. Number and accuracy of individual model parameters in the CIFAR-100 dataset

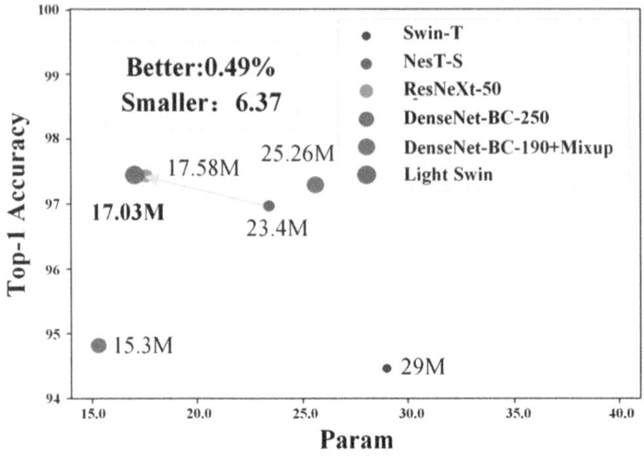

Fig. 9. Number and accuracy of individual model parameters in the CIFAR-10 dataset

To further validate the effectiveness of using InterWindow blocks to reduce computational complexity and improve the model's spatial modeling capability, we conducted experiments with a single GPU on the ImageNet-100 [13] test set. The experiments compared GFLOPs, prediction accuracy, and throughput (images/s) of the Swin-T and Light Swin models, as well as other state-of-the-art global and local attention-based models. The experimental results, as presented in Table 3, demonstrate that Light Swin achieves competitive performance with fewer parameters (17.03M) and lower computational cost (3.53G FLOPs) compared to both Transformer-based full-attention models (e.g., ViT-B, DeiT-S) and local-attention models. Specifically, the Light Swin model achieves a Top-1 accuracy of 87.04% and a throughput of 917.8 images/s, outperforming several alternatives in terms of efficiency and accuracy. When incorporating the sparse attention with l_∞, the Light Swin model achieves a further boost in performance, reaching a Top-1 accuracy of 88.28% with a throughput of 980.5 images/s while maintaining the same computational cost (3.53G FLOPs). Compared to Swin-T [3] and SwinV2-T [44], Light Swin not only reduces computational overhead but also achieves superior accuracy and throughput, confirming its effectiveness in balancing performance and efficiency.

Additionally, as illustrated in Fig. 10 the Light Swin model demonstrates optimal trade-offs among parameters, computational cost, and prediction accuracy compared to other Transformer-based global and local attention models. These results validate the strength of the proposed InterWindow blocks in enhancing performance under constrained computational budgets.

The INAT-18 dataset, as illustrated in Table 4, serves as a challenging benchmark for evaluating the Light Swin model and other state-of-the-art architectures. RDNet-T achieves the highest Top-1 accuracy of 77.0% but requires 5.0 GFLOPs with 24M parameters. Grafit ResNet-50 demonstrates lower computational cost (4.1 GFLOPs) but a reduced accuracy of 69.8%, despite having 26M parameters. DeiT-III-S offers modest accuracy (67.1%) while maintaining a relatively low computational cost (4.6 GFLOPs) and fewer parameters (22M). Light Swin achieves a competitive accuracy of 74.3%

with only 3.53 GFLOPs and 17.03M parameters, demonstrating its computational efficiency compared to other models. With an additional refinement using sparse attention with l_∞, the accuracy improves to 76.8%, maintaining the same lightweight design and computational cost (3.53 GFLOPs), improving the robustness of the model. The results validate the Light Swin model's balance between computational efficiency and accuracy, outperforming heavier models like ViT-B while offering competitive performance against RDNet-T. The significant improvement brought by the sparse attention with l_∞ variant highlights the model's robustness in optimizing sparse attention mechanisms for long-range feature extraction. In summary, the INAT-18 dataset demonstrates the Light Swin model's capability to achieve state-of-the-art performance with a lightweight architecture, making it well-suited for resource-constrained environments.

In order to evaluate the feature extraction ability of Light Swin in image prediction, two representative models, Swin-T [3] and NesT-T [19], are selected for comparison. The feature maps in the classification network are visualized in the form of heat maps of the model weights using Grad-CAM [20] technique, and the images in the test set of Imagenet-100 dataset are predicted. As illustrated in Fig. 11, both Swin-T [3] and NesT-T [19] are capable of focusing on the primary aspects of the target during the prediction process. However, they lack the ability to focus on the overall contour of the target. In contrast, Light Swin not only demonstrates a significant focus on the target as a whole but also has a more precise and prominent focus on its feature parts.

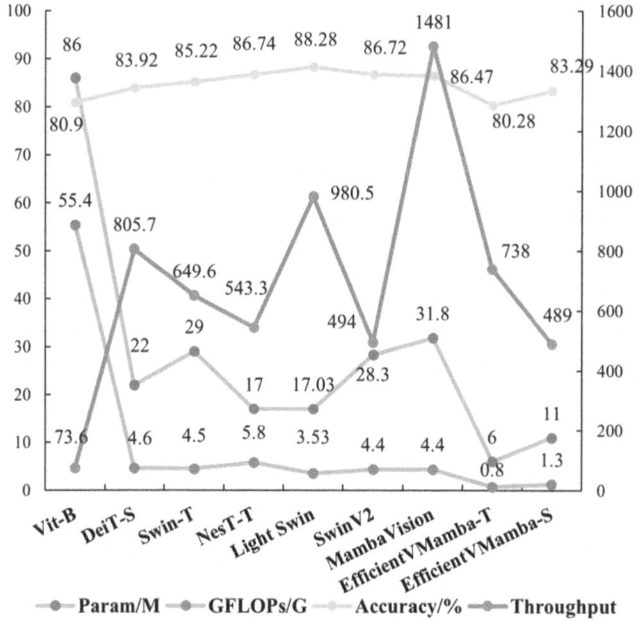

Fig. 10. Comparison of indicators (Param, GFLOPs, Accuracy, Throughput) of attention models in ImageNet-100 dataset

Table 3. The results of the Global Attention and Local Attention Model tests on the ImageNet-100 dataset

dataset	Arch. Base	Model	Image size	#Params	GFLOPs	Throughput	Top-1 acc
ImageNet-100	Transformer full-attention	ViT-B [2]	224^2	86M	55.4G	73.6	80.90
		DeiT-S [12]	224^2	22M	4.6G	805.7	83.92
		Twins-S [34]	224^2	24.1M	2.8G	897	85.90
		RegNetY-4G [47]	224^2	21M	4G	974	82.80
	Mamba-Based	MambaVision-T [31]	224^2	31.8M	4.4G	1481	86.47
		EfficientVMamba-T [33]	224^2	6M	0.8G	738	80.28
		EfficientVMamba-S [33]	224^2	11.0M	1.3G	489	83.29
	Transformer local-attention	Swin-T [3]	224^2	29M	4.5G	649.6	85.22
		SwinV2-T [32]	256^2	28.3M	4.4G	494	86.72
		NesT-T [19]	224^2	17M	5.8G	543.3	86.74
		Light Swin (ours)	224^2	17.03M	3.53G	917.8	87.04
		Light Swin with l_∞ (ours)	224^2	17.03M	3.53G	980.5	88.28

Table 4. The result of the Model tests on the INAT-18 dataset

dataset	Model	Image size	#Params	GFLOPs	Top-1 acc
INAT-18	RDNet-T[43]	224^2	24M	5.0G	77.0
	Grafit ResNet-50[44]	224^2	26M	4.1G	69.8
	DeiT-III-S[45]	224^2	22M	4.6G	67.1
	ViT-B[46]	224^2	86M	55.4G	74.1
	Light Swin (ours)	224^2	17.03M	3.53G	74.3
	Light Swin with l_∞ (ours)	224^2	17.03M	3.53G	76.8

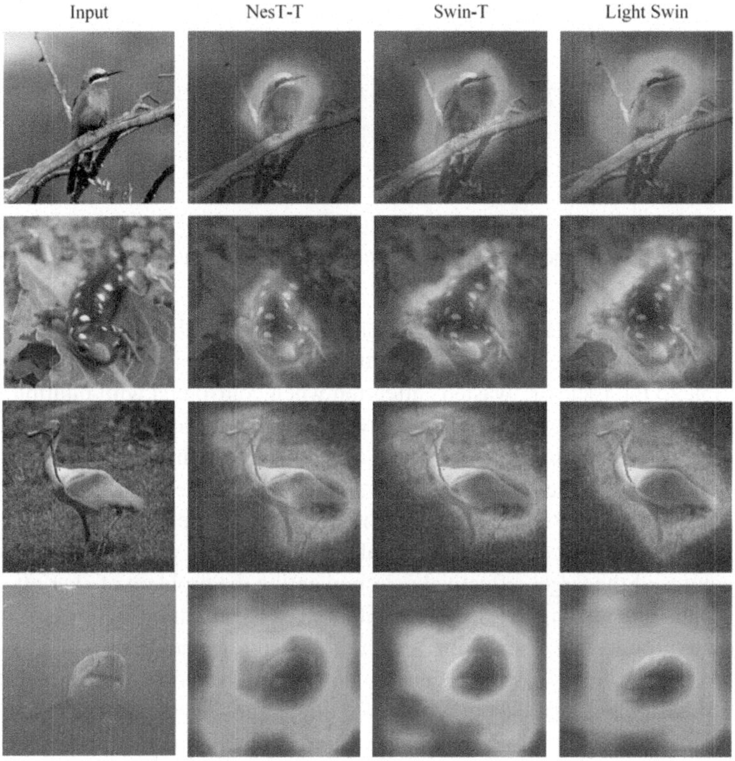

Fig. 11. Comparison of regions of attention across models

4.3 Ablation Experiment

This section examines the efficacy of implementing the l_∞ paradigm to achieve attentional sparsification for W-MSA and IW-MSA. To assess the impact of the l_∞ paradigm, ablation experiments are conducted concurrently for l_1, l_2, and l_∞, as well as for two distinct regular coefficients.

As illustrated in Fig. 12, the model prediction accuracy performs better on all three datasets when the hyperparameter λ is set to 1e-5. A comparison of the model prediction accuracies under the constraints of the l_1, l_2, and 1∞ paradigms reveals that sparsifying the attention distribution using the l_∞ paradigm yields superior results on all three datasets, with accuracies of 88.28%, 85.0%, and 97.45%, respectively.

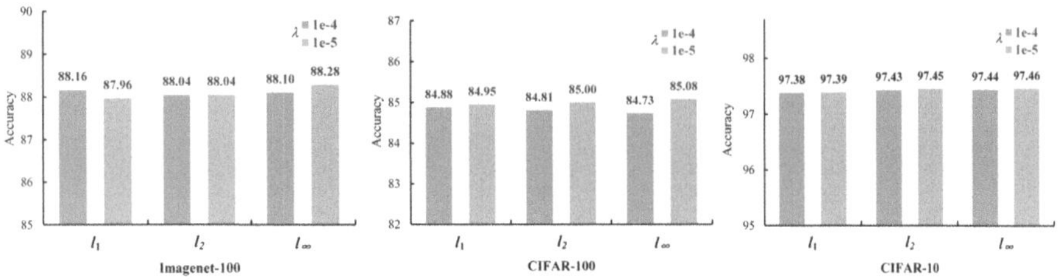

Fig. 12. Comparison of model prediction accuracies under different sizes of hyperparameter λ

As illustrated in Fig. 13, it can be observed that in the absence of paradigm constraints, the distribution of the peak of the attentional weights is more dispersed, which is susceptible to the issue of "wasted" attention when confronted with multiple targets. This occurs when attention is dispersed across multiple non-critical regions, resulting in the neglect of critical information. When confronted with multiple targets, it is common to encounter the issue of "wasted" attention, whereby attention is dispersed across multiple non-critical regions, resulting in the neglect of critical information. With the introduction of paradigm constraints, the peaks of the attention weights begin to focus on the key features of the target, effectively circumventing the problem of over-attention to the background and leading to a sparse distribution of the attention weights. A comparison of the attention weight distributions under three different paradigm constraints (l_1, l_2, and l_∞) reveals that the l_∞ paradigm constraint results in the most concentrated attention weight distribution, with peaks concentrated on the target. The l_∞ paradigm constraint does not focus on a single target to the exclusion of others due to over-regularization. However, it is able to attend to the salient features of each target simultaneously, which enhances the model's performance in scenarios involving multiple targets.

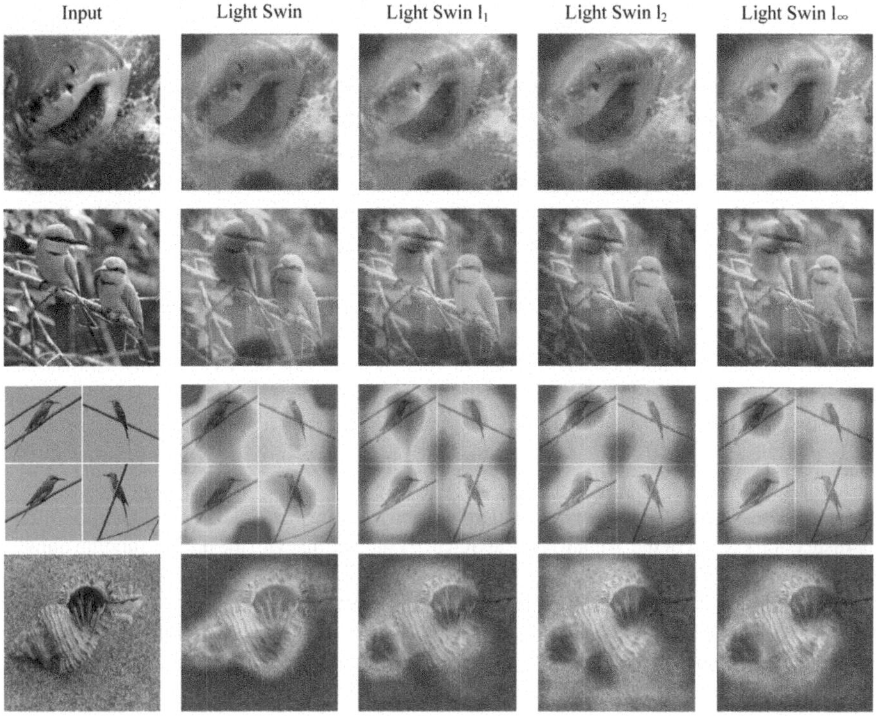

Fig. 13. Sparse attention weight distribution under different paradigms

4.4 Future Works

Light Swin model has been tested with ImageNet, CIFAR and INAT-18 dataset with good results. The architecture is also lightweight and fast in image processing. However, the use of InterWindow blocks has limitations and is currently only suitable for extracting image features at high latitudes and cannot directly form a backbone network; it still needs to be improved.

5 Conclusion

The main contributions of this work can be summarized as follows: The proposed InterWindow blocks enable efficient information exchange between non-adjacent regions while retaining local information extraction.; By focusing only on the most relevant parts of the input data, the sparse attention mechanism ensures comprehensive extraction of key information under limited computational resources. This significantly improves model efficiency and maintains robust performance; The proposed Light Swin model achieves an excellent balance between computational complexity and prediction accuracy, making it particularly suitable for resource-constrained environments.

InterWindow blocks may not fully utilize multi-scale information and may suffer when dealing with highly irregular or small objects in fine-grained tasks. In Sect. 3.3.2, we found that the InterWindow block was limited in use, so we did not completely abandon the original Swin-T model architecture, which also led to the shortcomings of the Swin model in the design of our model, such as the high computational cost of feature extraction at the bottom of the model and the loss of cross-window information. The sparse attention with l_∞ mechanism introduces an additional step to determine key-value pairs, which adds complexity to the model's implementation.

In the future, we will find a better sparse attention method to optimize our InterWindow block, so that it can reduce the computational complexity and improve the feature extraction effect in shallow networks, so as to improve the adaptability of the method.

Author Contributions. YX, TZ,YT,ZW HM and YL contributed to the study conception and design. Material preparation was performed by YX and TZ. Data collection and analysis were performed by HM, YT and ZW. The program code was written by YT. The first draft of the manuscript was written by YX and YL. All authors commented on previous versions of the manuscript. All authors read and approved the final manuscript.

Funding. This research was partially funded by the Metallurgical Materials and Quality Intelligent Inspection Project of the Wuhan Natural Science Foundation, with the project funding number (2024040701010055).

Data Availability. The ImageNet-100 in this study is a subset of the ImageNet-1k dataset from the 2012 ImageNet large Scale Visual Recognition Challenge and can be accessed at https://www.kaggle.com/datasets/ambityga/imagenet100. The INAT-18 datasets can be accessed at https://github.com/visipedia/inat_comp/tree/master/2018. Meanwhile, the CIFAR-100 and CIFAR-10 datasets can be accessed at https://www.cs.toronto.edu/~kriz/cifar.html.

Declarations.

Conflict of Interest. The authors declare that they have no conflict of interest or financial conflicts to disclose.

References

1. Vaswani, A., et al.: Attention is all you need. Adv. Neural Inf. Process. Syst. **30** (2017). https://doi.org/10.48550/arXiv.1706.03762
2. Dosovitskiy, A., et al.: An image is worth 16x16 words: transformers for image recognition at scale (2020). arxiv preprint arxiv:2010.11929
3. Liu, Z., et al.: Swin transformer: hierarchical vision transformer using shifted windows. In: Proceedings of the IEEE/CVF international conference on computer vision, (pp. 10012–10022) (2021). https://doi.org/10.48550/arXiv.2103.14030
4. Mehta, S., Rastegari, M.: Mobilevit: light-weight, general-purpose, and mobile-friendly vision transformer (2021). arxiv preprint arxiv:2110.02178
5. Ding, X., Zhang, X., Ma, N., Han, J., Ding, G., Sun, J.: Repvgg: Making vgg-style convnets great again. In Proceedings of the IEEE/CVF Conference on Computer Vision and Pattern Recognition (pp. 13733–13742) (2021). https://doi.org/10.48550/arXiv.2101.03697
6. Tan, M., Le, Q.: Efficientnet: Rethinking model scaling for convolutional neural networks. In International conference on machine learning (pp. 6105–6114). PMLR (2019). https://doi.org/10.48550/arXiv.1905.11946

7. Krizhevsky, A., Sutskever, I., Hinton, G.E.: ImageNet classification with deep convolutional neural networks. Commun. ACM **60**(6), 84–90 (2017). https://doi.org/10.1145/3065386
8. Simonyan, K., Zisserman, A.: Very deep convolutional networks for large-scale image recognition (2014). arXiv preprint arXiv:1409.1556
9. Szegedy, C., et al.: Going deeper with convolutions. In: Proceedings of the IEEE Conference on Computer Vision and Pattern Recognition, (pp. 1–9) (2015). https://doi.org/10.48550/arXiv.1409.4842
10. He, K., Zhang, X., Ren, S., Sun, J.: Deep residual learning for image recognition. In: Proceedings of the IEEE Conference on Computer Vision and Pattern Recognition (pp. 770–778) (2016). https://doi.org/10.48550/arXiv.1512.03385
11. Huang, G., Liu, Z., Van Der Maaten, L., Weinberger, K.Q.: Densely connected convolutional networks. In Proceedings of the IEEE Conference on Computer Vision and Pattern Recognition (pp. 4700–4708) (2017). https://doi.org/10.48550/arXiv.1608.06993
12. Touvron, H., Cord, M., Douze, M., Massa, F., Sablayrolles, A., Jégou, H.: Training data-efficient image transformers & distillation through attention. In: International Conference on Machine Learning (pp. 10347–10357). PMLR (2021). https://doi.org/10.48550/arXiv.2012.12877
13. Russakovsky, O., et al.: Imagenet large scale visual recognition challenge. Int. J. Comput. Vis. **115**, 211–252 (2015). https://doi.org/10.48550/arXiv.1409.0575
14. Krizhevsky, A., Sutskever, I., Hinton, G.E.: Imagenet classification with deep convolutional neural networks. Adv. Neural Inf. Process. Syst. **25** (2012)
15. Pinasthika, K., Laksono, B.S.P., Irsal, R.B.P., Yudistira, N.: SparseSwin: swin transformer with sparse transformer block. Neurocomputing **127433** (2024). https://doi.org/10.48550/arXiv.2309.05224
16. Mixup, S.A.: Boosting discriminative visual representa-tion learning with scenario-agnostic mixup. https://doi.org/10.48550/arXiv.2111.15454
17. Hu, H., Zhang, Z., Xie, Z., Lin, S.: Local relation networks for image recognition. In: Proceedings of the IEEE/CVF International Conference on Computer Vision (pp. 3464–3473) (2019). https://doi.org/10.48550/arXiv.1904.11491
18. Zhang, H., Cisse, M., Dauphin, Y.N., Lopez-Paz, D.: mixup: beyond empirical risk minimization (2017). arXiv preprint arXiv:1710.09412
19. Zhang, Z., Zhang, H., Zhao, L., Chen, T., Arik, S.Ö., Pfister, T.: Nested hierarchical transformer: towards accurate, data-efficient and interpretable visual understanding. In: Proceedings of the AAAI Conference on Artificial Intelligence (Vol. 36, No. 3, pp. 3417–3425) (2022). https://doi.org/10.48550/arXiv.2105.12723
20. Selvaraju, R. R., Cogswell, M., Das, A., Vedantam, R., Parikh, D., & Batra, D. (2017). Grad-cam: Visual explanations from deep networks via gradient-based localization. In Proceedings of the IEEE international conference on computer vision (pp. 618–626). https://doi.org/10.1109/ICCV.2017.74
21. Yuan, Y., Huang, L., Guo, J., Zhang, C., Chen, X., Wang, J.: Ocnet: object context network for scene parsing (2018). arXiv preprint arXiv:1809.00916
22. Xie, S., Girshick, R., Dollár, P., Tu, Z., He, K.: Aggregated residual transformations for deep neural networks. In Proceedings of the IEEE Conference on Computer Vision and Pattern Recognition (pp. 1492–1500) (2017). https://doi.org/10.48550/arXiv.1611.05431
23. Zhu, X., Hu, H., Lin, S., Dai, J.: Deformable convnets v2: more deformable, better results. In Proceedings of the IEEE/CVF Conference on Computer Vision and Pattern Recognition (pp. 9308–9316) (2019). https://doi.org/10.48550/arXiv.1811.11168
24. Dai, J., et al.: Deformable convolutional networks. In: Proceedings of the IEEE International Conference on Computer Vision (pp. 764–773) (2017). https://doi.org/10.48550/arXiv.1703.06211

25. LeCun, Y., Bottou, L., Bengio, Y., Haffner, P.: Gradient-based learning applied to document recognition. Proc. IEEE **86**(11), 2278–2324 (1998). https://doi.org/10.1109/5.726791
26. Raffel, C., et al.: Exploring the limits of transfer learning with a unified text-to-text transformer. J. Mach. Learn. Res. **21**(140), 1–67 (2020). https://doi.org/10.48550/arXiv.1910.10683
27. Bao, H., et al.: Unilmv2: Pseudo-masked language models for unified language model pre-training. In: International Conference on Machine Learning (pp. 642–652). PMLR (2020)
28. Hu, H., Gu, J., Zhang, Z., Dai, J., Wei, Y.: Relation networks for object detection. In: Proceedings of the IEEE Conference on Computer Vision and Pattern Recognition (pp. 3588–3597) (2018). https://doi.org/10.48550/arXiv.1711.11575
29. Beltagy, I., Peters, M.E., Cohan, A.: Longformer: the long-document transformer (2020). arXiv preprint arXiv:2004.05150
30. Kitaev, N., Kaiser, Ł., Levskaya, A.: Reformer: the efficient transformer (2020). arXiv preprint arXiv:2001.04451
31. Hatamizadeh, A., Kautz, J.: Mambavision: a hybrid mamba-transformer vision backbone (2024). arXiv preprint arXiv:2407.08083
32. Liu, Z., et al.: Swin transformer v2: Scaling up capacity and resolution. In: Proceedings of the IEEE/CVF Conference on Computer Vision and Pattern Recognition (pp. 12009–12019) (2022)
33. Pei, X., Huang, T., Xu, C.: Efficientvmamba: atrous selective scan for light weight visual mamba (2024). arXiv preprint arXiv:2403.09977
34. Chu, X., et al.: Twins: revisiting the design of spatial attention in vision transformers. Adv. Neural. Inf. Process. Syst. **34**, 9355–9366 (2021)
35. Qiu, J., Ma, H., Levy, O., Yih, S.W.T., Wang, S., Tang, J.: Blockwise self-attention for long document understanding (2019). arXiv preprint arXiv:1911.02972
36. Parmar, N., et al.: Image transformer. In: International Conference on Machine Learning (pp. 4055–4064). PMLR (2018). https://export.arxiv.org/pdf/1802.05751v3.pdf
37. Beltagy, I., Peters, M.E., Cohan, A.: Longformer: the long-document transformer (2020). arXiv preprint arXiv:2004.05150
38. Ding, J., et al.: Longnet: scaling transformers to 1,000,000,000 tokens (2023). arXiv preprint arXiv:2307.02486
39. Ho, J., Kalchbrenner, N., Weissenborn, D., Salimans, T.: Axial attention in multidimensional transformers (2019). arXiv preprint arXiv:1912.12180
40. Guo, D., Xu, C., Duan, N., Yin, J., McAuley, J.: Longcoder: a long-range pre-trained language model for code completion. In: International Conference on Machine Learning (pp. 12098–12107). PMLR (2023)
41. Ge, S., Zhang, Y., Liu, L., Zhang, M., Han, J., Gao, J.: Model tells you what to discard: adaptive kv cache compression for llms (2023). arXiv preprint arXiv:2310.01801
42. Anagnostidis, S., Pavllo, D., Biggio, L., Noci, L., Lucchi, A., Hofmann, T.: Dynamic context pruning for efficient and interpretable autoregressive transformers. Adv. Neural Inf. Process. Syst. **36** (2024)
43. Kim, D., Heo, B., Han, D.: Densenets reloaded: paradigm shift beyond resnets and vits. In: European Conference on Computer Vision (pp. 395–415). Springer, Cham (2025). https://arxiv.org/pdf/2403.19588v2
44. Touvron, H., Cord, M., Douze, M., Massa, F., Sablayrolles, A., Jégou, H.: Training data-efficient image transformers & distillation through attention. In: International Conference on Machine Learning (pp. 10347–10357). PMLR (2021). https://arxiv.org/abs/2012.12877
45. Touvron, H., Cord, M., Jégou, H.: Deit iii: Revenge of the vit. In: European Conference on Computer Vision (pp. 516–533). Cham: Springer Nature Switzerland (2022)
46. Touvron, H., Cord, M., El-Nouby, A., Verbeek, J., Jégou, H.: Three things everyone should know about vision transformers. In: European Conference on Computer Vision (pp. 497–515). Cham: Springer Nature Switzerland (2022). https://arxiv.org/abs/2203.09795

47. Radosavovic, I., Kosaraju, R.P., Girshick, R., He, K., Dollár, P.: Designing network design spaces. In: Proceedings of the IEEE/CVF Conference on Computer Vision and Pattern Recognition (pp. 10428–10436) (2020). https://arxiv.org/abs/2003.13678
48. Wang, Q., Wu, B., Zhu, P., Li, P., Zuo, W., Hu, Q.: ECA-Net: Efficient channel attention for deep convolutional neural networks. In: Proceedings of the IEEE/CVF Conference on Computer Vision and Pattern Recognition (pp. 11534–11542) (2020). https://ieeexplore.ieee.org/document/9156697

Integrating Viewpoint Planning and 3D Gaussian Splatting for Production Assembly Anomaly Automated Detection

Yanqi Niu[1], Zhixiang Feng[1], Pengjie Wang[1], Yicong Chen[2], Bingrui Chen[1], and Fuzhou Du[2](✉)

[1] School of Mechanical and Electrical Engineering, China University of Mining & Technology-Beijing, Beijing, China
niuyanqi@cumtb.edu.cn, {sqt2300401006,ZQT2400401044, zqt2300401002}@student.cumtb.edu.cn
[2] School of Mechanical Engineering and Automation, Beihang University, Beijing, China
chenyc9715@buaa.edu.cn, du_fuzhou@163.com

Abstract. The assembly quality of complex products such as spacecraft, ships, and large-scale equipment directly determines their performance and reliability. To overcome the limitations of manual inspection and the restricted viewpoints of traditional vision systems, this paper proposes an assembly anomaly detection method based on 3D Gaussian Splatting (3DGS) and robotic active perception. A CAD-guided adaptive viewpoint planning strategy is developed to enable coordinated multi-view data acquisition using an integrated manipulator-AGV system, ensuring comprehensive coverage of complex assemblies. The acquired multi-view images are reconstructed into a high-fidelity 3DGS model that serves as a digital reference template. Leveraging the real-time novel-view synthesis capability of 3DGS, image comparison is conducted under consistent illumination and geometry, effectively resolving the issues of viewpoint inconsistency and lighting mismatch in multi-view inspection. Compared with conventional Multi-View Stereo (MVS) and Neural Radiance Fields (NeRF) methods, the proposed framework achieves superior reconstruction accuracy and rendering efficiency. Experimental results demonstrate that the method can accurately detect typical assembly anomalies such as wrong part, missing part, and redundant components, while also supporting digital traceability and re-analysis of assembly processes. This study provides an efficient and scalable solution for high-precision visual inspection in complex industrial assembly environments.

Keywords: 3D Gaussian Splatting · Assembly anomaly detection · Viewpoint planning · Robotic active perception

1 Introduction

The assembly quality of complex products, such as spacecraft, ships, and large-scale equipment, directly determines their performance and safety. Owing to their large number of components, intricate structures, and massive dimensions, the assembly process

H. Wei and J. Niu (Eds.): COBOT 2025, CCIS 2893, pp. 176–197, 2027.
https://doi.org/10.1007/978-981-92-2074-8_11

of such products imposes stringent requirements on precision control and quality inspection [1]. Achieving precise, efficient, and traceable inspection in large-scale, multi-level assembly environments presents a key challenge for intelligent manufacturing.

Currently, assembly quality inspection of complex products still faces significant challenges. Common anomalies include incorrect assembly, missing parts, foreign objects, and surface defects. If not detected in time, these issues can seriously affect product quality and assembly reliability. In most cases, inspection still relies heavily on manual visual assessment, which depends on operator experience and suffers from strong subjectivity, low efficiency, and inconsistent standards, making it difficult to meet the high-precision and high-consistency requirements of complex products. In recent years, vision-based automated inspection methods have become a research hotspot. However, fixed-position visual systems are limited by viewing angles, lighting sensitivity, and insufficient coverage, making it difficult to achieve comprehensive inspection for large-scale and multi-layer assemblies. To overcome these limitations, researchers have combined mobile robotics with visual sensing technologies. By integrating SLAM-based localization, path planning, and visual servoing, automated multi-view data acquisition can be achieved within production environments, thereby improving inspection flexibility and automation. Nevertheless, most existing studies focus primarily on optimizing localization and path planning accuracy, while geometric consistency and completeness of inspection data remain insufficient. Due to variations in viewing geometry and illumination, images captured from different positions are difficult to align accurately and compare reliably, leading to false or missed detections caused by occlusion or viewpoint discrepancies.

To address these challenges, a new paradigm has emerged in recent years—constructing standard assembly templates through three-dimensional (3D) reconstruction, thereby shifting from "image-based inspection" to "model-based comparison." By reconstructing multi-view images of the assembly into a unified 3D model, accurate comparison can be performed under consistent geometric and illumination conditions, significantly enhancing detection accuracy. However, traditional reconstruction approaches, such as MVS [2] and NeRF [3], often struggle with large-scale industrial scenes involving reflective or complex surfaces, suffering from limited modeling precision, slow reconstruction speed, and low texture fidelity, which restrict their practical use in assembly inspection.

Recently, 3D Gaussian Splatting (3DGS) [4] has emerged as a promising solution to these issues. This method represents scenes explicitly using Gaussian distributions in 3D space, where each point is optimized in terms of position, scale, color, and opacity to efficiently encode geometric and radiometric properties. Unlike implicit volumetric representations, 3DGS avoids voxel sampling and volumetric integration, enabling both high reconstruction accuracy and real-time rendering efficiency. Its strengths in detail fidelity, lighting consistency, and scalability make it an ideal approach for constructing high-precision standard assembly templates.

To tackle the above problems, this paper proposes a 3D Gaussian Splatting and robot-based active perception framework for assembly anomaly detection. First, an adaptive viewpoint planning strategy based on CAD models is developed, considering both the geometric characteristics and visibility requirements of complex assemblies. Through coordinated optimization between a robotic arm and an Automated Guided Vehicle

(AGV), high-coverage, multi-view data acquisition is achieved, ensuring completeness and quality of reconstruction data at the source. Subsequently, a high-fidelity 3DGS model of the assembly is reconstructed from the collected images, and background removal and artifact suppression are performed to obtain a clean 3D template containing only the assembly body, thereby avoiding background-induced interference. Compared with traditional methods, 3DGS can efficiently reproduce realistic textures, reflections, and illumination characteristics of assemblies while maintaining high geometric precision. More importantly, its real-time novel view synthesis capability allows generation of high-quality images from arbitrary camera poses, fundamentally solving the problems of viewpoint inconsistency and lighting mismatch in multi-view inspection. Based on this capability, the proposed method further realizes same-view alignment and pixel-wise difference analysis between the 3DGS template images and on-site captured images, enabling robust detection of assembly anomalies under consistent viewing conditions. Finally, the method achieves accurate detection and localization of assembly defects like missing, misplaced, or extra parts. By doing so, it also generates high-fidelity 3D models as permanent digital assets, supporting assembly traceability and subsequent analysis. Experimental results demonstrate the effectiveness and robustness of the proposed approach in both 3D reconstruction quality and anomaly detection performance.

The main contributions of this paper are summarized as follows:

(1) For the first time, 3D Gaussian Splatting is introduced into the domain of complex assembly inspection, establishing a model-based automated anomaly detection framework that enables high-precision, photorealistic, and all-view template generation.
(2) An adaptive viewpoint planning algorithm is designed based on CAD models, integrating robotic arm–AGV collaboration to ensure geometric accuracy and texture consistency of 3DGS reconstruction.
(3) A digital assembly template with novel-view synthesis capability is constructed, supporting not only automated anomaly detection but also assembly traceability and data retention throughout the product lifecycle.

2 Related Work

The automation of assembly anomaly detection requires the seamless integration of several technologies: viewpoint planning, multi-view data acquisition, 3D reconstruction, and visual inspection. In recent years, extensive research efforts have been devoted to these areas, providing a solid foundation for the proposed method.

2.1 Viewpoint Planning and Multi-view Data Acquisition

In 3D reconstruction and industrial visual inspection, viewpoint planning for data acquisition directly affects model reconstruction quality and detection accuracy. Early studies largely relied on rule-based and geometric coverage strategies, selecting capture positions manually or semi-automatically. Guillemaut et al. [5] proposed a joint multilayer segmentation and reconstruction approach, using video sequences to achieve free-viewpoint 3D reconstruction; however, its full-coverage capability for complex industrial parts was limited. Hepp et al. [6] developed the Plan3D framework, optimizing viewpoints and trajectories for aerial multi-view capture, laying a foundation for efficient reconstruction of large-scale scenes.

In recent years, hierarchical and object-oriented viewpoint planning strategies have been introduced in industrial applications. Wang et al. [7] proposed a hierarchical viewpoint planning method for complex surfaces, ensuring high coverage and acquisition efficiency in multi-part and multi-surface scenarios. Li et al. [8] further developed an object-based viewpoint recommendation method, integrating 3D scene understanding with acquisition strategies to enable automated viewpoint selection. For mobile inspection platforms, Kong et al. [9] introduced station–viewpoint joint coverage path planning, enabling coordinated data collection with AGVs and robotic arms. Du et al. [10] incorporated CAD models and knowledge transfer to achieve cross-domain pose keypoint detection, providing prior support for automatic viewpoint selection. Wei et al. [11]demonstrated automatic viewpoint planning to accelerate multi-view 3D reconstruction in agricultural scenarios, confirming the direct impact of planning strategies on data diversity and reconstruction efficiency.

2.2 3D Reconstruction

Image-based 3D reconstruction methods have evolved from traditional geometric approaches to deep learning and Gaussian field representations. Hartung et al. [12] analyzed single-view reconstruction methods in industrial scenarios and highlighted their insufficient accuracy for complex parts and occluded regions. Huo and Yu [13] achieved 3D reconstruction of mechanical components from 2D images, but their performance was limited by texture and lighting conditions. Son et al. [14] combined laser scan data with CAD priors to reconstruct industrial instrumentation models with high precision, providing a reliable solution for structured data.

Deep learning methods have enhanced 3D scene representation through neural fields such as NeRF. Mildenhall et al. introduced neural radiance fields, enabling view synthesis and continuous spatial interpolation. Kerbl et al. developed a real-time 3D Gaussian splatting rendering approach, improving the efficiency of large-scale scene reconstruction. Sun [15] et al. applied 3D Gaussian Splatting to industrial component reconstruction and evaluated its precision, demonstrating advantages in high-accuracy reconstruction and texture consistency. Lim et al. [16] combined LiDAR data with 3DGS models to achieve multimodal initialization, improving reconstruction accuracy for complex surfaces. Qian et al. [17] proposed a compressed 3DGS model for large-scale UAV multi-view scene reconstruction, showing that efficiency can be maintained without compromising accuracy. Lu et al. [18] employed 3DGS for change detection in object rearrangement, indicating that 3D Gaussian representations can be directly applied to industrial inspection tasks.

2.3 Visual Inspection in Industrial Scenarios

Visual inspection in industrial scenarios typically involves detecting surface defects, assembly errors, and evaluating overall product quality. The development of these methods has progressed from traditional image processing to deep learning, and more recently, to approaches integrating three-dimensional geometric priors.

Traditional methods rely on rules, template matching, and hand-crafted features. Frustaci et al. [19] proposed a high-performance machine vision system for assembly inspection, which performs reliably in controlled environments. Sharifzadeh et al. [20] employed robot-mounted laser scanners for surface anomaly detection, enabling precise defect localization but requiring manual adjustment of scanning paths and thresholds.

Deep learning methods enhance adaptability and generalization through end-to-end training. Caeion et al. [21] utilized a Transformer architecture for end-to-end object detection. Tsai and Jen [22] developed an autoencoder-based anomaly detection approach. Liu et al. [23] applied GAN-based data augmentation to improve detection of rare defects. Peng et al. [24] predicted defect maps to generate realistic synthetic anomalies, achieving high-precision industrial inspection. Multimodal and unsupervised methods have also received attention. Zhao et al. [25] proposed a knowledge-driven generic framework adaptable to multiple tasks and scenarios. Božič et al. [26] evaluated the robustness of unsupervised methods. Cui et al. [27] systematically reviewed the potential and limitations of unsupervised industrial image inspection methods. Shao et al. [28] enhanced anomaly localization using a multi-scale prototype fusion network. Kaji et al. [29] introduced an in-situ deep learning detection method for laser-directed energy deposition.

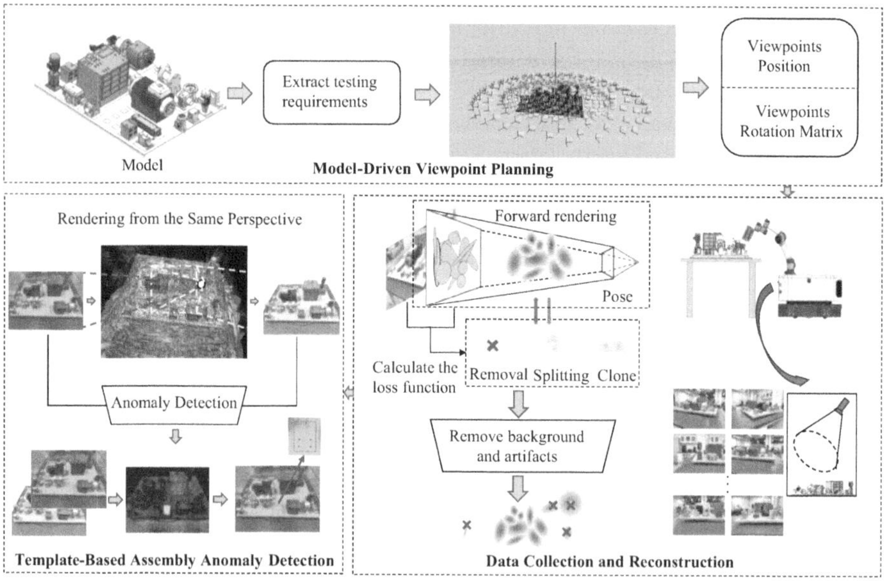

Fig. 1. System architecture of the 3DGS-driven robotic inspection Framework

Despite the improvements provided by deep learning and multimodal approaches, 2D images remain vulnerable to occlusions, reflections, and complex backgrounds. Multimodal methods often require additional equipment or complex scanning paths and lack efficient multi-view 3D fusion frameworks, limiting their robustness in scenarios involving multiple parts or stations. To address these limitations, this study proposes a multi-view assembly anomaly detection method based on 3D Gaussian Splatting. By combining high-precision 3D reconstruction with multi-view images, the approach achieves comprehensive, fast, and robust anomaly localization, filling a critical gap in existing research.

3 Methodology

3.1 System Overview

This paper proposes an assembly inspection method based on viewpoint planning, AGV–robot cooperative data acquisition, and 3D Gaussian Splatting. As shown in Fig. 1, the system first generates optimal observation viewpoints through a viewpoint planning algorithm, guiding the AGV and robotic arm to collaboratively perform multi-angle image acquisition. Then, the collected data are reconstructed with 3DGS, and background removal is applied to obtain a clean 3D model containing only the assembly body. Based on this model, template images from different viewpoints are rendered and compared with images of the real assembly scene. Finally, the proposed Robust Pixel-wise Difference and Geometric Aggregation (RPD-GA) method is used to achieve accurate detection of common assembly anomalies, including incorrect assembly, missing part, surface defects, and foreign objects. This approach establishes a complete closed-loop process from automated data acquisition to 3D reconstruction, virtual viewpoint generation, and anomaly detection, providing an efficient and scalable digital solution for assembly quality inspection.

3.2 Model-Driven Viewpoint Planning

In the 3D reconstruction of complex assemblies, viewpoint planning is a critical step to ensure reconstruction completeness and point cloud quality. To this end, this study adopts an adaptive multi-layer viewpoint planning strategy, which generates compact coverage regions based on the geometric features of the model and adaptively allocates observation viewpoints to balance reconstruction accuracy and acquisition efficiency. Since ellipsoids can tightly enclose the model and possess favorable mathematical parameterization properties, they are chosen as the bounding geometry of the model. Let the set of model vertices be $\mathcal{V} = \{v_i\}_{i=1}^{N}$. Through convex hull analysis, the vertex subset $\mathcal{H}$ is obtained. Then, the minimum volume enclosing ellipsoid (MVEE) is computed to determine the center c and the axis matrix A:

$$\mathcal{E} = \{x \in \mathbb{R}^3 | (x - c)^T A(x - c) \leq 1\} \tag{1}$$

The axes directions of the ellipsoid are determined by the eigenvectors of the matrix A, and the axis lengths are determined by its eigenvalues. Compared with cubic or spherical bounding shapes, this approach provides a tighter fit, effectively reducing redundant viewpoints caused by empty space while naturally encoding scale information for subsequent multi-layer distribution and adaptive viewpoint allocation.

Based on this, a multi-layer distribution strategy is adopted to generate viewpoints. Low-layer viewpoints are positioned close to the model to capture fine details and improve local geometric accuracy; mid-layer viewpoints cover areas occluded by the lower layer, increasing overall coverage; high-layer viewpoints observe the overall structure, ensuring effective sampling of distant and boundary regions. In this way, both reconstruction completeness and efficiency are considered. For each layer, the ellipsoid axes are scaled by a safety factor f_l:

$$r_i^{(l)} = r_i \cdot f_l, \quad f_l = f_0 + (l - 1) \cdot \Delta f \tag{2}$$

Here, f_0 denotes the number of viewpoints in the inner layer, and Δf represents the increment between layers. This design ensures that lower-layer viewpoints focus on fine details, while higher-layer viewpoints serve to complement occluded regions and provide overall coverage, thereby reducing local reconstruction gaps caused by occlusion at lower layers.

The number of viewpoints in each layer is adaptively allocated according to the model's surface area A_{est} and geometric complexity N_{hull}:

$$n_l = \left\lceil A_{\text{est}} \cdot V_{\text{base}} \cdot \left(1 + \frac{\log(1 + N_{\text{hull}}/N_0)}{5}\right) \cdot w_l \right\rceil \tag{3}$$

$$N_{\text{hull}} = |\text{ConvexHull}(\mathcal{V})| \tag{4}$$

$$\bar{r} = \frac{r_x + r_y + r_z}{3}, \quad A_{\text{est}} \approx 4\pi\bar{r}^2 \tag{5}$$

Here, V_{base} denotes the reference number of viewpoints per square meter, N_0 is the complexity threshold, and w_l represents the layer weight (e.g., 50% for the inner layer to preserve details, 30% for the middle layer to supplement coverage, and 20% for the outer layer in high-complexity models). The layer weights (50%/30%/20% for inner/middle/outer layers) are chosen to balance local detail preservation and overall acquisition efficiency. High-curvature or occluded regions require denser viewpoint coverage to constrain geometry and radiometric variation, whereas low-curvature regions yield diminishing returns with additional views. We conducted preliminary tuning across multiple representative assembly scenes, comparing several allocation schemes (e.g., 60/25/15, 50/30/20, 40/40/20, uniform). Results indicate that when the inner-layer weight falls below ~ 40%, local reconstruction fidelity and detection recall drop noticeably; conversely, allocating more than ~ 60% to the inner layer yields marginal accuracy gains but substantially increases acquisition redundancy and time. The 50/30/20 split provided the best trade-off in our tests—improving local PSNR and detection F1 while keeping collection time and redundancy at acceptable levels. We therefore adopt this distribution as the default, and provide sensitivity analysis in the supplementary material. The advantage of this adaptive strategy is that complex regions receive more viewpoints, ensuring uniform point cloud density and high local accuracy, while simple regions avoid redundant viewpoints, thereby improving reconstruction efficiency.

Fig. 2. Viewpoint planning performance of different models.

The specific viewpoint generation is based on a golden spiral distribution, ensuring uniform sampling over the ellipsoid surface:3

$$\theta_j = \arccos(\cos\theta_{\min} - h_j(\cos\theta_{\min} - \cos\theta_{\max})) \tag{6}$$

$$\phi_j = j \cdot \pi(3 - \sqrt{5}), \quad h_j = 1 - \frac{j + 0.5}{M} \tag{7}$$

Here, M denotes the total number of viewpoints in the layer, and (θ_j, ϕ_j) are the spherical coordinate angles. The schematic diagram is shown in Fig. 2.

Overall, this multi-layer adaptive viewpoint planning method achieves the goals of low redundancy, high coverage, and high accuracy. It effectively enhances the completeness and local precision of 3D reconstruction for complex assemblies, while providing a solid foundation for subsequent assembly anomaly detection and virtual assembly verification.

3.3 Data Collection and Reconstruction

After completing the multi-viewpoint planning based on scene geometric complexity, the system maps the planned viewpoints to actual acquisition stations that can be jointly covered by the mobile platform (AGV) and the robotic arm, achieving a seamless transition from virtual viewpoint planning to physical camera acquisition. As shown in Fig. 3, Under this approach, the AGV only needs to move to a limited number of stations, while the robotic arm performs local coverage, significantly reducing travel time, balancing efficiency and accuracy, and maintaining consistency between the virtual and

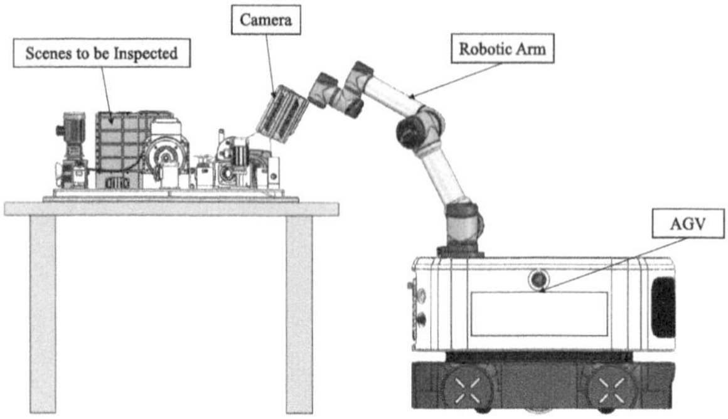

Fig. 3. Structure of the Data Acquisition System

physical setups. After each station completes data acquisition, the captured images and corresponding poses are uploaded to the reconstruction module for multi-view scene modeling, realizing a complete closed-loop process from planning to execution and reconstruction.

Based on the planned viewpoints, all viewpoints in the space are clustered and grouped. Clustering ensures local viewpoint coverage: the angular difference between viewpoints within each station does not exceed the camera's field of view, allowing the robotic arm to capture all assigned viewpoints at that station while reducing the AGV's global movements. By aggregating spatially proximate viewpoints into the same station, the AGV only needs to visit a limited number of stations, thereby decreasing travel time and path complexity. On this basis, spatial uniformity is also considered: each station is designed to cover different spatial regions as much as possible, ensuring overall acquisition uniformity and multi-angle coverage of the scene.

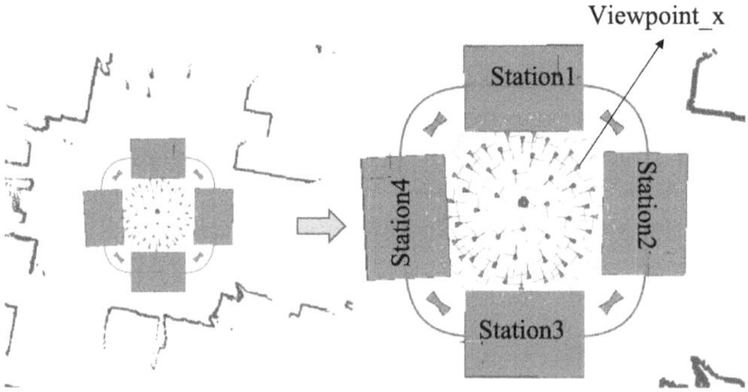

Fig. 4. Distribution of stations and viewpoints.

Let the poses of all planned viewpoints be denoted as:

$$(R_i, t_i)_{i=1}^{N} \tag{8}$$

Here, R_i denotes the rotation matrix, and t_i represents the position vector. A clustering criterion based on a combined distance of position and orientation is adopted:

$$D_{ij} = \alpha \left\| t_i - t_j \right\| + \beta \left(1 - \frac{\text{trace}(R_i R_j^\top)}{3} \right) \tag{9}$$

Here, α, β is a weighting coefficient used to balance the

influence of spatial distance and viewpoint difference on clustering. By applying threshold-based clustering, K station sets are obtained:

$$S_k = \{v_{k,1}, v_{k,2}, \ldots, v_{k,n_k}\} \tag{10}$$

Each set corresponds to a shooting station accessible by the AGV, and the viewpoints within the station are captured within the operational workspace of the robotic arm. A schematic of the stations and viewpoint planning is shown in the figure below. The planning diagram is shown in Fig. 4.

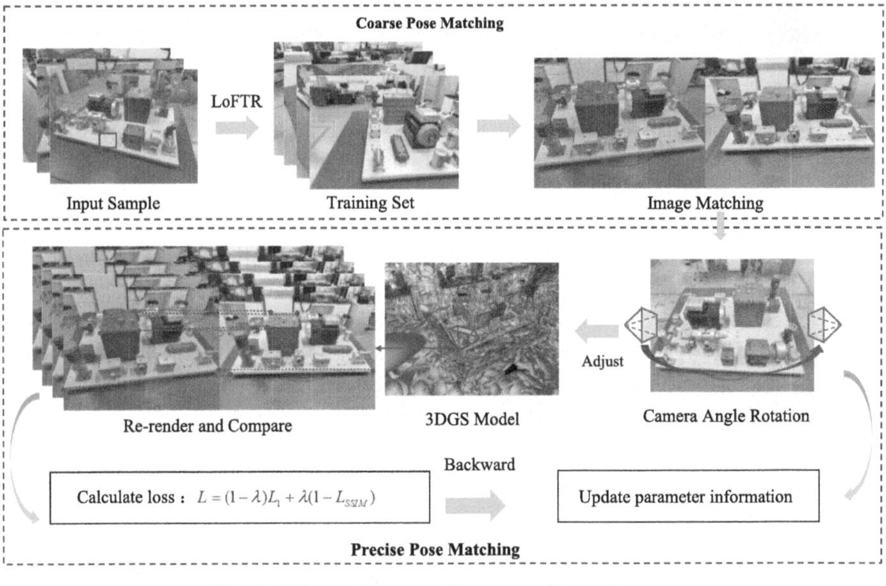

Fig. 5. Viewpoint pose alignment of input image.

At each station, the robotic arm sequentially moves to each target viewpoint. During execution, the system incorporates a two-level cooperative control strategy: the AGV performs station-level coarse positioning while the robotic arm executes precise local viewpoints within its operational workspace. To ensure geometric consistency between the planned virtual viewpoints and physical acquisition, hand-eye calibration is applied to determine the accurate transformation between the AGV base, robotic arm, and camera. In addition, online pose correction is performed at each station using visual markers to compensate for AGV navigation errors, and the robotic arm adjusts its end-effector trajectory to correct residual positioning and orientation deviations. This combined mechanism significantly reduces total data acquisition time while ensuring coverage and accuracy. After all viewpoints are captured, the collected images and corresponding poses are uploaded to the 3D reconstruction module for processing with 3D Gaussian Splatting (3DGS), producing a high-fidelity 3D template model. By integrating cooperative control and multi-stage motion error compensation, the system maintains precise alignment between virtual planning and physical execution, providing a reliable basis for subsequent anomaly detection.

3.4 Template-Based Assembly Anomaly Detection

The assembly inspection method based on the 3DGS template achieves a breakthrough in ensuring viewpoint consistency and lighting matching, making pixel-wise difference analysis feasible. However, to guarantee the robustness and practicality of the detection results, inherent environmental disturbances and high-frequency spurious differences in industrial scenes must be addressed, shown in Fig. 5. This paper introduces the Robust Pixel Difference and Geometric Aggregation (RPD-GA) method. By integrating 3D priors with multi-level geometric constraints, it precisely directs the detection focus toward anomalies that compromise structural integrity.

For each test image, we first analyze its visual information to determine the corresponding viewpoint in the reconstructed standard 3D assembly scene, and render a template image from that viewpoint. The test image is then compared with the rendered template at both pixel and structural levels, where the detected differences indicate potential anomaly regions. These difference regions are further localized and mapped back to the original image space to achieve accurate anomaly detection. In addition, during the construction of the 3D template, the background is removed and only the assembly object is preserved; therefore, the rendered template images also contain only the object of interest. This design allows the detection process to focus exclusively on the assembly component rather than background elements, effectively reducing interference and improving both accuracy and robustness of anomaly localization.

To enhance the geometric accuracy and surface consistency of the 3DGS-reconstructed model, we introduce an adaptive statistical filtering strategy. This method dynamically adjusts the neighborhood size and decision thresholds based on local geometric complexity. The core idea is to leverage local point density to refine noise identification criteria, effectively suppressing artifacts and blur while preserving critical fine details.

Specifically, for any point p_i in the point cloud, an adaptive-density neighborhood $\mathcal{N}(p_i)$ is first constructed, whose neighborhood size k_i is proportional to the local density of the point ρ_i. This ensures fine analysis in dense regions and enhanced robustness in sparse regions. The average Euclidean distance between the point and the points within its neighborhood is then computed as:

$$\overline{d_i} = \frac{1}{k_i} \sum_{p_j \in \mathcal{N}(p_i)} \|p_i - p_j\|_2 \tag{11}$$

Then, the global average distance and standard deviation of the entire point cloud are calculated:

$$\mu_d = \frac{1}{N}\sum_{i=1}^{N}\overline{d_i}, \quad \sigma_d = \sqrt{\frac{1}{N}\sum_{i=1}^{N}(\overline{d_i} - \mu_d)^2} \tag{12}$$

This enables adaptive dynamic adjustment of the filtering criterion. The condition for identifying outliers is defined as:

$$\overline{d_i} > \mu_d + \alpha_i \sigma_d \tag{13}$$

This criterion allows stricter thresholds in high-density regions to preserve fine details, while relatively relaxed thresholds in low-density regions enhance robustness. Through this density-aware adaptive mechanism, the system can automatically balance local smoothing with global consistency, significantly reducing artifacts and edge drifting caused by 3DGS rendering.

Meanwhile, to avoid the complexity and uncertainty of background segmentation in traditional visual inspection, we innovatively implement target-oriented geometric prior extraction after the 3DGS model optimization, The principle is shown in Fig. 6. The 3D Gaussian ellipsoids are used to perform foreground-background separation, resulting in a clean 3D template of the assembly free from background interference. Starting from this 3D template, any subsequent 2D images rendered from arbitrary viewpoints are inherently free from background distractions, fundamentally eliminating the influence of environmental lighting or background texture changes on anomaly detection. This design forms the cornerstone of the RPD-GA method's ability to achieve highly robust detection, precisely focusing the inspection on the assembly itself and greatly simplifying and enhancing the efficiency and reliability of subsequent anomaly analysis.

For any assembly image I_{test} captured during the inspection stage, its camera pose is typically unknown or only roughly estimated. To find the corresponding viewpoint in the 3D template, this work draws on the *SplatPose* [30]. Method proposed by Kruse et al., and leverages the differentiable rendering characteristics of the 3D Gaussian scene to construct a pose alignment method based on the clean model. Building on the original technique, we reconstructed the data interface and rendering module of *SplatPose,* allowing pose optimization to be directly applied in the Gaussian point parameter space, without requiring voxel field integration or radiance field sampling. This enhances both the convergence speed and accuracy of pose optimization, as illustrated in the figure.

During the pose optimization stage, the complete model (including background) is used in the matching iterations:

$$T^* = \arg\min_{T} L(R(G_{full}; T), I_{obs}) \tag{14}$$

This allows full utilization of the background to constrain geometric consistency. After obtaining the optimal pose T^*, the background-free model G_{full} is used for the final rendering to produce an aligned, clean image without background interference:

$$I_{align} = R(G_{full}, T^*) \tag{15}$$

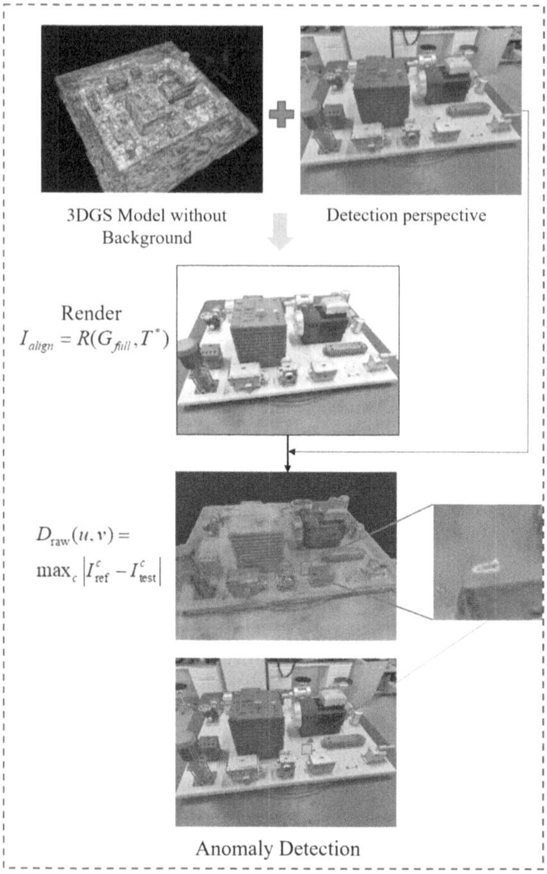

Fig. 6. Principle diagram of anomaly detection.

In the 2D difference analysis stage, this prior knowledge is utilized for difference computation and background masking. First, the absolute difference between the on-site image I_{test} and the template rendering I_{ref} is computed as D_{raw}. Then, all difference regions corresponding to the white background are precisely masked out, producing the masked difference map D_{masked}:

$$
\begin{aligned}
D_{raw}(u, v) &= \max_c \left| I^c_{ref} - I^c_{test} \right| \rightarrow \\
D_{masked}(u, v) &= D_{raw}(u, v) \cdot \mathbb{I}(\max(I_{ref}, I_{test}) \leq \tau_{white})
\end{aligned}
\tag{16}
$$

where $\mathbb{I}(\cdot)$ is the indicator function, and τ_{white} denotes the background threshold. This innovative 3D–2D collaborative mechanism ensures that the detection process remains entirely focused on the assembly object itself, effectively eliminating the interference of environmental noise.

After eliminating environmental interference, we turn to address the issue of spurious anomalies along object boundaries. Even under perfect alignment, the contour edges of the assembly may produce elongated, high-intensity pseudo-differences due to pixel sampling or minor geometric residuals. These pseudo-differences, characterized by high connectivity and large area, are prone to false detection. To solve this problem, the RPD-GA method introduces a *morphological erosion constraint*. We apply the morphological erosion operation on D_{masked}:

$$D_{\text{eroded}} = D_{\text{masked}} \ominus K_{\text{Erosion}} \tag{17}$$

This erosion operation leverages geometric priors to effectively eliminate narrow boundary noise while preserving larger, more connected core defect regions. On this basis, an adaptive boundary determination is further applied to prevent single anomalies from being mistakenly divided into multiple detections due to uneven distribution. This process is crucial for distinguishing true structural defects from non-structural sampling noise. Finally, through the above procedures, accurate and robust localization of assembly errors is achieved, with the improvement illustrated in the Fig. 7.

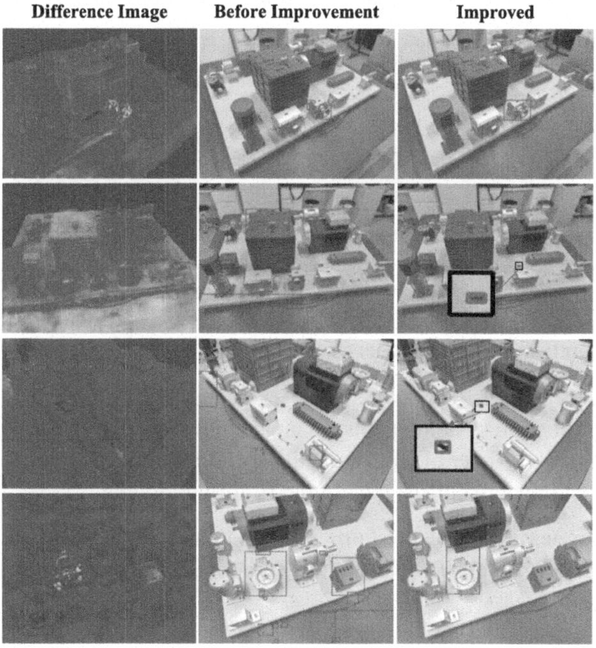

Fig. 7. Comparison of results

4 Experiments

To verify the effectiveness of the assembly inspection method and the RPD-GA method proposed in this paper, experiments were conducted in a real assembly environment. The main objectives of the experiments are to evaluate two aspects of performance: first, the three-dimensional reconstruction effect, which examines the ability of the viewpoint planning and AGV–robot arm collaborative acquisition strategy to achieve high precision reconstruction of the assembly body from different viewpoints; second, the anomaly detection capability, which verifies the accuracy and robustness of the RPD-GA method based on the comparison between 3DGS-rendered templates and real assembly images in identifying common assembly anomalies such as wrong part, missing parts, surface defects, and foreign objects. The experiments quantitatively analyzed and visually validated the system performance through a series of controlled variables and real acquisition data, providing a basis for subsequent method optimization and practical application.

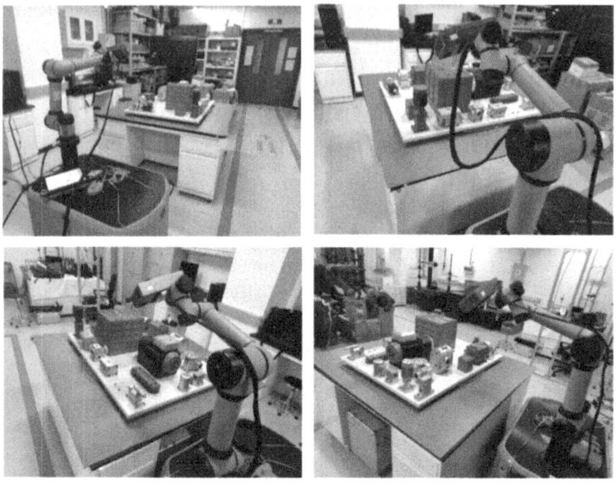

Fig. 8. Schematic diagram of experimental data acquisition

4.1 1Reconstruction Quality Experiment

Experimental Objective. This experiment evaluates the proposed AGV–robot arm collaborative acquisition system against manual capture to assess its impact on 3DGS reconstruction quality. Under identical conditions, we compare the geometric accuracy, texture fidelity, and overall consistency of the resulting models.

Experimental Scheme. To ensure fairness in comparison, the experiments were conducted in the same assembly environment, with both acquisition methods capturing the same number of images:

Manual acquisition (control group): The operator manually adjusted the camera position to capture the assembly object from different angles. This method lacks systematic viewpoint distribution planning and is prone to concentration or uncovered occlusion areas.

Proposed method (experimental group): The AGV equipped with a robotic arm moved to planned positions, and the robotic arm automatically adjusted the camera pose according to the viewpoint planning algorithm to achieve multi-view coverage of the assembly object illustrated in the Fig. 8. This strategy comprehensively considers the geometric complexity and visibility of the assembly, aiming to maximize information coverage and ensure uniform viewpoint distribution.

The same number of image data obtained by the two methods were respectively input into the 3DGS model for reconstruction, with identical training parameters and hardware environments. In addition, to evaluate reconstruction accuracy, an independent validation image set of 150 images (not used in training) was additionally collected as real image references. For each validation image, the corresponding viewpoint in both reconstructed models was rendered, and the generated reconstructed rendering images were compared with the real validation images to assess the realism and consistency of the reconstruction results.

The reconstruction quality was evaluated by the similarity between the original images and the rendered images, using the following three commonly used quantitative metrics: Peak Signal-to-Noise Ratio (PSNR), Structural Similarity Index (SSIM), and Learned Perceptual Image Patch Similarity (LPIPS). Higher PSNR and SSIM values and lower LPIPS values indicate higher similarity between the reconstructed and real images, and thus better reconstruction quality.

Table 1. Results analysis comparison table

Data	PSNR	SSIM	LPIPS
Control Group	24.5796	0.7854	0.0843
Experimental Group	29.4236	0.9222	0.0582

Experimental Results and Analysis. The table presents a comparison of reconstruction quality among the three data acquisition methods with the same number of images. As shown in the Table 1, the proposed AGV-robotic arm collaborative acquisition strategy significantly outperforms manual acquisition in terms of reconstruction accuracy and texture consistency. Specifically, PSNR and SSIM increased by approximately 20.9% and 17.4%, respectively, while LPIPS decreased by about 30.9%. These results indicate that the proposed viewpoint planning effectively enhances image diversity and coverage, thereby improving the convergence and completeness of the 3DGS model during training.

From a visual perspective (as shown in the Fig. 9), the model generated by the proposed method exhibits a smoother surface, with reflective areas and edge details better preserved. In contrast, the reconstruction results from manual acquisition show noticeable texture tearing and occlusion artifacts (Fig. 10). Moreover, we recorded the reconstruction time for each scene, since the automated acquisition system ensures more uniform image overlap between viewpoints, the overall reconstruction time is slightly reduced (see Table 2).

Fig. 9. Comparison of reconstruction results

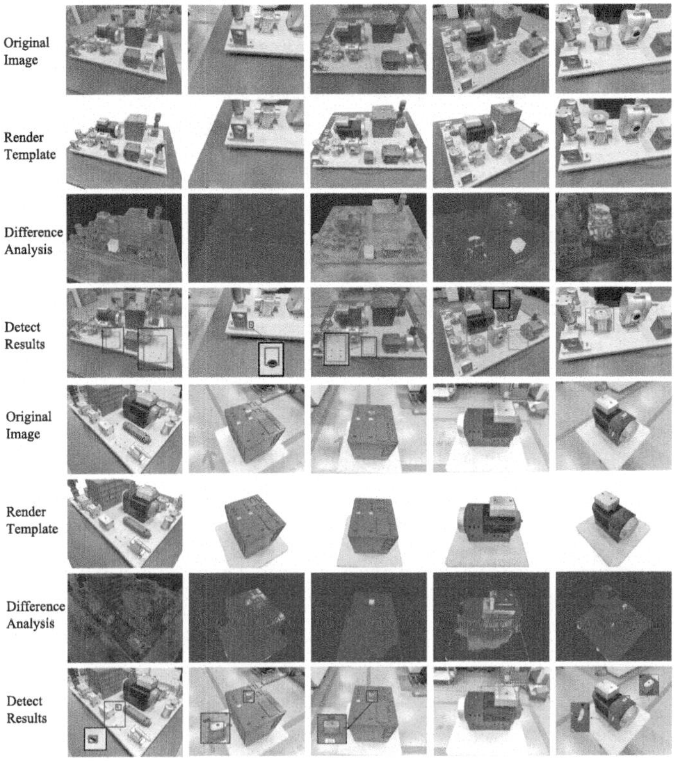

Fig. 10. Assembly defect detection results

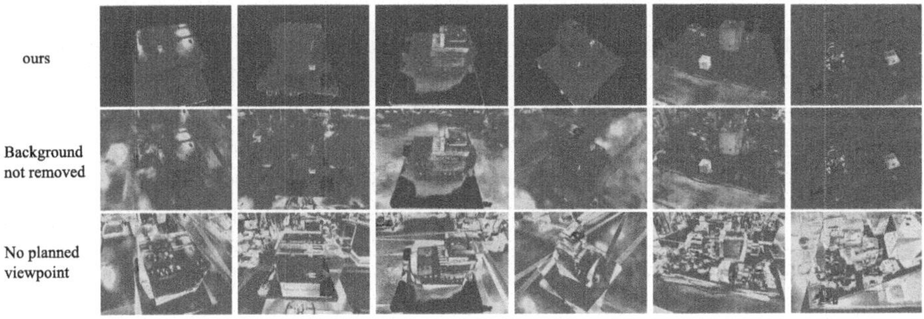

Fig. 11. Analysis of differences between different methods

Table 2. Reconstruction Schedule

Method	Scene 1	Scene 1	Scene 1
Original method	20 min 18 s	17 min 35 s	17 min 10 s
Our method	22 min 54 s	19 min 25 s	19 min 04 s

4.2 Abnormal Detection Results Analysis

To validate the feasibility and detection performance of the proposed 3DGS-based assembly anomaly detection method in real-world assembly scenarios, this study focuses on assessing its capability and robustness in identifying typical assembly anomalies, including wrong part, missing part, surface defects, and extraneous parts.

Table 3. Test Results Statistical Analysis Table.

Anomaly Type	Total	Number of detected	Accuracy	False rate	Missed rate	Average time
wrong part	50	52	96%	4%	0	0.1s
missing Part	50	52	96%	4%	0	0.1s
redundant part	50	55	91%	9%	0	0.1s

94%, with a missed detection rate below 6%. The detailed detection results are shown in Table 3. These results validate the effectiveness and robustness of the 3DGS-based anomaly detection framework in complex assembly environments and demonstrate its potential for further development.

After verifying the effectiveness of the proposed method, we conducted additional ablation studies to further demonstrate its advantages. Specifically, we compared four settings: (1) our full method using viewpoint-matched template images, (2) viewpoint-matched images without background removal, (3) viewpoint-matched images reconstructed without viewpoint planning, and (4) the test image itself. As shown in the Fig. 11, our method produces a clear and accurate difference map that precisely localizes the anomalous region. In contrast, the method without background removal suffers from severe background interference, introducing irrelevant differences that degrade the detection accuracy. The method without viewpoint planning performs the worst: due to poor reconstruction quality and the inability to obtain a viewpoint that matches the test image, a large amount of noise is introduced, making the resulting difference map unreliable for anomaly detection. In summary, our method achieves high-quality 3D reconstruction, enables accurate template generation for each test viewpoint, and provides robust and precise anomaly localization.

5 Conclusion

This study focuses on the digital reconstruction of assembly environments and the detection of assembly anomalies, and proposes an automated detection method based on 3D Gaussian Splatting 3DGS. Starting from real assembly sites, the research achieves autonomous multi-angle data acquisition through viewpoint planning and AGV-robotic arm collaborative capture. The 3DGS technique is then employed to reconstruct the assembly environment with high fidelity, generating virtual 3D models that preserve physical consistency and controllable lighting conditions.

Methods On this basis, a Robust Pixel Difference and Geometric Aggregation (RPD-GA) method is proposed, integrating 3D geometric priors with 2D difference analysis to achieve high-precision detection of assembly errors. This approach effectively removes background interference and illumination-induced pseudo-differences, demonstrating strong detection performance and robustness for common assembly anomalies such as wrong part, missing parts, surface defects, and extraneous components. The results validate the feasibility and effectiveness of 3DGS-based anomaly detection in practical industrial environments.

Overall, this research establishes a complete closed-loop workflow from autonomous viewpoint planning, multi-source data acquisition, and 3D reconstruction to anomaly detection, providing a scalable digital solution for intelligent assembly quality inspection. By introducing 3DGS as a lightweight and high-fidelity scene representation, the study bridges critical steps from data acquisition to detection analysis, laying a foundation for rapid modeling and online inspection in complex assembly scenarios.

However, there remain areas for further improvement. In particular, handling highly reflective and semi-transparent surfaces poses significant challenges for 3DGS-based reconstruction. These materials cause unstable photometric observations due to specular highlights, light transmission, and view-dependent appearance, which violate the multi-view consistency assumptions required by Gaussian parameter optimization. As a result, the estimated opacity, covariance, and color parameters may oscillate or converge to incorrect values, leading to blurred geometry, floating artifacts, or incomplete surface reconstruction. Moreover, existing density regularization strategies in 3DGS are not designed to separate true surface signals from environment-dependent reflections, further exacerbating reconstruction errors.

Beyond this, the efficiency of real-time detection still needs enhancement, and anomaly prominence analysis under complex occlusions requires the integration of advanced deep learning and physical modeling techniques. Future research will therefore focus on improving real-time performance, developing reflection-aware error modeling, enhancing dynamic assembly process monitoring, and achieving deeper integration with industrial control systems, further advancing the practical deployment of 3DGS-based intelligent assembly inspection in industrial environments (Table 4).

**For the convenience of readers and to help them better understand our workflow and the underlying code, we provide here the parameters used in our system along with their corresponding values.*

Table 4. Inference hyperparameter table

Viewpoint planing	fov	55°
	width	1920
	height	1280
	Near	0.01
	far	1000
	overlap	0.96
	safety	1.05
3DGS	nb_neighbors	40
	std_ratio	1
Detetion	DIFF_THRESHOLD	80
	WHITE_THRESHOLD	240
	AREA_THRESHOLD	20
	EDGE_CROP_PIXELS	10
	EROSION_KERNEL_SIZE	3
	HEATMAP_GAMMA	1

6 Disclosure of Interests

The authors have no competing interests to declare that are relevant to the content of this article.

References

1. Liu, Y., Li, S., Wang, J., Zeng, H., Lu, J.: A computer vision-based assistant system for the assembly of narrow cabin products. Int. J. Adv. Manuf. Technol. **76**(1), 281–293 (2015)
2. Wang, T., Gan, V.J.L.: Enhancing 3D reconstruction of textureless indoor scenes with IndoReal multi-view stereo (MVS). Autom. Constr. **166**, 105600 (2024)
3. Mildenhall, B., Srinivasan, P.P., Tancik, M., et al.: Nerf: representing scenes as neural radiance fields for view synthesis. Commun. ACM **65**(1), 99–106 (2021)
4. Kerbl, B., Kopanas, G., Leimkühler, T., et al.: 3D Gaussian splatting for real-time radiance field rendering. ACM Trans. Graph. **42**(4), 139:1-139:14 (2023)
5. Guillemaut, J.Y., Hilton, A.: Joint multi-layer segmentation and reconstruction for free-viewpoint video applications. Int. J. Comput. Vision **93**(1), 73–100 (2011)
6. Hepp, B., Nießner, M., Hilliges, O.: Plan3d: viewpoint and trajectory optimization for aerial multi-view stereo reconstruction. ACM Trans. Graph. (TOG) **38**(1), 1–17 (2018)
7. Wang, S., Tong, Y., Shang, X., et al.: Hierarchical viewpoint planning for complex surfaces in industrial product inspection. IEEE/ASME Trans. Mechatron. **29**(5), 3289–3299 (2023)
8. Li, K., Wu, Y., Xue, Y., et al.: Viewpoint recommendation based on object-oriented 3D scene reconstruction. IEEE Trans. Multimed. **23**, 257–267 (2020)

9. Kong, F., Du, F., Zhao, D.: Station-viewpoint joint coverage path planning towards mobile visual inspection. Robot. Comput. Integr. Manuf. **91**, 102821 (2025)
10. Du, F., Kong, F., Zhao, D.: A knowledge transfer method for unsupervised pose keypoint detection based on domain adaptation and CAD models. Adv. Intell. Syst. **5**(2), 2200214 (2023)
11. Wei, K., Liu, S., Chen, Q., et al.: Fast Multi-View 3D reconstruction of seedlings based on automatic viewpoint planning. Comput. Electron. Agric. **218**, 108708 (2024)
12. Hartung, J., Dold, P.M., Jahn, A., et al.: Analysis of AI-based single-view 3D reconstruction methods for an industrial application. Sensors **22**(17), 6425 (2022)
13. Huo, J., Yu, X.: Three-dimensional mechanical parts reconstruction technology based on two-dimensional image. Int. J. Adv. Rob. Syst. **17**(2), 1729881420910008 (2020)
14. Son, H., Kim, C., Kim, C.: 3D reconstruction of as-built industrial instrumentation models from laser-scan data and a 3D CAD database based on prior knowledge. Autom. Constr. **49**, 193–200 (2015)
15. Sun, G., Liu, D., Yang, Z., et al.: 3D reconstruction and precision evaluation of industrial components via Gaussian splatting. Comput. Graph., 104422 (2025)
16. Lim, H., Chang, H., Choi, J.B., et al.: LiDAR-3DGS: LiDAR reinforcement for multimodal initialization of 3D Gaussian splats. Comput. Graph., 104293 (2025)
17. Qian, J., Yan, Y., Gao, F., et al.: C3DGS: compressing 3D Gaussian model for surface reconstruction of large-scale scenes based on multi-view UAV images. IEEE J. Sel. Top. Appl. Earth Obs. Remote Sens., (2025)
18. Lu, Z., Ye, J., Leonard, J.: 3dgs-cd: 3d Gaussian splatting-based change detection for physical object rearrangement. IEEE Robot. Autom. Lett. (2025)
19. Frustaci, F., Spagnolo, F., Perri, S., et al.: Robust and high-performance machine vision system for automatic quality inspection in assembly processes. Sensors **22**(8), 2839 (2022)
20. Sharifzadeh, S., Biro, I., Lohse, N., et al.: Abnormality detection strategies for surface inspection using robot mounted laser scanners. Mechatronics **51**, 59–74 (2018)
21. Caeion, N., Massa, F., Synnaeve, G., et al.: End-to-end object detection with transformers. In: European conference on computer vision. Cham: Springer International Publishing, pp. 213–229 (2020)
22. Tsai, D.M., Jen, P.H.: Autoencoder-based anomaly detection for surface defect inspection. Adv. Eng. Inform. **48**, 101272 (2021)
23. Liu, R., Liu, W., Zheng, Z., et al.: Anomaly-GAN: a data augmentation method for train surface anomaly detection. Expert Syst. Appl. **228**, 120284 (2023)
24. Peng, T., Zheng, Y., Zhao, L., et al.: Industrial product surface anomaly detection with realistic synthetic anomalies based on defect map prediction. Sensors **24**(1), 264 (2024)
25. Zhao, D., Kong, F., Lv, N., et al.: A common knowledge-driven generic vision inspection framework for adaptation to multiple scenarios, tasks, and objects. Sensors **24**(13), 4120 (2024)
26. Božič, J., Fučka, M., Zavrtanik, V., et al.: Robustness of unsupervised methods for image surface-anomaly detection. Pattern Anal. Appl. **28**(2), 99 (2025)
27. Cui, Y., Liu, Z., Lian, S.: A survey on unsupervised anomaly detection algorithms for industrial images. IEEE Access **11**, 55297–55315 (2023)
28. Shao, H., Peng, J., Shao, M., et al.: Multi-scale prototype fusion network for industrial product surface anomaly detection and localization. IEEE Sens. J., (2024)
29. Kaji, F., Nguyen-Huu, H., Budhwani, A., et al.: A deep-learning-based in-situ surface anomaly detection methodology for laser directed energy deposition via powder feeding. J. Manuf. Process. **81**, 624–637 (2022)
30. Kruse, M., Rudolph, M., Woiwode, D., et al.: Splatpose & detect: pose-agnostic 3d anomaly detection. In: Proceedings of the IEEE/CVF Conference on Computer Vision and Pattern Recognition, pp. 3950–3960 (2024)

Active Control of an Unmanned Bicycle Lean Angle Based on Variable Domain Fuzzy PID

Nannan Wang(✉), Yigang Fan, and Jiahao Lu

School of Mechanical and Electrical Engineering, Shandong Jianzhu University, Jinan, People's Republic of China
wangnannan@sdjzu.edu.cn

Abstract. For unmanned bicycle, active control is the key to maintain its balance. To achieve optimal control of the bicycle's lean angle, this paper proposed a variable domain fuzzy PID control method. First of all, based on the concept of centrifugal force, we derive the centrifugal force torque equation and torque equilibrium equation for bicycle steering, and establish the dynamic equation for the equilibrium state of a bicycle. Then, by analyzing system stability through transfer functions and combining handlebar steering control methods, the conditions required for system stability are obtained. Based on the lateral lean angle of the bicycle body, a variable domain fuzzy PID control scheme is designed. Finally, MATLAB was used to perform lateral lean angle simulation analysis on the bicycle using PID control, fuzzy PID control, and variable domain fuzzy PID control, respectively. The results showed that when the control parameters K_p, K_i and K_d were set to 3.5, 0.02 and 0.03 respectively, compared with traditional PID and fuzzy PID controllers, the fuzzy PID control algorithm with variable domain improves the stability of the bicycle system by nearly 60% and shortens the stabilization time by 40%. The results validated the effectiveness of the method, providing a new control method for autonomous vehicle control that has the potential to be applied in fields such as intelligent transportation systems.

Keywords: unmanned bicycle · dynamic · fuzzy PID · variable domain · lean angle

1 Introduction

Driverless transportation has a bright future in the context of the rise of artificial intelligence and self-driving technology, with drones, driverless cars, and driverless bicycles. However, compared to other unmanned vehicles, the stability control of the self-driving bicycle faces unique challenges. The phenomenon of low-speed instability and high-speed self-stability in bicycle systems, coupled with their nonlinear and time-varying mechanical characteristics, has become a

H. Wei and J. Niu (Eds.): COBOT 2025, CCIS 2893, pp. 198–214, 2027.
https://doi.org/10.1007/978-981-92-2074-8_12

key focus of modern dynamics research. Among these challenges, one of the key issues in maintaining the balance of unmanned cycling is the precise control of the body lean angle, which is also the focus of this paper [26–28]. Therefore, this paper investigates the control of the lean angle of the bicycle's body.

Current research on intelligent self-driving bicycles by most scholars primarily focuses on dynamic modeling and stability control. Regarding physical explanations for bicycle stability control, three mainstream perspectives are widely accepted [30]: (1) Centrifugal force effect; (2) Gyroscopic effect; (3) Caster effect. In terms of dynamic modeling, existing methodologies mainly encompass Newton-Euler equation, Euler-Lagrange equation, and so on... Boyer et al. [2] systematically derived a 3D nonlinear dynamics model of the Whipple bicycle using geometric mechanics methods. Xiong et al. [30] innovated parametric reduction via Gibbs-Appell formalism, effectively condensing the 9-degree-of-freedom (DOF) system into a 3-DOF core dynamic representation. Zhang et al. [32] constructed a four-rigid-body nonlinear dynamic model using Kane's method, systematically elucidating how structural parameters (wheelbase, center-of-mass position, fork angle) govern self-balancing mechanisms, with experimental validation confirming model fidelity. Schwab et al. [23,24]extended the classical Whipple model to incorporate passive rider postures, quantitatively differentiating lateral stability impacts between forward-leaning and upright riding configurations. Hess et al. [11] parameterized rider dynamics as gain-filter-preview time structures, pioneering quantitative metrics for human-vehicle interaction. Chen et al. [4] developed nonlinear coupling models integrating active upper-body leaning, achieving coordinated steering-lean control via model predictive strategies.

The nonlinear and strongly coupled dynamic characteristics inherent in bicycle simplified single-track models have prompted researchers to develop diverse stability control strategies. Existing control methods mainly include PID control, LQR control, fuzzy control, model predictive control, adaptive sliding mode control, and so on... Dao et al. [5] demonstrated Sliding Mode Control's (SMC) robustness in bicycle stabilization, while emphasizing the critical balance between tracking precision and chattering suppression in switching function design. Bravo et al. [3] conducted comparative experimental analyses of PID, LQR, and computed torque control, providing multi-faceted insights into nonlinear balancing strategies. Li et al. [20] integrated riders' experience rider experience with adaptive fuzzy PID to mitigate initial parameter sensitivity. Kim et al. [16,17]coupled flywheel inertial moment modulation with PID algorithms, effectively suppressing motor speed drift and energy consumption. Zhao et al. [33] reduced parametric tuning complexity through genetic algorithm optimization. Bravo's et al. [3] experiments showed that when bicycle speed decreased from 15 km/h to 5 km/h, the tracking error of fixed-parameter PID increased from 0.6° to 4.6°, indicating significant performance degradation. Jiang et al. [13] developed Takagi-Sugeno fuzzy control with stability domain constraints, achieving optimal vehicle balance through model predictive integration. Hashemnia et al. [9] designed Lyapunov-stable fuzzy controllers enabling wide-range roll angle tracking (0°-35°) under diverse initial conditions. Li et al. [18] implemented four-wheel steering coordination via fuzzy torque distribution.

However, while the majority of the aforementioned scholars' dynamic modelling and vehicle-occupant interaction models are accurate, from a control perspective, they exhibit high complexity, numerous design parameters, and generate vast computational data volumes, rendering them impractical to implement. Existing research still exhibits shortcomings in dynamically adjusting characteristics and fuzzy rules. Fuzzy PID control heavily relies on the optimal design of the fuzzy rule base and membership functions; improper selection of domain scaling factors may lead to diminished control accuracy. From a modelling perspective, the simplified dynamic model designed herein exhibits high compatibility with fuzzy PID control, simultaneously satisfying real-time control requirements while preserving the key state variables essential for control. From a control perspective, facing complex road conditions and parameter uncertainties, the fixed model parameters of fuzzy PID limit its dynamic response and robustness, often leading to overshoot and steady-state error under complex operating conditions. The variable domain fuzzy PID controller overcomes these limitations of conventional fuzzy PID control by enabling adaptive control, offering a novel strategy for robotic stability control.

This paper presents the design of a variable-domain fuzzy PID controller, which addresses the issue of low control accuracy of the bicycle's body lean angle. The second section of the paper establishes the dynamic and control models of the bicycle system based on its structural characteristics. The third section introduces the design of the PID, fuzzy PID, and variable-domain fuzzy PID controllers. The fourth section carry out MATLAB simulations of the three controllers for the bicycle. The fifth section concludes with a summary of our work.

2 Establishment of Unmanned Bicycle Dynamic Model

2.1 Establishment of the Coordinate System

The Carvallo-Whipple bicycle is taken as an research object which is schematically shown in Fig. 1. The bicycle structure is mainly composed of four rigid bodies: front frame H, rear frame B, front wheel F and rear wheel W. Point P represents the rear wheel-ground contact point, while point Q denotes the front wheel-ground contact point. G represents the center of mass of the bicycle, h is the distance from the center of mass to the ground, a is the distance from the projection of G on the horizontal plane to point P. b is the wheelbase of the bicycle, corresponding to the distance between point P and Q, c is the distance between the contact point Q of the front wheel with the ground and the intersection of the steering axis with the ground.

The following coordinate systems are defined for the bicycle in Fig. 1: the inertial coordinate system O-XYZ, with O as the origin. The line connecting points P and Q defines the X-axis, oriented from point P to Q. The Y-axis is defined perpendicular to the X-axis through point P. The Z-axis is perpendicular to the XY-plane with its positive direction pointing downward. A body-fixed coordinate frame o-xyz is established, where point o coincides with contact point P between the rear wheel and ground in this configuration. ϕ is the head angle of the bicycle, representing the lean angle measured in the vertical direction, θ is the lean angle of the bicycle and δ is the angle of the front handlebar.

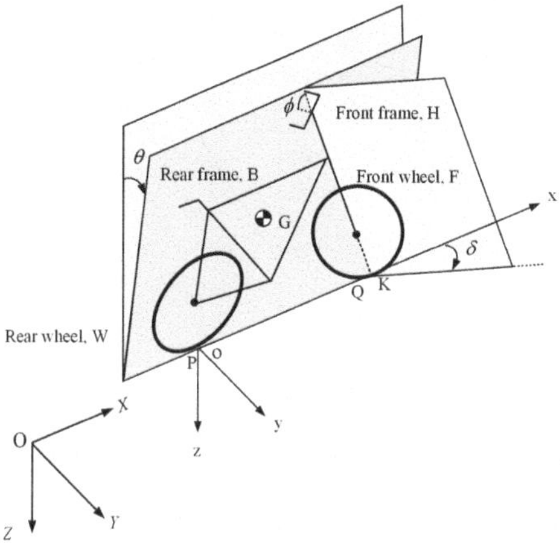

Fig. 1. Schematic of the bicycle structure.

2.2 Bicycle System Dynamic Model

Acturally, bicycle is a complex system with multiple degrees of freedom, in order to simplify the model and focus on the core dynamics of body inclination control, factors such as tire slip, road disturbance need to be ignored when performing dynamic analysis. Therefore, the following assumptions are made for the mechanical analysis of the bicycle shown in Fig. 1: (1) A bicycle is regarded as a rigid body with four hinged joints. (2) The effect of ground friction is neglected. (3) In the process of movement, no lateral slip occurs in the front and rear wheels.

The centrifugal force effect perspective posits that the centrifugal force generated during bicycle cornering counterbalances the overturning effect of gravity. Since centrifugal force is related to vehicle speed and steering, its effect resembles that of the external force F_y, this perspective can leverage the inherent characteristics of the bike combined with active control strategies for more effective control [10]. Consequently, the centrifugal force effect perspective is adopted for conducting dynamic analysis of bicycles.

As shown in Fig. 2, under the conditions of lean angle θ and front handlebar angle δ, the bicycle body moves in a uniform circular motion, bicycle itself is subjected to gravitational force, with its gravitational moment denoted as M_g. To maintain balance, an external force F_y is assumed to applied at the bicycle's

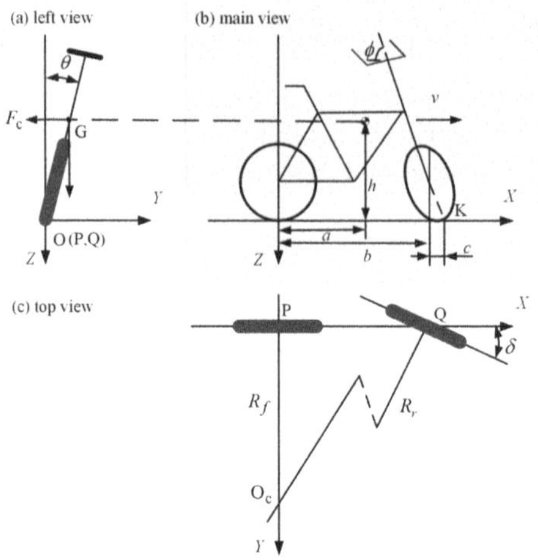

Fig. 2. Bicycle Parameters Diagram.

center of gravity, with the moment of external force is M_y. The moment balance equation for the bicycle in the stationary state is:

$$M_g = -mgh \sin\theta = h' F_y \cos\theta = M_y \tag{1}$$

where m is the total bicycle mass, h' is the height of the reaction moment application point above the ground.

From the perspective of centrifugal force, and in conjunction with Zhao's [10] results, the centrifugal moment generated by the bicycle's steering includes contributions from the centrifugal force, gravitational force, and components related to the handlebar's angular velocity. Let the center of the turning radius be denoted as O_c, and the turning radius of the bicycle be R, front wheel turning radius R_r, rear wheel turning radius R_f. Assuming that the turning radius R is much larger than the wheelbase b, and when the frame angle is θ, the turning radius R is assumed to be the geometric distance from the focal point of the ellipse to the nearest point on the circumference. According to the assumption, the turning radius R of the bicycle is the same as that of the rear wheel R_f and front wheel R_r. The turning radius of the bicycle body is represented as:

$$R = \frac{b\cos\theta}{\tan\delta \sin\phi} \tag{2}$$

when δ approaches 0, $\tan\delta \approx \delta$, the turning radius of the bicycle body simplifies to:

$$R = \frac{b\cos\theta}{\delta \sin\phi} \tag{3}$$

As show in Fig. 2, The centrifugal force acting on the bicycle is defined as F_c. In accordance with Eq. 3, the centrifugal force can be expressed as:

$$F_c = mR\omega^2 = \frac{mv^2}{R} = \frac{mv^2\delta\sin\phi}{b\cos\theta} \tag{4}$$

In the process of steering, the center of gravity is slightly lowered due to the influence of the front fork structure, with the lowered height denoted as ΔG. Incorporating the angular velocity-related component of handlebar steering, the equation for the centrifugal moment M_c generated by bicycle steering is given by:

$$M_c = \frac{mahv\sin\phi\dot{\delta}}{b} + F_c(h\cos\theta - \Delta G) \tag{5}$$

Substituting Eq. 4 into Eq. 5 and expanding, the complete form of the centrifugal moment is expressed as:

$$M_c = \frac{mahv\sin\phi\dot{\delta} + (mhv^2\sin\phi - macg\sin^2\phi)\delta}{b} \tag{6}$$

The Eq. 4 for the moment of inertia of the bicycle I is as follows:

$$I = mh^2 \tag{7}$$

Consequently, the dynamic equation for the equilibrium state of the bicycle is obtained:

$$M_c - M_g = I\ddot{\theta} \tag{8}$$

where $\ddot{\theta}$ is the angular acceleration of the bicycle's body lean angle. By substituting Eq. 6 and Eq. 7 into Eq. 8 and expanding, we obtain the dynamic equilibrium equation of the bicycle:

$$mh^2\ddot{\theta} - mgh\sin\theta = \frac{mah\sin\phi\dot{\delta}}{b} + \frac{(mhv^2\sin\phi - macg\sin^2\phi)\dot{\delta}}{b} \tag{9}$$

2.3 Bicycle System Control Model

Assuming that the roll angle θ of the bicycle is extremely small ($\sin\theta \approx \theta$)during motion. Then the dynamic equation of the bicycle is simplified as:

$$mh^2\ddot{\theta} - mgh\theta = \frac{mah\sin\phi\dot{\delta}}{b} + \frac{(mhv^2\sin\phi - macg\sin^2\phi)\dot{\delta}}{b} \tag{10}$$

By applying the Laplace transform to Eq. 10, the transfer function of the system is given by:

$$G(s) = \frac{\theta(s)}{\delta(s)} = \frac{sahv\sin\phi + \left(hv^2\sin\phi - acg\sin^2\phi\right)}{b\left(S^2h^2 - gh\right)} \tag{11}$$

In Eq. 11, when the handlebar steering angle δ is held constant, a stable bicycle roll angle θ can be achieved through dynamic regulation. By specifying a fixed handlebar steering angle δ_l and its corresponding stabilized roll angle θ_l, the ratio θ_l/δ_l is defined as the variable K_l , formulated as:

$$K_l = \frac{\theta_l}{\delta_l} = \lim_{s \to 0} G(s) = -\frac{hv^2 \sin\phi - acg \sin^2\phi}{gbh} \tag{12}$$

In Eq. 12, apart from the inherent parameters of the bicycle, there is only one variable, the speed v. When the speed v is determined, K_l is a constant.

According to stability criterion theory, K_l and speed v are linearly correlated, which means that the steering control method only takes effect when $K_l < 0$. Consequently, this method requires a critical minimum speed condition, formulated as:

$$v > \sqrt{\frac{acg \sin\phi}{h}} \tag{13}$$

3 Design of Fuzzy PID Controller with Dynamic Domain Adjustment

Intelligent control of the bicycle's body lean angle is crucial for achieving self-balancing. The bicycle system is complex, exhibiting characteristics like nonlinearity, strong coupling, and instability, making real-time control and accurate mathematical modeling challenging. To overcome these challenges, this paper develops an adaptive fuzzy PID controller with domain adjustment capabilities, employing a scaling factor for dynamic parameter optimization. This approach provides robust flexibility and adaptability, enabling effective control of the bicycle's body lean angle [8,22,25].

3.1 Fuzzy PID Controller Design

The core process of fuzzy PID control consists of four main steps: (1) fuzzification, (2) fuzzy reasoning, (3) defuzzification, and (4) parameter adjustment. The fuzzy PID controller uses two input variables: the deviation e and the rate of change of the deviation ec. These variables are processed by the fuzzy rule base in the fuzzy controller for fuzzy reasoning. The result is an adjustment to the PID parameters—specifically, the correction values ΔK_p, ΔK_i and ΔK_d, which are the output variables of the fuzzy logic controller.

The mathematical model for a PID controller is:

$$u(t) = K_p e(t) + K_i \int e(t)\,dt + K_d \frac{de(t)}{dt} \tag{14}$$

where $u(t)$ is the control variable, $e(t)$ is the difference between the expected value and the actual value, K_p is the proportional gain, K_i is the integral gain and K_d is the derivative gain.

The PID parameter adjustment formula is given by:

$$\begin{cases} K_p = K_{p0} + \Delta K_p \\ K_i = K_{i0} + \Delta K_i \\ K_d = K_{d0} + \Delta K_d \end{cases} \tag{15}$$

where K_{p0}, K_{i0} and K_{d0} are the preset values of PID control parameters.

Based on the dynamic control equations of bicycle balance, the selection of fuzzy domains needs to be combined with the system control model of bicycles, comprehensively considering the quantification range of inputs and outputs and parameter selection. The fuzzy domain of the input variables is set to [-6, 6], while the basic and fuzzy domains of the output variables K_p, K_i and K_d are set to [2.8, 3.7], [0, 0.04] and [0.01, 0.09] respectively. The proportional factors for the input variables and the quantization factors for the output variables are both set to 1. The fuzzy linguistic subsets for the input and output variables of the fuzzy control system are defined as: NB, NM, NS, ZO, PS, PM, PB. The difference in membership functions can affect the control performance. In this paper, both the input and output membership functions are defined as Gaussian functions, and their membership functions are shown in Fig. 3.

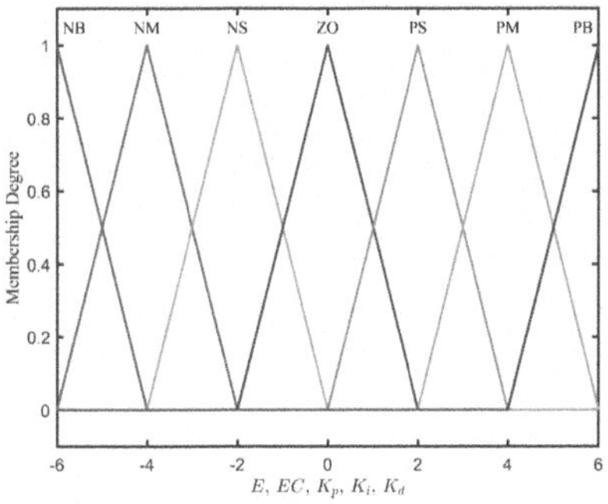

Fig. 3. Membership Function and Fuzzy Rule Diagram.

Fuzzy rules constitute the core component of a fuzzy controller, and their design critically determines the control system's performance. These rules are typically formulated based on expert knowledge and empirical data, then refined through iterative practical adjustments to optimize system behavior. The resulting fuzzy rule control system is presented in Tables 1,2,3.

On the base of the obtained membership functions and fuzzy rules, the output surfaces for proportional, integral, and derivative actions are determined according to the outlined procedure, as shown in Fig. 4.

Finally, defuzzification is conducted using the center of mass method. The defuzzification equation is expressed as:

$$f = \frac{\sum_{i=1}^{P} C^*(i)\mu_{C^*}(i)}{\sum_{i=1}^{P} \mu_{C^*}(i)} \tag{16}$$

where C^* is the fuzzy set, f is the precise value after defuzzification.

Table 1. Fuzzy control rules for ΔK_p

e	*ec*						
	NB	NM	NS	ZO	PS	PM	PB
NB	PB	PB	PM	PM	PS	ZO	ZO
NM	PB	PB	PM	PS	PS	ZO	NS
NS	PM	PM	PM	PS	ZO	NS	NS
ZO	PM	PM	PS	ZO	NS	NM	NM
PS	PS	PS	ZO	NS	NS	NM	NM
PM	PS	ZO	NS	NM	NM	NM	NB
PB	ZO	ZO	NM	NM	NM	NB	NB

Table 2. Fuzzy control rules for ΔK_i

e	*ec*						
	NB	NM	NS	ZO	PS	PM	PB
NB	NB	NB	NM	NM	NS	ZO	ZO
NM	NB	NM	NS	NS	ZO	PS	PS
NS	NM	NM	NS	ZO	PS	PM	PM
ZO	NM	NS	ZO	PS	PS	PM	PB
PS	ZO	ZO	PS	PS	PM	PB	PB
PM	ZO	ZO	PS	PM	PM	PB	PB
PB	NB	NB	NM	NM	NS	ZO	ZO

3.2 Design of Variable Domain Fuzzy PID Controller

Principle of Variable Domain Fuzzy Control. The principle of variable domain fuzzy control enhances performance by dynamically adjusting the controller's domain. Without altering the fuzzy rules, the domain shrinks as the error decreases and expands as the error increases, thereby enhancing control accuracy. The process of the initial domain scaling transformation is shown in Fig. 5.

Assuming that the initial domain of the input error e and the error rate ec are $[-E, E]$and $[-EC, EC]$, respectively, and the initial domain of the output variable is $[-U, U]$, the domain formula containing the scaling factors is given by:

$$\begin{cases} X(e) = [-\alpha(e)E, \alpha(e)E] \\ X(ec) = [-\alpha(ec)EC, \alpha(ec)EC] \\ Y(y) = [-\beta(y)U, \beta(y)U] \end{cases} \tag{17}$$

where $\alpha(e)$ and $\alpha(ec)$ are the scaling factors of the input variable, $\beta(y)$ is the scaling factor of the output variable, $X(e)$, $X(ec)$, and $Y(y)$ are the transformed input and output domains.

Table 3. Fuzzy control rules for ΔK_d

e	ec						
	NB	NM	NS	ZO	PS	PM	PB
NB	PS	NS	NB	NB	NB	NM	PS
NM	ZO	NS	NM	NM	NS	NS	ZO
NS	ZO	NS	NS	NS	NS	NS	ZO
ZO	ZO	ZO	ZO	ZO	ZO	ZO	ZO
PS	PB	NS	PS	PS	PS	PS	PB
PM	PB	PM	PM	PM	PS	PS	PB
PB	PS	NS	NB	NB	NB	NM	PS

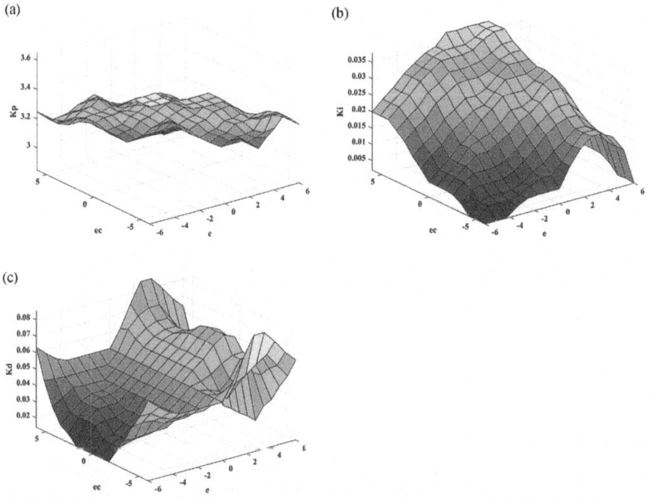

Fig. 4. (a) Output Surface Diagrams of Proportional Coefficient, (b) Output Surface Diagrams of Integral Coefficient, (c) Output Surface Diagrams of Differential Coefficient.

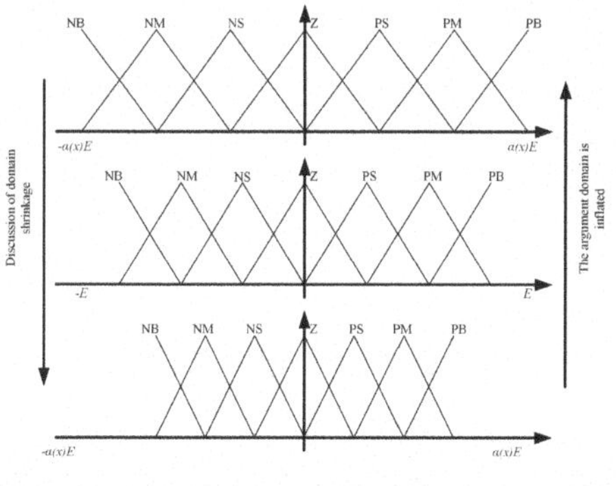

Fig. 5. Schematic Diagram of Universe of Discourse Variation.

Based on the above analysis, the schematic diagram of the variable domain PID controller is shown in Fig. 6.

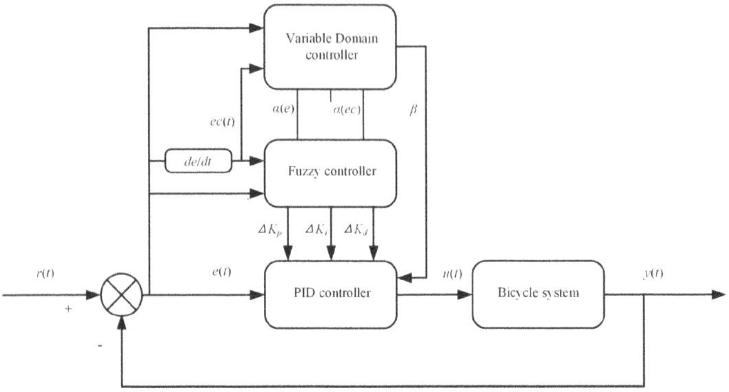

Fig. 6. Schematic Diagram of Variable Universe of Discourse Fuzzy PID Control System Principle.

Determination of Scaling Factors. Selection of domain scaling factors directly affects control accuracy. Therefore, the scaling factors must adhere to the principles of duality, zero-avoidance, monotonicity, coordination, and normality. The selection of scaling factors primarily relies on three methodological approaches: function-based, fuzzy inference-based, and error grading-based strategies. However, since fuzzy inference-based scaling factors require substantial computation and have poor real-time performance, this paper adopts an exponential function, based on a functional form, as the scaling factor. The commonly used functions for the scaling factors of the input and output domains include:

$$\begin{cases} \alpha(x) = 1 - \lambda e^{-kx^2}, \lambda \in (0,1),\ k > 0 \\ \beta(y) = K_1 \sum\limits_{i=1}^{n} p_i \int_0^y e_i(\tau) P d\tau + \beta(0) \end{cases} \tag{18}$$

Based on the selection principles of the scaling factor function, while considering system stability, the selection range for the parameters λ and k is controlled within [0, 1]. After multiple experimental attempts, the final values of λ and k are determined to be 0.6 and 0.5 respectively. Specifically expressed as:

$$\begin{cases} \alpha(e) = 1 - 0.6e^{-0.5x^2} \\ \alpha(ec) = 1 - 0.6e^{-0.5x^2} \end{cases} \tag{19}$$

Output scaling factor is determined by analyzing the effect of the input scaling factor on the system ΔK_p, ΔK_i and ΔK_d control. The results show that the monotonicity of K_p and K_d should be consistent with the monotonicity of $|e|$, while the monotonicity of K_i should be opposite to that of $|e|$. After multiple experimental validations, the following output scaling factor was selected:

$$\beta_p = 0.5|e|, \quad \beta_i = \frac{1}{|e| + 0.6}, \quad \beta_d = 5|e| \tag{20}$$

where β_p, β_i, and β_d are the output domain scaling factors for ΔK_p, ΔK_i and ΔK_d respectively.

4 Results and Discussion

4.1 Analysis of Bicycle Control System

We use the SMLRO electric-assisted mountain bike system to implement the variable domain fuzzy PID control method mentioned in this paper. The parameters of the bicycle are shown in Table 4.

Table 4. Bicycle Parameters

Parameter	Value	Unit
m	32.7	Kg
h	0.65	m
λ	70	degree
a	0.40	m
b	1.1	m
c	0.09	m

4.2 Simulation Tests and Comparisons of Control Methods

To evaluate the effectiveness and reliability of the variable-domain fuzzy PID control for the bicycle system, simulations were conducted using MATLAB/Simulink. Comparing the three modes of PID control, fuzzy PID control and variable domain fuzzy PID control. The simulation diagram of the bicycle lateral lean angle system and the variable domain submodule are shown in Fig. 7 and Fig. 8.

After incorporating the variable domain fuzzy PID controller into the original system, the bicycle body tilt control system was simulated using MATLAB. A step signal with an amplitude of 5° was selected as the input signal, yielding the simulation curves shown in Fig. 9 and Fig. 10. As depicted in Fig. 9, the variable domain fuzzy PID controller enhanced the system's response speed, eliminated oscillations, and achieved a stable state with minimal overshoot in a shorter time. Figure 10 indicates that in the low-frequency range, the variable-domain fuzzy PID exhibits virtually no fluctuations and demonstrates excellent stability. The

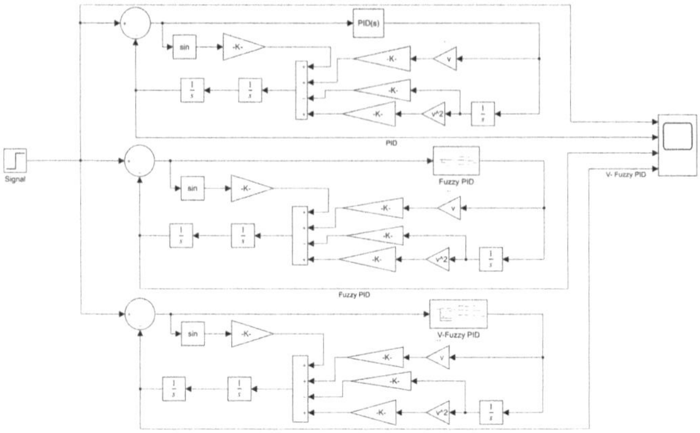

Fig. 7. Simulation diagrams of PID, Fuzzy PID, and variable universe Fuzzy PID controllers.

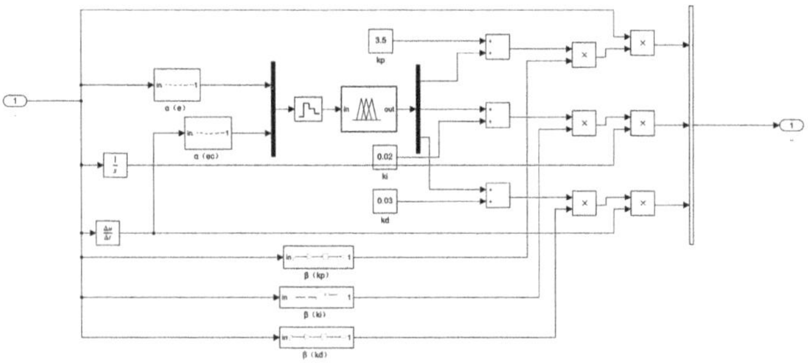

Fig. 8. Block diagram of variable universe Fuzzy PID control module.

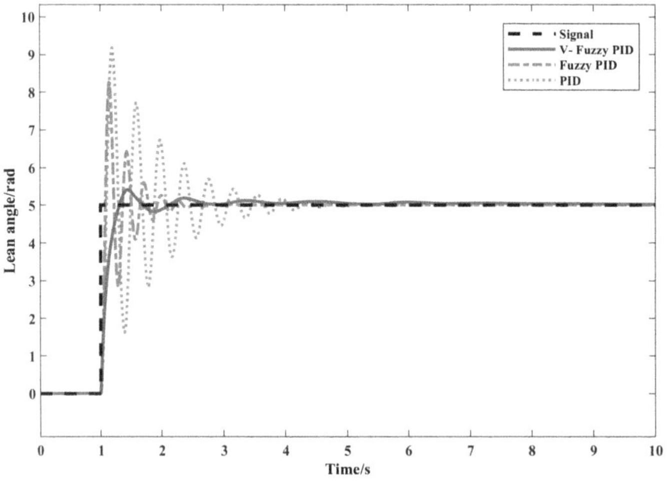

Fig. 9. Performance Comparison Curves of Three Different Control Methods.

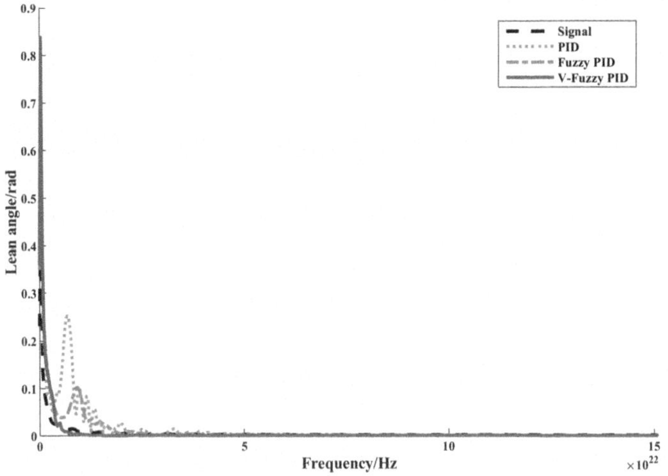

Fig. 10. Fourier Transform of Lean Angel.

comparison results between variable domain fuzzy PID control and traditional PID/fuzzy PID control are shown in Table 5. While PID and fuzzy PID controls stabilize the system, they exhibit significant overshoot and oscillations. In contrast, variable domain fuzzy PID control offers superior response speed and accuracy, greatly enhancing the system's disturbance rejection and adaptability to changes.

Table 5. Comparison of Three Control Methods for Vehicle Body Lean Angle Control

Control Method	Overshoot	Settling Time
PID	4%	5 s
Fuzzy PID	3.2%	3 s
VD-Fuzzy PID	0.5%	3 s

Simulation and comparative analysis demonstrate that PID and fuzzy PID controllers feature simple structures and ease of implementation. However, they suffer from issues such as large steady-state errors, significant overshoot, and prolonged settling times (exceeding 5 s). Furthermore, their parameter tuning relies on manual intervention and involves high computational complexity. In contrast, variable-domain fuzzy PID control demonstrates strong adaptability, not only reducing overshoot and shortening response time but also exhibiting superior stability, thereby enabling excellent control of the unmanned bicycle.

5 Conclusions

This paper addresses the issues of low control accuracy and poor stability in bicycle body lean angle control. It analyses the mathematical models of bicycle self-balancing characteristics and body structure, and proposes an improved bicycle body lean angle control method based on variable domain fuzzy PID control. First, based on the concept of centrifugal force, derive the centrifugal torque equation and torque equilibrium equation for bicycle steering and establish the dynamic equation governing the bicycle's state of equilibrium. Then, analyze system stability through transfer functions and handlebar steering control methods to obtain the necessary conditions for system stability. Next, design a variable-domain fuzzy PID control scheme based on the bicycle's body lean angle. Finally, develop a control system using MATLAB/Simulink and simulate its response characteristics and adaptability. Simulation results demonstrate that when the K_p value is 3.5, the K_i value is 0.02, and the K_d value is 0.03, the proposed variable-domain fuzzy PID control method demonstrates significant advantages over traditional PID and fuzzy PID controllers in terms of response speed, disturbance rejection capability, and control performance for the nonlinear dynamics of bicycles.

References

1. Ahmadi, A., Abdul Rahim, H., Abdul Rahim, R.: Optimization of a self-tuning pid type fuzzy controller and a pid controller for an inverted pendulum. J. Intell. Fuzzy Syst. **26**(4), 1987–1999 (2014)
2. Boyer, F., Porez, M., Mauny, J.: Reduced dynamics of the non-holonomic whipple bicycle. J. Nonlinear Sci. **28**(3), 943–983 (2018)
3. Bravo M, D.A., Rengifo R, C.F., Díaz O, J.F.: Comparative analysis between computed torque control, lqr control and pid control for a robotic bicycle. In: 2019 IEEE 4th Colombian Conference on Automatic Control (CCAC), pp. 1–6. IEEE (2019)
4. Chu, T., Chen, C.: Modelling and model predictive control for a bicycle-rider system. Veh. Syst. Dyn. **56**(1), 128–149 (2018)
5. Dao, T.K., Chen, C.K.: Sliding-mode control for the roll-angle tracking of an unmanned bicycle. Veh. Syst. Dyn. **49**(6), 915–930 (2011)
6. Das, S., Pan, I., Das, S., Gupta, A.: A novel fractional order fuzzy pid controller and its optimal time domain tuning based on integral performance indices. Eng. Appl. Artif. Intell. **25**(2), 430–442 (2012)
7. Guo, Z., Zhu, F., Zhao, P., Chen, H.: Ba-optimized variable domain fuzzy pid control algorithm for water and fertilizer ratio control system in cotton field. Processes **12**(6), 1202 (2024)
8. Han, J.: From pid to active disturbance rejection control. IEEE Trans. Ind. Electron. **56**(3), 900–906 (2009)
9. Hashemnia, S., Shariat Panahi, M., Mahjoob, M.J.: Unmanned bicycle balancing via lyapunov rule-based fuzzy control. Multibody Syst. Dyn. **31**(2), 147–168 (2014)
10. He, J., Zhao, M., Stasinopoulos, S.: Constant-velocity steering control design for unmanned bicycles. In: 2015 IEEE International Conference on Robotics and Biomimetics (ROBIO), pp. 428–433. IEEE (2015)

11. Hess, R., Moore, J.K., Hubbard, M.: Modeling the manually controlled bicycle. IEEE Trans. Syst. Man Cybern.-Part A: Syst. Humans **42**(3), 545–557 (2012)
12. Ji, G., et al.: Research on variable universe fuzzy pid control for semi-active suspension with cdc dampers based on dynamic adjustment functions. Sci. Rep. **14**(1), 3442 (2024)
13. Jiang, S., Wu, G., Li, Y., Mao, L., Zhang, D.: Intelligent vehicle path tracking and stability cooperative control strategy based on stable domain. SAE Int. J. Veh. Dyn. Stabil. NVH **8**(10-08-04-0025), 445–465 (2024)
14. Jin, X., Chen, K., Zhao, Y., Ji, J., Jing, P.: Simulation of hydraulic transplanting robot control system based on fuzzy pid controller. Measurement **164**, 108023 (2020)
15. Keo, L., Masaki, Y.: Trajectory control for an autonomous bicycle with balancer. In: IEEE/ASME International Conference on Advanced Intelligent Mechatronics, pp. 676–681. IEEE (2008)
16. Kim, H.W., An, J.W., dong Yoo, H., Lee, J.M.: Balancing control of bicycle robot using pid control. In: 2013 13th International Conference on Control, Automation and Systems (ICCAS 2013), pp. 145–147. IEEE (2013)
17. Kim, Y., Kim, H., Lee, J.: Stable control of the bicycle robot on a curved path by using a reaction wheel. J. Mech. Sci. Technol. **29**(5), 2219–2226 (2015)
18. Li, B., Yu, F.: Design of a vehicle lateral stability control system via a fuzzy logic control approach. Proc. Inst. Mech. Engineers, Part D: J. Automob. Eng. **224**(3), 313–326 (2010)
19. Li, H., Hui, Y., Ma, J., Wang, Q., Zhou, Y., Wang, H.: Research on variable universe fuzzy multi-parameter self-tuning pid control of bridge crane. Appl. Sci. **13**(8), 4830 (2023)
20. Li, Y., Ren, X., Liu, J.: A new fuzzy control and dynamic modeling of bicycle robot. In: 2012 4th International Conference on Intelligent Human-Machine Systems and Cybernetics, vol. 2, pp. 53–58. IEEE (2012)
21. Mohindru, P.: Review on pid, fuzzy and hybrid fuzzy pid controllers for controlling non-linear dynamic behaviour of chemical plants. Artif. Intell. Rev. **57**(4), 97 (2024)
22. Monje, C.A., Vinagre, B.M., Feliu, V., Chen, Y.: Tuning and auto-tuning of fractional order controllers for industry applications. Control. Eng. Pract. **16**(7), 798–812 (2008)
23. Schwab, A.L., Meijaard, J.P.: A review on bicycle dynamics and rider control. Veh. Syst. Dyn. **51**(7), 1059–1090 (2013)
24. Schwab, A.L., Meijaard, J.P., Kooijman, J.D.: Lateral dynamics of a bicycle with a passive rider model: stability and controllability. Veh. Syst. Dyn. **50**(8), 1209–1224 (2012)
25. Shah, P., Agashe, S.: Review of fractional pid controller. Mechatronics **38**, 29–41 (2016)
26. Singhania, S., Kageyama, I., Karanam, V.M.: Steering control to balance a motorcycle at low speeds based on riders' input. Proc. Inst. Mech. Engineers, Part D: J. Automob. Eng. **234**(12), 2892–2904 (2020)
27. Tan, J., Gu, Y., Liu, C.K., Turk, G.: Learning bicycle stunts. ACM Trans. Graph. (TOG) **33**(4), 1–12 (2014)
28. Vu, N.K., Nguyen, H.Q.: Balancing control of two-wheel bicycle problems. Math. Probl. Eng. **2020**(1), 6724382 (2020)
29. Wang, J.J.: Simulation studies of inverted pendulum based on pid controllers. Simul. Model. Pract. Theory **19**(1), 440–449 (2011)

30. Xiong, J., Wang, N., Liu, C.: Stability analysis for the whipple bicycle dynamics. Multibody Syst. Dyn. **48**(3), 311–335 (2020)
31. Zhang, J., Zhao, T., Guo, B., Dian, S.: Fuzzy fractional-order pid control for two-wheeled self-balancing robots on inclined road surface. Syst. Sci. Control Eng. **10**(1), 289–299 (2022)
32. Zhang, Y., Zhao, G., Li, H.: Multibody dynamic modeling and controlling for unmanned bicycle system. ISA Trans. **118**, 174–188 (2021)
33. Zhao, B., Wang, H., Li, Q., Li, J., Zhao, Y.: Pid trajectory tracking control of autonomous ground vehicle based on genetic algorithm. In: 2019 Chinese Control And Decision Conference (CCDC), pp. 3677–3682. IEEE (2019)
34. Zhao, C., Wang, D., Xue, W.: Beyond linear limits: design of robust nonlinear pid control. Automatica **173**, 112075 (2025)

Trajectory Correction for Omnidirectional Wheeled Mobile Chassis of Educational Robots: A Low-Complexity Kinematic Compensation Method

Dianjun Wang[1], Guanglei Li[2], Ya Chen[3](✉), Shujing Liu[1,2,3], and Peng Wang[1,2,3]

[1] School of Mechanical Engineering & Hydrogen Energy Research Centre, Beijing Institute of Petrochemical Technology, No.19 Qingyuan North Street, Huangcun Town, Daxing District, Beijing 102617, China

[2] School of Mechanical Engineering, Beijing Institute of Petrochemical Technology, No.19 Qingyuan North Street, Huangcun Town, Daxing District, Beijing 102617, China

[3] School of Information Engineering, Beijing Institute of Petrochemical Technology, No.19 Qingyuan North Street, Huangcun Town, Daxing District, Beijing 102617, China

chenya@bipt.edu.cn

Abstract. Small omnidirectional wheeled mobile robots (OWMRs) have gained widespread adoption in educational contexts owing to their lightweight construction and superior maneuverability. However, the inherent slippage characteristic of omnidirectional wheel configurations, coupled with inertial disturbances induced by the manipulator during start-stop sequences and motion phases, results in substantial trajectory deviations of the omnidirectional wheeled chassis. This not only impairs the accuracy of kinematics teaching demonstrations but also compromises students' comprehension of core theoretical principles. To mitigate this problem, this paper first develops and solves the forward and inverse kinematic models of the OWMR, laying a theoretical foundation for the subsequent development of the compensation strategy. Building on this framework, a simplified kinematic compensation approach is proposed, wherein correction coefficients—ascertained via experimental tests—are embedded into the chassis control program to specifically counteract trajectory deviations arising from slippage and inertial disturbances during the robot's start-stop and motion cycles. Finally, comparative experiments are carried out on a standard laboratory surface to evaluate the performance of the proposed compensation strategy against the uncompensated control approach. Experimental findings demonstrate that the proposed method markedly reduces the robot's trajectory deviation, accompanied by a considerable enhancement in trajectory tracking precision. This study offers a practical and low-complexity solution for the precise control of OWMRs in educational settings, thereby effectively enhancing the efficacy of kinematics teaching demonstrations.

Keywords: Omnidirectional wheeled mobile robot (OWMR) · Kinematic compensation · Trajectory tracking · Educational robotics

H. Wei and J. Niu (Eds.): COBOT 2025, CCIS 2893, pp. 215–233, 2027.
https://doi.org/10.1007/978-981-92-2074-8_13

1 Introduction

With the rapid evolution of robotics technology, robot education has become a core vehicle for cultivating innovative practical capabilities and engineering thinking, while high-performance educational robot platforms serve as crucial support for ensuring teaching quality. Small-scale Omnidirectional Wheeled Mobile Robots (OWMRs), leveraging their lightweight structure and flexible omnidirectional mobility, have been widely adopted in educational scenarios such as kinematics principle demonstrations, students' hands-on training, and robot competitions. This type of robot can intuitively present three-degree-of-freedom (3DOF) kinematic characteristics, building an efficient bridge between theoretical knowledge and engineering practice, and significantly enhancing students' engagement and cognitive efficiency. However, the mechanical structural characteristics of omnidirectional wheels (e.g., discontinuous contact at the wheel-ground interface) and inertial disturbances generated during the start-stop and movement phases of the manipulator collectively lead to inevitable wheel slip [1–3]. This slip directly causes trajectory deviations in OWMRs, which not only undermine the accuracy of kinematics teaching demonstrations and fail to accurately reproduce the expected results of theoretical derivations but also may mislead students' understanding of core principles, severely impairing the effectiveness of practical teaching. Consequently, it has become a major technical bottleneck restricting the in-depth application of OWMRs in educational settings.

To address the trajectory deviation compensation problem of OWMRs, the academic community has conducted a series of studies, and existing solutions can be categorized into three types [4–10]: The first is the slip state estimation and compensation method, whose core idea is to real-time estimate the slip state and design targeted control strategies to offset trajectory disturbances. For instance, Ref. [4] proposed a disturbance-rejection tracking control strategy based on an Extended State Observer (ESO), which estimates the total disturbance including uncertainties, external perturbations, and slip effects, then uses the backstepping method to design a control law for compensation and employs a second-order tracking differentiator to solve the problem of computational complexity explosion. Finally, based on cascade theory analysis, it is verified that the closed-loop system satisfies Input-to-State Stability (ISS). Simulation results show that this scheme achieves higher tracking accuracy than traditional PID controllers, with errors converging to a small neighborhood around zero, and the ESO can effectively reconstruct velocity information and accurately estimate disturbances. Ref. [5] developed an RLS-based slip estimation and compensation scheme, which fuses data from MPU-9250 multi-axis sensors and motor encoders based on the kinematic and dynamic models of a quarter-car, real-time estimates the slip ratio of four wheels via the Recursive Least Squares (RLS) algorithm, designs compensation strategies for linear, translational, and diagonal motion modes, and dynamically adjusts wheel speeds using a dsPIC30F6010A microcontroller. Experimental results indicate that for the chassis in diagonal motion and cross walking, the start-stop position deviations are reduced from 1.52 m and 1.56 m to 0.03 m, respectively, achieving a significant improvement in trajectory tracking accuracy. Ref. [6] proposed a fuzzy adaptive PID method that combines the stability of PID control with the adaptability of fuzzy control. It obtains motion error e, slip coefficient μ, and tilt angle λ through geometric modeling, then real-time adjusts PID parameters based on

Mamdani-type reasoning. Experiments demonstrate that this method improves the anti-slip stability and anti-tilt stability of curved trajectories by 38% and 19%, respectively, and those of circular trajectories by 17% and 10%. Additionally, the curved trajectory error converges within 1 s (compared to 3 s for traditional PID), showing remarkable optimization of motion performance. However, despite their excellent compensation effects, the above three methods require the design of complex controllers and algorithms, making them difficult to adapt to low-cost educational robots. The second type is the sensor fusion method for improving pose accuracy, which reduces measurement errors caused by slip by fusing multi-source sensor data. Ref. [7] designed a nonlinear controller based on Trajectory Linearization Control (TLC), adopting a dual-loop control architecture (the outer loop adjusts the pose, and the inner loop tracks the body rate), fuses data from on-board encoders and vision systems to reduce measurement errors caused by slip, and introduces a time-varying bandwidth command shaping filter to avoid control saturation. Experimental results show that under fixed controller parameters, this scheme can accurately track various 3DOF trajectories with a position tracking error of ≤ 1 cm and an orientation angle error of $\leq 6°$, achieving a good tracking effect. Ref. [8] implemented error compensation using an Extended Kalman Filter (EKF), fusing data from a compass (heading angle), gyroscope (yaw rate), and encoder (wheel speed), and real-time estimates the robot's position and dynamically adjusts the path to offset slip disturbances according to the process of "state prediction - Kalman gain calculation - state update". MATLAB simulation results show that the robot's trajectory deviation is significantly reduced, and the final position is closer to the target. However, these two methods require additional sensors, and the control logic of Ref. [7] is relatively complex, making it more suitable for industrial scenarios and difficult to meet the low-cost and easy-to-operate requirements of educational robots. The third type is the mechanical structure optimization method, which improves wheel-ground contact stability by modifying the omnidirectional wheel structure [9, 10]. Nevertheless, such schemes are still in the experimental stage and have not yet achieved market-oriented application, resulting in limited practical reference value for educational scenarios.

In summary, existing compensation schemes either suffer from complex control/algorithms or high hardware costs, or lack engineering application conditions, all of which are difficult to meet the core requirements of "low cost, easy operation, and high reliability" for OWMRs in educational scenarios. Based on this, this paper conducts targeted research to address the precise control demands of OWMRs in educational settings: Firstly, establish and solve the forward/inverse kinematic models of OWMRs to provide a theoretical foundation for trajectory deviation compensation; secondly, propose a simplified kinematic compensation method, calibrate adaptive correction coefficients through experiments, embed them into the chassis control program, and specifically offset trajectory deviations caused by slip and inertial disturbances (without the need for complex algorithms or additional sensors); Finally, the effectiveness of the proposed scheme is verified through comparative experiments conducted on a standard laboratory floor.

2 Kinematics Analysis

2.1 Mechanical System Scheme

The mechanical structure of the composite robot's experimental prototype is divided into two core components: the robotic arm and the chassis. Specifically, the robotic arm adopts a 6-axis configuration modeled after the PUMA560 [11], while the chassis employs a four-wheel omnidirectional drive design [12]. The motion schematics of the robotic arm and chassis are illustrated in Figs. 1 and 2, respectively.

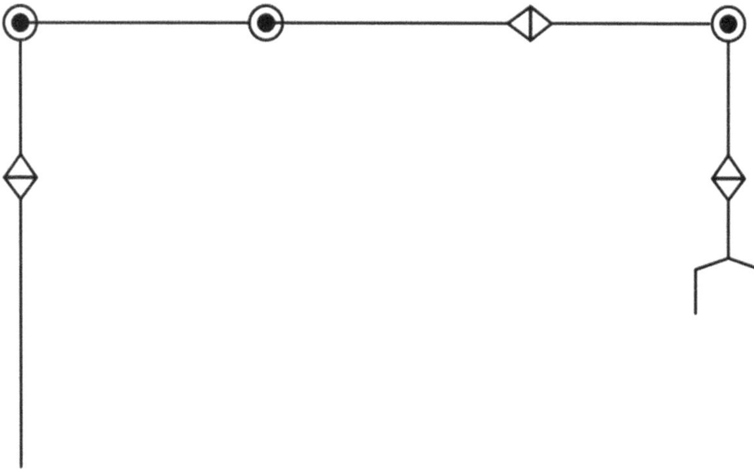

Fig. 1. Kinematic Motion Diagram of the 6-DOF Robotic Arm (PUMA560-like Configuration)

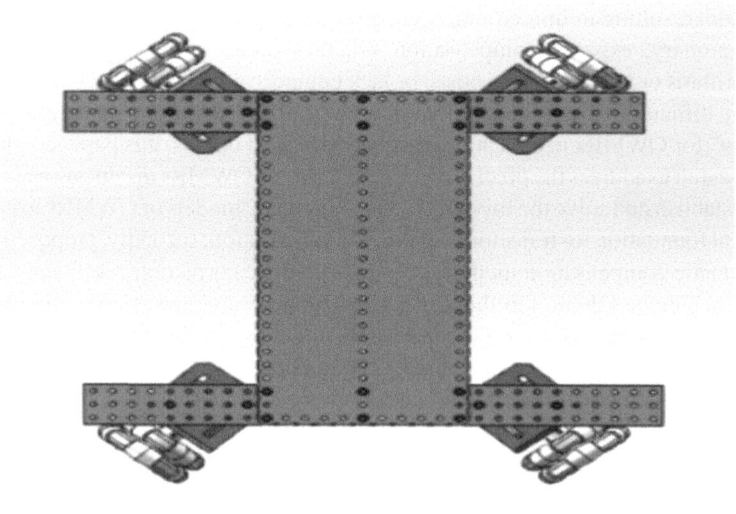

Fig. 2. Schematic of the 4-Wheel Omnidirectional Drive Chassis for the Educational Robot

2.2 Forward Kinematics Analysis of Robotic Arm

First, an improved linkage coordinate system was established for the robotic manipulator; subsequently, a linkage parameter table was constructed [13]. The detailed configuration of the linkage coordinate system and the specific contents of the linkage parameter table are presented in Fig. 3 and Table 1, respectively.

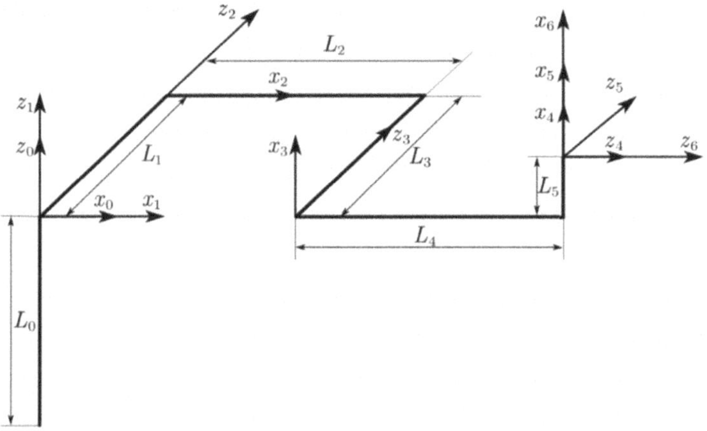

Fig. 3. Linkage Coordinate System of the 6-DOF Robotic Arm (PUMA560-like Configuration)

Table 1. Linkage Parameter Table for the 6-DOF Robotic Arm (PUMA560-like Configuration)

i	α_{i-1}	a_{i-1}	d_i	θ_i
1	0	0	0	θ_1
2	−90°	0	L_1	θ_2
3	0	L_2	$-L_3$	θ_3
4	−90°	L_5	L_4	θ_4
5	90°	0	0	θ_5
6	−90°	0	0	θ_6

Based on the forward kinematics formula:

$$
{}_{i}^{i-1}T = \begin{bmatrix} cos\theta_i & -sin\theta_i & 0 & a_{i-1} \\ sin\theta_i cos\alpha_{i-1} & cos\theta_i cos\alpha_{i-1} & -sin\alpha_{i-1} & -sin\alpha_{i-1} d_i \\ sin\theta_i sin\alpha_{i-1} & cos\theta_i sin\alpha_{i-1} & cos\alpha_{i-1} & cos\alpha_{i-1} d_i \\ 0 & 0 & 0 & 1 \end{bmatrix} \tag{1}
$$

where α_{i-1} denotes the twist angle of the $(i-1)$-th link, a_{i-1} denotes the length of the $(i-1)$-th link, d_i denotes the offset of the i th link, and θ_i denotes the joint angle of the i-th joint.

The homogeneous transformation matrix for the forward kinematics of the robotic arm is expressed as follows:

$$ {}_6^0T = {}_1^0T{}_2^1T{}_3^2T{}_4^3T{}_5^4T{}_6^5T = \begin{bmatrix} r_{11} & r_{12} & r_{13} & P_x \\ r_{21} & r_{22} & r_{23} & P_y \\ r_{31} & r_{32} & r_{33} & P_z \\ 0 & 0 & 0 & 1 \end{bmatrix} \quad (2) $$

where r_{ij} (where $i, j = 1, 2, 3$) denote the elements of the rotation matrix for the transformation from the base coordinate system to the end-effector coordinate system, and P_x, P_y, P_z denote the position coordinate components of the end-effector.

2.3 Inverse Kinematics Analysis of Robotic Arm

The inverse kinematics solution for the robotic arm of the composite robot experimental prototype was derived using the variable separation method. Specifically, equations containing only individual joint angles (θ_1, θ_2, θ_3, θ_4, θ_5, and θ_6) were first separated, following which the values of θ_1 to θ_6 were solved sequentially [14]. The detailed solution process is presented as follows:

$$ \theta_1 = atan2(P_y, P_x) - atan2\left(L_1 - L_3, \pm\sqrt{P_x^2 + P_y^2 - (L_1 - L_3)^2}\right) \quad (3) $$

$$ \theta_2 = atan2(-L_5P_z - L_4P_ys_1 - L_4P_xc_1, L_5P_ys_1 - L_4P_z + L_5P_xc_1) - \theta_3 \quad (4) $$

$$ \theta_3 = atan2(L_5, L_4) - atan2\left(K, \pm\sqrt{L_4^2 + L_5^2 - K^2}\right) \quad (5) $$

$$ \theta_4 = atan2(r_{23}c_1 - r_{13}s_1, -r_{13}c_1c_{23} - r_{23}s_1c_{23} + r_{33}s_{23}) \quad (6) $$

$$ \theta_5 = atan2(p, q) \quad (7) $$

$$ \theta_6 = atan2(e, r) \quad (8) $$

wherein:

$$ K = \frac{P_x^2 + P_y^2 + P_z^2 + 2L_1L_3 - L_1^2 - L_2^2 - L_3^2 - L_4^2 - L_5^2}{2L_2} $$

$$ p = r_{13}(c_1c_4c_{23} + s_1s_4) + r_{23}(c_4s_1c_{23} - c_1s_4) + r_{33}c_4s_{23} $$

$$ q = r_{13}c_1s_{23} + r_{23}s_1s_{23} + r_{33}c_{23} $$

$$ e = r_{11}(c_4s_1 - c_1c_{23}s_4) - r_{21}(c_1c_4 + s_1s_4c_{23}) + r_{31}s_4s_{23} $$

$$ r = r_{11}(c_5(s_1s_4 + c_1c_4c_{23})) - r_{31}(s_5c_{23} + c_4c_5s_{23}) - r_{21}(c_5(c_1s_4 - c_4s_1c_{23}) + s_1s_5s_{23}) $$

where s_i, c_i, s_{ij}, c_{ij} denote $sin(\theta_i)$, $cos(\theta_i)$, $sin(\theta_i + \theta_j)$, $cos(\theta_i + \theta_j)$, respectively, where $(i, j = 1, 2, 3, 4, 5)$. The remaining symbols have the same meanings as those in Eqs. (1) and (2).

2.4 Forward Kinematics Analysis of a Mobile Chassis

First, the world coordinate system *XOY* and the chassis coordinate system *X'O'Y'* are established, as depicted in Fig. 4.

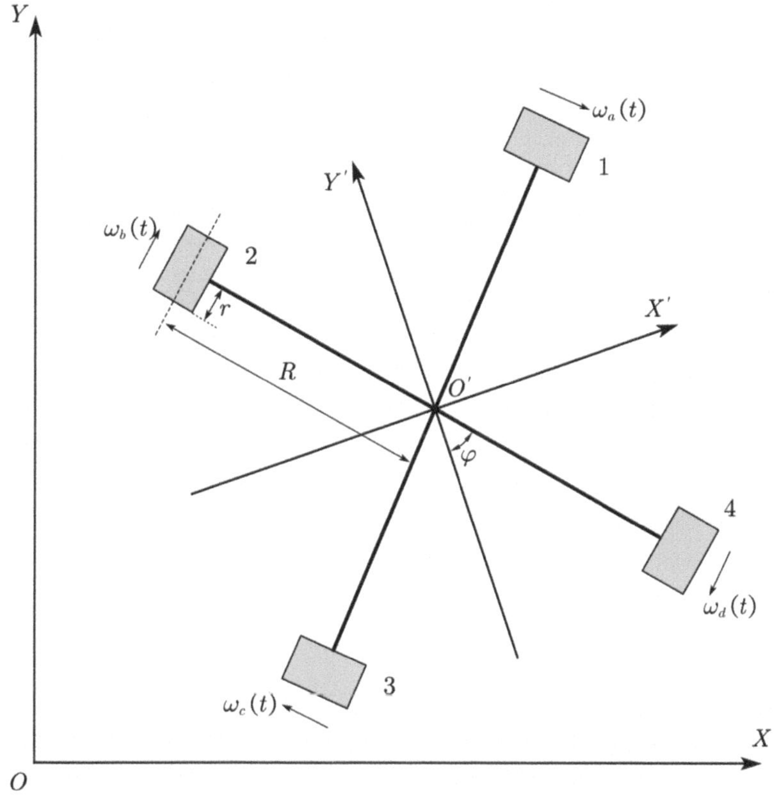

Fig. 4. Pose Schematic of the 4-Wheel Omnidirectional Mobile Chassis

Then, the velocity $V_{X'}(t)$ of the mobile chassis along the X' direction, the velocity $V_{Y'}(t)$ along theY'direction, and the rotational velocity $\omega_z(t)$ around the chassis center are respectively expressed as follows:

$$V_{X\prime}(t) = \frac{1}{4}(\omega_a(t)rcos(\varphi) + \omega_b(t)rcos(\varphi) - \omega_c(t)rcos(\varphi) - \omega_d(t)rcos(\varphi)) \tag{9}$$

$$V_{Y\prime}(t) = \frac{1}{4}(-\omega_a(t)rsin(\varphi) + \omega_b(t)rsin(\varphi) + \omega_c(t)rsin(\varphi) - \omega_d(t)rsin(\varphi)) \tag{10}$$

$$\omega_z(t) = \frac{r}{4R}(-\omega_a(t) - \omega_b(t) - \omega_c(t) - \omega_d(t)) \tag{11}$$

where $\omega_i(t)$, where $(i = a, b, c, d)$ denotes the angular velocity of the four wheels, R denotes the distance from the chassis center to the wheel mounting position, and φ denotes the wheel mounting angle (the angle with the $Y\prime$-axis).

Then, the pose of the mobile chassis in the *XOY* coordinate system at time *t* is given by:

$$X_t(t) = \int_0^t \left[V_{X'}(t)cos\left(\int_0^t \omega_z(t)dt \right) - V_{Y'}(t)sin\left(\int_0^t \omega_z(t)dt \right) \right] dt \tag{12}$$

$$Y_t(t) = \int_0^t \left[V_{X'}(t)sin\left(\int_0^t \omega_z(t)dt \right) + V_{Y'}(t)cos\left(\int_0^t \omega_z(t)dt \right) \right] dt \tag{13}$$

$$\theta_t(t) = \int_0^t \omega_z(t)dt \tag{14}$$

Therefore, the homogeneous transformation matrix from the world coordinate system *XOY* to the coordinate system *X'O'Y'* of the mobile chassis is as follows:

$$_{O'}{}^{O}T = \begin{bmatrix} cos(\theta_t(t)) \; -sin(\theta_t(t)) \;\; 0 \; X_t(t) \\ sin(\theta_t(t)) \;\; cos(\theta_t(t)) \;\; 0 \; Y_t(t) \\ 0\,0 \qquad\qquad 1\,0 \\ 0\,0 \qquad\qquad 0\,1 \end{bmatrix} \tag{15}$$

2.5 Inverse Kinematics Analysis of Moving Chassis

Initially, a coordinate system dedicated to the mobile chassis of the composite robot is constructed, as depicted in Fig. 5.

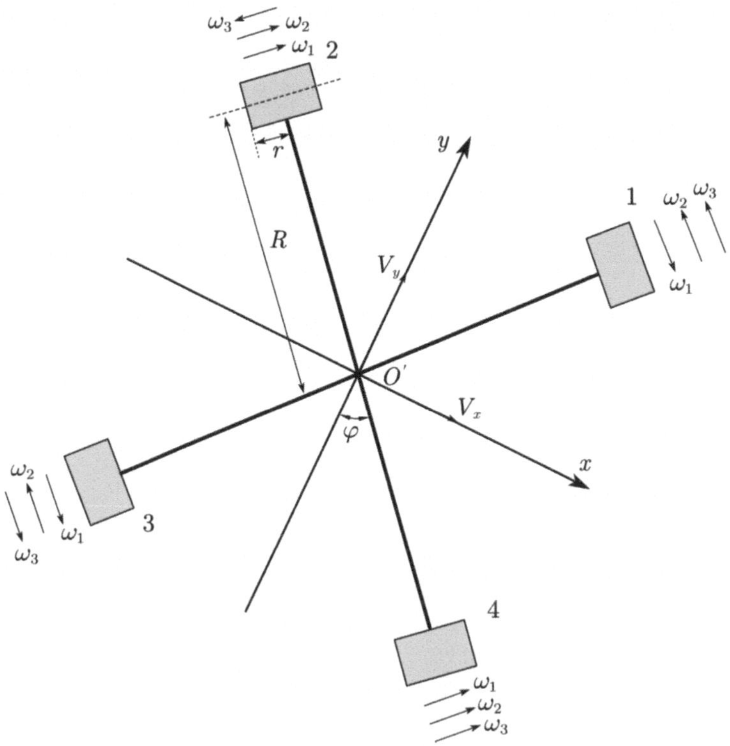

Fig. 5. Inverse Kinematics Analysis Diagram of the 4-Wheel Omnidirectional Mobile Chassis

When the mobile platform is in general motion, its motion can be decomposed into linear motion along the x-axis and y-axis, as well as rotational motion around its center. The counterclockwise direction is defined as the positive direction, from which the following can be derived:

$$\begin{bmatrix} V_X \\ V_Y \\ \omega_Z \end{bmatrix} = \begin{bmatrix} V_X \\ 0 \\ 0 \end{bmatrix} + \begin{bmatrix} 0 \\ V_Y \\ 0 \end{bmatrix} + \begin{bmatrix} 0 \\ 0 \\ \omega_Z \end{bmatrix} \tag{16}$$

where V_X, V_Y, and ω_Z denote respectively the translational velocity along the x-axis, the translational velocity along the y-axis, and the angular velocity about the coordinate origin.

In the figure, ω_i (where $i = 1, 2, 3$) respectively represents the angular velocity of the four wheels corresponding to three decomposed motion states of the mobile chassis: linear motion along the x-axis, linear motion along the y-axis, and rotation around its own center. For each individual decomposed motion state, the angular velocities of the four wheels are equal; the positive direction of the angular velocity is defined as the direction where the thumb of the right-hand rule points outward along the wheel center.

$$V_x = \omega_1 rcos(\varphi) \tag{17}$$

$$V_y = \omega_2 r sin(\varphi) \tag{18}$$

$$\omega_z = \frac{\omega_3 r}{R} \tag{19}$$

So we can invert this for $\omega_1, \omega_2, \omega_3$:

$$\omega_1 = \frac{V_x}{rcos(\varphi)} \tag{20}$$

$$\omega_2 = \frac{V_y}{rsin(\varphi)} \tag{21}$$

$$\omega_3 = \frac{\omega_z R}{r} \tag{22}$$

So we can find the relationship between $\omega_a, \omega_b, \omega_c, \omega_d$ and $V_x, V_y, , \omega_z$:

$$\omega_a = \omega_1 - \omega_2 - \omega_3 = \frac{V_x}{rcos(\varphi)} - \frac{V_y}{rsin(\varphi)} - \frac{\omega_z R}{r} \tag{23}$$

$$\omega_b = \omega_1 + \omega_2 - \omega_3 = \frac{V_x}{rcos(\varphi)} + \frac{V_y}{rsin(\varphi)} - \frac{\omega_z R}{r} \tag{24}$$

$$\omega_c = -\omega_1 + \omega_2 - \omega_3 = -\frac{V_x}{rcos(\varphi)} + \frac{V_y}{rsin(\varphi)} - \frac{\omega_z R}{r} \tag{25}$$

$$\omega_d = -\omega_1 - \omega_2 - \omega_3 = -\frac{V_x}{rcos(\varphi)} - \frac{V_y}{rsin(\varphi)} - \frac{\omega_z R}{r} \tag{26}$$

2.6 Forward Kinematics Analysis of Composite Robot

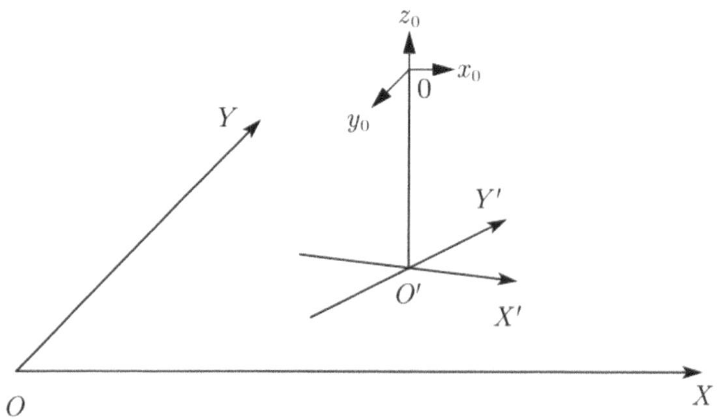

Fig. 6. Global Coordinate System of the Composite Robot (World, Chassis, and Robotic Arm Base Coordinates)

The mobile chassis has 3 degrees of freedom (DOFs), including 2 translational DOFs and 1 rotational DOF. Due to the high coupling degree of this configuration, its kinematic analysis process is relatively complex. To simplify the calculation, the motion of the mobile chassis can be decoupled into 2 translational motions and 1 rotational motion [15]. As shown in Fig. 6, XOY is the world coordinate system, $X'O'Y'$ is the mobile chassis coordinate system, and $x_0y_0z_0$ is the robotic arm base coordinate system; among them, the robotic arm base coordinate system only has a translational transformation relative to the mobile chassis coordinate system along the Z-axis (with a translation distance of l_0), and its homogeneous transformation matrix can be expressed as:

$$
{}_0^{O'}T = \begin{bmatrix} 1\,0 & 0\,0 \\ 0\,1 & 0\,0 \\ 0\,0 & 1\;l_0 \\ 0\,0 & 0\;1 \end{bmatrix} \tag{27}
$$

Based on Eqs. (2), (15), and (27), the overall forward kinematic homogeneous transformation matrix is derived as follows:

$$
{}_6^{O}T = {}_{O'}^{O}T\,{}_0^{O'}T\,{}_6^{0}T \tag{28}
$$

2.7 Motion Error Compensation for the Mobile Chassis

Affected by the constraints of the experimental site and the load deviation caused by the inertia of the robotic arm, the mobile chassis of the composite robot exhibits significant motion errors during linear movement, thus necessitating motion compensation. Given the symmetry in the motion of the mobile chassis, compensation only needs to be implemented for linear motion in a single direction. It is assumed that in Fig. 5, the angular velocity of wheel 1 is $-\widehat{\omega}_1$ and that of wheel 3 is $\widehat{\omega}_1$ (The positive direction of the wheel is consistent with that described in Sect. 2.5), while the angular velocity of wheel 2 is $\widehat{\omega}_2$ and that of wheel 4 is $-\widehat{\omega}_2$. Under these conditions, the theoretical motion direction of the mobile chassis is illustrated in Fig. 7.

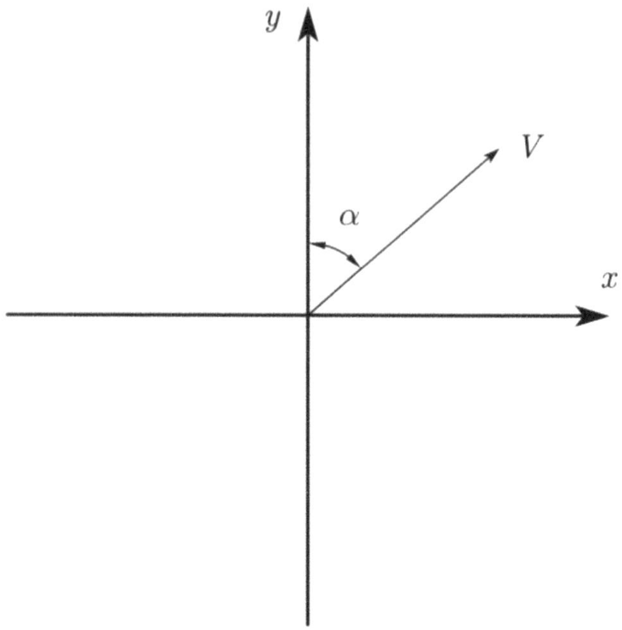

Fig. 7. Actual Motion Direction of the Omnidirectional Wheel Chassis

The angle α in the figure can be calculated by the following formula:

$$\alpha = tan^{-1}(cot(\varphi)\frac{1 - \frac{\widehat{\omega}_1}{\widehat{\omega}_2}}{1 + \frac{\widehat{\omega}_1}{\widehat{\omega}_2}}) \tag{29}$$

wherein

$$-|\widehat{\omega}_2| \leq \widehat{\omega}_1 \leq |\widehat{\omega}_2|$$

$$\widehat{\omega}_2 > 0$$

where φ has the same meaning as that in Eqs. (9) and (10).

However, due to the effects of wheel slippage and the motion inertia of the robotic arm, the actual motion trajectory of the mobile chassis will deviate, and its actual motion trajectory is shown in Fig. 8.

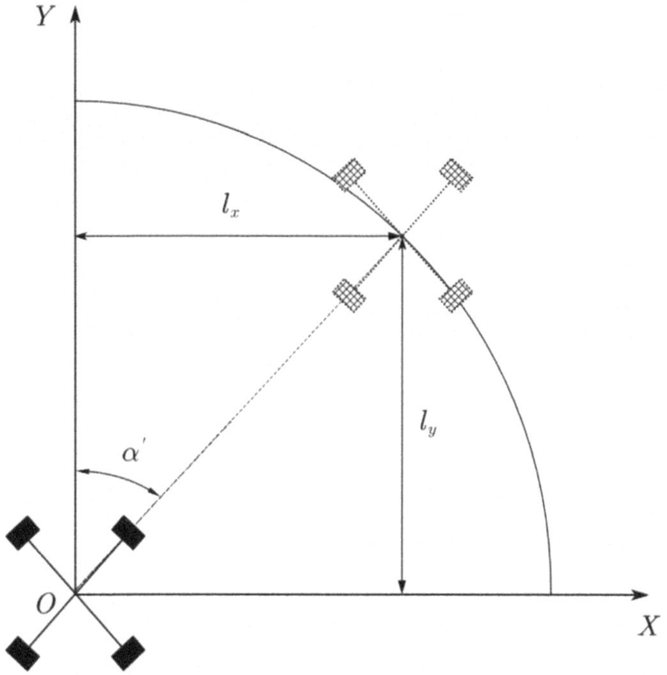

Fig. 8. Actual Motion Trajectory of 4-Wheel Omnidirectional Chassis (Slip & Inertial Disturbances; Dashed Line)

In Fig. 9, l_x and l_y denote the distances from the center of the chassis to the horizontal axis and vertical axis, respectively, and R is the distance from the center of the chassis to the coordinate origin. The actual angle $\alpha\prime$ can be calculated by the following formula:

$$\alpha' = \begin{cases} sin^{-1}\left(\frac{l_y}{R}\right), \frac{\hat{\omega}_1}{\hat{\omega}_2} \geq 0 \\ cos^{-1}\left(\frac{l_x}{R}\right), \frac{\hat{\omega}_1}{\hat{\omega}_2} < 0 \end{cases} \tag{30}$$

Let the deviation coefficient be $a = \frac{\alpha\prime}{\alpha}$ (where $\alpha \neq 0$, and for $\alpha = 0$, it is calculated using linear interpolation), then the value of a can be calculated by Eqs. (29) and (30).

$$a = \frac{\alpha'}{\alpha} = \begin{cases} \dfrac{sin^{-1}\left(\frac{l_y}{R}\right)}{tan^{-1}(cot\varphi)\frac{1-\frac{\hat{\omega}_1}{\hat{\omega}_2}}{1+\frac{\hat{\omega}_1}{\hat{\omega}_2}})}, \frac{\hat{\omega}_1}{\hat{\omega}_2} \geq 0 \\ \dfrac{cos^{-1}\left(\frac{l_x}{R}\right)}{tan^{-1}(cot(\varphi)\frac{1-\frac{\hat{\omega}_1}{\hat{\omega}_2}}{1+\frac{\hat{\omega}_1}{\hat{\omega}_2}})}, \frac{\hat{\omega}_1}{\hat{\omega}_2} < 0 \end{cases} \tag{31}$$

3 Motion Compensation Experiment of the Mobile Chassis

3.1 Introduction to the Experimental Platform

The composite robot platform adopted in this experiment is mainly composed of two parts: a robotic arm and a mobile chassis. Its core hardware configuration is as follows: In the joint drive system of the robotic arm, joints 2 and 3 use 180° servos with a torque of 35 kg.cm, while joints 1, 4, 5, and 6 are equipped with 270° servos with a torque of 12 kg.cm, so as to meet the load requirements and motion range specifications of different joints. The drive unit of the mobile chassis adopts 42 stepper motors to ensure precise control of the chassis motion. The control core of the entire system is an Arduino Mega 2560 microcontroller, which is responsible for coordinating the collaborative operation of the robotic arm and the mobile chassis. The above hardware configuration provides a stable hardware foundation for the motion control of the composite robot and subsequent error compensation experiments.

3.2 Experimental Methods and Procedures

As shown in Fig. 9, the mobile chassis starts from the coordinate origin and moves linearly to the end point on a circular arc with radius R. Due to the difficulty in directly measuring the actual angle $\alpha\prime$, its value is determined indirectly by measuring l_x and l_y. In the experiment, the ratio $\frac{\widehat{\omega_1}}{\widehat{\omega_2}}$ is set to take the values of $\frac{3}{4}, \frac{1}{2}, \frac{1}{4}, 0, -\frac{1}{4}, -\frac{1}{2}, -\frac{3}{4}$, and -1 (When $\frac{\widehat{\omega_1}}{\widehat{\omega_2}} = 1$, the theoretical angle $\alpha = 0$, and it is calculated using linear interpolation in this case.). The mobile chassis is activated separately under each of the above ratio values. After the mobile chassis reaches the end point, the corresponding l_x and l_y values are measured. Subsequently, each $\frac{\widehat{\omega_1}}{\widehat{\omega_2}}$ ratio and its corresponding l_x and l_y values are substituted into Eq. (31) to calculate the compensation coefficient a, and the results are filled into Tables 2 and 3 respectively.

Table 2. Correction Coefficient α for Wheel Angular Velocity Ratio $\frac{\omega_1}{\omega_2} \geq 0$

$\frac{\omega_1}{\omega_2}$	α	l_y	α'	a
$\frac{3}{4}$	9.02°	0.66 m	12.71°	1.41
$\frac{1}{2}$	20.31°	0.81 m	15.66°	0.77
$\frac{1}{4}$	33.68°	1.28 m	25.26°	0.75
0	48°	2.22 m	47.73°	0.99

Table 3. Correction Coefficient α for Wheel Angular Velocity Ratio $\frac{\omega_1}{\omega_2} < 0$

$\frac{\omega_1}{\omega_2}$	α	l_x	α'	a
$-\frac{1}{4}$	61.62°	1.02 m	70.12°	1.14
$-\frac{1}{2}$	73.29°	0.63 m	77.88°	1.06
$-\frac{3}{4}$	82.67°	0.48 m	80.79°	0.98
-1	90°	0.40 m	82.34°	0.91

After calibrating the aforementioned correction coefficients through experiments, the remaining ones can be derived using linear interpolation. To verify the effectiveness of the proposed compensation strategy in correcting trajectory deviations, comparative experiments were further conducted in a standard experimental site. The experimental setup is as follows: First, the chassis of the omnidirectional wheeled mobile robot (OWMRs) was positioned across the yellow line on the ground, with its geometric center aligned with the intersection of the ground tile gaps—this served as the movement starting point. Meanwhile, the manipulator was placed in an upwardly extended state to simulate the inertial disturbances generated during the chassis movement (as shown in Fig. 9).

Fig. 9. Initial Pose of the Composite Robot at the Motion Experiment Starting Point

Subsequently, the robot was controlled to move along the yellow line to the fourth intersection of tile gaps, which was designated as the endpoint. Among them, Figs. 10 and 11 respectively illustrate the actual deviations of the chassis from the endpoint under the uncorrected and corrected conditions.

Fig. 10. The pose of the robot at the endpoint in the absence of correction coefficients

Fig. 11. The pose of the robot at the endpoint with correction coefficients

Based on Eq. 30, the angle $\alpha\prime$ under the uncorrected and corrected conditions mentioned above can be determined, with the results presented in Table 4.

Table 4. The values of angle α' under the uncorrected and corrected conditions

Project	α	α'
Without the correction value	0°	8.63°
With the correction value	0°	1.27°

3.3 Experiment Conclusion

Experimental results show that without applying the correction coefficient, the deviation angle of the omnidirectional wheeled chassis is 8.63° with a deviation rate of 18% (considering the scenario where Wheels 1 and 3 are completely ineffective, the deviation rate is calculated based on a reference of 48°). When the correction coefficient is applied, the deviation angle of the omnidirectional wheeled chassis is 1.27° with a deviation rate of 2.65%. Thus, after adopting the proposed compensation strategy, the trajectory deviation of the omnidirectional wheeled mobile robots (OWMRs) is reduced by 85.28%, and this difference is visually illustrated in Figs. 10 and 11. This result confirms that the proposed method can effectively suppress trajectory deviations caused by wheel slippage and inertial disturbances of the manipulator. These experimental data verify that this strategy can meet the trajectory accuracy requirements in educational scenarios, providing experimental evidence for the reliable application of OWMRs in kinematics teaching demonstrations and student practices.

4 Conclusions

This study addresses the trajectory deviation issue of omnidirectional wheeled mobile robots (OWMRs) in educational scenarios—a deviation induced by the coupling effect of omnidirectional wheel slippage and inertial disturbances from the manipulator. To tackle this, a kinematic theoretical framework for the composite robot is constructed, and a simplified kinematic compensation strategy is proposed.

First, using methods such as homogeneous transformation matrices and dual-coordinate system transformation, the forward/inverse kinematic models of the manipulator and the omnidirectional wheeled chassis are derived separately, and then integrated to obtain the overall forward kinematic model of the composite robot. Subsequently, correction coefficients under different wheel angular velocity ratios are experimentally calibrated, and linear interpolation is adopted to ensure the continuity of compensation. An experimental platform with Arduino Mega 2560 as the control core was built; comparative tests in standard environments show that the proposed strategy significantly reduces the robot's trajectory deviation, effectively suppresses errors caused by wheel slippage and manipulator inertial disturbances, and fully meets the trajectory accuracy requirements of educational scenarios.

This study provides key theoretical support and experimental evidence for the reliable application of OWMRs in educational scenarios such as kinematics teaching demonstrations and student practices, enriching the error compensation technology system in the field of educational robotics. Future work will further optimize the compensation strategy by incorporating compensation for the coupling of rotational and linear motions, expand experimental validation in complex interference conditions and multi-task collaboration scenarios, and promote the in-depth application and performance enhancement of OWMRs in the educational domain.

References

1. Fu, X., Yang, J., Wang, S.: Research on structural layout and control of omnidirectional wheeled mobile robots(in Chinese). In: Shenyang Municipal Party Committee of the Communist Party of China, Shenyang Municipal People's Government (eds.) Proceedings of the 14th Shenyang Academic Annual Conference (Science, Engineering, Agriculture and Medicine), pp. 560–565. School of Electrical Engineering, Shenyang University of Technology, Shenyang (2017)
2. Balakrishna, R., Ghosal, A.: Modeling of slip for wheeled mobile robots. IEEE Trans. Robot. Autom. **11**(1), 126–132 (1995)
3. Williams, et al.: Dynamic Model With Slip for Wheeled Omnidirectional Robots. IEEE Transactions on Robotics & Automation (2002)

4. Wang, C., Wang, D., Peng, Z.: Extended state observer based anti-disturbance tracking control for omnidirectional mobile robots subject to uncertainties and wheel skidding. IEEE (2021). https://doi.org/10.1109/CACRE52464.2021.9501304
5. Chen, P.J., Yang, S.Y., Chen, Y.P., et al.: Slip estimation and compensation control of omnidirectional wheeled automated guided vehicle. Electronics **10**(7), 840 (2021). https://doi.org/10.3390/electronics10070840
6. Cao, G., Zhao, X., Ye, C., et al.: Fuzzy adaptive PID control method for multi-mecanum-wheeled mobile robot. J. Mech. Sci. Technol. **36**(4), 1 (2022). https://doi.org/10.1007/s12206-022-0337-x
7. Liu, Y., Zhu, J.J., Ii, R.L.W., et al.: Omni-directional mobile robot controller based on trajectory linearization. Robot. Auton. Syst. **56**(5), 461–479 (2008). https://doi.org/10.1016/j.robot.2007.08.007
8. Phunopas, A., Inoue, S.: Motion improvement of four-wheeled omnidirectional mobile robots for indoor terrain. J. Robot. Netw. Artif. Life **4**(4), 275 (2018). https://doi.org/10.2991/jrnal.2018.4.4.4
9. Fu, J., Hill, R.C.: Development of a Multi-Stage Control Strategy for Omnidirectional Mobility in the Presence of Slip. In: 61st IEEE International Midwest Symposium on Circuits and Systems. IEEE (2018). https://doi.org/10.1109/MWSCAS.2018.8624090
10. Long, S., Terakawa, T., Yogou, R.K.M.: Kinetic analysis of active omni wheel with barrel-shaped rollers for avoiding slippage and vibration. J. Mech. Robot. Trans. ASME **16**(5), (2024)
11. Chen, H., Li, L.S.: Research on forward and inverse solution of PUMA560 Robot based on MATLAB. Manuf. Autom. **40**(12), 34–36 (2018)
12. Guo, H.Y.: Research and Design of Modular Material Transfer Platform Based on Omnidirectional Wheel. D, Jiangsu University of Science and Technology, Zhenjiang (2022)
13. Wei, F.X., Sun, M.G., Zhou, H.Y.: Based on MATLAB analysis and verification of the six axis of mechanical arm kinematics. J. Jilin Inst. Chem. Technol. **38**(11), 59–62+79 (2021)
14. Zhu, Q.H.: Six Axis Industrial Robot Motion Planning Study. D, Nanjing University of Information Science and Technology, Nanjing (2022)
15. Sgao, H.: Research on Design, Positioning and Path Planning of Wheeled Mobile Manipulator. D, Xi'an University of Technology, Xi'an (2017)

A Smart Blind Assistance System Based on MobileViT

Xinshuo Wang[1], Pengcheng Wang[1], and Chun Zhao[1,2(✉)]

[1] Honor College, Beijing Information Science and Technology University, Beijing, China
[2] School of Computer, Beijing Information Science and Technology University, Beijing, China
zhaochun@bistu.edu.cn

Abstract. To address the challenges faced by visually impaired individuals in information acquisition and environmental perception in daily life by designing and implementing a human-machine collaborative blind assistance system. This paper presents a smart blind assistance system based on MobileViT. The system includes a set of functions such as road recognition, deviation warning, and obstacle notification. Furthermore, a volunteer assistance platform is developed to ensure visually impaired users receive timely and accurate human guidance in complex environments. Experimental results demonstrate that the system significantly enhances the safety and independence of visually impaired individuals during travel.

Keywords: visual impairment assistance · road deviation detection · remote assistance

1 Introduction

Visual impairment severely affects people's quality of life. Globally, there are approximately 285 million individuals with visual impairments, including 39 million who are completely blind [1]. Visually impaired individuals face numerous challenges in daily travel, such as recognizing obstacles, determining directions, and avoiding dangers. Traditional assistive tools, such as guide dogs and white canes, have many limitations. For example, guide dogs require long training periods, are scarce, and involve high costs, while white canes have limited functionality and bulky designs. With the advancement of computer vision and artificial intelligence technologies, intelligent assistive devices for the visually impaired have gradually become a research topic [2]. However, most existing intelligent assistive systems rely on a single technical approach and exhibit poor environmental adaptability, often failing to provide comprehensive and reliable assistance in complex environments.

Typical systems—such as Microsoft Seeing AI, OrCam MyEye, and the WeWALK smart cane—perform well in specific scenarios. Microsoft Seeing AI

H. Wei and J. Niu (Eds.): COBOT 2025, CCIS 2893, pp. 234–252, 2027.
https://doi.org/10.1007/978-981-92-2074-8_14

identifies objects and text using smartphone cameras, OrCam MyEye reads textual information in real time, and the WeWALK smart cane integrates ultrasonic sensors for obstacle detection. However, these systems suffer from low recognition accuracy and poor environmental adaptability in complex, dynamic real-world environments.

In terms of human-assisted solutions, applications such as Be My Eyes connect visually impaired users with volunteers for remote support. However, the platforms lack effective integration with smart hardware devices, making it difficult to ensure comprehensive travel assistance [3,4].

Gurari et al. propose the VizWiz Grand Challenge [5], establishing the first large-scale visual question answering dataset by the real needs of visually impaired users, which provides a valuable research platform for integrating smart question answering with assistive applications. Nevertheless, research that organically combines artificial intelligence with human assistance remains relatively limited.

Since 2023, research on human–machine collaborative assistive systems for blind and low-vision users has accelerated, highlighting hybrid pipelines that integrate on-device perception with cloud or human-in-the-loop support. Recent surveys synthesize AI-centric and communication-enhanced approaches, reporting growing adoption of lightweight deep models, multimodal sensing, and privacy-aware data exchange [6,7]. Concurrent studies explore deep transfer learning for robust navigation and object awareness under real-world constraints [8], and wearable assistance that combines real-time vision with large vision–language models to deliver contextual guidance and dialog-based interaction [9]. These trends substantiate our dual-layer design that couples automated detection with remote human assistance.

This paper proposes a blind assistance system that integrates hardware intelligent detection and artificial remote assistance, presents a human-machine collaboration mode, and provides a dual guarantee mechanism for visually impaired people.

The main contributions of this paper are as follows:

- A novel human–machine collaboration model that combines intelligent hardware detection with remote human assistance, compensating for the shortcomings of single-technology approaches and providing enhanced perception through multimodal vision models in critical scenarios.
- Adoption of the MobileViT model in the real-time decision system, balancing computational efficiency and inference speed.
- Introducing a trip publication, volunteer selection, and credit evaluation mechanism, enabling precise and timely human assistance based on geographical and temporal factors.

2 System Design and Methodology

The key technologies of this work are as follows:

Design and implementation of a road deviation detection system based on computer vision and deep learning. The system employs the BiSeNet V2 semantic segmentation model combined with geometric center estimation [10] [11] to achieve real-time and accurate pedestrian pathway recognition. Development of a real-time decision-making system based on image classification models, with a comparative performance analysis of five models including ResNet50, EfficientNet–B0, DenseNet121, MobileNetV3, and MobileViT [12–16]. Design of a multifunctional intelligent hardware device, integrating a Raspberry Pi 5B with a high-definition camera, GNSS positioning module, ultrasonic sensors, and a voice interaction unit. The device supports multiple functions such as road deviation detection, obstacle alert, and navigation assistance. Construction of a volunteer-based remote assistance platform, utilizing WebSocket and multi-handshake mechanisms to achieve low-latency bidirectional communication [17–19]. A Flutter-based mobile client enables volunteers to browse user trips and provide real-time guidance, thereby realizing a human–machine collaborative assistance model. Design of a family-binding and safety zone monitoring mechanism, enhancing system safety and reliability. When users approach complex traffic conditions, the system can invoke multimodal vision AI models (e.g., traffic light recognition, scene description, text reading) on demand to enhance environmental perception.

2.1 System Architecture

This study adopts a three-layer architecture consisting of the hardware terminal, software platform, and cloud server. The design draws upon existing research and implementation experience from domestic and international assistive technologies for the visually impaired as shown in Fig. 1.

Hardware Terminal Layer. Composed of fundamental components such as development boards, sensors, and batteries, this layer is responsible for acquiring and processing environmental information and generating corresponding voice prompts.

Software Platform Layer. Provides a volunteer-assisted remote support platform and a family care system. While enabling human-in-the-loop assistance for visually impaired users, it also provides family members with real-time information such as the user's location.

Cloud Server Layer. Built on Huawei Cloud, this layer handles data storage and command forwarding. It employs the WebSocket protocol to achieve low-latency communication between hardware and software terminals.

The key functional modules of the system include a road deviation detection module, real-time decision-making module, navigation assistance module, remote assistance module, and family binding module. Information exchange

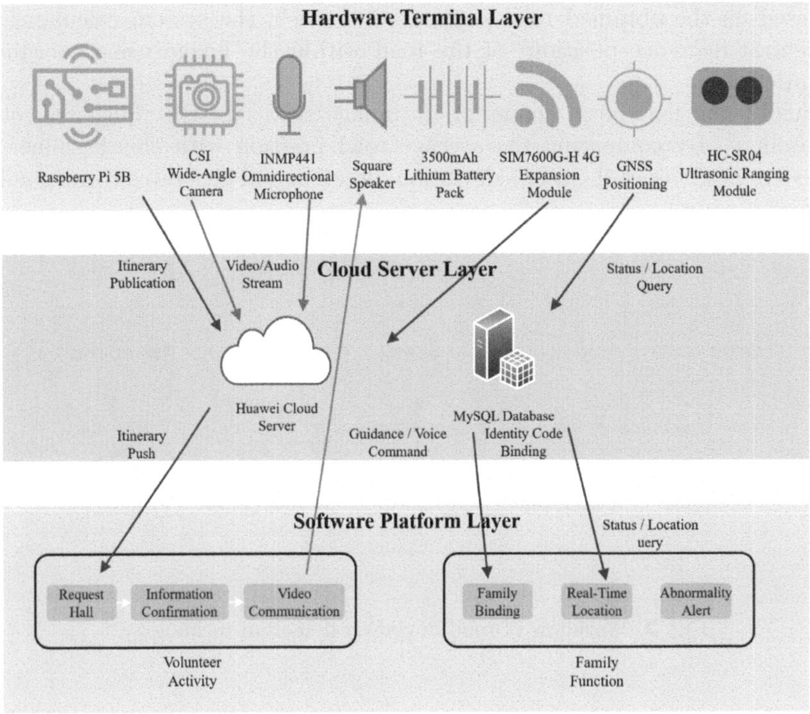

Fig. 1. System Overall Architecture.

between modules occurs through data buses and network communication, forming an integrated system that provides comprehensive travel assistance for visually impaired users.

2.2 Road Deviation Detection Technology

Road deviation detection is a key component ensuring travel safety for visually impaired users. Its primary objective is to accurately identify pedestrian road areas within images and compute the positional deviation relative to the user, providing explicit directional adjustment guidance. The proposed method adopts the BiSeNet V2 (Bilateral Segmentation Network) semantic segmentation model, combined with coordinate-based computation and offset judgment logic, to achieve an automated end-to-end process—from image acquisition to directional sound feedback, as shown in Fig. 2. The technical workflow is as follows [10]:

1. The camera first captures the forward-facing road image.
2. The BiSeNet V2 model performs semantic segmentation to classify pixels at the image level.

3. Based on the obtained road segmentation mask, the system calculates the "central reference position" of the road within the image using coordinate statistics.
4. Finally, the horizontal center of the image serves as the "ideal trajectory baseline". By comparing the average road position with this baseline, the system determines the user's deviation direction and generates corresponding voice prompts.

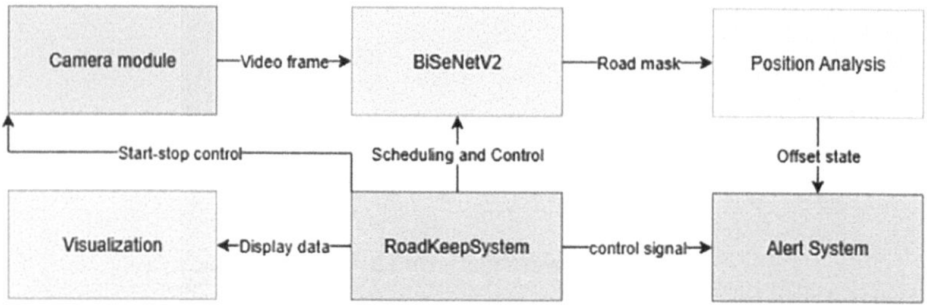

Fig. 2. Workflow of road deviation detection technology.

2.3 Real-Time Decision-Making Method

To meet the requirements of real-time performance and reliability in guidance scenarios, we compare five models: ResNet50, EfficientNet-B0, DenseNet121, MobileNetV3, and MobileViT. The selection covers accuracy–efficiency trade-offs under wearable-device constraints: ResNet50 serves as a standard accuracy benchmark, DenseNet121 and EfficientNet-B0 reflect parameter efficiency and scalability, and MobileNetV3 together with MobileViT ensure feasibility for real-time edge deployment. A custom dataset is constructed with five classes: stairs, intersection, pole, normal road, and obstacle, containing a total of 841 images. All models are trained and validated using the same specifications (5 classes, 320×240 pixels), and were tested on the same hardware environment.

Images were acquired with a wearable camera across diverse scenes, covering indoor and outdoor environments; daytime, evening, and night conditions; clear, overcast, and rainy weather; varied crowd densities; and both static and dynamic motion. Frames were sampled at fixed intervals and near-duplicate sequences were filtered to minimize redundancy. Class definitions were specified a priori: Intersection required visible junction geometry and potential traffic conflict; Normal road demanded a clear pedestrian path without salient obstacles; Obstacle included stationary and moving impediments such as poles and pedestrians; Stairs and Pole followed their literal presence with sufficient visual evidence. Borderline cases were adjudicated via category precedence to ensure consistent labeling. Labels were produced by two independent annotators with conflict resolution by a third reviewer, and inter-annotator agreement was assessed on a held-out subset using Fleiss' kappa to confirm consistency across classes.

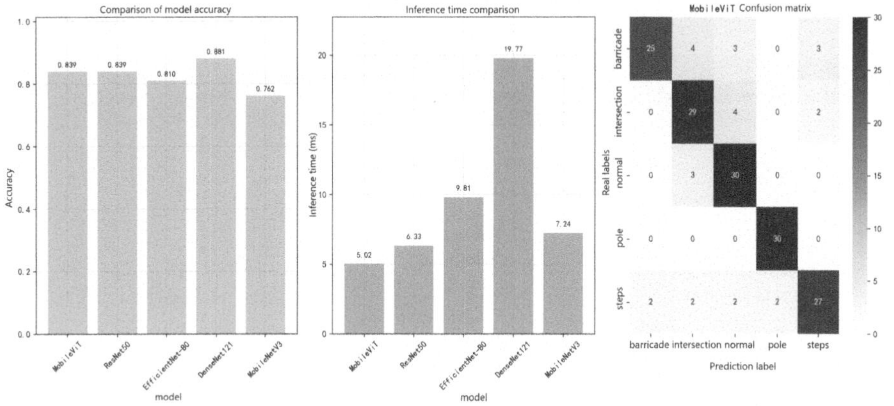

Fig. 3. Model Performance Comparison.

As shown in Fig. 3, using accuracy and per-frame inference latency as evaluation metrics, DenseNet121 achieves the highest accuracy (0.881), while MobileViT demonstrates superior real-time performance with an inference delay of only 5.02 ms/frame (199 FPS)—significantly outperforming ResNet50 (6.33 ms/frame, 158 FPS), MobileNetV3 (7.24 ms/frame, 138 FPS), EfficientNet-B0 (9.81 ms/frame, 102 FPS), and DenseNet121 (19.77 ms/frame, 50.6 FPS).

Considering both accuracy and latency, MobileViT was selected as the primary visual backbone of the guidance system for its high responsiveness and throughput while maintaining acceptable recognition performance. The choice is made to optimize the accuracy–latency–energy trade-off under edge-device constraints rather than to assert absolute accuracy superiority over other backbones. The real-time decision-making system serves as the core decision engine of the entire guidance framework, performing rapid visual analysis and generating user feedback strategies based on recognition results. By distinguishing among various scenarios of stairs, intersections, poles, normal roads, and obstacles, we establish a hierarchical risk evaluation and response mechanism:

1. Low-risk scenarios (normal road): The system maintains silent monitoring to reduce power consumption.
2. Medium-risk scenarios (stairs, poles, obstacles): The system triggers basic voice alerts to warn the user of potential hazards ahead.
3. High-risk scenarios (intersections): The system activates an advanced decision mode, invoking a large-scale vision model for refined analysis.

2.4 Remote Assistance Module Design

The remote assistance module is a critical subsystem that connects visually impaired users with volunteers through the cloud server, enabling a human–machine collaborative assistance mode.

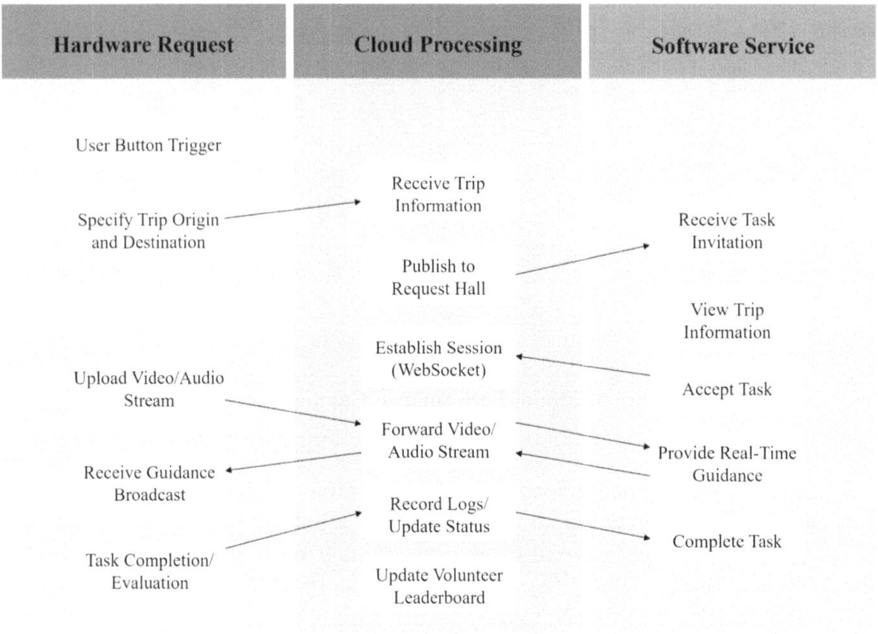

Fig. 4. Timing design of remote assistance module.

As shown in Fig. 4, the remote assistance module is divided into three sides.

Hardware Request Side. Occurs on the device side and includes functions such as trip planning, request publishing, and real-time communication. Users can set destinations via voice commands, after which the system automatically plans routes and provides navigation. During navigation, users may choose to publish trip information to the volunteer hall to request remote support.

Cloud Service Side. The cloud server manages user–volunteer matching, data storage, and video transmission. An intelligent matching algorithm based on geographic proximity and time scheduling prioritizes nearby volunteers to enhance response speed. The system also implements mechanisms such as volunteer credit evaluation and service duration ranking to ensure service quality and encourage participation.

Software Service Side. The Request Hall serves as the main functional module of the mobile application, offering trip browsing, task acceptance, and remote guidance features. Volunteers can review trip requests, including departure, destination, and time, and choose to assist based on their availability. Once a volunteer accepts a task, a real-time video connection is established, allowing the volunteer to access the user's camera feed and provide live voice guidance. The application also includes modules such as My Focus, My Assistance History, and My Family, helping volunteers identify familiar users and build long-term support relationships.

3 System Implementation

3.1 Road Deviation Detection Algorithm

The proposed system adopts a "deep semantic segmentation ± geometric center estimation" strategy to monitor, in real time, the relative deviation of the guiding device from the pedestrian pathway.

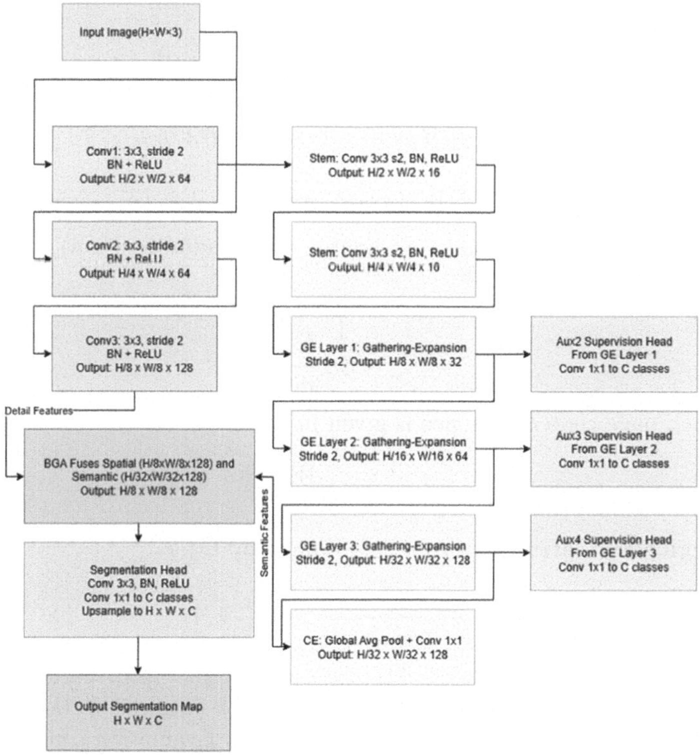

Fig. 5. Flowchart of BiSeNet V2.

As shown in Fig. 5, consecutive image frames captured by the front camera are first processed through the lightweight real-time segmentation network BiSeNet V2, generating a binary road mask of size $H \times W$.

$$road\_mask(i,j) \in 0,1, i \in [1,H], j \in [1,W] \tag{1}$$

where $road\_mask(i,j) = 1$ indicates that pixel (i,j) belongs to the road, and $road\_mask(i,j) = 0$ represents non-road regions. This segmentation output serves as a stable input for subsequent geometric analysis.

In the post-processing stage, only the lower half of the image is selected as the Region of Interest (ROI) to emphasize near-field road information and suppress instability caused by perspective distortion, occlusion, or noise in the distant field. The system computes the proportion of road pixels within this ROI, as shown in Eq. 2.

$$r_{road} = \frac{\sum_{i=h/2}^{H} \sum_{j=1}^{W} road\_mask(i,j)}{(H/2) \times W} \tag{2}$$

and compares it with a predefined threshold $\tau_{road} = 0.4$ to determine whether the device is currently "on the road", as shown in Eq. 3

$$is\_on\_road(bool) = (r_{road} > \tau_{road}) \tag{3}$$

To quantify the lateral position, the algorithm extracts all road pixel indices from the bottom image row $i = H$, which is shown in Eq. 4.

$$\Omega = \{\, j \in [1,W] \mid road\_mask(H,j) = 1 \,\} \tag{4}$$

If $|\Omega| > 0$, the average road coordinate (road center position) is defined in Eq. 5.

$$x_{road} = \frac{1}{|\Omega|} \sum_{j \in \Omega} j \tag{5}$$

and the image center reference is given in Eq. 6.

$$x_{center} = \frac{W}{2} \tag{6}$$

The normalized lateral deviation is obtained by Eq. 7.

$$d = \frac{x_{road} - x_{center}}{x_{center}}, d \in [-1,1] \tag{7}$$

To determine the fixed thresholds τ_{road} and τ_{dev}, a grid search and sensitivity analysis were conducted on the validation set ($\tau_{road} \in [0.20, 0.60]$ with a step of 0.05; $\tau_{dev} \in [0.05, 0.20]$ with a step of 0.01). Evaluation considered road detection accuracy and directional prompt accuracy against human-annotated ground truth, together with the false-positive rate; an additional constraint on prompt frequency was applied to preserve usability. Under the criterion of maximizing accuracy subject to a bounded false-positive rate, the thresholds were set to $\tau_{road} = 0.40$ and $\tau_{dev} = 0.10$. Further analysis indicated stable performance when $\tau_{road} \in [0.35, 0.45]$ and $\tau_{dev} \in [0.08, 0.12]$, supporting the robustness of the selected values.

Which eliminates resolution-related bias and facilitates consistent thresholding across different devices. When $|d| < \tau_{dev} = 0.1$, the device is considered centered on the road. Otherwise, the deviation direction is determined by the sign of $d > 0$ indicates that the device is shifted left relative to the road center (requiring a right correction), while $d < 0$ indicates a rightward shift (requiring a left correction). If no road pixels are found in the bottom row ($|\Omega| = 0$), the direction is set to unknown and a general deviation warning is triggered.

The warning module combines both r_{road} and d indicators to generate specific correction prompts, with a minimum time interval to suppress excessive alerts. Throughout the process, semantic segmentation constitutes the primary computational load, while post-processing involves only indexing and mean-value operations with a time complexity of $O(W)$. The entire pipeline achieves real-time performance exceeding 30 FPS on a Raspberry Pi 5B.

This design ensures both high detection accuracy and computational efficiency, while maintaining engineering deployability. Moreover, it provides flexibility for future improvements—such as replacing the geometric center with connected-component analysis, applying smoothing filters, or introducing adaptive thresholds—to further enhance robustness.

3.2 Real-Time Decision Module

The module uses MobileViT to establish a closed-loop decision-making pipeline consisting of rapid recognition, risk assessment, hierarchical alerting, and on-demand enhancement. MobileViT is a hybrid mobile vision architecture that integrates lightweight convolutional layers with Transformer modules. Its core concept is to extract local textures and edge information efficiently through shallow inverted residual blocks and depthwise separable convolutions, while partitioning mid-to-high-level feature maps into patches and applying lightweight self-attention to capture long-range dependencies and global context. This fusion of local and global representations endows the network with Transformer-level perceptual capabilities while maintaining computational efficiency—making it well-suited for real-time inference on resource-constrained edge devices.

As shown in Fig. 6, camera input is first standardized (resized to 320×240 and normalized to the training distribution), then passed through the MobileViT model for forward inference, producing five semantic environment categories—stairs, intersection, pole, normal, and obstacle—along with confidence scores. The classification head uses global average pooling followed by a linear layer, achieving a balance between inference speed and discriminative power. At the decision layer, risk assessment and hierarchical responses are performed based on class labels and confidence levels. For normal cases, the system maintains silence or low-intensity prompts to reduce user disturbance. For medium-risk

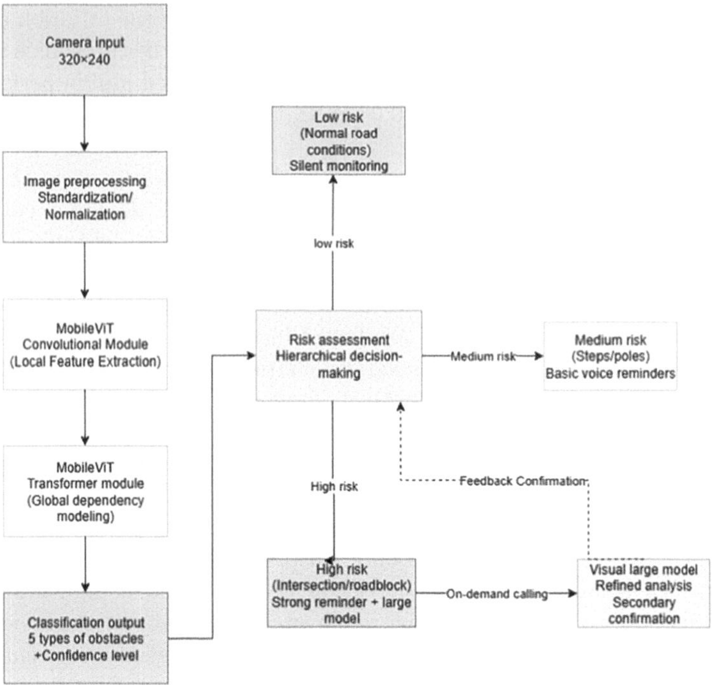

Fig. 6. Flowchart of Decision-Making Module.

scenarios such as stairs and poles, it triggers audio or vibration alerts to prompt deceleration and avoidance. For high-risk scenarios like intersections, it issues strong warnings with directional guidance, invoking an upper-level vision model for secondary confirmation. This mechanism ensures reliability under complex conditions while keeping latency and power consumption within acceptable limits.

3.3 Remote Assistance Interface

According to the design in Sect. 2, we construct the shell of the Blind Assistance System using 3D printing. Meanwhile, relevant software is deployed.

As shown in Fig. 7, the hardware architecture is built around a Raspberry Pi 5B as the main controller, equipped with 8 GB RAM and running a customized Linux operating system. The visual module employs a CSI wide-angle camera (resolution 2592×1944) supporting up to 1080p video capture. Communication and positioning are handled by a SIM7600G-H 4G HAT, which supports 4G/3G/2G connectivity and GNSS-based geolocation—enabling precise positioning for navigation and family monitoring functions. An HC-SR04 ultrasonic ranging sensor (accuracy up to 0.3 cm) is mounted downward to maintain high short-range accuracy and low latency, providing emergency obstacle alerts as

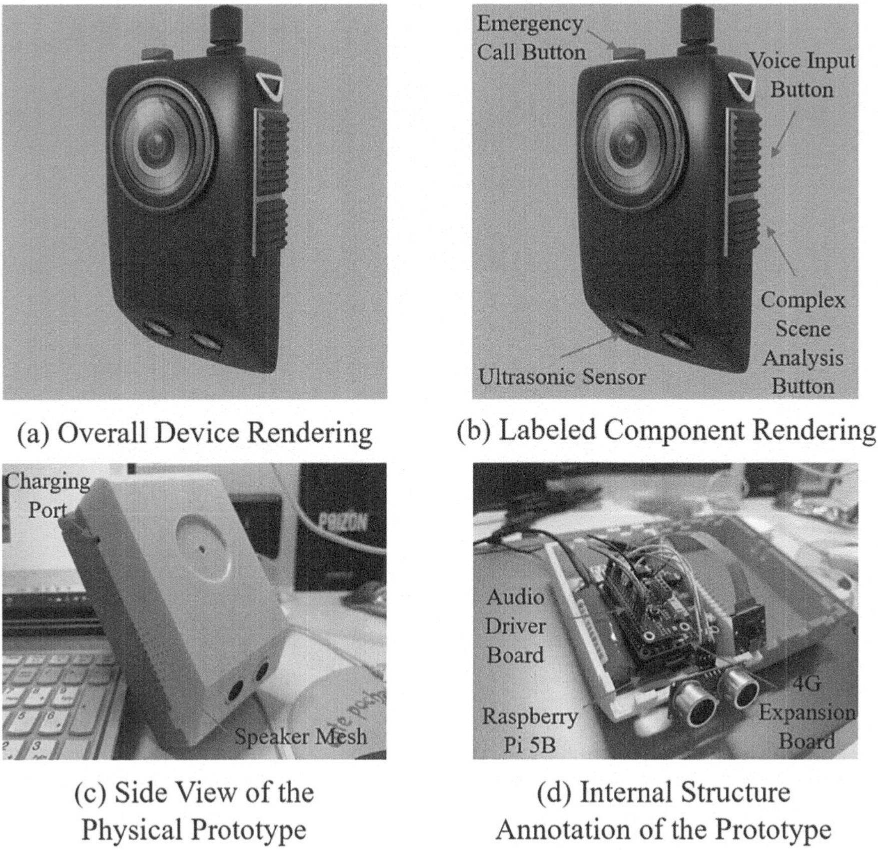

Fig. 7. Hardware Prototype of proposed Blind Assistance System.

the final safety safeguard for users. For human–device interaction, the hardware includes an INMP441 omnidirectional microphone and a small square speaker. The user can adjust volume via a rotary potentiometer or connect to external Bluetooth audio.

In addition, to enhance system security and reliability, a family binding function is implemented.

Family members of visually impaired users can bind their accounts to the device using a unique device ID within the mobile application. When a trip is initiated, the system automatically notifies the bound family members, allowing them to view the user's real-time location and surrounding environment. They can provide remote assistance or intervene in person when necessary. The family binding module also supports safety zone configuration and anomaly detection. If a user leaves the designated safe area or the system detects abnormal behavior (e.g., prolonged immobility or sudden acceleration), an automatic alert is sent to the family members. This feature provides an additional layer of security

(a) Remote Assistance Platform UI Design

(b) Implementation of Remote Assistance Platform

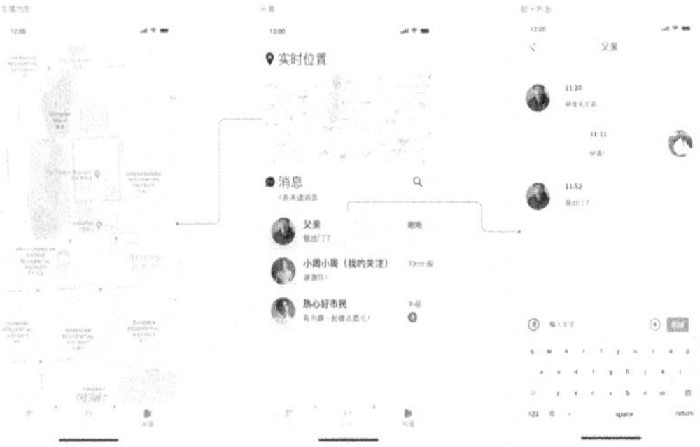

(c) Family Binding UI Design

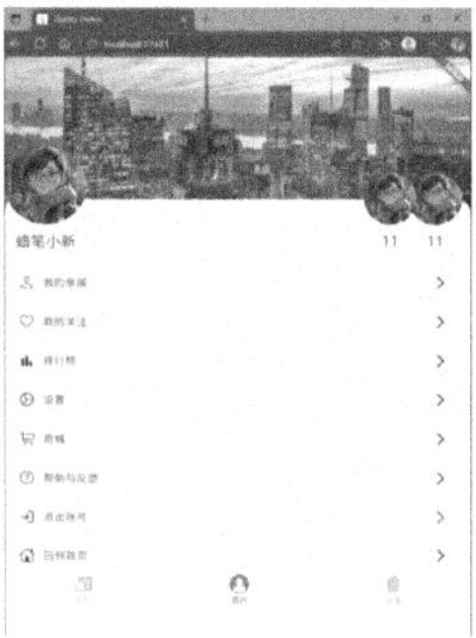

(d) Implementation of the Bound Homepage

Fig. 8. Implementation of the Mobile Application.

for visually impaired users and alleviates family concerns. As shown in Fig. 8, the software platform serves both volunteers and family members: the former engages in task participation and real-time assistance, while the latter monitors user status and provides feedback.

4 Verification and Discussion

4.1 Verification of Road Deviation Detection

To evaluate the performance of the proposed road deviation detection algorithm, experiments were conducted on the self-built Urban-Sidewalk-500 dataset. This dataset contains 500 images collected under typical urban sidewalk environments. Evaluation metrics include road recognition accuracy, deviation detection accuracy, and system response time. Experimental results show that under standard lighting conditions, the segmentation model achieved an accuracy of

96.8%, with an average system response time of 120 ms. Compared with existing road detection algorithms, our method achieves a better balance between computational efficiency and accuracy, performing particularly well on resource-constrained embedded devices. A comprehensive evaluation of the BiSeNet V2 model was conducted on the customized sidewalk dataset using four different evaluation strategies: Single Scale (SS), Single Scale Crop (SSC), Multi-Scale Flip (MSF), and Multi-Scale Flip Crop (MSFC).

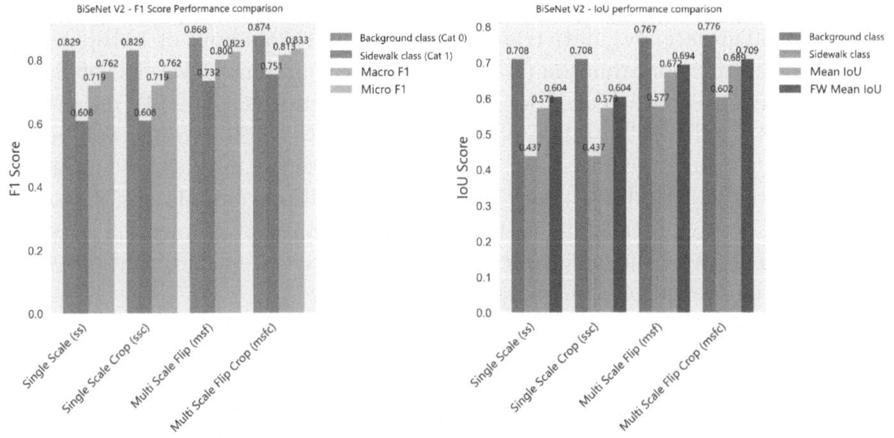

Fig. 9. Comparison of Segmentation Technology Performance.

After 700 epochs of training, as shown in Fig. 9, the results indicate that BiSeNet V2 demonstrates robust performance in this binary semantic segmentation task. Under the Micro F1 metric, the MSFC strategy achieved the best result (0.832558), representing a 9.2% improvement over the single-scale baseline. For Mean IoU, the MSFC strategy also yielded the highest score (0.688807), highlighting the positive impact of data augmentation on segmentation precision. From a class-level perspective, the background category exhibited stable performance across all strategies, with F1 scores ranging between 0.829302–0.873772, indicating the model's strong ability to recognize dominant background regions. In contrast, the sidewalk category also performed well, achieving an F1 score of 0.751384 under the MSFC strategy. This demonstrates that complex data augmentation enhances segmentation accuracy for minority classes. These findings indicate that training optimization significantly improves model performance. The background class accounted for 61.5% of pixel distribution, while the sidewalk class accounted for 38.5%, suggesting the model effectively balances major and minor class representation.

4.2 Verification of Remote Assistance

To validate the feasibility of the remote assistance platform, we conducted three rounds of functional repeat tests and five rounds of performance stress tests, covering a total of 200 connection requests. Key evaluation metrics included connection establishment latency (sampled every 100 ms), stability of the three-way handshake mechanism, and continuity of voice forwarding (tested with 5-minute randomly interrupted speech samples). Under the experimental environment (Snapdragon 888 CPU, 8 GB RAM), the average connection establishment time ranged from 1.2 to 1.8 s, with 92% of connection requests entering communication within 1.5 s. In a simulated 10-client concurrent access scenario, the platform maintained stable message handling and state transitions across all communication stages—message request (100% success rate), receiver acknowledgment (<50 ms latency), data transmission (stable throughput of 2 Mbps), and session termination confirmation (<30 ms). Regarding latency, WebSocket-based real-time transmission achieved an average end-to-end delay of 142 ms over a 100 Mbps LAN, with 95% of data points ranging between 86–198 ms, well below the 200 ms threshold required for real-time audio-video interaction (industry standard <300 ms). Under network fluctuation conditions (simulated 5% packet loss, ±80 ms latency jitter), the system maintained a 98.7% connection survival rate. The three-way handshake mechanism ensured state synchronization, with synchronization deviation remaining below 10 ms.

4.3 Performance Analysis of Real-Time Decision-Making

To validate the effectiveness of the proposed real-time decision-making module, we conducted a comprehensive analysis focusing on backbone selection and system robustness under complex environmental conditions.

Backbone Selection Analysis. We evaluated five candidate models, including ResNet50, EfficientNet-B0, DenseNet121, MobileNetV3, and MobileViT, under identical training and hardware protocols. The experimental results reveal a distinct accuracy–latency trade-off: while DenseNet121 achieves the highest absolute accuracy, it incurs significant latency penalties unsuitable for real-time guidance. Conversely, MobileViT demonstrates a superior balance, residing on the Pareto frontier with accuracy comparable to lightweight CNNs but at substantially lower latency. Consequently, MobileViT is selected as the core decision engine, prioritizing response speed and energy efficiency for wearable deployment without compromising essential recognition capabilities.

Robustness Evaluation. Reliability in dynamic real-world scenarios is critical for blind assistance. We assessed the model's robustness against common perturbations, including brightness and contrast shifts (±30%), Gaussian noise ($\sigma \in [0.01, 0.05]$), motion blur, and partial occlusions (10–25%). The system exhibited bounded performance degradation across these variations. Notably, MobileViT's attention mechanism provided superior resilience in scenes requiring global context (e.g., intersection identification) compared to pure CNN baselines. When integrated with the fail-safe prompting policy governed by τ_{road} and τ_{dev}, the decision-making module maintains reliable guidance performance across diverse lighting, weather, and occlusion conditions.

4.4 Power Consumption

Power consumption was characterized across three operational modes using an inline 5V power meter at 1 Hz under consistent peripheral settings, following a 60 s warm-up period. Modes were defined as: Offline (decision-making and road deviation detection only), Online (volunteer remote assistance only), and Hybrid (both systems concurrently). For each mode, average power (W) and current (A) were recorded, and estimated runtime was computed as battery energy (Wh) divided by average power. Measurements were performed under identical ambient conditions; transient spikes during connection establishment were excluded from averages to reflect steady-state behavior.

Assuming a 4000 mAh battery (nominal 3.7 V, $\approx$ 14.8 Wh) and an 85% conversion efficiency to 5 V, the available energy is $\approx$ 12.6 Wh. Under this assumption, the average power and estimated runtime for the three modes are summarized below (Table 1).

Table 1. Average power and estimated runtime (battery 4000 mAh; available energy $\approx$ 12.6 Wh)

Mode	Avg Power (W)	Estimated Runtime (h)
Offline	6.1	2.06
Online	5.2	2.42
Hybrid	8.2	1.53

4.5 System Performance Evaluation and Analysis Based on User Questionnaire

After completing the initial implementation of the hardware, cloud platform, and software components, we conducted a comprehensive assessment of the system's operational behavior.

From a performance perspective, under a single-server setting the prototype concurrently supports multiple session connections, with CPU and memory utilization remaining within acceptable bounds. The multi-handshake mechanism together with the finite-state-machine design yields robust communication; no deadlocks or message losses were observed.

In terms of security, a custom JSON protocol constrains data exchange and ensures message-format consistency and parseability. In future iterations, JWT-based authentication and HTTPS transport will be introduced to further strengthen the system's security posture.

Regarding usability, we organized a small-scale volunteer study and administered the System Usability Scale (SUS) questionnaire (see Appendix), which contains 10 items on a five-point Likert scale. The study involved 20 participants: 17 able-bodied volunteers (primarily university students aged 18–28) and

3 visually impaired participants (family members of the research team). Given the limited sample size and the small proportion of visually impaired users, the results primarily reflect general usability, while broader validation within the target population remains necessary. We summarize the SUS outcomes with a box plot and a histogram (see Fig. 10).

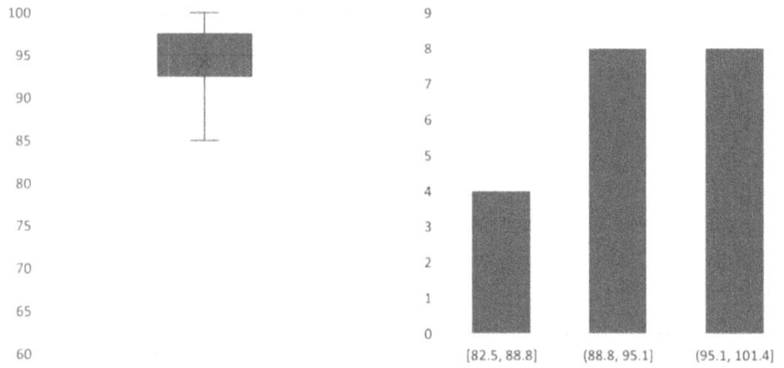

(a) Box Plot of SUS Scores (b) Histogram of SUS Scores

Fig. 10. System Usability Evaluation Results.

Figure 10(a) shows that the SUS scores concentrate in the 85–100 range, with a median of approximately 95 and a narrow interquartile span, indicating high consistency among user ratings and no substantial outliers. Figure 10(b) further reveals the distribution characteristics: most scores cluster at 95 and above, with only a small fraction falling within the 85–90 interval. Notably, the minimum score exceeds 80 and no extreme low values or marked outliers are observed, suggesting stable usability performance across user groups. Relative to the SUS benchmark of 68, the system attains an excellent level of overall usability, implying that the prototype has essentially met practical requirements in terms of ease of use.

Overall, the prototype exhibits favorable technical feasibility and user experience, laying a solid foundation for subsequent feature refinement and scaled deployment.

5 Conclusion

This paper presents an intelligent assistive system for the visually impaired based on the concept of human–machine collaboration. By integrating intelligent hardware-based perception with remote human assistance, the system provides comprehensive travel support and safety assurance for visually impaired users. The core innovation of this work lies in the organic integration of road deviation detection technology and the remote assistance platform, forming a dual-layer safeguard mechanism that combines technical automation and human guidance.

Experimental results demonstrate that the proposed system achieves high performance in road recognition accuracy, deviation detection precision, and user satisfaction. It effectively enhances both the safety and independence of visually impaired individuals during travel. Moreover, the human–machine collaboration model not only compensates for the limitations of purely technical approaches but also creates social value by encouraging volunteer participation, thereby contributing to inclusive technological development. In future work, we plan to further optimize system performance, expand application scenarios, and promote broader deployment of the system, aiming to contribute to the construction of a more accessible and inclusive social environment.

Acknowledgments. We thank the funding provided by the "XingGuang Fund" (XG2025PT99) project of Beijing Information Science and Technology University in 2025, and the "Innovation and Entrepreneurship Training Program for College Students" project of Beijing Information Science and Technology University.

Appendix

(See Table 2).

Table 2. SUS Questionnaire Items and Scoring

No.	Item	Score
Q1	I think that I would like to use this system frequently.	1 2 3 4 5
Q2	I found the system unnecessarily complex.	1 2 3 4 5
Q3	I thought the system was easy to use.	1 2 3 4 5
Q4	I think that I would need the support of a technical person to be able to use this system.	1 2 3 4 5
Q5	I found the various functions in this system were well integrated.	1 2 3 4 5
Q6	I thought there was too much inconsistency in this system.	1 2 3 4 5
Q7	I would imagine that most people would learn to use this system very quickly.	1 2 3 4 5
Q8	I found the system very cumbersome to use.	1 2 3 4 5
Q9	I felt very confident using the system.	1 2 3 4 5
Q10	I needed to learn a lot of things before I could get going with this system.	1 2 3 4 5

(1 = Strongly disagree, 5 = Strongly agree)

References

1. Elmannai, W., Elleithy, K.: Sensor-based assistive devices for visually-impaired people: current status, challenges, and future directions. Sensors **17**(3), 565 (2017)
2. Tapu, R., Mocanu, B., Zaharia, T.: DEEP-SEE: joint object detection, tracking and recognition with application to visually impaired navigational assistance. Sensors **17**(11), 2473 (2017)
3. Amore, F., Silvestri, V., Guidobaldi, M., et al.: Efficacy and patients' satisfaction with the ORCAM MyEye device among visually impaired people: a multicenter study. J. Med. Syst. **47**(1), 11 (2023)
4. Be My Eyes: Accessibility technology for blind & low-vision people. https://www.bemyeyes.com/. Accessed 27 Nov 2025
5. Gurari, D., Li, Q., Stangl, A.J.: VizWiz grand challenge: answering visual questions from blind people. In: Proceedings of the IEEE Conference on Computer Vision and Pattern Recognition (CVPR), pp. 3608–3617 (2018)
6. Lavric, A., Beguni, C., Zadobrischi, E., Căilean, A.M., Avătămăniţei, S.A.: A comprehensive survey on emerging assistive technologies for the visually impaired: AI and visible light communications. Sensors **24**(15), 4834 (2024)
7. Pancholi, S., Wachs, J.P., Duerstock, B.S.: Use of artificial intelligence techniques to assist individuals with physical disabilities. Ann. Rev. Biomed. Eng. **26**, 1–24 (2024). https://doi.org/10.1146/annurev-bioeng-082222-012531
8. Adam, M.M.S., Aljehane, N.O., Alzahrani, M.Y., et al.: Leveraging assistive technology for visually impaired people through optimal deep transfer learning based object detection model. Sci. Rep. **15**, 30113 (2025)
9. Baig, M.S.A., Gillani, S.A., Shah, S.M., Aljawarneh, M., Khan, A.A., Siddiqui, M.H.: AI-based wearable vision assistance system for the visually impaired: integrating real-time object recognition and contextual understanding using large vision-language models (2024). arXiv:2412.20059
10. Yu, C., Gao, C., Wang, J., et al.: BiSeNet V2: bilateral network with guided aggregation for real-time semantic segmentation. Int. J. Comput. Vision **129**(11), 3051–3068 (2021)
11. Yang, S., Li, Y., Liu, Y.: Lane detection based on instance segmentation of bisenet v2 backbone network. Appl. Artif. Intell. **36**(1), 2085321 (2022)
12. Howard, A., Sandler, M., Chu, G.: Searching for MobileNetV3. In: Proceedings of the IEEE/CVF International Conference on Computer Vision (ICCV) (2019)
13. Mehta, S., Rastegari, M.: MobileViT: light-weight. in: general-purpose, and mobile-friendly vision transformer (2021). arXiv:2110.02178
14. He, K., Zhang, X., Ren, S., Sun, J.: Deep residual learning for image recognition. In: Proceedings of the IEEE Conference on Computer Vision and Pattern Recognition (CVPR), pp. 770–778 (2016)
15. Tan, M., Le, Q.: EfficientNet: rethinking model scaling for convolutional neural networks. In: Proceedings of the International Conference on Machine Learning (ICML), pp. 6105–6114. PMLR (2019)
16. Huang, G., Liu, Z., Van Der Maaten, L., Weinberger, K.Q.: Densely connected convolutional networks. In: Proceedings of the IEEE Conference on Computer Vision and Pattern Recognition (CVPR), pp. 4700–4708 (2017)
17. Jones, M., Bradley, J., Sakimura, N.: JSON Web Token (JWT). RFC 7519 (2015)
18. Kreps, J., Narkhede, N., Rao, J.: Kafka: a distributed messaging system for log processing. In: Proceedings of NetDB (2011)
19. Williams, J.: RabbitMQ in Action: Distributed Messaging for Everyone. Simon and Schuster (2012)

Development of a Wall-Climbing Mobile Manipulator for Concrete Bridge Tower Inspection

Wang Peng[1], Lyu Guizhi[2(✉)], Li Guohong[3], Lu Feng[1], and Dai Shenglong[1]

[1] Beijing Institute of Aeronautical Materials, Beijing 100095, China
[2] School of Mechanical and Electronical Engineering, Heze University, Heze 274015, China
lvguizhi@126.com
[3] Hong Kong-Zhuhai-Macao Bridge Authority, Zhuhai 519000, China

Abstract. A wall-climbing mobile manipulator designed for the inspection of concrete bridge towers has been introduced. Based on an analysis of the deficiencies exhibited by existing wall-climbing robots in the inspection of concrete bridge towers, a solution of a composite robot in the form of a wall-climbing mobile manipulator, which is compatible with multiple inspection devices, is proposed. The overall design scheme is described, encompassing a detailed account of the mechanical structure of the developed composite wall-climbing robot and the modification process undertaken for the AUBO collaborative robotic. The overall control architecture of the composite robot is provided, with particular emphasis on the robot localization achieved through multi-sensor fusion, as well as its navigation planning and logical obstacle avoidance architecture. Finally, the specific parameters of the developed wall-climbing composite robot (WCCR) are presented. Taking the Qingzhou Bridge Tower of the Hong Kong-Zhuhai-Macao Bridge (HZMB) as an example, it is demonstrated that the developed composite robot can meet the inspection requirements for concrete wall surfaces of bridge towers.

Keywords: Wall-climbing composite robot (WCCR) · Negative-pressure adhesion · Structural design · Control architecture · Retrofitting of collaborative robot

1 Introduction

Currently, existing manual inspection methods for bridge towers have been inefficient and inaccurate, posing challenges for close-up inspections and presenting high safety risks. To overcome these challenges, a variety of intelligent devices have been developed for bridge inspection in recent years, such as mobile robots, wall-climbing robots, and unmanned aerial vehicles (UAVs) [1].UAVs offer the advantage of carrying inspection equipment like high-definition cameras to perform patrol inspections of bridge structures under favorable weather conditions. Nevertheless, they are limited by poor

H. Wei and J. Niu (Eds.): COBOT 2025, CCIS 2893, pp. 253–268, 2027.
https://doi.org/10.1007/978-981-92-2074-8_15

endurance, the potential for camera field-of-view occlusion during flight, and the inability to conduct close-range inspections. To compensate for the deficiencies of UAV-based inspection methods, wall-climbing robots have been successively developed for bridge structure inspection. Wheel-type [2] or leg-type [3] wall-climbing robots utilizing permanent-magnet adhesion are primarily employed for inspecting bridge steel structures. Annular wall-climbing robots equipped with high-definition cameras can carry out close-range inspections of bridge pier cracks [4]. However, these robots, which adopt a clamping-support structure, are relatively bulky and heavy, posing challenges in terms of installation and deployment.

Negative pressure adhesion wall-climbing robots, which are capable of carrying inspection devices and possess strong wall-adaptable, have been developed for the inspection of various concrete-structured infrastructure [5–10]. Yet, when inspecting the surfaces of concrete structures, the currently developed wall-climbing robots typically carry only one type of non-destructive testing equipment, such as cameras, LiDAR, or ultrasonic devices. This lack of compatibility with multiple inspection devices simultaneously, coupled with insufficient operational flexibility, restricts their effectiveness.

Given that high bridge towers require not only the inspection of important concrete structural surfaces but also the use of other inspection tools to assess performance indicators beyond appearance defects, a more versatile solution is needed. Drawing inspiration from mobile manipulators, a 6-degree-of-freedom (DOF) robotic arm is mounted on a wall-climbing robot. This configuration enables the robot to be compatible with a range of inspection devices, including high-definition cameras, infrared cameras, steel corrosion meters, and rebound hammers. As a result, it can conduct precise close-range inspections of concrete bridge towers, significantly expanding the application scope and methods of traditional wall-climbing robots in concrete wall operations.

In this context, the main contributions of this paper are as follows: (1) It introduces a developed negative pressure adhesion wall-climbing mobile manipulator-type composite robot specifically designed for the inspection of concrete bridge towers. (2) It details the design structure of the wall-climbing robot body and the modification process of the Cobot. (3) It presents the composition architecture of the robot control system, along with the positioning, navigation, and trajectory planning schemes. (4) It conducts an application verification of the developed WCCR on the Qingzhou Bridge Tower of the HZMB.

2 Overall System Design Scheme

2.1 Overall Composition

To enhance the operational flexibility and applicability of wall-climbing robots, enabling them to be compatible with multiple inspection tools and ensuring that the inspection area could cover as many positions as possible on the surface of concrete bridge towers, a wall-climbing mobile manipulator-type composite robot capable of free movement on the bridge tower surface has been designed, as shown in Fig. 1. The WCCR primarily consists of a negative-pressure adhesion wall-climbing robot carrier and a 6 DOF cobot mounted on it. The end-effector of the robotic arm can not only be compatible with various devices to expand the inspection capabilities of the system but also enable the inspection tools to reach relatively narrow spaces, such as the area below the lock cables, thereby expanding the operational range of the system on the surface of concrete bridge towers.

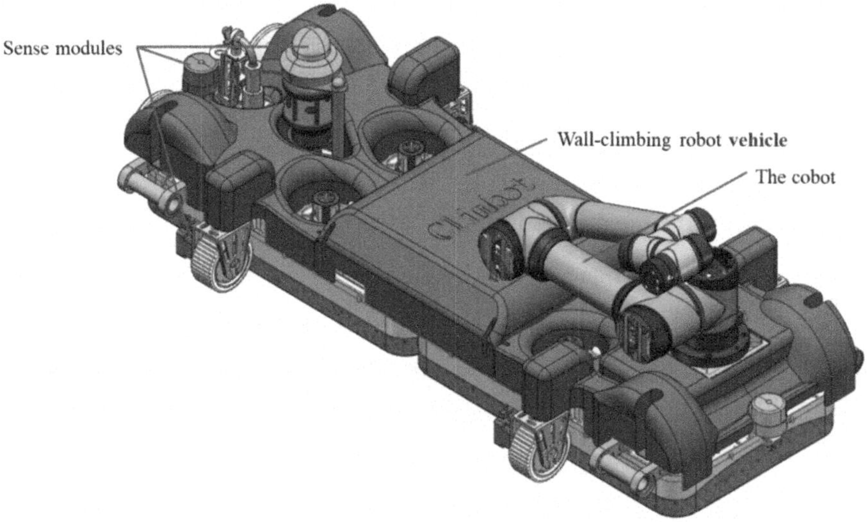

Fig. 1. Overall structure of the wall-climbing composite robot

2.2 Portable Inspection Devices

The WCCR is capable of using multiple inspection tools. Through the cobot, it can carry inspection equipment such as high-definition cameras, high-definition video cameras, infrared thermal imagers, and steel corrosion detectors, as shown in Fig. 2. These devices are used to detect defects in concrete bridge tower structures, including cracks, bubbles, holes, pitted surfaces, and internal steel bar corrosion.

Consumer-grade cameras and video cameras have been selected as the image and video data acquisition devices. The main consideration is that these two types of image acquisition devices need to have adaptive functions for parameters such as light and focal length. This is to meet the requirement that when installed at the end-effector of the cobot, they can obtain high-definition images and videos of suspected defects from different postures and distances.

(a) (b) (c) (d)

(a) Canon GFX100S camera (b) BMPCC 6k video camera (c) TI65s Infrared Thermal Imager (d) Reinforcement Rust Detection Meter

Fig. 2. Equipment for concrete defect inspection

2.3 Planning for the Layout on the Tower Top

Two sets of winders are arranged on the platform at the top of the bridge tower when the inspection operation is carried out with the WCCR. A winder is used to retract and release the power supply cable. The other winder is used to retract and release anti-fall rope for safety. The operation diagram is shown in Fig. 3. The figure shows the top of the Qingzhou Tower of the HZMB. Based on the original facilities, a WCCR implementation plan has been designed.

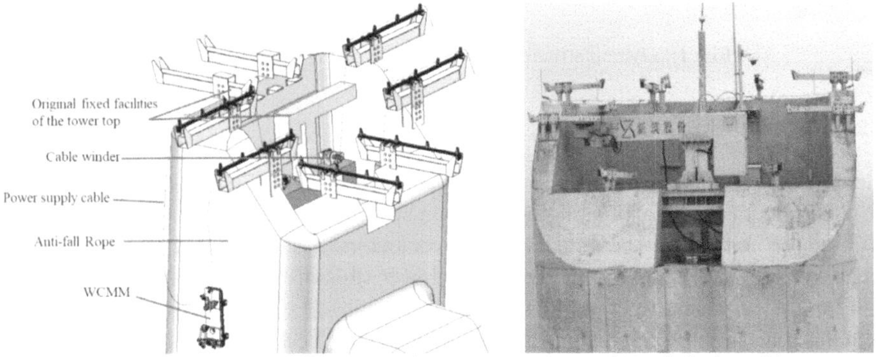

(a) Implementation Plan (b)Realistic view of the tower top

Fig. 3. Schematic diagram of the WCCR implementing work tasks

3 Structural Design

3.1 Structure of the Wall-Climbing Robot

The installation of the cobot and inspection tools necessitates that the wall-climbing robot's body possesses a relatively high load-bearing capacity. The realization of the heavy load function mainly depends on the design of the negative pressure adhesion module of the wall-climbing robot. Flexible movement on the bridge tower wall is achieved through four sets of active wheel movement modules, as shown in Fig. 4.

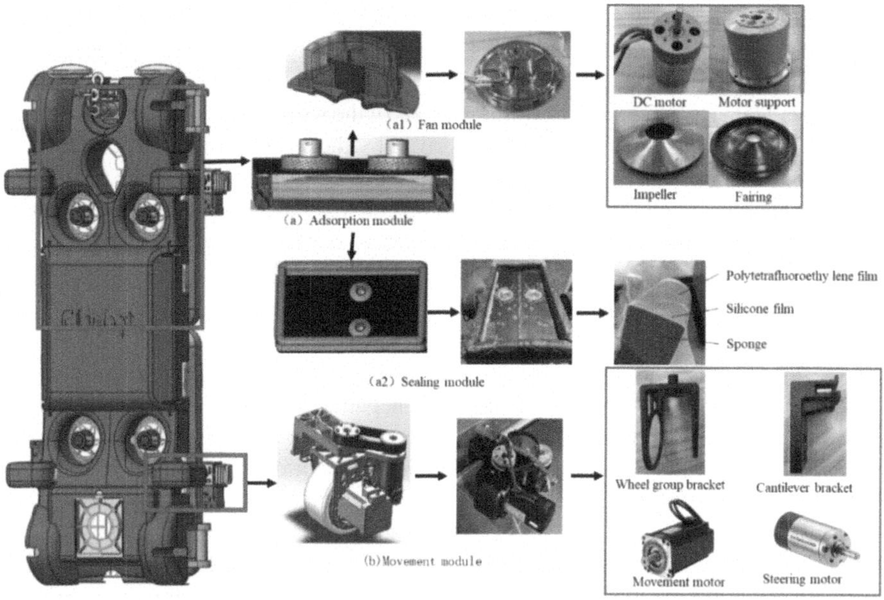

Fig. 4. Design of the wall-climbing robot

The wall-climbing robot carrier adopts a dual-chamber and four fans negative pressure adhesion method. The design of the dual-chamber structure enhances the reliability of the robot's adhesion, reducing the risk of accidents caused by pressure loss in a single adhesion chamber when the wall-climbing robot performs tasks such as crossing obstacles. The design of two fans per chamber improves the adhesion reliability of the adhesion chamber, reducing the risk of pressure loss due to electrical failures of a single fan and other reasons. Each adhesion module consists of two fan modules and one sealing module, as shown in Fig. 4(a). The fan module is composed of parts such as a fairing, a fan impeller, and an adhesion motor. The sealing module is composed of parts such as a sealing frame and a sealing ring. The sealing module forms a closed sealing chamber by fitting the sealing ring on the sealing frame against the wall surface. The DC motor in the adhesion module drives the impeller to extract air from the sealing chamber, creating a pressure difference between the inside and outside of the top partition of the sealing chamber, which enables the robot to adhere to the wall surface.

The adhesion chambers of the wall-climbing robot carrier use a combination of highly elastic and wear-resistant materials for sealing. During the movement of the wall-climbing robot on the wall surface, the sealing ring of the adhesion module is required to be in close contact with the wall surface while also allowing relative sliding. While controlling the leakage amount of the sealing chamber to achieve good sliding characteristics, a sealing ring with good adaptability, wear resistance, and a relatively low friction coefficient with the wall surface has been developed by combining highly elastic and wear-resistant materials.

The wall-climbing robot carrier uses four omnidirectional wheels for movement. The four sets of omnidirectional wheel modules are symmetrically arranged on the robot body, reducing the applied power of each drive module. Each active wheel movement module is an omnidirectional wheel set, as shown in Fig. 4(b). The omnidirectional wheel set is fixed to the robot body frame through a cantilever bracket. Each omnidirectional wheel set contains two motors. The rolling motion of the wheel is driven by the movement motor, and the in-place steering motion is driven by the steering motor. The routing of the power and control cables for the two motors in the wheel set enters the fixed shaft sleeve through the wheel set bracket, preventing entanglement caused by the movement of the wheels. The use of four omnidirectional wheel sets enhances the flexibility of the wall-climbing robot's movement on the wall surface.

3.2 Modification of the Cobot Controller

Since the wall-climbing robot carrier adopts a negative-pressure adhesion method, its adhesion capacity is limited. To adapt to different bridge tower inspection tasks and enhance the accessibility of inspection equipment on the bridge tower surface, considering the load capacity of the negative pressure wall-climbing robot and the self-weights of the inspection equipment and repair tools, the Aubo i5 cobot, which can achieve 6 DOF motion, is selected as the arm to be mounted on the negative pressure wall-climbing robot. However, the existing robotic arms cannot meet the application requirements in target detection scenarios. It is necessary to conduct systematic optimization and modification on the collaborative arm based on the size, structure, and load-bearing capacity of the negative-pressure wall-climbing robot, as well as the actual process conditions during the robot's maintenance operations.

The Aubo i5 cobot is composed of three main parts: the robotic arm body, the control cabinet, and the teaching pendant. The end effector of the robotic arm body contains a set of 12/24 V power supply interfaces that can be used to power the inspection equipment. To ensure the retention of the collaborative arm's original functional and technical parameter advantages, such as load capacity and reachability, the arm's main body structure remains unchanged.

The existing teaching pendants are commonly used for convenient programming applications in industrial settings, but they do not align with the target scenario of bridge tower inspection and can thus be optimized and removed. The WCCR system adopts an Ethernet control method, removing the body teaching pendant and its associated communication and power supply cables. The functions of the teaching pendant are realized remotely through the upper computer program.

To meet the characteristics of high adaptability, high safety, openness, and compatibility, the existing Aubo collaborative robot controller is designed with relatively large volume and weight, which fails to meet the requirements for mounting it on a wall-climbing robot carrier. To achieve the integration of the negative pressure wall-climbing robot and the robotic arm, according to the actual process during the robot's operation and considering the size, structure, and load capacity of the negative pressure wall-climbing robot body, the robotic arm has been optimized and modified. On one hand, the external I/O interface board employs data shielding, eliminating the original control cabinet enclosure and unused interface boards. On the other hand, the layout control of the PCB board has been restructured, resulting in a control box with higher layout integration and more compact and reasonable component placement. The modification process is shown in Fig. 5.

The original size of the control cabinet is 380 mm × 350 mm × 258 mm, and its weight is 15 kg. It is bulky and heavy. The weight of the modified robotic arm control box is 4 kg, and its size is 264 mm × 234 mm × 162 mm, which is significantly reduced compared to the original one. This makes it more convenient to be mounted on the designed negative pressure wall-climbing robot.

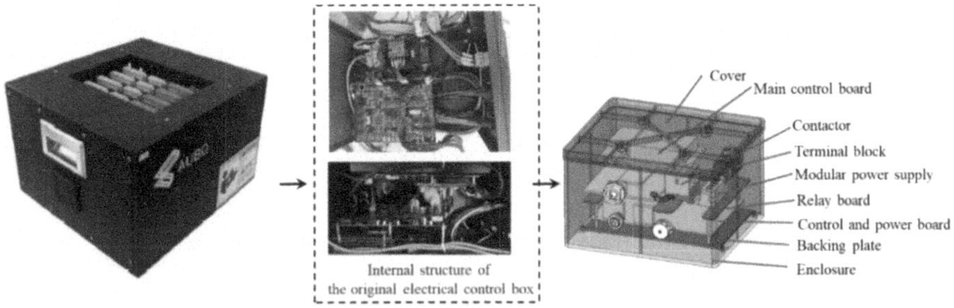

Fig. 5. Retrofit design of the cobot controller

4 Composition of the Control System, Positioning, and Navigation Planning

4.1 Control Architecture of the WCCR

The WCCR control system mainly consists of four parts: a main industrial personal computer (IPC), sensing systems, a cobot control system, and a wall-climbing robot control system, as illustrated in Fig. 6. The upper computer can communicate with the IPC installed with the Robot Operating System (ROS) through a dedicated wireless network. The IPC communicates with the motion controller of the wall-climbing robot and the controller of the cobot respectively via the EtherCAT bus.After the upper computer assigns an inspection task, the IPC sends control commands to the motion controller. These commands control the adhesion module motors for the robot to adhere to the bridge tower surface, the movement module hub motors for the robot to move on the wall, and the movement module steering motors for the robot to change its moving direction.

In accordance with the requirements for robot positioning, orientation, navigation, and ensuring its safety during movement, the robot body is equipped with visual sensors such as monitoring cameras and pan-tilt dome cameras, as well as positioning and orientation sensors including a differential positioning module, an inertial measurement unit (IMU), and steering encoders. Since slippage between the wheels and the wall surface is inevitable when the wall-climbing robot moves on the wall, a passive wheel odometer is specifically added. This odometer can calculate the robot's travel distance based on the wheel circumference in combination with an absolute encoder.

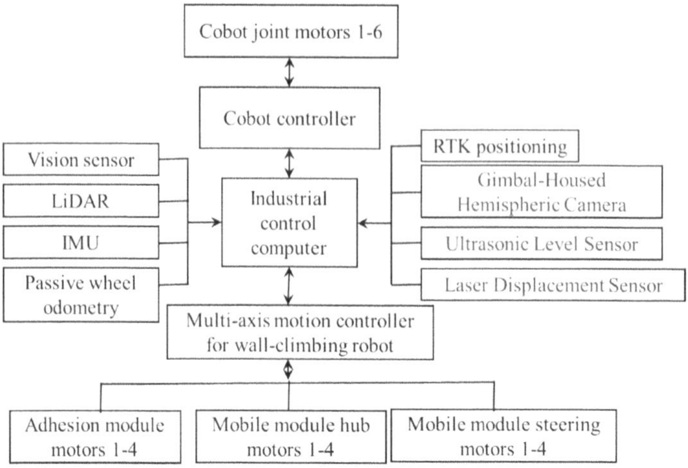

Fig. 6. The wall-climbing hybrid robot control block diagram

4.2 Positioning Based on Multi-sensor Fusion

The positioning of the wall-climbing robot body is a prerequisite for accurately feedbacking the locations of bridge tower defects. A multi-sensor combination positioning approach is adopted to determine the precise position of the robot body on the bridge tower. Firstly, rough positioning is carried out using the data returned by the laser rangefinders installed around the wall-climbing robot body to ensure that the positioned location is within the operational radius of the robotic arm for inspection.

The combination of odometer displacement and the yaw angle from the Inertial Measurement Unit (IMU) is used as the initial estimate of pose change. This is then combined with the calculation results from Real-Time Kinematic (RTK, a carrier-phase differential technique) for Kalman filtering, as shown in Fig. 7. Through multi-sensor data fusion, the inherent constraints of single-odometer system positioning are eliminated, and the impact of cumulative errors on the positioning results is reduced. Meanwhile, the laser rangefinders and vision units continuously match, extract, and record surrounding obstacles such as stay cables in real time.

Due to the offset errors generated by the output signals of inertial sensors, which accumulate over time, according to the error characteristics of inertial sensors, the Levenberg-Marquardt (LM) nonlinear method is used to fit the data and establish a mathematical model of inertial sensor errors, and then the errors are calibrated. Subsequently, an Extended Kalman Filter (EKF) algorithm is employed to fuse the sensor data and compensate for the errors, obtaining the optimal estimate of the robot's pose. After 37 tests of positioning accuracy with caliber rule in 5 different days, we found the combined errors are between 13.2 cm and 14.8 cm. Finally, through the robot pose control algorithm, high-precision continuous autonomous cruising of the robot is achieved.

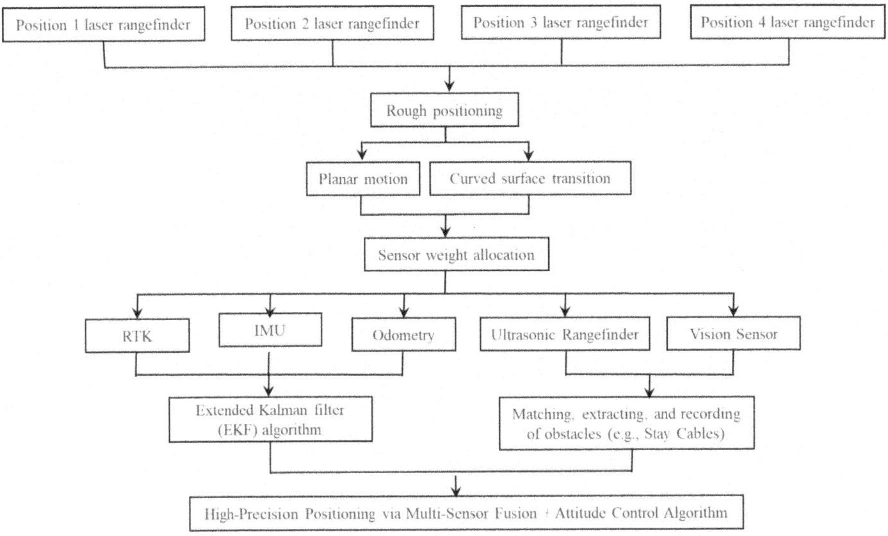

Fig. 7. Block diagram of positioning and attitude control

4.3 Inspection Trajectory Planning

Given that the structure and dimensions of bridge towers are relatively fixed, cruise routes can be designed based on the tower dimensions with reference to positioning sensors, and the robot can follow these routes for inspection. At the end of each route inspection, the robot rotates a full circle in place to ensure that its forward direction still utilizes the visible field of view provided by the imaging sensors. Since, except at the edges, there are generally no significant obstacles, protrusions, or depressions on the surface of a typical bridge tower, the WCCR system can first be moved to a fixed position on the bridge tower surface as the initial starting point. The forward direction is set as +y, the upward direction of the bridge tower as +z, and the x-direction is determined according to the right-hand rule. In this way, the positional information returned by the odometer and inertial sensors is relative to the initial starting point, and the angular velocity information represents the current attitude of the system. The robot conducts a comprehensive inspection of the concrete surface of the bridge tower. When concrete lesions are detected, the RTK (Real-Time Kinematic) carrier-phase differential technology and the odometer model data of the negative-pressure inspection robot are used to record the location information of the lesions. During the cruise, fall-prevention and collision-avoidance sensors come into effect in real time, and the real-time obstacle-avoidance logic is shown in Fig. 8.

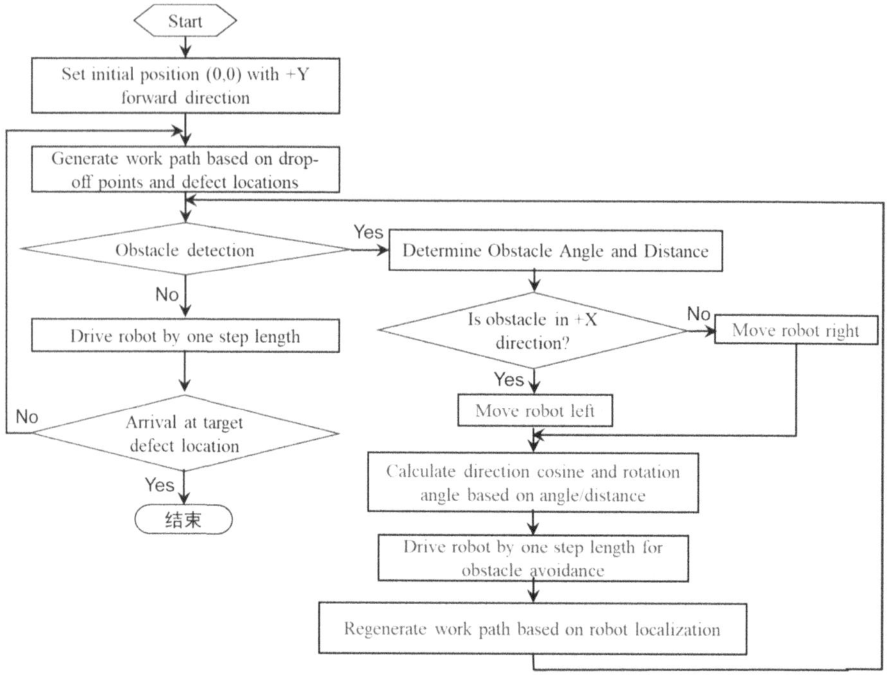

Fig. 8. Real-time obstacle avoidance logic framework

Taking the Qingzhou Bridge Tower of the HZMB as an example, its inspection planning is shown in Fig. 9. Figure 9(a) illustrates the anchor cable structure of the bridge tower, with the tower being symmetrically divided along the anchor cable structure. The robot travels half the circumference of the bridge tower each time, as shown in Fig. 9(b). Taking the path from A_0 to D_0 on the left side as an example, the robot first moves downward vertically from the top of A_0 to perform inspection tasks; after completion, it moves horizontally to B_0 at the same height and performs tasks; then it moves vertically downward to C_0 and performs tasks; finally, it starts moving horizontally to D_0 at the same height, completing one cycle. Figure 9(c) shows the planned walking trajectory of the robot.

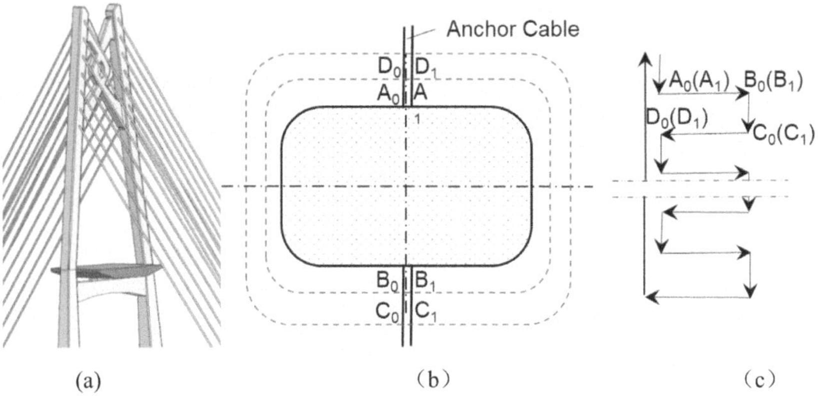

Fig. 9. Path planning for the tower along the anchor cable structure

5 Prototype and Experiments

5.1 Prototype Parameters

The designed WCCR is depicted in Fig. 10, with detailed specifications listed in Table 1.

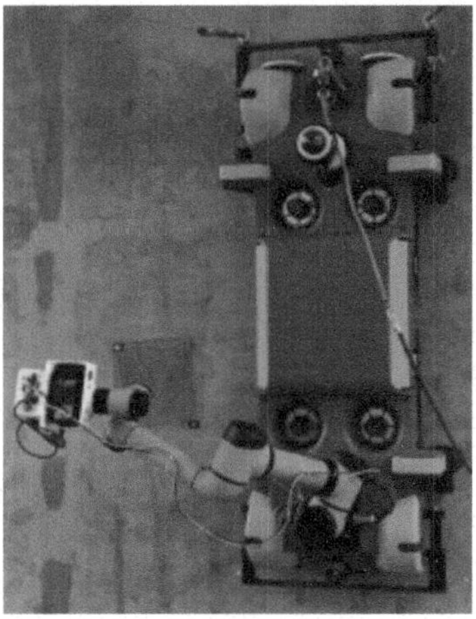

Fig. 10. Composition of the WCCR

Table 1. Technical parameters of the WCCR

Wall-climbing robot	Item	Values	Cobot	Item	Values

(*continued*)

Table 1. *(continued)*

	Weight (kg)	≤ 60		Weight (kg)	≤ 25
	Load (kg)	≥ 30		Load (kg)	≥ 5
	Dimensions (mm^3) length $\times$ width $\times$ height	$1600 \times 810 \times 525$		Operating radius(mm)	≥ 850
	Speed(m/min)	≥ 15		DOF	6

The end-effector of the cobot in the WCCR system can be connected via flanges, enabling it to be compatibly equipped with inspection devices such as the high-definition camera, high-definition video camera, infrared thermal imager, and steel bar corrosion detector shown in Fig. 1, as depicted in Fig. 11. The system uses the Intel UP Xtreme Board as its core. Functional circuit boards are designed to interact with the robot's industrial personal computer (IPC). Through this interaction, they can control functions such as the shutter and zoom of the camera and video camera. Meanwhile, they provide services including power supply, communication, and data forwarding for the inspection devices.

(a) (b) (c) (d)

(a) Installation of the camera

(b) Installation of the video camera

(c) Installation of the infrared thermal imager

(d) Installation of the reinforcement rust detection meter

Fig. 11. Integrated installation of various detection devices at the end of the cobot

5.2 Experimental Validation

To test the practicality of the WCCR, defect inspections were conducted on the Qingzhou tower of the HZMB, as shown in Fig. 12.

Fig. 12. Application validation of Qingzhou bridge tower

After installing the selected inspection equipment onto the end - effector of the composite robot's robotic arm, inspections were carried out on the Qingzhou Tower respectively. The high - definition video camera with 20 million pixels and a frame rate of no less than 24 FPS was mainly used for routine inspections of areas with a large number of defects. The GFX100S high - definition camera with over 100 million pixels was employed for detailed inspections of suspected defect areas identified during the routine inspections. The infrared thermal imager served as a supplementary means for surface defect detection. The steel bar corrosion detector was mainly used to detect the corrosion condition of steel bars inside the concrete. So far, more than 400 high-definition inspection images of the concrete wall surface of the HZMB have been accumulated. The processing results of collected image from the Qingzhou Tower, recognized by YOLO algorithm, are shown in Fig. 13 and Table 2, and the inspection and monitoring scenarios of the infrared thermal imager obtained from the laboratory are presented in Fig. 14. Table 3 provides some of the data information collected by the steel bar corrosion detector on the test platform. All the inspection results obtained from he Qingzhou Tower indicate that it is currently in a good service condition.

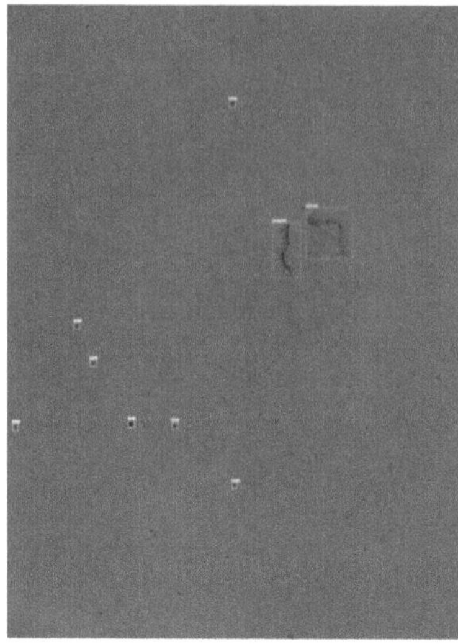

Fig. 13. Individual images collected from Qingzhou tower

Fig. 14. Application of infrared thermal imaging in surveillance

Table 2. Identify Result (with red star) of collected image from the Qingzhou Tower

Serial	defectName	defectLength (mm)	defectWidth (mm)	defectArea (mm)	defectPerimeter (mm)
1	Peeroff	5.872254	1.929396	9.389334	15.603299

Table 3. Inspection data obtained by the reinforcement rust detection meter

Serial	open circuit potential (V)	medium resistance (Ω·cm^2)	polarization resistance (Ω·cm^2)	corrosion current density (uA/cm^2)	corrosion rate (um/a)
1	−0.935	411.1	1153	22.53	267.2
2	−0.958	413.4	257.1	101.0	1198
3	−1.023	458.3	382.4	67.97	805.9
4	−1.012	169.6	193.2	134.5	1595
5	−1.046	63.14	1296	20.05	237.7
6	−1.003	457.4	1085	23.96	284.1
7	−1.088	468.8	1854	14.02	166.2

6 Conclusion

This paper has presented the development of a composite wall-climbing robot. It provides an approach-type inspection method which can provide more detecting instruments and high positional accuracy, compare with the Unmanned Aerial Vehicle. The robot is capable of carrying different inspection devices weighing up to 5 kg to inspect the concrete bridge towers of large bridges, and it can also be used for inspecting other large concrete structures with similar configurations.

The paper analyzes the shortcomings of existing wall-climbing robots in inspecting large concrete structures on concrete surfaces. It details the structure of the wall-climbing robot within the developed composite robot, the control modifications of the cobot, as well as the overall control architecture, and positioning and navigation methods. Additionally, it describes the application and verification of this robot on the Qingzhou Bridge Tower of the HZMB.

Acknowledgements. This work was supported by a grant from the National Key R&D Program of China (No. 2019YFB1600700).

References

1. Tian, Y., Chen, C., Sagoe-Crentsil, K., Zhang, J., Duan, W.: Intelligent robotic systems for structural health monitoring: applications and future trends. Autom. Constr. **139**, 104273 (2022)
2. Wang, R., Kawamura, Y.: Development of climbing robot for steel bridge inspection. Ind. Robot Int. J. **43**(4), 429–447 (2016)
3. Qin, Y., Dong, S., Pang, R., Xia, Z., Yang, J.: Design and kinematic analysis of a wall-climbing robot for bridge appearance inspection. IOP Conf. Earth Environ. Sci. **638**(1), 120 (2021)
4. Jang, K., An, Y., Kim, B., Cho, S.: Automated crack evaluation of a high-rise bridge pier using a ring-type climbing robot. Comput. Aided Civ. Infrastruct. Eng. **36**(1), 14–29 (2020)

5. Hernando, M., Gómez, V., Brunete, A., Gambao, E.: CFD modelling and optimization procedure of an adhesive system for a modular climbing robot. Sensors **21**, 1117 (2021)
6. Parween, R., Wen, T.Y., Elara, M.R.: Design and development of a vertical propagation robot for inspection of flat and curved surfaces. IEEE Access **9**, 26168–26176 (2021)
7. Qin, G., Li, C., Wu, H., Ji, A.: Wall-climbing mobile robot for inspecting DEMO vacuum vessel. Appl. Sci. **12**(9260), 1–17 (2022)
8. Kim, D., Kim, Y., Noh, K., Jang, M., Kim, S.: Wall climbing robot with active sealing for radiation safety of nuclear power plants. Nucl. Sci. Eng. **7**, 1–13 (2020)
9. Chen, N., Shi, K., Li, X.: Theoretical and experimental study and design method of blade height of a rotational-flow suction unit in a wall-climbing robot. Trans. ASME **12**(8), 045002 (2020)
10. Yang, L., Li, B., Feng, J., Yang, G., Chang, Y., Jiang, B., Xiao, J.: Automated wall-climbing robot for concrete construction inspection. J. Field Robot. **9**, 1–20 (2022)

Section II

Trajectory Planning and Stability Optimization of an AGV–SCARA System for Automated Gauge Block Verification

Siyuan Liu[1], Jingjing Fan[2], Xiaodong Wang[1], and Yi Luo[1](✉)

[1] Dalian University of Technology, School of Mechanical Engineering, Dalian 116024, China
luoy@dlut.edu.cn

[2] Changcheng Institute of Metrology & Measurement, Beijing 100095, China

Abstract. Gauge blocks are essential physical standards for length metrology. Their accuracy directly impacts product dimension control and overall manufacturing quality. The system features a hybrid robot setup, integrating an Automated Guided Vehicle (AGV) and cobot, working alongside SCARA robot. To explore the vibration and dynamic characteristics of AGVs during transport, a dynamic model of the differential-driven AGV is developed to analyze how the turning radius and acceleration affect the vehicle's stability. An improved A*-APF path planning algorithm that incorporates dynamic constraints is proposed to enhance path smoothness and handling stability. For the trajectory planning of SCARA robot, the RRT* algorithm is enhanced by parent node reselection and path pruning. Furthermore, by applying B-spline interpolation, low-pass filtering, and fifth-order non-uniform B-spline curve interpolation, we optimize the trajectory smoothness. An experimental system for gauge block handling and verification is built to validate the proposed approach. Experimental results demonstrate that the optimized AGV path planning reduces vibration amplitude and inertial impact, with robot handling accuracy reaching $\pm$ 0.10 mm. These results verify the system's feasibility for high-precision automated handling and verification of gauge blocks.

Keywords: Gauge block verification · AGV path planning · Robot trajectory optimization · Dynamic constraints

1 Introduction

Gauge blocks are core physical standards for length value transmission and calibration of length-measuring instruments, and their detection accuracy directly determines product dimension control and manufacturing quality [1]. In traditional verification processes, gauge blocks undergo multiple steps such as cleaning, isothermal conditioning, and repeated measurements. The entire process is highly dependent on manual operations, which is time-consuming and susceptible to human factors, making it difficult to meet the metrological requirements of high precision and efficiency [2]. Therefore, researching automated gauge block verification systems that adopt collaborative operations of AGVs

H. Wei and J. Niu (Eds.): COBOT 2025, CCIS 2893, pp. 271–289, 2027.
https://doi.org/10.1007/978-981-92-2074-8_16

and robots to realize automatic handling, pose adjustment, and verification of gauge blocks throughout the process is an important direction to improve the reliability of length metrology. This approach not only reduces human errors but also significantly enhances verification efficiency and accuracy.

Given that tungsten steel gauge blocks feature high density and large mass in the high-precision handling process, they are prone to generating significant inertial impact and vibration during transportation. Therefore, the path and trajectory planning as well as stability of AGVs and robots exert a crucial impact on handling accuracy and repeatability.

In the field of AGV path planning, classical graph search algorithms include the A* algorithm and Dijkstra algorithm. These algorithms quickly search for a path from the start to the end point through heuristic functions or breadth-first search strategies. They can efficiently plan paths in simple environments but tend to fall into local optima in complex, obstacle-dense environments, making them unsuitable for scenarios with high real-time requirements [3, 4]. With technological development, optimization theory-based algorithms such as genetic algorithms and particle swarm optimization algorithms have emerged. These algorithms search for optimal paths in the solution space by simulating biological behaviors or selection, crossover, and mutation operations in evolutionary processes, exhibiting global search capabilities. However, their slow convergence rate and complex parameter adjustment processes [5, 6] limit their application in practical industrial scenarios. In recent years, machine learning algorithms such as deep learning and reinforcement learning have been applied to AGV path planning. Deep learning algorithms, such as Convolutional Neural Networks (CNNs) and Recurrent Neural Networks (RNNs), enable AGVs to perceive complex environments and plan paths through learning from large amounts of environmental data. Reinforcement learning algorithms allow AGVs to continuously learn through interaction with the environment and adjust strategies based on reward mechanisms to obtain optimal paths [7–9]. Nevertheless, these algorithms involve high training costs and long model training times, which affect their accuracy and generalization ability. Most existing algorithms focus on obstacle avoidance and achieving the globally shortest path, with insufficient consideration of motion stability and dynamic constraints during AGV operation. In scenarios like gauge block verification that demand extremely high handling stability, vibrations during AGV operation may cause positional deviations or damage to gauge blocks, thereby affecting detection accuracy.

In terms of robot trajectory planning and smoothing, the traditional potential field method constructs a potential field model, treating the robot's movement as motion within the potential field. It prioritizes obstacle avoidance and quickly searches to avoid obstacle spaces [10]. However, this method is prone to local minima, preventing the robot from finding the optimal path. Traditional Rapidly-exploring Random Tree (RRT) series algorithms quickly search for a feasible path by randomly sampling in the state space and expanding the tree structure, but their paths are often insufficiently smooth and require further optimization [11–13]. Although these algorithms improve operational efficiency, they pay insufficient attention to the continuity of acceleration during robot movement. During direction changes, start-up, or shutdown phases of the robotic arm, abrupt changes in acceleration can cause impact vibrations, directly affecting the positioning accuracy of the end effector. For high-precision operations like gauge block verification, even minor vibrations may increase detection errors.

In the field of precision measurement automation, several advanced automated measuring devices and systems have been applied in industrial production, such as coordinate measuring machines and laser interferometers, enabling high-precision measurement of workpiece dimensions. In the field of multi-robot collaborative handling, numerous research achievements have also been made. By establishing multi-robot collaboration models, functions such as task allocation, path planning, and motion coordination have been realized, improving handling efficiency and flexibility. However, in the integrated application of precision measurement automation and multi-robot collaborative handling, issues related to stability and accuracy remain [14–16]. The collaborative operation of AGVs and robots requires high stability and precision, but existing collaborative methods tend to exhibit inconsistent coordination when dealing with complex tasks and dynamic environments. When verification tasks change or temporary obstacles appear in the verification environment, the collaborative strategies of AGVs and robots cannot be adjusted in a timely manner, leading to delays or errors in handling and inspection processes and affecting the accuracy and efficiency of gauge block verification.

To address these issues, this paper establishes an automated gauge block detection system with collaborative operation of AGVs and robots, and proposes a modified A*-APF path planning algorithm based on dynamic constraints. Kinematic and dynamic models of the AGV differential drive system are developed to quantitatively analyze the influence of turning radius and acceleration on operational vibration, with optimal parameters integrated into the path planning process as constraints. Compared with traditional path planning algorithms, the proposed method not only improves path smoothness but also ensures AGV operational stability. In terms of robot trajectory planning, compared with conventional approaches, the proposed trajectory planning method significantly enhances planning efficiency and shortens path length by improving the parent node reselection and pruning-reconnection mechanisms of RRT*. By combining multiple smoothing technologies and fifth-order non-uniform B-spline interpolation, high-order continuity of joint position, velocity, and acceleration is achieved to avoid impact and jitter. A hierarchical strategy guarantees global optimal search and local optimization, thereby improving trajectory accuracy and stability. Finally, by constructing an automated gauge block verification scenario and conducting both simulations and experiments, it is verified that the proposed method effectively improves the accuracy and stability of gauge block handling, providing a theoretical basis and technical support for the field of high-precision automated handling operations.

2 Path Planning and Stability Optimization Method of Composite Robot

2.1 System Composition and Equipment Division of Labor

To achieve the handling and pose adjustment throughout the entire process of automated gauge block verification, a hybrid robot verification system is designed and constructed, as shown in Fig. 1. The hybrid robot consists of an AGV and a collaborative robot, where the collaborative robot accomplishes stable handling operations of the gauge block tray between the cleaning station, verification station, and packaging station. In the verification station phase, a SCARA robot is adopted to grasp individual sets of gauge blocks and perform pose transformation.

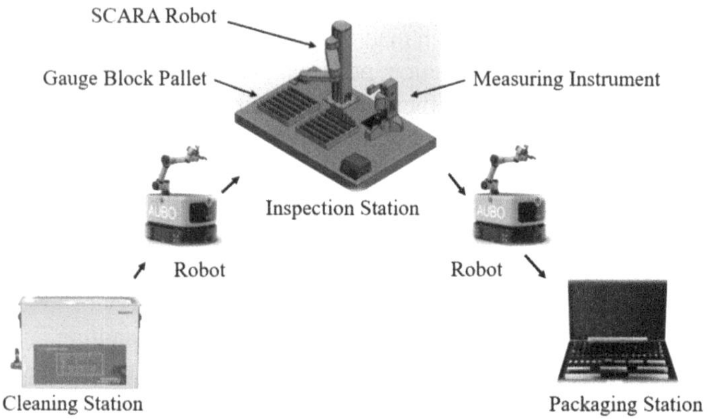

Fig. 1. Flow chart of automated gauge block verification

2.2 AGV Path Planning and Dynamic Stability Modeling

In the automated gauge block verification system, vibrations and inertial impacts may occur during AGV transportation, which can affect the positioning accuracy and handling stability of gauge blocks. There are two main factors contributing to vibrations: first, dynamic factors—differences in turning radius and inertial impacts during acceleration/deceleration stages can cause vehicle instability. Second, in terms of path planning algorithms, unsmooth paths can lead to high-frequency vibrations of the vehicle during operation. Therefore, ADAMS simulation is first performed on the AGV to simulate and analyze its motion states under different turning radii and acceleration conditions, optimizing dynamic operation parameters. Based on this, an improved A*-APF path planning algorithm under dynamic constraints is proposed to enhance stability during transportation.

The AGV adopts two-wheel differential drive; thus, in its dynamic model, the vehicle body consists of a chassis, four caster wheels, and two left/right driving wheels. To simulate the load state in actual handling scenarios, a concentrated load equivalent to a mass of 200 kg is applied at the center of mass on the top surface of the AGV body, simulating the collaborative robot and its upper components. Meanwhile, to simulate the inertial characteristics and vibration response of the gauge block tray during AGV operation, a solid sphere with mass equivalent to that of the gauge block tray is constructed and placed on the vehicle body, as shown in Fig. 2(a). The sphere is fixed by four sets of spring-damper mechanisms, where the spring stiffness is set to 12.3 N/mm and the preload is set to 0 N. By monitoring and analyzing the deformation and stress state of the spring-damper mechanisms during AGV motion, the inertial impacts and vibration characteristics borne by the gauge block tray are obtained.

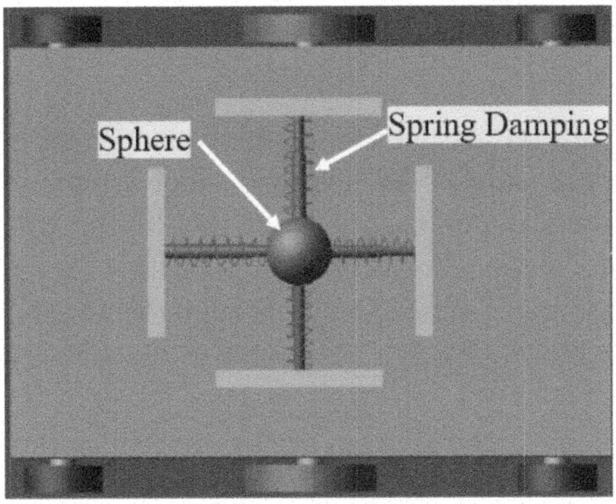

(a) Schematic of the spring-damper mechanism

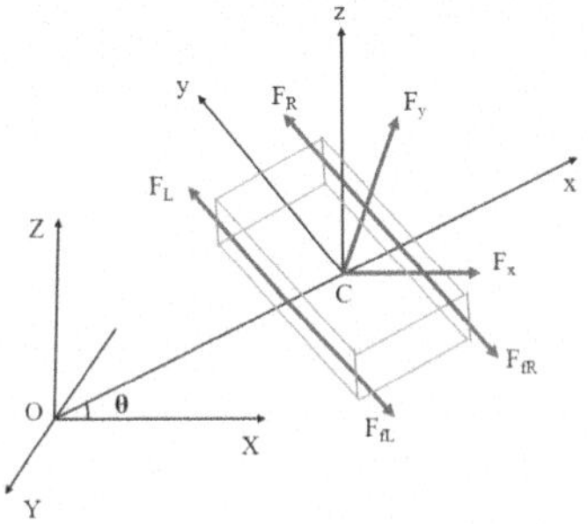

(b) Dynamic modeling

Fig. 2. AGV model construction

For the two-wheel differential AGV, its motion can be decomposed into translational motion in the vehicle body coordinate system and rotational motion around the center of mass. Therefore, the Newton-Euler equations are applied to establish the dynamic model of the AGV. The inertial coordinate system $\{O - XYZ\}$ and the vehicle body coordinate system $\{C - xyz\}$ are established respectively, as shown in Fig. 2(b). Here, m is the mass of the AGV; F_L and F_R are the driving forces acting on the left and right wheels, respectively; F_{fL} and F_{fR} are the ground friction forces acting on the left and right wheels, respectively; and M_z is the torque generated by the friction forces. This yields the dynamic relationship as shown in Eq. (1):

$$\begin{cases} m\ddot{x} = F_L cos\theta + F_R cos\theta - F_{fx} \\ m\ddot{y} = F_L sin\theta - F_R sin\theta - F_{fy} \\ M_z = \frac{1}{2}L(F_R - F_L) - M_{fz} \end{cases} \tag{1}$$

Model parameters are constructed in ADAMS based on the dynamic equations. The rubber wheel material is defined via the contact module in ADAMS: the contact stiffness between the rubber wheel and the ground is 5×10^6N/m; the damping coefficient is set to 5×104N·s/m, a typical value for AGVs; the force exponent, which characterizes the nonlinear characteristics of rubber wheel contact, is set to 1.5; the allowable penetration depth refers to the maximum distance that the wheel can "penetrate" the ground in the simulation. Considering that the diameter of the rubber wheel is 300 mm, it is set to 2 mm in the simulation.

Turning Radius and Angular Acceleration Analysis. When the AGV performs a turning maneuver, the speed ratio of its inner and outer driving wheels increases significantly, which destabilizes the contact between the driving wheels and the ground. This leads to irregular fluctuations in local friction at the contact interface, and such fluctuations are transmitted to the vehicle body through the wheels, ultimately inducing vibrations in the vehicle body—this is detrimental to the stable transportation of the gauge block tray.

To investigate the influence of different speed ratios under varying turning radii on vehicle stability, a variable m is defined, which represents the ratio of the rotational speed of the left driving wheel to that of the right driving wheel. Through kinematic analysis, the relationship between m, the turning radius R, and the wheelbase $L = 630mm$ of the driving wheels is derived as follows:

$$R = \frac{L}{2}\left(1 + \frac{2}{m - 1}\right) \tag{2}$$

Based on the turning radii R that the AGV may encounter during motion, the corresponding speed ratios are derived from Eq. (2). Simulation analysis yields Fig. 3, which illustrates the influence of turning radii corresponding to different speed ratios m on the traction force of the AGV's driving wheels.

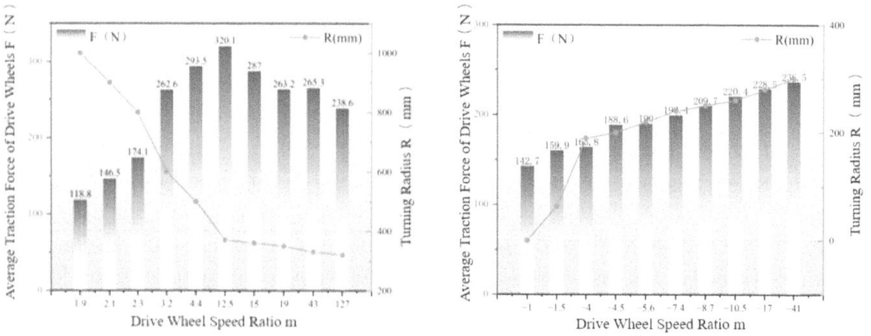

(a) Speed ratio m is positive (b) Speed ratio m is negative

Fig. 3. AGV model construction

Figure 3 shows the trend of driving wheel traction force varying with turning radius: when the turning radius is close to 370 mm, i.e., $L/2$, , the traction force reaches its peak. The peak traction force will intensify friction fluctuations at the contact interface between the driving wheels and the ground, inducing vehicle body vibrations and thus affecting the stable transportation of the gauge block tray. To ensure transportation stability, the operating condition with a turning radius of 370 mm (i.e., $L/2$) should be prioritized to avoid in actual path planning, and steering parameters should be adjusted to reduce vibration interference.

To analyze the influence of driving wheel acceleration on the operational stability of the AGV body, parameter settings are shown in Fig. 4(a), aiming to cover the angular acceleration range that may occur during actual AGV operation, so as to systematically evaluate the impact of different accelerations on vehicle body vibration and tray stability. Through simulation, the AGV is controlled to complete acceleration-deceleration motion, and the force response of the spring-damper mechanism connected to the equivalent tray (solid sphere) above the vehicle body is monitored in real time, as shown in Fig. 4(b).

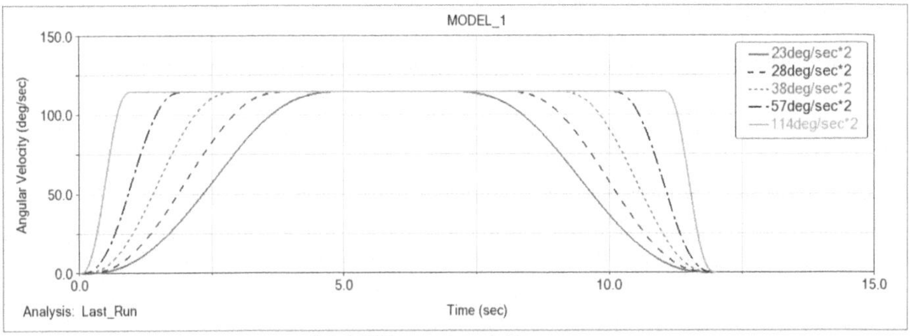

(a) AGV speed variation curve

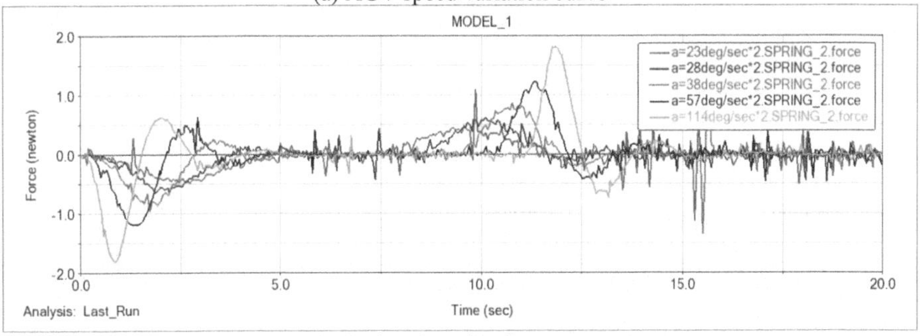

(b) Spring damping force curve

Fig. 4. AGV's motion and mechanical response under different driving wheel angular accelerations

The results show that as the amplitude of acceleration increases, the overall amplitude of the spring-damper force presents an upward trend. Under small accelerations, the high-frequency fluctuations of the force curve are more pronounced, characterized by high fluctuation frequency and large instantaneous amplitude changes. Mechanical transmission path analysis indicates that these high-frequency fluctuations are transmitted to the vehicle body through the spring-damper structure, causing vehicle body vibrations. These vibrations further act on the tray, increasing the micro-displacements of the gauge blocks.

Through multi-group data simulations, when the vehicle acceleration is controlled within the range of 0.08 m/s^2 to 0.12 m/s^2, it can not only constrain the deformation of the spring-damper within an acceptable range that does not affect the gauge block posture but also effectively suppress vehicle vibration caused by high-frequency fluctuations in stress. This effectively controls the repetitive positioning error of the gauge block during handling within 1.98 ± 0.02 mm, thereby providing a guarantee for the stable transfer of gauge blocks between stations and ensuring that the positioning accuracy of subsequent verification processes is not interfered by the transfer process.

Path Planning Algorithm Based on Dynamic Constraints. After completing the dynamic stability analysis of the AGV under different turning radii and driving wheel angular accelerations, to further enhance the smoothness of its path planning and operational stability, an improved A*-APF path planning algorithm under dynamic constraints is constructed based on the dynamic constraints derived earlier.

By deleting redundant nodes, only the currently known optimal path for each state is retained (based on the criterion of minimizing f(n) or g(n)), which improves the efficiency of path planning. On the other hand, an inflection point penalty coefficient ω is introduced into the traditional heuristic function to dynamically adjust the balance weight of inflection points in the path. Under the same map conditions, five different ω values (0.5, 1.0, 1.5, 2.0, and 2.5) are selected for testing respectively. The results show that if ω is too small, the suppression of inflection points is insufficient; if ω is too large, the path becomes excessively long. When ω is 1.5, the number of path inflection points is reduced by 37%, which effectively improves motion smoothness.

In the improvement of the artificial potential field method, aiming at the problem that the traditional method is prone to path interruption caused by the balance between attractive force and repulsive force, the repulsive field model is optimized:

$$F_{rep-obj-obs} = Eta_{rep-obj} \times \left(\frac{1}{dist(J,1)} - \frac{1}{d_0}\right) \times \frac{dist(N,1)}{dist(J,1)^2} \tag{3}$$

$dist(N,1)$ (distance from the vehicle to the target) is introduced as an adjustment factor. In this way, when the vehicle is approaching the target, even if it is close to an obstacle, the repulsive force will be adjusted under the influence of the distance to the target, avoiding the path interruption issue.

In addition, to incorporate the dynamic analysis results into path planning, dynamic constraints on velocity and acceleration as shown in Eqs. (4–7) are introduced. The AGV monitors the distance to obstacles in real time and dynamically adjusts its operating velocity and acceleration according to the distance range.

$$v = \begin{cases} v_{min} & if\ d \leq d_{min} \\ v_{min} + (v_{max} - v_{min}) \cdot \frac{d-d_{min}}{d_{sup}-d_{min}} & if\ d_{min} < d \leq d_{\text{safe}} \\ v_{max} & otherwise \end{cases} \tag{4}$$

$$\Delta v = v_{\text{desired}} - v \tag{5}$$

$$a_{actual} = min\left(\frac{\|\Delta v\|}{\Delta t}, a_{max}\right) \cdot \frac{\Delta v}{\|\Delta v\|} \tag{6}$$

$$v_{new} = v + a_{actual} \cdot \Delta t \tag{7}$$

wherein: d: real-time distance between the AGV and the nearest obstacle; d_{min}: minimum safe distance, below which the AGV must decelerate to v_{min}; d_{safe}: deceleration threshold, above which the AGV can operate at full speed v_{max}; v_{min}: minimum speed, ensuring safety when approaching obstacles; v_{max}: maximum speed, the highest speed in open areas; v_{desired}: desired velocity calculated by the speed control formula; a_{max}: maximum allowable acceleration; Δt: control cycle.

Hazardous zone ($d \leq d_{min}$): Forced low speed to avoid collisions.
Transition zone ($d_{min} < d \leq d_{safe}$): Velocity increases linearly with distance, balancing safety and efficiency.
Safe zone ($d > d_{safe}$): Full-speed operation to improve efficiency.

When approaching an obstacle within the constraint range, the AGV decelerates with a preset constrained acceleration to reduce collision risks; when moving away from obstacles, it accelerates smoothly with an adaptive acceleration. This ensures safety while reducing vibrations caused by abrupt velocity changes, ultimately achieving collaborative optimization of path planning and dynamic characteristics, and providing reliable motion guarantees for high-precision gauge block handling. Figure 5 shows the pathfinding process of the dynamic constraint algorithm based on A*-APF: (a) Expanding initial nodes based on the heuristic function; (b) Repulsive force surging under the influence of dynamic obstacles, changing the path direction; (c) The adjustment factor dynamically regulating the balance between obstacle avoidance and guidance in long-distance paths; (d) Attractive force dominating to converge the direction toward the endpoint.

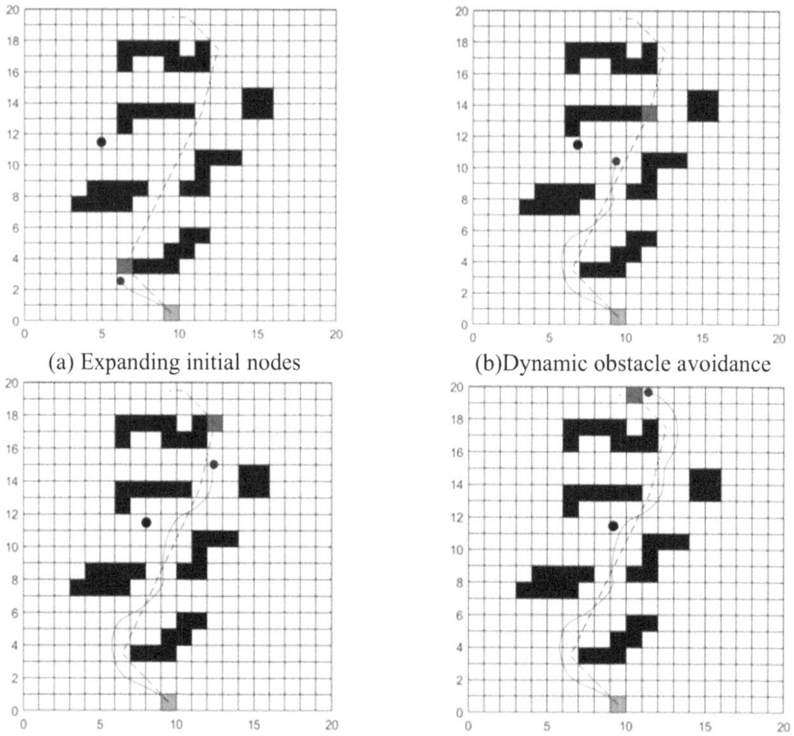

(a) Expanding initial nodes (b)Dynamic obstacle avoidance

(c) Balance regulation (d) Path convergence

Fig. 5. Pathfinding process of the improved A*-APF algorithm under dynamic constraints

2.3 Hierarchical Optimization Strategy for Robotic Arm Trajectory

The purpose of the SCARA robot is to transfer gauge blocks from the tray to the detection platform for inspection. Unlike AGVs, the instability factors of the robot in this process originate from end-effector trajectory fluctuations and joint vibrations of the SCARA robot. Therefore, a hierarchical trajectory optimization strategy is proposed. First, the improved RRT* algorithm is applied for parent node reselection and path pruning to optimize the rationality of the global path. Second, in the Cartesian space, B-spline interpolation and fourth-order low-pass filtering are used to smooth discrete trajectories, eliminating zigzag fluctuations and high-frequency burrs. Finally, in the joint space, fifth-order non-uniform B-splines (NURBS) interpolation is adopted to ensure the continuity of joint position, velocity, and acceleration, reducing instantaneous impacts during start-stop or direction changes.

Improved RRT* Algorithm. Parent node reselection and path pruning optimizations are performed on the traditional RRT* algorithm, as shown in Fig. 6. When the node corresponding to path $d3$ is taken as the parent node of X_{new}, its total path cost is lower than that of the original parent node. Therefore, the node of path $d3$ is determined as the new parent node of X_{new} to reduce the local path cost. After completing the parent node reselection, further path pruning is performed: if the cost of the new path is less than the original path cost of the neighboring node, the parent node of this neighboring node is updated to X_{new}.

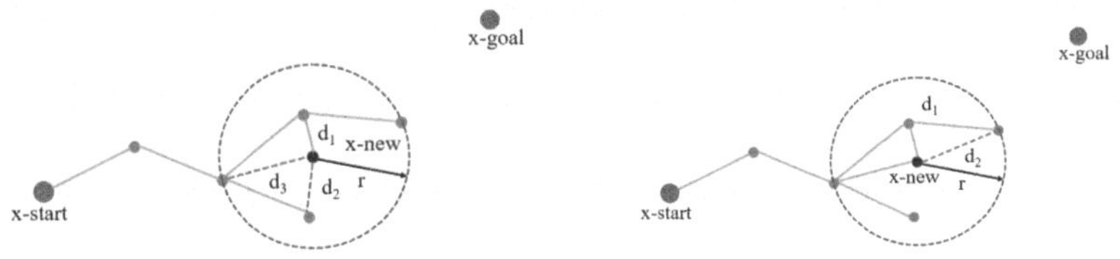

Fig. 6. Improved RRT* algorithm

As shown in Eq. (8), to improve search efficiency and guidance, a goal bias strategy is introduced. After algorithm initialization, the goal bias probability *Pb* is set within a certain range, and meanwhile, the goal bias probability is dynamically adjusted according to collision occurrences during the growth step of the node tree.

$$Pa < Pb < Pc \tag{8}$$

As shown in Fig. 7, a 50 × 50 × 50 three-dimensional space based on the coordinate system {O-XYZ} is constructed in MATLAB software, with the same obstacle configuration. Both the traditional RRT* algorithm and the improved RRT* algorithm are run 30 times in the obstacle map, and the average values are calculated. The average path length of the improved algorithm is 73, with a time consumption of 0.14 s. Compared with the traditional RRT*, due to the higher guidance of the improved RRT* algorithm and fewer search nodes, its average path length is shortened by 23.7%, and its running time is reduced by 94%. This verifies that the improved algorithm has higher search efficiency and stronger guidance compared to the traditional RRT* algorithm in obstacle maps.

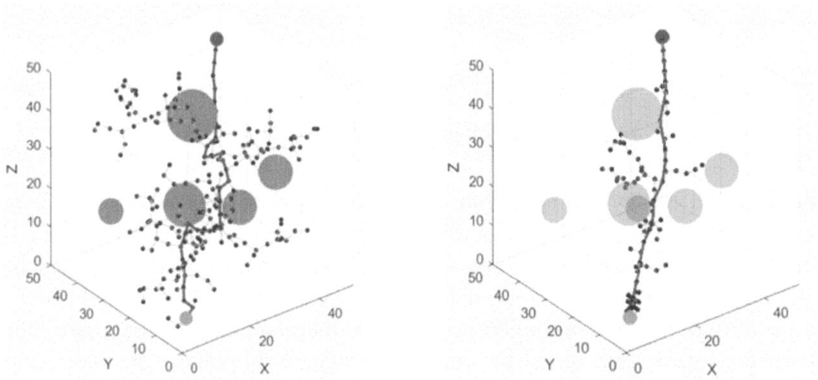

(a) Traditional RRT* algorithm (b) Improved RRT* algorithm

Fig. 7. Simulation results of 3D map

Trajectory Smoothing Processing in Cartesian Space. In the Cartesian space, b-spline interpolation is used for fitting and interpolating the original discrete path points, expanding these discrete points into high-density path points through interpolation. This improves path resolution to avoid path discretization caused by excessively large spacing between discrete points.

A fourth-order low-pass filter is introduced, adopting a zero-phase processing strategy of "forward filtering + reverse filtering". Bidirectional filtering is used to offset phase deviations, optimizing residual zigzag fluctuations, local burrs, and high-frequency jitter in the path.

A collision detection function and a weighted optimization mechanism are introduced. A multi-objective optimization function is constructed by setting smoothing weights and path preservation weights, avoiding safety risks caused by excessive deviation of the smoothed path from the original obstacle avoidance path.

As shown in Fig. 8, compared with the original discrete path, the zigzag fluctuations and high-frequency burrs in the path optimized by the above methods have been effectively suppressed. The path curve exhibits continuous and smooth characteristics, which effectively avoids impact vibrations during robot movement and provides stable support for Cartesian space motion trajectories in high-precision gauge block handling.

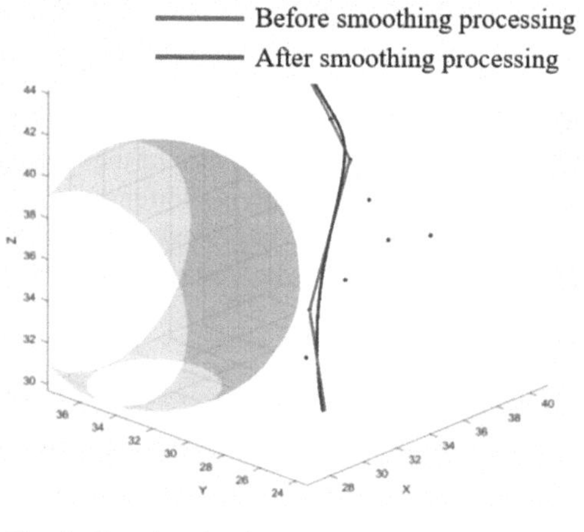

Fig. 8. Results of trajectory smoothing processing

Joint Space Trajectory Interpolation. To ensure the continuity of joint acceleration and reduce impact during start-stop or direction changes, fifth-order NURBS curves are used for interpolation. a NURBS curve consists of control points, weight factors, and a knot vector. Control points determine the position of the curve but generally do not lie on the curve. Weight factors are equivalent to the gravitational influence of control points; the larger their values, the closer the curve is to the control points. The knot vector is defined as $U = [u_0, u_1 ... u_n ... u_{n+m+1}]$, and the variation of the NURBS curve can be controlled through the knot vector.

Interpolation is performed based on the time and position node sequence $(\{P_i, t_i\}, i = 1,2,3...m)$, where the expression of the K-th order non-uniform B-spline curve is shown in Eq. (9):

$$p(u) = \sum_{i=0}^{n} d_i N_{i,k}(u) \tag{9}$$

In Eq. (9), K is the degree of the B-spline; d_i are control points; n denotes the number of control points, with a value of m − 1 + k; $N_{i,k}(u)$ is a function defined on the non-uniformly distributed knot vector U, where the starting and ending knots have a multiplicity r of $K + 1$. The basis functions of the NURBS curve can be derived from the Cox-deBoor recurrence formula as follows:

$$\begin{aligned} & N_{i,0}(u) = \begin{cases} 1, \ u_i \le u \le u_{i+1} \\ 0, \qquad other \end{cases} \\ & N_{i,k}(u) = \frac{u-u_i}{u_{i+k}-u_i} N_{i,k-1}(u) + \frac{u_{i+k+1}-u}{u_{i+k+1}-u_{i+1}} N_{i+1,k-1}(u) \\ & define = \tfrac{0}{0} = 0 \end{aligned} \tag{10}$$

For the fifth-order NURBS curve, to ensure that the trajectory passes through each specified path point, the knots $ui + 5(i = 1,2 \ldots m)$ need to be paired sequentially with the corresponding control points P_i. Here, the total number of knot vectors is m + 2k, and the number of control points is m − 1 + k. The knot vector must have sixth-order multiplicity at both the start and end, with interpolation times of 0 and 1 at the start and end, respectively. Thus, the vector U = [0,0, 0,0, 0,0, u_7, u_8..., u_{m+k-1}, 1,1, 1,1, 1,1].

Figure 9 shows the comparison results of joint angle, joint velocity, joint acceleration, and joint jerk curves obtained after interpolation using third-order NURBS and fifth-order NURBS, respectively. The fifth-order NURBS curve, with more control points, can generate smoother curves, ensuring continuous and smooth acceleration. Moreover, the increased number of control points provides higher degrees of freedom and flexibility, enabling better fitting of complex curve shapes. This effectively reduces impact vibrations generated by the robot during direction changes, start-ups, or shutdowns, which could otherwise affect the positioning accuracy of the end effector.

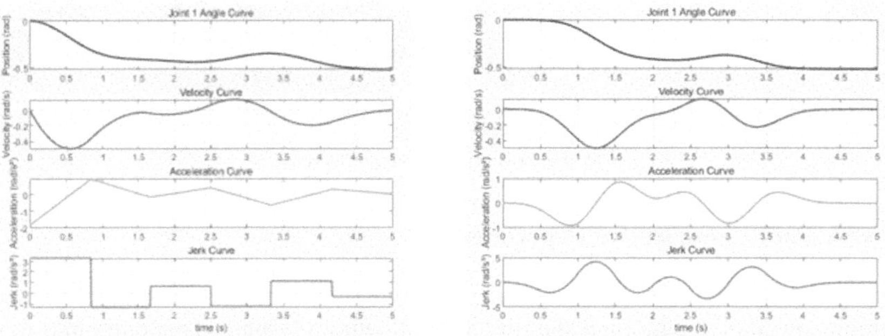

(a) Third-order NURBS interpolation (b) Fifth-order NURBS interpolation

Fig. 9. Interpolation results

3 Experiments and Results

3.1 Experimental Platform and Test Scenarios

To verify the feasibility of the previously proposed two-wheel differential AGV path planning, robot trajectory optimization, and gauge block handling stability optimization methods, an experimental scenario as shown in Fig. 10 was established. In accordance with the gauge block verification process, the initial station, cleaning station, detection station, and packaging station in Fig. 10(a) were sequentially set up. To simulate the complex environment in production, multiple groups of fixed columns were arranged to simulate equipment bases, with the spacing between obstacles controlled within the range of 2.0 m to 3.0 m. The AGV adopted AUBO-ARM300 and was equipped with an AUBO-i10 robot. For the SCARA robot at the detection station in Fig. 10(b), HITBOT Z-Arm 4160N0-A0M2 was selected, featuring a repetitive positioning accuracy of ± 0.02 mm. It grips preset single sets of gauge blocks in the gauge block tray according to the optimized RRT* path and fifth-order NURBS trajectory. The gauge blocks are made of tungsten steel with a density of 0.0145 g/mm^3, cross-sectional dimensions of 9 mm × 35 mm, and heights ranging from 0.5 mm to 100 mm, totaling 83 blocks. The measurement platform is equipped with a HEIDENHAIN CT2500 grating displacement sensor to detect the length deviation, flatness, and parallelism of the gauge blocks, with an accuracy of up to ± 0.1 μm. The detection data is displayed via host computer software, and the rationality of the proposed research methods is verified by monitoring the real-time operating status data of the AGV and the real-time detection data of the gauge blocks.

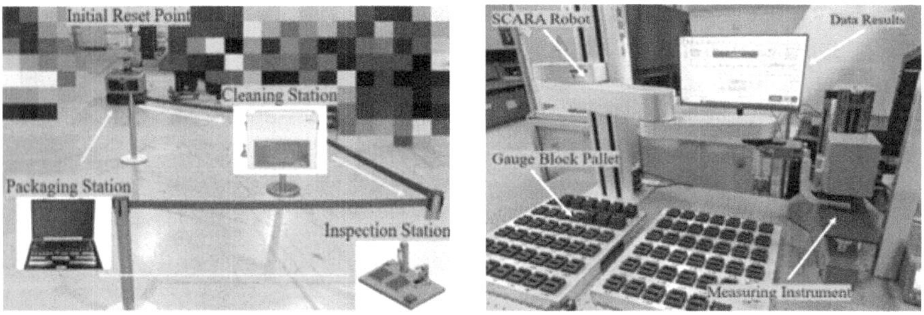

(a) Gauge block handling test scenario (b) Verification station test scenario

Fig. 10. Experimental scenarios

3.2 AGV Path Comparative Experiment

To verify the optimization effect of the improved A*-APF algorithm under dynamic constraints on AGV operational stability, the following settings were configured in the experimental scenario: the AGV's laser radar was set with a scanning frequency of 10 Hz, a detection radius of 15 m, and a scanning angle of 360°. The global AGV driving dynamic constraints were defined as: $d_{safe} = 1.5m, d_{min} = 0.8m, v_{min} = 0.1m/s, v_{max} = 0.3m/s, a_{max} = 0.1m/s2$. Under the condition of ensuring the same path conditions, a comparative analysis was conducted between the traditional path planning algorithm and the dynamic constraint path planning algorithm. A vibration sensor mounted inside the AGV was used to real-time monitor the stability during AGV operation, resulting in the vibration amplitude and inertial impact curves of the AGV during operation as shown in Fig. 11.

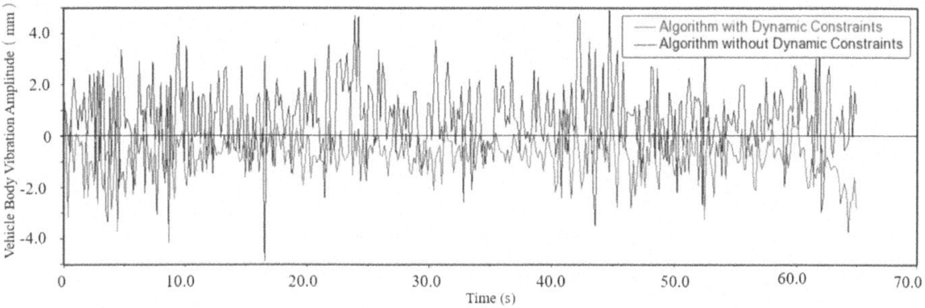

(a) AGV vibration amplitude curve during operation

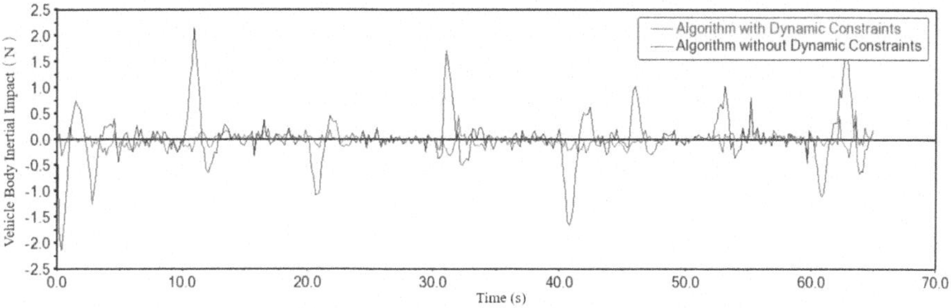

(b) AGV inertial impact curve during operation

Fig. 11. Data Comparison between Traditional Algorithm and Dynamic Constraint Algorithm

From the curve characteristics, it can be seen that the traditional path planning algorithm, due to the lack of dynamic constraints, leads to surging driving wheel traction during vehicle operation—caused by factors such as turning or abrupt speed changes. This further results in excessively high vibration amplitude and peak inertial impact of the vehicle body, making the overall stability of the vehicle body difficult to control. In contrast, the dynamic constraint path planning algorithm reasonably restricts the turning radius threshold and dynamically adjusts the acceleration during acceleration and deceleration processes. This effectively reduces the vibration amplitude and peak inertial impact of the vehicle body during operation, improving the stability of the vehicle body.

Table 1. AGV operation data.

Algorithm	Average vibration amplitude (mm) ± SD	Peak average vibration amplitude (mm) ± SD	Average inertial impact (N) ± SD	Peak average inertial impact (N) ± SD
Traditional path planning algorithm	3.17 ± 0.35	6.12 ± 0.58	1.01 ± 0.12	2.15 ± 0.23
Dynamic constraint path planning algorithm	1.98 ± 0.22	3.31 ± 0.32	0.25 ± 0.04	0.75 ± 0.09

To further clarify the specific performance differences between the two algorithms, ten groups of comparative experiments were conducted under the same experimental conditions, and the experimental data were quantitatively analyzed as shown in Table 1. From the perspective of vibration performance, the average vibration amplitude of the dynamic constraint path planning algorithm decreased from 3.17 mm (of the traditional algorithm) to 1.98 mm, and the peak average vibration amplitude reduced from 6.12 mm to 3.31 mm. Meanwhile, the standard deviation of the average vibration amplitude and the standard deviation of the peak average vibration amplitude also decreased significantly. From the perspective of inertial impact performance, the average inertial impact of the dynamic constraint path planning algorithm dropped from 1.01 N (of the traditional algorithm) to 0.25 N, and the peak average inertial impact decreased from 2.15 N to 0.75 N. Furthermore, the standard deviation of the average inertial impact and the standard deviation of the peak average inertial impact were greatly reduced. After t-test, the p-values of all comparative indicators are less than 0.001, indicating statistically significant effects. This demonstrates that the dynamic constraint path planning algorithm has significant advantages in suppressing vibration and inertial impact during AGV operation, thereby improving the stability of gauge block handling.

3.3 Consistency Analysis Between Robot Path Simulation and Actual Trajectory

To verify the actual effect of the robot trajectory optimization scheme, real-time tracking and recording were performed for the full process of the robot gripping gauge blocks from the gauge block tray, completing posture transformation, and placing them on the detection platform. A total of 83 gauge blocks of different specifications were handled respectively, and the results are shown in Fig. 12. The actual operation path of the robot is highly consistent with the simulation-optimized path. During the movement of each joint of the robot, there is no local vibration or joint singularity, and the positional accuracy error of the robot when placing the gauge blocks is controlled within ± 0.10 mm.

(a) Grip the gauge block and place it on the detection platform

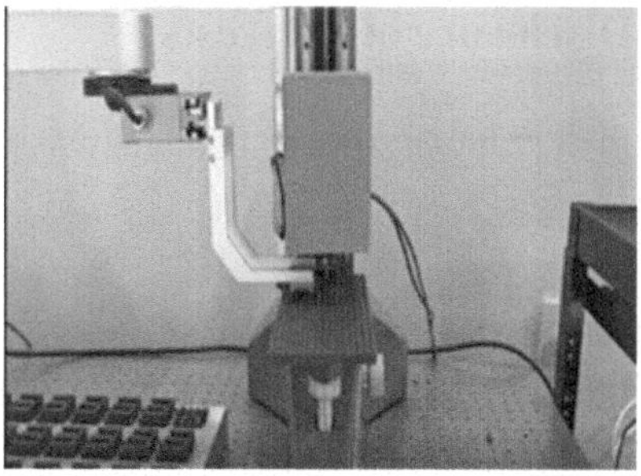

(b) Adjust the detection pose of the gauge block using the robotic arm

(c) Completion of single set of gauge block inspection

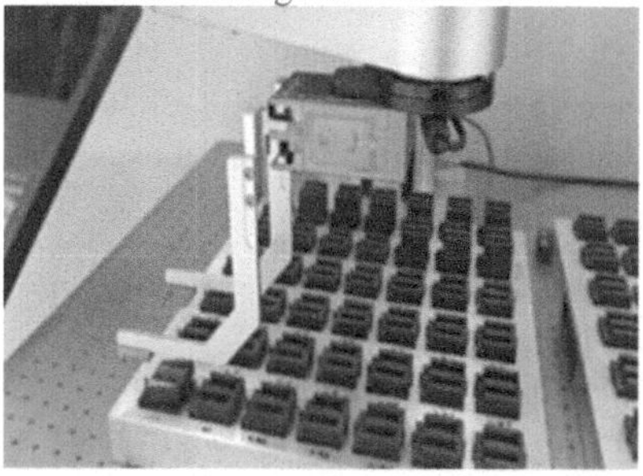

(d) Place the gauge block back into the tray and proceed to the next set of inspection

Fig. 12. Gauge block gripping-inspection-replay full-process experiment

4 Conclusions

This study presents an automated gauge block verification system that integrates operation of Automated Guided Vehicles (AGVs), cobot and SCARA robot, enabling the automation of gauge block handling, pose adjustment, and verification.

In AGV handling, we introduced an improved A*-APF path planning algorithm under dynamic constraints. Experimental results demonstrate that, compared to traditional methods, the optimized path planning reduces the AGV's average vibration amplitude to 1.98 mm, decreases peak inertial impact to 0.75 N, reduces path inflection points by approximately 37%, and shortens total handling time by about 30%. These improvements contribute to enhanced handling stability and efficiency.

For robot trajectory planning, we enhanced the RRT* algorithm, incorporating trajectory smoothing and fifth-order NURBS interpolation to ensure smooth and stable motion of the robot. The experimental results show that the positional accuracy of gauge block clamping and placement is within ± 0.10 mm, with an average error rate of 7.4%.

This research demonstrates that the proposed system achieves improvements in both stability and efficiency for automated gauge block handling. The findings provide technical references for the design and optimization of precision automated handling systems, with potential applications in precision manufacturing and inspection tasks.

5 Disclosure of Interests

The authors have no competing interests to declare that are relevant to the content of this article.

Acknowledgments. This study was funded by 2022 Hybrid Robot Program (grant No. TC220H05D).

References

1. Ahmed, K.M., AlTalhh, Y.A., AlZamil, A.N., et al.: Development of novel interferometric system for short and long gauge block calibration. Meas. Sci. Technol. **35**(12), 126015 (2024)
2. Sanjid, M.A.: A novel method to calibrate standard angle gauge blocks. Measurement **229**, 114496 (2024)
3. Ahmed, G., Sheltami, T., Yasar, A.: Optimal path recommendation in dynamic traffic networks using the hybrid Tabu-A* algorithm. Transp. Res. Part E Logist. Transp. Rev. **204**, 104414 (2025)
4. Liu, Q., Liu, Q., Chen, L.: Intelligent scheduling and emergency management for mining areas integrating improved bidirectional A* algorithm path planning. Appl. Earth Sci. **134**(3), 172–183 (2025)
5. Wang, W., Tian, J.: Dynamic UAV path planning in mountainous terrain utilizing an arithmetic optimization algorithm incorporating adaptive thermal conduction search and elite population genetic strategies. Aerosp. Sci. Technol., 110950 (2025)
6. Van Nguyen, L., Kwok, N.M., Ha, Q.P.: Fermat-Weber location particle swarm optimization for cooperative path planning of unmanned aerial vehicles. Appl. Soft Comput. **167**, 112269 (2024)
7. Gao, H., Wang, J., Wang, X., et al.: Deep reinforcement learning coupled with topological scene graph for dynamic path planning of autonomous bulldozer in complex earthwork construction. Autom. Constr. **181**, 106617 (2026)
8. Zhang, Z., Fu, H., Yang, J., et al.: Deep reinforcement learning for path planning of autonomous mobile robots in complicated environments. Complex Intell. Syst. **11**(6), 277 (2025)
9. Choi, S.C., Lee, Y., Cho, S.W.: Reinforcement learning-integrated evolutionary algorithm for enhanced unmanned aerial vehicle coverage path planning. Swarm Evol. Comput. **97**, 102051 (2025)
10. Fang, Z., Liang, X.: Intelligent obstacle avoidance path planning method for picking manipulator combined with artificial potential field method. Ind. Robot Int. J. Robot. Res. Appl. **49**(5), 835–850 (2022)
11. Li, Y., Chen, N.: Improved RRT algorithm for robotic arm path planning based on reward strategy. Proc. Inst. Mech. Eng. C J. Mech. Eng. Sci. **239**(17), 6983–6995 (2025)
12. Cao, M., Mao, H., Tang, X., et al.: A novel RRT*-Connect algorithm for path planning on robotic arm collision avoidance. Sci. Rep. **15**(1), 2836 (2025)
13. Gao, Q., Yuan, Q., Sun, Y., et al.: Path planning algorithm of robot arm based on improved RRT* and BP neural network algorithm. J. King Saud Univ. Comput. Inf. Sci. **35**(8), 101650 (2023)
14. Wu, Y., Li, H.: Multi-robot collaborative obstacle avoidance based on hierarchical path planning methods. Eng. Appl. Artif. Intell. **163**, 112796 (2026)

15. Zhao W, Zhang J, Liu Y, et al. High accuracy measurement path planning of robot optical measurement system for free-form surfaces based on convex optimization. Measurement, 2025: 118639
16. Li, C., Yao, Y., Liu, X., et al.: Neighborhood-perturbed and nonlinearly enhanced whale optimization for high-precision trajectory tracking and stability control in four-steering-wheel AGVs. Int. J. Electr. Power Energy Syst. **172**, 111048 (2025)

Research on Local Path Planning for Mobile Robots in Unstructured Environments

Yang Meng[1], Pengju Wang[1], Honglei Mu[2], Shuoshi Wang[3], and Mingtao Yao[4(✉)]

[1] Beihang University, Beijing 100191, China
{mengyang_buaa,pengjuwang}@buaa.edu.cn
[2] Tsinghua University, Beijing 100084, China
[3] Beijing Luhang Yilong Intelligent Technology Co., Ltd., Beijing 100083, China
wss@ckzone.cn
[4] Beijing University of Chemical Technology, Beijing 100029, China
yaomingtao@buct.edu.cn

Abstract. Aiming at the problems of low global path tracking accuracy, dynamic obstacle avoidance lag and insufficient vehicle kinematic constraint adaptability of the traditional dynamic window approach (DWA) in unstructured environments, a multi-dimensional improved DWA algorithm is proposed. Methodologically, the turning radius constraint cost function is constructed based on the Ackermann steering model, and the over-limit path generation is suppressed by an exponential penalty term; the spatio-temporal trajectory prediction mechanism is designed, and a linear extrapolation model is used to predict the trajectory of the dynamic obstacle and compute the collision risk, so as to enhance the dynamic obstacle avoidance ability; a hierarchical dynamic weight adjustment strategy is proposed, and the weight allocation is optimized in real-time according to the distance of the goal and the proximity of obstacles; a fractional-order PID controller is introduced to optimize trajectory tracking, and the Grünwald-Letnikov discretization method is used to reduce the overshooting and steady state error. The results of simulation and real-vehicle experiments show that the motion time of the improved algorithm is reduced by 11.3%, the path length is shortened by 8.1%, and the number of iterations is reduced by 12.2% in the scenario without dynamic obstacles, and the path length is further reduced by 6.1% and the number of iterations is reduced by 10.0% in the scenario with dynamic obstacles. The conclusion shows that the improved algorithm effectively improves the feasibility of path planning, dynamic obstacle avoidance efficiency and trajectory tracking accuracy, and is suitable for localized path planning of mobile robot in complex dynamic unstructured environments.

Keywords: mobile robot · local path planning · unstructured environments · improved dynamic window approach

H. Wei and J. Niu (Eds.): COBOT 2025, CCIS 2893, pp. 290–302, 2027.
https://doi.org/10.1007/978-981-92-2074-8_17

1 Introduction

With the rapid development of automation and information technology, mobile robots are widely used in many fields such as logistics, agriculture and public transportation. Meanwhile, the accuracy and real-time performance of local path planning algorithms in dynamic and complex environments face higher challenges. The Dynamic Window Approach (DWA) is an algorithm for path planning in mobile robots, particularly suited for navigation challenges in dynamic environments. Its objective is to flexibly plan the robot's path within such environments to adapt to constantly changing obstacles and environmental conditions [1]. The DWA algorithm achieves real-time responsiveness and adaptability by dynamically adjusting the robot's velocity and angular velocity in real time to accommodate the evolving environment.

In recent years, improvement studies for DWA have shown a multi-dimensional development trend. Ndidiamaka Adiuku et al. [2] proposed a hybrid algorithm that combines RRT and DWA, and verified the safety and real-time performance of the algorithm, but the computational overhead of multi-sensor fusion limits its application in resource-constrained platforms. Liu Yang et al. [3] design the adaptive DWA based on the riskiness function, and optimize the dynamic obstacle collision risk by quantifying the risk of dynamic obstacles. Dandan Hu et al. [4] reconstruct the evaluation function of an off-axle full trailer system to achieve the minimum obstacle avoidance distances. Wang et al. [5] proposed the DAV_DWA algorithm, which uses the dynamic and static dual obstacle evaluation function to distinguish the obstacle avoidance priority of dynamic and static obstacles, and combines with the virtual target method to solve the local extremum problem. Y Ming et al. [6] introduce an energy consumption model of the drive motor, and extends the evaluation function of the DWA to generate a low-energy consumption smoothing path.

The trajectory tracking control needs to ensure driving stability and comfort while avoiding obstacles. He Yang et al. [7] fused the speed obstacle model with fuzzy MPC to achieve better performance of the car at high speeds, but their controller did not consider the decision conflict problem of cooperative obstacle avoidance with multiple obstacles. Zhang zhineng et al. [8] introduced spline curve parameterization of candidate path clusters in dynamic path planning and optimized the tracking error by combining with a fractional-order PID controller, but the generation of their candidate path clusters relied on fixed parameters and lagged in the response to sudden obstacles. Chen Shuping et al. [9] design a hierarchical MPC architecture for four-wheel independent drive electric vehicles to coordinate the trajectory tracking with the stability of traversing sway, and realize the high-precision control with the center-of-mass deviation angle of less than 2.32° under the extreme conditions, but the computational complexity of the algorithm is high, which makes it difficult to satisfy the demand of millisecond real-time performance. Zhai Li et al. [10] used Bessel curves for smoothing after generating the initial obstacle avoidance path through the potential field method, but their potential field method still has the problem of local minima, and the Bessel curves have limited real-time adjustment ability for dynamic obstacles.

In this paper, a multidimensional improved DWA algorithm is proposed, aiming to solve the above problem. First, a kinematic constraint cost function is constructed based on the Ackermann steering model, which is combined with an instantaneous turning radius penalty term to avoid generating an overrun path. Second, the dynamic obstacle avoidance mechanism incorporating spatio-temporal trajectory prediction improves the dynamic environment adaptability by linearly extrapolating the obstacle trajectory and calculating the collision risk. A hierarchical dynamic weight adjustment strategy is proposed to optimize the weight allocation in real time according to the target distance and obstacle proximity. Finally, a fractional-order PID controller is introduced to optimize trajectory tracking, and its performance advantage over traditional PID is verified through comparative experiments.

2 Improved Dynamic Window Approach

The Dynamic Window Approach (DWA) selects the optimal path by using a scoring function in the velocity sampling space. This algorithm sets boundaries for velocity sampling based on robot performance, forming a dynamic window. It does this by modeling and scoring all possible velocity combination paths to pick the highest scoring path as the next direction of travel for the robot. The trajectory consists of multiple segments affected by linear and angular velocities and is computed considering.

In this paper, the traditional DWA algorithm has poor global path point tracking and poor dynamic obstacle avoidance in the process of utilization. An improved DWA algorithm is proposed in terms of steering limitation, spatio-temporal trajectory prediction and path tracking optimization.

2.1 Vehicle Kinematics Model

Ackermann geometry is a common method of vehicle steering in which the inside wheel has a smaller turning radius than the outside wheel when turning. The model is shown in Fig. 1.

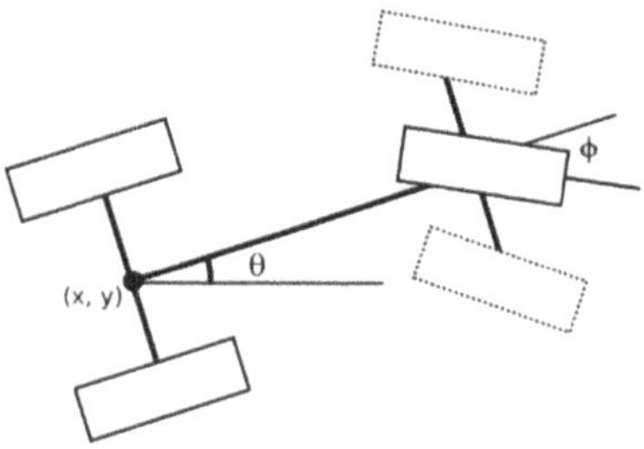

Fig. 1. Ackermann steering model

After simplifying the kinematic model of Ackermann trolley, the control quantities are usually chosen as the rear wheel speed v and the front wheel angle ϕ, the state quantities are chosen as the coordinates in the world coordinate system (x, y) and the heading angle θ, and the vehicle wheelbase l. The kinematic model of Ackermann steering can be expressed as Eq. 1.

$$\begin{cases} \dot{x} = v\cos(\theta) \\ \dot{y} = v\sin(\theta) \\ \dot{\theta} = \frac{v}{l}\tan(\phi) \end{cases} \tag{1}$$

where $\dot{x}$, $\dot{y}$ denotes the linear velocity components of the vehicle in the x direction and y direction respectively at the current moment, and $\dot{\theta}$ denotes the angular velocity of the vehicle at the current moment.

2.2 B. Kinematic Constraint Optimization and Dynamic Obstacle Avoidance Enhancement

The traditional Dynamic Window Algorithm (DWA) mainly considers three cost terms of target proximity, speed maximization and obstacle avoidance in path planning, but does not consider vehicle kinematic characteristics. For vehicle models with minimum turning radius constraints, there will be some scenarios in which unexecutable paths may be generated. In this paper, we introduce the turning radius constraint cost function to ensure that the generated paths satisfy the vehicle kinematic characteristics; and we introduce the spatio-temporal collision risk cost to optimize the ability of avoiding dynamic obstacles.

A new steering radius cost term is added to the cost function, and for the candidate velocity v and angular velocity ω, the instantaneous turning radius R can be expressed as Eq. 2.

$$R = \left|\frac{v}{\omega}\right|, \omega \neq 0 \tag{2}$$

An exponential penalty term is introduced to design the steering cost. When the instantaneous turning radius R is less than the sum of the minimum turning radius and the safety threshold ($R_{\min} + r$), the exponential penalty term is introduced as shown in Eq. 3.

$$\begin{cases} k_{turn} \cdot e^{\alpha(R_{\min} - R)},\ R < R_{\min} + r \\ 0,\ R \geq R_{\min} + r \end{cases} \tag{3}$$

where k_{turn} is the weighting coefficient; α is the adjustment coefficient, which can be adjusted according to the vehicle maneuverability. This function makes the closer R is to $R_{\min}$, the more significant the cost increase is, so as to inhibit the generation of paths with too small turning radius.

Introducing a spatio-temporal trajectory prediction mechanism for dynamic obstacles. The spatio-temporal trajectory prediction mechanism for dynamic obstacles aims to predict the movement path of moving obstacles in the future period [11]to help vehicles perform real-time obstacle avoidance and path planning. The algorithm in this paper adopts a linear extrapolation model, which assumes that the dynamic obstacle moves at a constant velocity, and the future position of the obstacle is projected from the initial position of the obstacle and the predicted duration. Define the spatio-temporal collision risk cost function as shown in Eq. 4.

$$C_d = w_o \cdot \sum_{t=0}^{T_p} \frac{1}{\|p_r(t) - p_d(t)\| - d_s} \tag{4}$$

where T_p is the predicted duration; $p_r(t)$ and $p_d(t)$ denote the predicted positions of the vehicle and the dynamic obstacle at the time of t, respectively; and d_s denotes the safe distance between the vehicle and the dynamic obstacle. If the distance between the predicted trajectory and the obstacle is less than the safe distance, the trajectory candidate is directly rejected.

2.3 Dynamic Weight Adjustment

This algorithm introduces a dynamic weight adjustment mechanism on the basis of the traditional DWA framework, which adaptively adjusts the weight allocation of goal orientation and obstacle avoidance by real-time sensing of environmental features. The weight adjustment rules are defined as follows:

2.3.1 Target Proximity Weight Enhancement

When the vehicle distance from the target is less than the threshold D_g, the target weights are dynamically increased in proportion to the distance as shown in Eq. 5.

$$w_g = w_{g\_base} + (w_{g\_max} - w_{g\_base}) \times (1 - \frac{d_g}{D_g}) \tag{5}$$

where d_g is the current Euclidean distance to the target, w_{g_max} is the maximum target weight and w_{g_base} is the base target weight.

2.3.2 Obstacle Avoidance Enhancement in the Vicinity of Obstacles

When the detected obstacle distance is less than the safety range D_o, the obstacle avoidance weights are linearly enhanced as shown in Eq. 6.

$$w_o = w_{o\_base} + (w_{o\_max} - w_{o\_base}) \times (1 - \frac{d_o}{D_o}) \tag{6}$$

where d_o is the nearest obstacle distance, w_{o_max} is the maximum obstacle avoidance weight, and w_{o_base} is the base obstacle avoidance weight.

2.3.3 Steering Constraint Cost Enhancement

To prevent the vehicle from performing steering beyond its kinematic capability, the weight coefficients of steering cost are dynamically optimized as shown in Eq. 7.

$$k_{turn} = \kappa \cdot \frac{R_{min}}{max(R, \in)} \tag{7}$$

where κ is the tuning coefficient and $\in$ is the minimal amount to avoid division by zero.

2.4 D. Fractional Order PID Optimized Trajectory Tracking

Traditional PID controllers often cause trajectory oscillations or response hysteresis due to integral saturation and differential noise in the dynamic window method [12]. In this paper, a fractional order PID (FOPID) controller is introduced with the transfer function shown in Eq. 8.

$$C(s) = K_p + K_i s^{-\lambda} + K_d s^{\mu} \tag{8}$$

where $\lambda \in (0, 1)$ and $\mu \in (0, 1)$ are the fractional-order integral and differential orders, respectively, the adaptability to non-smooth error signals is enhanced by extending the order degrees of freedom of integral and differential.

Based on the Grünwald-Letnikov fractional order calculus definition, the discretized differential and integral terms are approximated as Eqs. 9 and 10.

$$D^{\mu} e(t_k) \approx \frac{1}{h^{\mu}} \sum_{j=0}^{N} (-1)^j \binom{\mu}{j} e(t_{k-j}) \tag{9}$$

$$I^{\lambda} e(t_k) \approx h^{\lambda} \sum_{j=0}^{N} \binom{-\lambda}{j} e(t_{k-j}) \tag{10}$$

where h is the sampling period and N is the memory length. By maintaining a historical error queue, the controller retains the number of recent error samples, which is realized to change the computational complexity and fractional-order memory characteristics.

The optimization effect of comparing fractional order PID control and PID control for trajectory tracking is shown in Fig. 2.

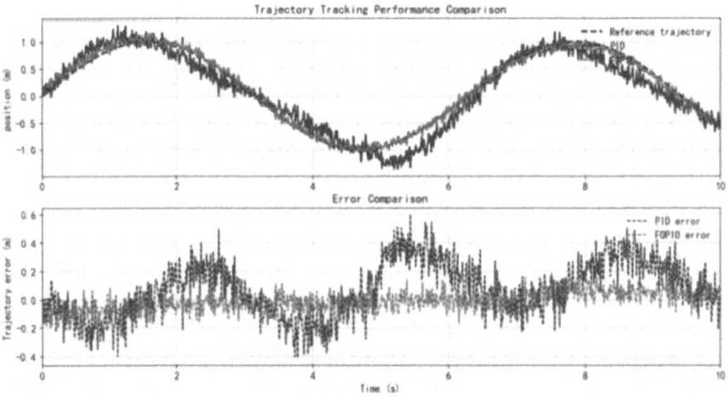

Fig. 2. Comparison of trajectory tracking effects between fractional-order PID and PID

The black dotted line in the figure represents the reference trajectory, the red realization is the tracking trajectory of fractional-order PID control and the blue realization is the tracking trajectory of PID control. It is found that the PID control has a large overshooting problem and generates a large error during sudden changes, with the maximum error reaching 0.6 m. The fractional-order PID control has a faster response and smaller steady-state error, with the maximum error of only 0.2 m. The fractional order PID control has a larger error than the fractional-order PID control.

3 Simulation and Experiment

The real mobile robot platform is based on Songling's scout mini mobile platform, and the real-vehicle experiment picture is shown in Fig. 3.

Fig. 3. Real Mobile Robot Platform

Based on the traditional DWA algorithm, steering optimization FOPID-DWA algorithm, steering and dynamic obstacle avoidance optimization FOPID-DWA algorithm path planning diagrams obtained from the two simulation scenarios of no moving obstacles and moving obstacles are shown in Figs. 5, 6, 7, 8 and 9, in which the green solid line indicates the predicted trajectory of the algorithm, the red solid line indicates the vehicle driving trajectory, the black circle represents the static obstacle, and the red circle represent dynamic obstacles. The evaluation indexes are movement time, path length and iteration number, the shorter the time means the higher the optimization efficiency; the lower the iteration number means the higher the execution efficiency of the algorithm; the shorter the path length means the better the optimization performance.

In the scene without moving obstacles, the traditional DWA algorithm and steering optimization FOPID-DWA algorithm are compared, and the effects are shown in Figs. 4 and 5, and the evaluation indexes are shown in Table 1.

In the scenario without moving obstacles, the steering-optimized FOPID-DWA algorithm reduces the motion time by 11.3%, the path length by 8.1%, and the number of iterations by 12.2% compared to the traditional DWA algorithm.

In the scene with moving obstacles, the traditional DWA algorithm, steering optimization FOPID-DWA algorithm and steering and dynamic obstacle avoidance optimization FOPID-DWA algorithm are compared, the effect is shown in Figs. 6, 7 and 8, and the evaluation indexes are shown in Table 2.

In the scenario with moving obstacles, compared to the steering optimization FOPID-DWA algorithm, the steering and dynamic obstacle avoidance optimization FOPID-DWA algorithm further shortens the path length by 6.1% and reduces the number of iterations by 10.0% with close movement time.

This study focuses on unstructured environments. To meet experimental requirements and make full use of existing conditions, the experiments were conducted in an

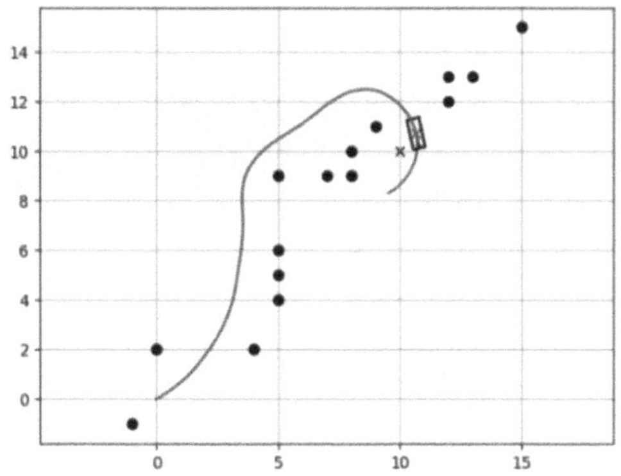

Fig. 4. Traditional DWA Algorithm

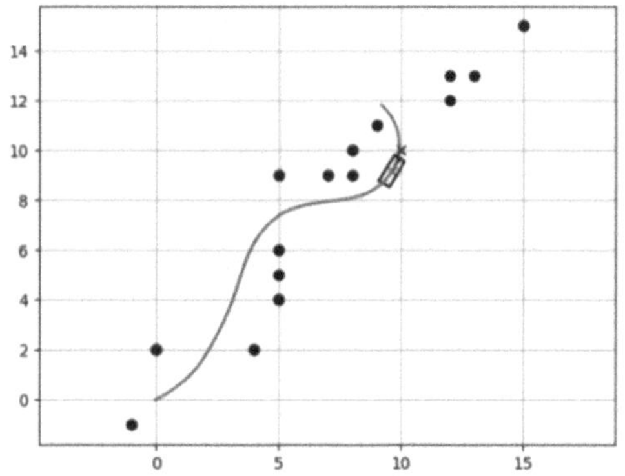

Fig. 5. Steering Optimization FOPID-DWA Algorithm

Table 1. Evaluation metrics of two algorithms for scenes without moving obstacles

Evaluation metrics	DWA Algorithm	Steering Optimization FOPID-DWA Algorithm
Path length(m)	231.7	212.9
Iterations	698	613
Movement time (s)	78.6	69.7

indoor setting. The experimental scene was relatively complex, featuring multiple obstacles of varying sizes and shapes. Figure 9 shows a panoramic view of the experimental environment.

As shown in Fig. 10, the blue point is selected as the navigation starting point and the red point as the navigation endpoint.

When reaching the position shown in Fig. 11, the mobile robot encountered an unknown obstacle marked by the yellow triangle. The steering and dynamic obstacle

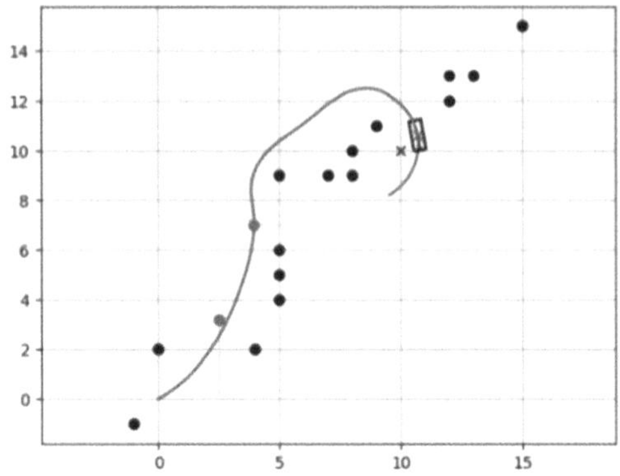

Fig. 6. Traditional DWA Algorithm

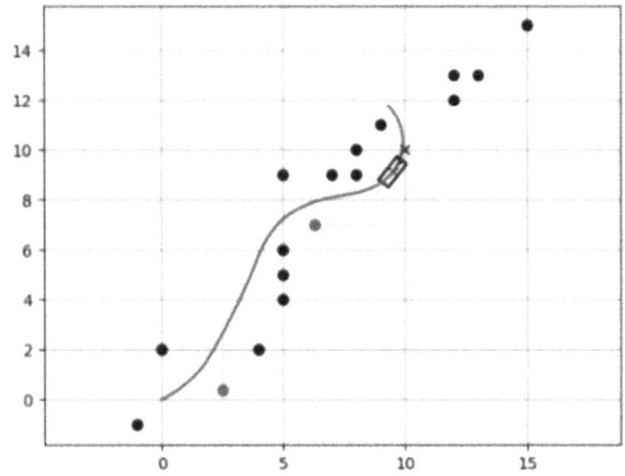

Fig. 7. Steering Optimization FOPID-DWA Algorithm

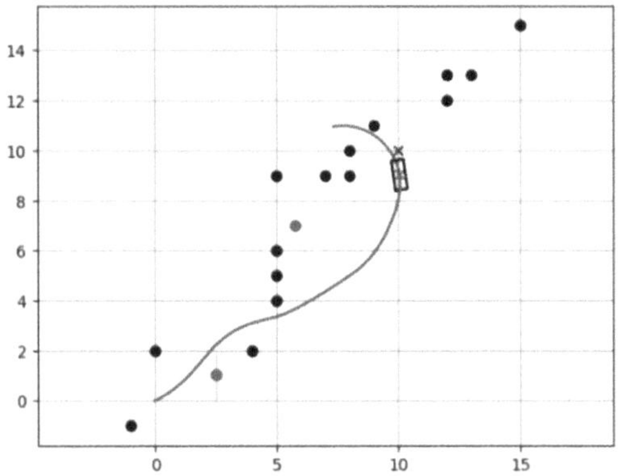

Fig. 8. Steering and Dynamic Obstacle Avoidance Optimization FOPID-DWA Algorithm

Table 2. Evaluation metrics of three algorithms for scenes with moving obstacles

Evaluation metrics	DWA Algorithm	Steering optimization FOPID-DWA algorithm	Steering and dynamic obstacle avoidance optimization FOPID-DWA algorithm
Path length(m)	236.2	214.9	201.8
Iteration number	719	627	564
Movement time (s)	82.9	74.8	74.9

Fig. 9. Panoramic View of Unstructured Environments

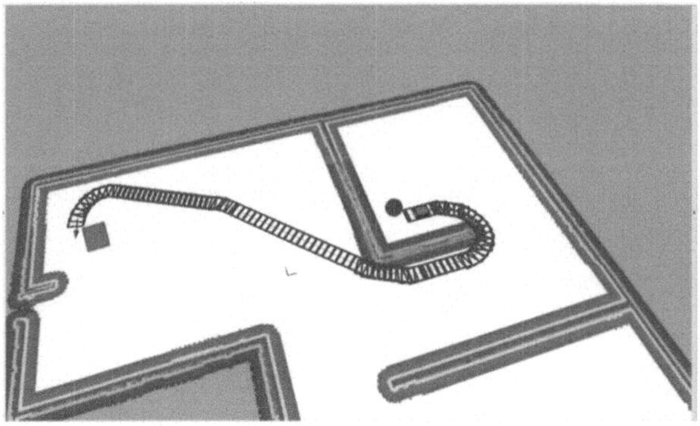

Fig. 10. Global path

avoidance optimization FOPID-DWA algorithm accurately identified it and altered the original trajectory curve depicted in Fig. 10. The local path planning, indicated by the red line in Fig. 11, represents the path planned by the algorithm upon encountering obstacles in an unknown environment.

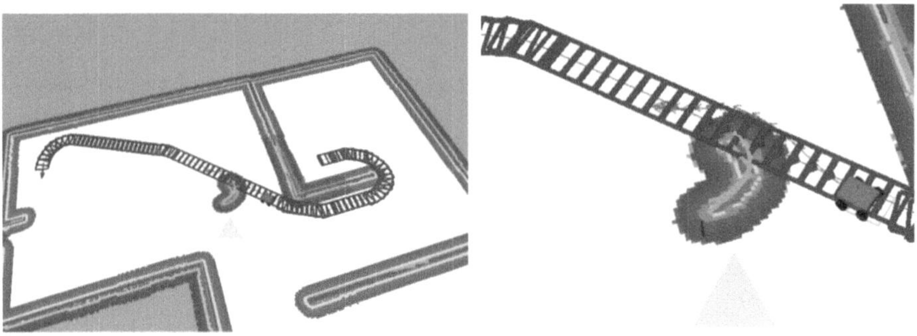

Fig. 11. Encounter unknown obstacles

When encountering a cylindrical unknown obstacle, the mobile robot adopts the new path planned by the steering and dynamic obstacle avoidance optimization FOPID-DWA algorithm, as shown by the red line in Fig. 12.

Fig. 12. Avoid cylindrical obstacle

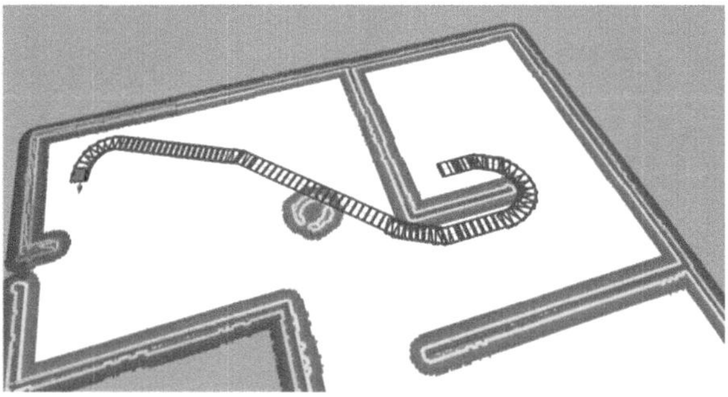

Fig. 13. Navigation complete

As shown in Fig. 13, the mobile robot departs from its starting position. Its initial route is the global path connected by the blue box. During navigation, sensors detect an unrecorded obstacle. The mobile robot then employs the steering and dynamic obstacle avoidance optimization FOPID-DWA algorithm to plan a new path (represented by the thin red line, indicating the newly planned local path) to bypass the obstacle, ultimately reaching the navigation endpoint successfully.

4 Conclusion

In this paper, under the unstructured environments an improved dynamic window method for mobile robots is proposed for overcoming the shortcomings of the traditional DWA in local path planning. Through simulation and experiments, the results consistently demonstrate that the improved algorithm can efficiently track the global path and achieve dynamic obstacle avoidance, significantly outperforming traditional DWA.

(1) Kinematic constraint optimization: Based on the Ackermann steering model, a new turning radius constraint cost function is introduced to prevent generating paths beyond the vehicle's maneuvering capabilities. This addresses the issue of path infeasibility in traditional DWA algorithms caused by ignoring minimum turning radius constraints.
(2) Enhanced dynamic obstacle avoidance: Through a spatiotemporal trajectory prediction mechanism and collision risk cost function, the algorithm's ability to avoid dynamic obstacles has been effectively improved.
(3) Dynamic weight adjustment: The algorithm adaptively adjusts the weight distribution for target proximity, obstacle avoidance, and steering constraints based on environmental characteristics, enhancing its flexibility and robustness. Experiments demonstrate that this mechanism significantly optimizes path selection when approaching obstacles or targets.
(4) Enhanced trajectory tracking capability: The adoption of fractional-order PID controllers optimizes trajectory tracking, reducing overshoot and steady-state error while improving path execution smoothness and precision.

Acknowledgement. This paper is supported by Fundamental Research Funds for the Central Universities (buctrc202204).

This work has been developed at the institute of advanced materials processing and intelligent engineering, college of mechanical and electrical engineering, Beijing University of Chemical Technology, China.

References

1. Fox, D., Burgard, W., Thrun, S.: The dynamic window approach to collision avoidance. IEEE Robot. Autom. Mag. **4**(1), 23–33 (1997)
2. Adiuku, N., Avdelidis, P.N., Tang, G., et al.: Improved hybrid model for obstacle detection and avoidance in robot operating system framework (Rapidly Exploring Random Tree and Dynamic Windows Approach). Sensors **24**(7), 2262 (2024)
3. Yang, L., Li, C., Wang, Q., et al.: Adaptive dynamic windowing approach based on risk degree function. Trans. Inst. Meas. Control. **46**(8), 1542–1550 (2024)

4. Dandan, H., Jinju, Z., Guochen, N.: Active obstacle avoidance based on improved dynamic window approach for off-axis full-trailer vehicles. J. Beijing Univ. Aeronaut. Astronaut. 1–17 (2025)
5. Xiaochan, W., Zihan, Q., Zhenyu, Y., et al.: Local path planning for agricultural robots based on DAV_DWA. Trans. Chin. Soc. Agric. Mach. **56**(02), 105–114 (2025)
6. Ming, Y., Haigang, D., Xianying, F., et al.: Improved dynamic windows approach based on energy consumption management and fuzzy logic control for local path planning of mobile robots. Comput. Ind. Eng. **187**, 109767 (2024)
7. Yang, H., Gang, L.: Research on trajectory planning control method of intelligent vehicle based on velocity obstacle model. Acta Armamentarii **46**(4), 397–410 (2025)
8. Zhineng, Z., et al.: Study on local path planning and tracking algorithm of intelligent vehicle in complex dynamic environment. China J. Highway Transport **35**(9), 372–386 (2022)
9. Shuping, C., Zhiguo, Z., Kun, Z.: Coordinated control of trajectory tracking and yaw stability of four-wheel-independent-drive autonomous electric vehicle. J. Tongji Univ. (Nat. Sci.) **52**(S1), 185–196 (2024)
10. Li, Z., Xueying, Z., Xian, Z., et al.: Local dynamic obstacle avoidance path planning algorithm for unmanned vehicles based on potential field method. Trans. Beijing Inst. Technol. **42**(7), 696–705 (2022)

Survey on Intelligent Driving Technologies for Humanoid Robots

Hongxin Xu[1], Hongxing Wei[2], Guizhen Yu[1], Zhuofei Ding[1], and Xudong Ji[1](✉)

[1] Hefei Innovation Research Institute of Beihang University, Hefei 230012, China
jixd@buaa.edu.cn
[2] Beihang University, Beijing 100191, China

Abstract. Intelligent driving for humanoid robots has emerged as a novel interdisciplinary domain integrating robotics, perception, cognitive decision-making, and autonomous driving. Its primary goal is to enable humanoid robots—with human-like morphology and motion capabilities—to perform or assist in vehicle driving tasks under complex and dynamic environments. Unlike conventional autonomous driving systems constrained by task-specific hardware and rule-based models, humanoid intelligent driving features enhanced adaptability in human–machine interaction and strong generalization across diverse driving scenarios. This paper reviews the conceptual foundations, technological framework, and recent advances in intelligent driving for humanoid robots. Core technologies, including multimodal perception fusion, human-like driving behavior modeling, hierarchical task planning, cooperative control mechanisms, and simulation-to-reality verification platforms, are systematically analyzed. Remaining challenges such as motion coordination, control precision, real-time responsiveness, and safety assurance are highlighted. Finally, the paper outlines prospective applications in intelligent transportation, special operations, and human–robot collaboration, and discusses future trends emphasizing the integration of artificial intelligence with robot dynamics modeling and adaptive multi-task learning. The review aims to provide systematic insights and guidance for advancing humanoid robot intelligent driving research and applications.

Keywords: Humanoid Robots · Intelligent Driving · Humanoid intelligent driving · Cooperative Control

1 Introduction

With the deep integration of artificial intelligence, robotics, and intelligent driving technologies, the traditional human-centric driving paradigm is rapidly shifting toward automation and intelligence [1]. However, existing autonomous driving systems continue to face substantial challenges in autonomous decision-making within unstructured environments, complex human–machine interaction, and emergency response under extreme operating conditions [2]. Along with the wave of the new generation of large language models (LLMs), such as DeepSeek [3] Chat-4 [4] and other innovative breakthroughs. Embodied intelligence is an important branch of artificial intelligence. Emphasizing

H. Wei and J. Niu (Eds.): COBOT 2025, CCIS 2893, pp. 303–322, 2027.
https://doi.org/10.1007/978-981-92-2074-8_18

that agents can exhibit intelligence through direct interaction with the environment, with little or no reliance on explicit symbolic reasoning, embodied robots and embodied intelligence show broad application prospects in open fields such as home services, autonomous navigation, and inspection and rescue [5]. Against this backdrop, Intelligent driving for humanoid robots refers to the capability of an intelligent robot with a human-like physique to achieve autonomous operation and navigation of a car in both structured and unstructured real-world scenarios, through the fusion of multimodal perception. as a new driving mode featuring biological structure adaptability and universal execution capability [6, 7] has attracted widespread attention from research institutions and industry both at home and abroad [8].

Traditional robots rely on preprogrammed models and fixed procedures to perform specific tasks [9]. In contrast, embodied intelligent robots integrate artificial intelligence with robotics. At the current stage, along with deep learning [10]. Its innovative advancements have enabled humanoid robots to self-learn, achieving perception of the surrounding environment as well as planning and control of intelligent vehicles. Compared with traditional robots, humanoid robots feature human-like skeletal structures, end effectors, and degree-of-freedom configurations [11]. They possess the potential to perform physical driving tasks such as steering wheel control, accelerator/brake pedal operation, and gear shifting through "human-like manipulation". Compared with driverless vehicles based on drive-by-wire chassis or autonomous driving platforms with fixed structures, the humanoid robot autonomous driving system can directly "take over" traditional vehicles without modification. It boasts the advantage of replacing humans in completing driving tasks, along with stronger cross-platform adaptability, portability, and the ability to imitate human behaviors in emergency scenarios [12]. In environments with clear rules such as highways and closed parks, technologies have evolved from traditional neural networks and computer vision to today's machine learning and deep learning. This capability serves as the technical requirement and foundation for ensuring the safety, efficiency, and reliability of autonomous vehicles [13]. However, in urban roads, disaster scenarios, or military missions characterized by complex structures and frequent dynamic changes, as well as in unstructured environments such as rural roads and off-road terrains [14]. Navigating in areas without clear boundaries or predictable conditions still poses numerous challenges, and unmanned vehicle systems that rely on rule bases and deep models continue to face multiple bottlenecks. Traditional autonomous driving algorithms exhibit limited robustness and flexibility in handling emergencies, human-vehicle collaborative interactions, and the understanding of unstructured semantics [15]. In contrast, humanoid robots, equipped with human-like cognitive structures, physical manipulation capabilities, and interaction potential, serve as an important bridge between intelligent systems and the physical world through their human-like operation methods and environmental adaptability. They are regarded as one of the key supporting technologies for the integrated development of "intelligent driving—smart transportation—robots" in the future [16]. The main challenge lies in integrating human-like logic and driving risk constraints based on behavioral decision-making outcomes. This integration is crucial for aligning the cognitive level, actions, memory, and reasoning of autonomous driving closely with human driving needs, habits, and styles [17].

Although some studies in recent years have focused on the feasibility verification and system integration of "humanoid robot driving"—such as experiments on robot steering wheel control based on motion planning and control [18]. Application of humanoid foot control algorithms in pedal operation [19]. However, on the whole, humanoid robot autonomous driving technology is still in the early exploration stage. It still has broad application prospects in unstructured roads, and in terms of the control model accuracy of autonomous driving robots [20]. The system integration of autonomous driving technology and robot manipulation [21]. Interactive Intelligence and Safety Redundancy [22] and other aspects have not yet formed a mature system. Therefore, this review comprehensively sorts out and conducts an in-depth analysis of the research status of humanoid robot autonomous driving, starting from the concept, key technologies, current status and problems, existing research foundations, and future development directions of humanoid robot autonomous driving [23].

2 Concept of Humanoid Robot Autonomous Driving

2.1 Humanoid Robot

Human exploration and imagination of robots began to take shape as early as the 20th century. Playwright Karel Čapek first proposed the term "Robot" in his 1920 play [24]. This creation also marked the origin of the term. In 1950, Alan Turing, the father of artificial intelligence, put forward the forward-looking concept of "embodied intelligence" in his paper Computing Machinery and Intelligence [25]. He emphasized that intelligent behavior must be acquired through interaction between an agent of corresponding form (i.e., a robot with a physical body) and the environment. This theory laid an important ideological foundation for the subsequent development of humanoid robots. Meanwhile, with the vigorous rise and in-depth development of the Third Industrial Revolution—especially the leapfrog progress in the processing capacity of electronic computers and the emergence of new engineering materials—many limitations in early computing, control, and physical structures were overcome. This provided key support for robots to formally move from exploratory prototypes to practical applications. The combination of theoretical and technical foundations officially initiated the development course of humanoid robots.

In 1972, WABOT-1, the world's first full-scale humanoid intelligent robot, was born [26]. It could carry objects and was equipped with a basic sensory system (such as vision and hearing), but its functions were limited. This marked the starting phase of the development of modern humanoid robots. From 1986 to 2010, led mainly by Japanese companies, humanoid robots evolved into integrated mechanical devices equipped with subsystems such as perception, control, and execution. ASIMO (Honda) was capable of running [27]. It made significant progress in motion control and action coordination, such as being able to go downstairs. NAO (SoftBank) can communicate with humans through voice, vision and other means, and has basic human-computer interaction capabilities [28]. The launch and industrial application of the first generation of humanoid robots laid the foundation for the subsequent high-dynamic and intelligent development of humanoid robots.

In recent years, with the development of fields such as artificial intelligence, machine learning, and large - model technology, humanoid robots have been able to better understand and adapt to complex environments through technologies such as deep learning, natural language processing, and image perception and recognition [29]. It can perform higher - level tasks. For example, Boston Dynamics' Atlas is capable of completing high - complexity movements such as backflips [30]. For example, Boston Dynamics' Atlas is capable of performing high - complexity movements such as backflips [31]. With the continuous innovation and improvement of humanoid robot technology, costs have gradually decreased and the technology has become mature. Humanoid robots have gradually moved from experimental scenarios to industrial settings and practical applications, entering the stage of commercialization and industrialization—such as industrial production, home services, medical care, and educational assistance. The most representative humanoid robot enterprises in this stage are Tesla and Unitree Robotics [32]. It has demonstrated broad commercial application scenarios.

While robots are developing, autonomous driving is also thriving. As one of the cutting-edge fields in robotics, autonomous driving is undergoing profound evolution. Compared with pure autonomous driving, humanoid robot autonomous driving has advantages in its natural adaptation to human environments, multi-tasking capabilities, and continuously enhanced intelligence. For example, centralized robots in unstructured scenarios sometimes not only act as drivers but also complete complex tasks in disaster or dangerous contexts—such as moving debris and clearing operations [33]. This makes them more flexible than traditional autonomous driving systems in complex scenarios (such as logistics, security, and rescue), while also having stronger task transfer capabilities and generalization capabilities. Exploring the integration of humanoid robots and autonomous driving technology is expected to provide a new solution for autonomous driving in complex environments.

2.2 Humanoid Robot Intelligent Driving

Humanoid robot autonomous driving refers to the process in which robots with human-like morphology and motor capabilities simulate human driving behaviors, manipulate existing vehicles through physical interaction, and realize autonomous or semi-autonomous driving tasks. Unlike traditional autonomous driving systems that rely on high-precision maps and end-to-end modeling, humanoid robot autonomous driving emphasizes "embodied execution" and "structural compatibility," enabling it to complete driving tasks in the original environment without special modifications to the vehicle. In 2012, the Robotics Challenge (DRC) initiated by the Defense Advanced Research Projects Agency (DARPA) became an important turning point in the research of humanoid robots [34]. The background of this challenge stems from the 2011 Fukushima nuclear accident in Japan, which prominently exposed the functional shortcomings of robots in the emergency response to nuclear leaks. In disaster scenarios such as high temperatures and high radiation, robots played an extremely limited role—prompting DARPA to aim at advancing the practical capabilities of robots in complex accident environments through the competition [35]. The challenge required humanoid robots to replace humans in performing tasks in disaster settings, including emergency response operations such as using tools to break through walls, climbing ladders, and navigating complex terrains. Among these requirements, the driving capability of humanoid robots emerged as one of the key assessment areas of the DRC. This not only provided valuable experience for the future application of humanoid robot autonomous driving on both structured and unstructured roads (even in disaster environments for hazardous tasks) but also verified the feasibility and market value of integrating humanoid robots into future autonomous driving modules (Fig. 1).

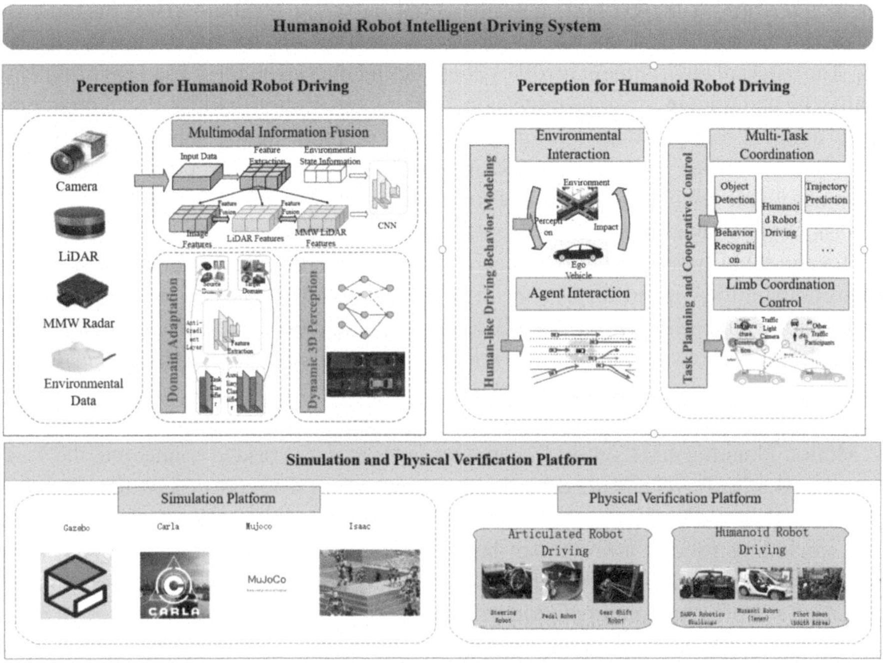

Fig. 1. Humanoid Robot Intelligent Driving Technology System

As noted by Paolillo et al. In their research [36] disaster rescue sites often involve the direct operation of traditional human-operated equipment (such as excavators, drilling machines, transport vehicles, etc.). Due to their complex designs and diverse structures, these devices are difficult to achieve unified automated transformation. Over the past decade, a paradigm shift has occurred in manual rescue monitoring operations, with robot teams being adopted to replace manual rescue efforts. In unstructured environments, villages, or muddy roads, robot intelligent driving has demonstrated new advantages compared to traditional autonomous driving, which can effectively reduce casualties in disaster rescue scenarios. Humanoid robots can directly adaptto existing vehicle control interfaces, avoiding the need for high-cost vehicle reconstruction, and possess a high level of scene generalization ability and deployment flexibility. The realization of a humanoid robot intelligent driving system requires the construction of a highly integrated, multi-level collaborative technical architecture. This architecture can be divided into four levels, which progressively achieve a complete chain from environmental perception to autonomous decision-making.

Environmental Perception and Understanding Layer: The environmental perception of traditional intelligent driving relies on the integration of conventional visual neural networks (CNN, RNN) with autonomous driving technologies, focusing more on the vehicle's perception of the external environment to enable driving on structured roads. However, the humanoid robot intelligent driving system needs to pay attention to both the internal environment of the vehicle and the external environment simultaneously. The internal environment includes the identification of positions such as instrument panels and seats in the cockpit, how the robot correctly positions itself on the seat and maintains a sitting posture, and the recovery after missing the accelerator pedal [37]. The external environment involves constructing the surrounding road environment, identifying and tracking various targets on the road, etc. Robots build a global cognition of the dynamic environment through the fusion of multi-modal sensors (integrating monocular/binocular cameras, LiDAR, millimeter-wave radar, etc.), providing a real-time and reliable environmental model for upper-level decision-making [38].

Task Decision-Making Layer: Its core task is to generate executable behavioral strategies for the robot based on multi-source perception information, environmental status, and in combination with preset task objectives [39]. This layer needs to integrate semantic understanding, intention reasoning, and task management logic through a modular architecture. It comprehensively considers various types of information such as visual, tactile, and radar signals, coordinates the work priorities of each subsystem, and dynamically balances safety and efficiency to ensure a safe and efficient driving path.

Motion Planning and Control Layer: This layer acts as a bridge connecting the Task Decision-Making Layer and the robot's physical actuators. The core task within this layer is to convert high-level driving decisions into low-level control commands for the robot's body (including robotic arms, torso, head, etc.), which specifically encompass path generation, posture adjustment, and reactive balance [36]. This includes the coordinated control of grasping and manipulation actions. In driving tasks, the robot is required to execute commands such as precise control of the steering wheel by the robotic arm, coordinated control of the acceleration and deceleration pedals by the feet, and dynamic maintenance of the torso posture. The coordinated execution of such multi-dimensional operation commands imposes requirements and challenges on the stability and degrees of freedom of the robot's physical structure. While ensuring the stable operation of the robot, it is also necessary to take into account disturbance factors such as vehicle vibrations and terrain slopes, so as to guarantee the robustness of the humanoid robot's intelligent driving control and the safety of driving.

Autonomous Learning and Adaptation Layer: To address the diverse vehicle types and unstructured road conditions in real-world environments, humanoid robots need to possess long-term autonomous learning capabilities and the ability to quickly adapt to new environments, ensuring stable operation even in sudden complex traffic scenarios. During interactions with vehicles and the environment, the robot leverages mechanisms such as imitation learning of navigation behaviors, reinforcement learning, or few-shot learning [40]. To optimize key parameters in the perception-decision-control loop of intelligent driving either online or offline, thereby enhancing generalization ability in unknown environments. When the robot encounters unfamiliar vehicle models or changes in interface layouts, the system should quickly adjust operational strategies based on its own perception and learning to ensure the fluency and safety of the humanoid robot's driving behaviors.

3 Key Technologies of Humanoid Robot Autonomous Driving

Humanoid robot autonomous driving is a complex interdisciplinary field covering various key technologies such as perception, decision-making, execution, and system integration. This review summarizes the current research achievements in humanoid robot autonomous driving. Combined with the current wave of next-generation large language models, this section will conduct an in-depth analysis of the core key technologies of humanoid robot autonomous driving from aspects including multimodal perception, decision-making and execution, and the multi-system integration of robot systems and autonomous driving systems.

3.1 Multimodal Perception for Humanoid Robot Driving

Perception is the core supporting module of the humanoid robot autonomous driving system. Compared with traditional autonomous driving, there is an essential difference in the design logic of their perception modules: the latter constructs a multimodal environmental perception framework centered on the"vehicle", while the former needs to take the"human-like driver" as a reference to meet the dual requirements of in-depth environmental understanding and expanded interactive functions. The multimodal perception of humanoid robot autonomous driving combines visual perception, inertial perception, and force perception [41]. On one hand, it requires humanoid robots to achieve real-time acquisition and understanding of multiple dimensions such as vehicle status, road environment, and traffic participants. On the other hand, an additional in-vehicle perception subsystem needs to be constructed. Through multi-sensor fusion including visual perception and tactile perception, the robot's pose estimation and behavior prediction are realized, enabling the humanoid robot to achieve precise positioning of instrument panels and control devices as well as human-robot interaction.

Single-Modal Information Perception: The research summary of humanoid robot autonomous driving combined with deep - learning - based object detection algorithms by Yadav s.t [42] in autonomous driving is as follows [43]. Firstly, the commonly - used data - collection sensors are systematically classified, and the most widely - used 3D detection datasets and evaluation indicators are introduced. The perception is classified into single - modal information perception and multi - modal information fusion perception [44]. Algorithms are classified according to the type of input data. Single-modal information perception includes methods based on vision, point clouds, Transformer, Mamba, and hybrid point clouds [45]. Integrating humanoid robots with traditional vision and point cloud technologies can enhance the accuracy and safety of environmental information acquisition for intelligent driving technologies in unstructured and high-risk environments.

Multimodal Information Fusion: When humanoid robots perform driving tasks, they need to integrate information from multimodal sensors such as vision, force, inertia, and touch to achieve a unified cognition of the environment and vehicle status. The HRP-2 robot developed by Paolillo et al. realizes the recognition of road boundaries and lane lines through a monocular camera [46] and used the optical flow method combined with IMU data to estimate the motion states of itself and the vehicle, providing basic data for path planning and decision-making. Rasmussen et al., on the other hand, configured a forward camera and additional in-vehicle cameras [47]. They are used to identify the positions and states of internal vehicle control devices (such as the steering wheel, pedals, etc. inside the vehicle) and integrate multi-point force sensors to perceive contact intensity, enabling tactile interaction. In real-world dynamic environments, effective fusion of these multimodal inputs requires technical approaches such as temporal synchronization, spatial alignment, and sensor calibration to cope with uncertainties like illumination changes, motion blur, sudden disturbances, and pedestrian intrusions. More advanced systems have moved beyond relying on a single modality. Such as Musashi developed by the University of Tokyo [48]. It is equipped with a diverse network of tactile and pressure sensors, integrating force sensing, position sensing, visual perception, and internal state perception capabilities to achieve a natural motion closed loop from environmental contact (such as tactile feedback when grasping objects) to action adjustment (such as changing grip strength according to pressure distribution). This multimodal fusion-based perception hierarchy design provides the robot with a human-like interactive decision-making path of"contact-perception-adjustment". The key lies in achieving real-time and robust fusion of multi-source information under dynamically changing conditions.

Dynamic 3D Perception: Dynamic 3D perception technology refers to a technology that uses various sensors and algorithms to obtain real-time 3D information about the environment around the vehicle, and to accurately perceive and track the position, shape, motion state, etc., of dynamic objects. For example, the BEV (Bird's-Eye View) perception technology [49] Such technologies as LiDAR maps, etc. 3D perception provides key support for humanoid robots in spatial positioning and obstacle avoidance in the surrounding environment during driving. Compared with fixed-position sensors, humanoid robots, as embodied intelligent agents, can continuously perceive and adapt to environmental changes through human-like behaviors such as"learning" and"imitation" during vehicle operation. To achieve precise perception for intelligent driving, robots need to complete self-localization in the vehicle coordinate system and pose estimation of the robotic arm. This not only requires analyzing the geometric structure of the surrounding 3D environment, such as key objects like pedals and steering wheels, but also further realizing precise alignment by constructing 3D point cloud models [50]. In the system proposed by Rasmussen et al., using data information provided by two cameras, the robot can establish a spatial model of the vehicle's internal layout and calculate its relative position, thereby dynamically adjusting its sitting posture and operation strategies [51]. However, in the scenario of embodied intelligent driving, robots still need to face additional dynamic perception challenges. Due to the impact of road conditions such as road bumps, the body movements of humanoid robots will introduce additional dynamic factors that cause perception deviations. It is necessary to ensure the robot's perception accuracy through mechanisms such as pose estimation, coordinate transformation, and motion compensation. For example, in situations of insufficient light or strong jitter, methods such as filtering and image enhancement can be used to improve the stability of visual perception, or sensor channels with higher stability can be prioritized to ensure the reliability of overall perception.

Domain Adaptation: Unlike autonomous driving itself, traditional intelligent driving systems usually focus on adaptation to fixed vehicle models. However, in unstructured scenarios, humanoid robots can more easily integrate into the environment to accomplish diverse tasks. They have inherent advantages in experimental scenarios involving cross-vehicle models and cross-environments—specifically, domain adaptation and generalization capabilities [52]. Therefore, having domain adaptation capability is crucial for achieving universal driving. When faced with different vehicle brands, operational layouts, and component sizes, the perception system needs to quickly learn and adapt to new vehicle configurations, and establish mapping relationships with specific operations. The Yuanhang Zhang team [53] proposed a dual-agent reinforcement learning framework named FALCON, which decomposes the whole-body control of humanoid robots into a"lower-body stable motion agent" and an"upper-body force compensation manipulation agent". The lower-body agent ensures stable movement under external force disturbances, while the upper-body agent accurately tracks the position of the end-effector through implicit adaptive force compensation. The joint operation of these two agents has significantly enhanced the robot's adaptability, providing a new idea for the adaptive design of robots in real-world unstructured scenarios.

3.2 Decision-Making and Humanoid Robot Intelligent Driving

The decision-making and execution module for humanoid robot intelligent driving is responsible for converting perception results into specific operational actions through its own"learning" process—specifically, the manipulation and motion control of the steering wheel, accelerator pedal, etc., within the next k time steps.

Human-like Driving Behavior Modeling: In traditional intelligent driving systems, driving tasks are often modeled as trajectory planning and path tracking problems, which involve road conditions, other traffic participants, and so on. However, in humanoid robot systems, driving behaviors need to be mapped into action sequences that conform to the logical structure of the human-like body. These sequences not only include basic actions such as dual-arm grasping and steering wheel rotation, but also sequential actions like stepping on the accelerator and brake pedals, and adjusting the posture in the cab. The team led by Dejan Milojevic proposed a resource-efficient, task-driven collaborative design framework for perception and decision-making in autonomous vehicle mobile robots, named CODEI [54]. The HA-VLN model proposed by Carnegie Mellon University [55] By integrating the 3D human motion state model with human-like intelligent agents, the agents can think like humans, predict future action trajectories, and adjust paths accordingly. Lin [56] The ADAPT environmental perception model for mobile robots was constructed, which connects the robot's actions with language to achieve effective association between verbal instructions and vision. The Musashi robot, developed by the University of Tokyo, is designed to support multi-action collaborative tasks such as"simultaneous operation of the steering wheel with both arms, control of the accelerator pedal, and maintenance of driving posture". Its musculoskeletal system adopts a bionic drive design, which can adjust stiffness according to task requirements by imitating the contraction characteristics of human muscles. Its human-like shoulder joint features a scapula structure that slides on the rib cage, enabling a wide range of movements, improving the continuity and stability of control, and initially realizing human-like driving behavior modeling in terms of structure.

Task Planning and Collaborative Control: In driving tasks, humanoid robots need to plan the overall vehicle tasks and achieve collaborative control between various parts of the robot, as well as between the robot and humans. The system must not only perform single actions but also possess the capability of multi-channel concurrent task scheduling and control [57]. For example, during steering, it needs to simultaneously adjust pedal input and maintain a stable body posture.

Paolillo et al.'s framework provides three driving strategies: teleoperation, full autonomous driving by robots, and human-robot hybrid control. Human supervisors can choose to let robots take on all or part of the driving tasks as needed. This design is essentially a task-level collaborative control strategy, ensuring that humans can take over driving at any time, thereby improving the safety and flexibility of the task. The Sohn team [58] The auxiliary system they developed enables operators to predict vehicle routes based on the current vehicle state, calculate the optimal path using obstacle avoidance and driving distance data, and display it on the user control interface. They also conducted a comparative analysis of driving patterns under different operation modes (such as full human driving and full remote control) to evaluate the performance of this control strategy, and verified its effectiveness by driving vehicles in real road environments. Hao Mao [59] They proposed the Constraint-Aware Diffusion Guidance (CoDiG) model, which is applied to race car-level humanoid robot intelligent driving. By tracking the real-time generated trajectories of CoDiG, the robot achieves efficient obstacle avoidance on race car-level roads. Yuqing Guo et al. proposed a unified Optimal Control Problem (OCP) framework to describe general trajectory planning and collaborative control schemes. It aims to establish a unified task planning solution among different planners on unstructured roads, regardless of the driving mode. For current research, the humanoid driving system has initially achieved human-robot switching and closed-loop control capabilities, but task planning is limited to simple paths and obstacle avoidance, and a multi-agent collaboration framework for robots has not yet been formed.

3.3 Humanoid Robot Intelligent Driving Platform

The humanoid robot intelligent driving platform serves as the integrated carrier of the entire human-robot driving system, hosting multiple modules such as perception, execution, control, communication, and power supply. Current research mainly progresses along two parallel tracks: simulation platforms and physical verification platforms.

Simulation Platform: In existing research, before implementing humanoid robot driving, simulation platforms are mostly used for algorithm verification and debugging. Currently, in humanoid robot driving research, research teams generally utilize general robot simulation environments such as Gazebo and MuJoCo to conduct studies on vehicle dynamics model construction, robot dynamics characteristic analysis, and sensor data simulation. In the simulation environment, researchers can efficiently verify the control strategies of driving algorithms without relying on physical hardware for repeated debugging. By reproducing real driving scenarios in a virtual environment, simulation platforms reduce costs such as hardware wear and tear, site occupation in physical testing, and minimize risks like equipment collisions that may occur in actual operations, providing a reliable performance prediction basis for subsequent physical verification. In Rasmussen's experiment, after physically testing the methods for steering and speed control, the team collected over five kilometers of data from campus environments and golf courses to verify algorithms on the Gazebo simulation platform, and conducted simulation tests on the accuracy of obstacle detection, environment modeling, path planning, and trajectory tracking. In practical research, many teams also adopt a multi-platform integration approach, using different simulation platforms at different stages. For example, during the adjustment phase of robot motion and control algorithms, the MuJoCo simulation platform can accurately simulate complex contact dynamics processes such as the interaction between the robot's arm and the steering wheel, and the interaction between the foot and the pedal. When testing visual perception algorithms and driving decision-making systems, CARLA's built-in diverse elements such as vehicle models, pedestrians, and traffic lights can simulate complex traffic scenarios. This multi-platform collaboration approach can give full play to the advantages of each platform, improving research efficiency and the reliability of results.

Physical Verification Platform: The physical verification platform is crucial for promoting the implementation of humanoid robot driving systems. Testing in real environments can verify the system's perceptual robustness, execution accuracy, and platform reliability. Currently, various humanoid robot platforms have been used for driving task verification[60]. The Kawaharazuka team conducted research based on the Musashi muscle muscle-driven muscle-driven robot from Japan's WARI Laboratory [58]. The experiment utilized the ultra-compact electric vehicle COMS (Chotto Odekake Machimade Suisui, which means"short-distance convenient mobility vehicle") series, with the test site located at the Kashiwa Campus of the University of Tokyo. Components required for the experiment, including a power supply and a Wi-Fi router, were integrated into the trunk of this small car. As an electric vehicle, all the power required by the humanoid robot in the future can be directly supplied by the COMS model. The Jeongmin Moon team [61] They proposed an Edge-Driven Robot Platform (EDRP) and a Local Dynamic Map Platform (LDMP) based on 5G mobile edge computing and Kubernetes, which effectively address the issues of computing power strain and asynchronous delays caused by processing large volumes of data during deep learning. Teams including Sohn's utilized an upgraded DRC-Hubo + robot in the DRC Robotics Challenge, with improvements made to its body and perception head. After the introduction of the driving assistance system, operators trained on the system, in collaboration with full auxiliary functions, achieved a significant reduction in decision-making time when the vehicle was avoiding obstacles.

Physical validation platforms have successfully demonstrated the feasibility of humanoid robot autonomous driving. The Musashi team's COMS platform relies on high-precision vision-force perception fusion, offering stable control suitable for structured scenarios. The Jeongmin Moon team's EDRP and LDMP platforms utilize 5G edge computing for perception, enabling fast response and strong load capacity adapted to dynamic unstructured environments. Meanwhile, DRC-Hubo +, integrated with a driving assistance system, performs well in unstructured road conditions. This survey aims to establish unified evaluation dimensions including robustness, computational cost, and adaptability by defining quantitative metrics such as environmental adaptability, response real-time performance, and system generalization capability. This framework will systematically assess the characteristics and applicable scenarios of different validation platforms, providing clear direction for the future of humanoid robot intelligent driving.

Compared with in-depth modifications to vehicle hardware, a common feature of the aforementioned platforms is that they can perform driving operations without installing additional components on the vehicle. This characteristic directly enhances the cross-platform versatility of the system. Future research can further improve the hardware design of humanoid robots for driving (such as lightweight wrists, adaptive pedal contact, etc.) and human-robot collaboration strategies to enhance safety and operational efficiency, thereby achieving more comprehensive intelligent driving functions.

4 Current Status and Problems

Currently, humanoid robot intelligent driving, as a cutting-edge research direction at the intersection of unmanned driving and humanoid robot technology, has realized the execution of basic driving tasks in controlled environments such as structured roads and urban roads. This provides initial feasibility for the future popularization of humanoid robot intelligent driving in unstructured scenarios such as mine transportation and emergency rescue. For instance, humanoid robots can perceive and make decisions based on data provided by cameras, lidar, etc., to drive vehicles along the road center and perform operations such as acceleration, steering, and braking. Musculoskeletal robots with advanced structures can leverage the advantages of their human-like structure, demonstrating flexible and more precise operational performance in using both arms to operate the steering wheel, maintaining driving posture, and even performing behaviors beyond the scope of driving. For example, the Yue Dong team [62] They proposed a model-based dual-arm steering wheel control method, which integrates the dual-arm robot and the steering wheel into a system through a quadratic programming (QP) controller, effectively reducing steering delays. This research has been verified on real vehicles and real dual-arm robot platforms. However, most of these research results have been validated in idealized or simplified scenarios (such as closed parks and fixed routes), and there remains a significant gap in environmental adaptability compared with the complex requirements of actual road driving scenarios.

From a developmental perspective, the insufficient environmental adaptability of robots essentially stems from the shortcomings of humanoid robots in generalization ability and adaptive capacity. Firstly, there remains a significant gap in the physical structure between humanoid robots and human drivers—although current humanoid robots mimic human physical structures and behavioral patterns, their own structures are limited by the development of current hardware technology. However, the structure of the robots themselves is constrained by the current state of hardware technology. Studies have found that most robots still adopt rigid drive systems. The Lingling Tang team[63] Their rigid multi-body model with a kinematic closed-loop has improved gait symmetry and full-body coordinated gaits. However, robots with this structure have a limited range of joint motion, resulting in insufficient stability and flexibility when faced with complex driving operations (such as rapid large-angle steering of the steering wheel, fine adjustment of brake depth, etc.). However, the leg structure of current robots exhibits weak adaptability to different pedals, and foot reset capability and stability are difficult to guarantee on bumpy roads. Even for the musculoskeletal robot Musashi, which has a superior structure, the team still faces many challenges in steering wheel operation. Unlike the cross-arm steering method often used in actual human driving, this system can only perform push-pull steering at present, making it difficult to complete large-angle movements. Moreover, Deficiencies in hardware directly constrain the operational dimensionality and robustness of robots, making it difficult for them to cope with the variable in-vehicle environment and precise control demands in real-world driving. Such as mechanical structure interference and collisions are prone to occur during operation. In addition, the non-grasping contact posture causes its finger structure to be frequently stuck by the steering wheel, requiring manual reset. In terms of lower limb control, almost all current robots only achieve control of a single pedal (accelerator). The leg structure has weak adaptability to the pedal heights of different vehicle models. Meanwhile, under bumpy road conditions, the autonomous reset function of the feet still requires further testing to ensure stability.

As a prerequisite for humanoid robots to achieve autonomous driving, the capability of the perception system directly determines the accuracy of the entire system's cognition of the external environment and the timeliness of its responses. Most systems only use a monocular camera (or a simple combination of sensors) to perform perception tasks when observing the external environment. This configuration not only lacks data support in night scenes but also suffers from problems such as high noise and low robustness in scenarios with frequent occlusions or high information density. In test environments that are open and lack other traffic participants, simple perception mechanisms have basically met the needs; however, in unstructured roads and actual road applications, We must take into account the implications of practical hardware issues for the safety of intelligent driving. A single visual method is difficult to stably identify lane lines, traffic lights, pedestrians, and other traffic participants, and cannot handle complex scenarios such as multi-lane roads and intersections [64]. More critically, most current systems have not established a sound dynamic target prediction mechanism, resulting in limited ability to respond to possible emergencies in road environments, making it difficult to achieve truly safe and efficient driving.

Although multiple studies have explored the initial capabilities of humanoid robots in driving tasks, most work remains in the technical verification stage, lacking universal evaluation and standardized testing of system performance. In many research papers with arbitrary methodologies, experimental results are mostly presented through qualitative descriptions or a small number of metrics. Especially in terms of perceptual robustness and multi-target collaborative control, a unified evaluation index system has not yet been established for quantification. This evaluation method weakens the persuasiveness of results and also restricts cross-platform and cross-system performance comparison and iterative optimization. The Kawaharazuka team proposed that"the time required to complete a specific route" could be used as a performance metric. However, test routes also need to consider multiple indicators for standardization; without structured design and support from an index system, they cannot effectively facilitate horizontal evaluation. For example, whether the route includes subtasks such as lane changing, U-turns, and pedestrian avoidance; whether the evaluation considers operational stability on ramps or non-horizontal environments, etc. On the other hand[65], Musashi, as the first humanoid robot to complete driving operations in a vehicle using only its arms without the assistance of fixtures, demonstrates extremely high adaptability of bionic human-like structures. However, defining the standardization of robot intelligent driving is also a challenge to be explored. Establishing standardized testing procedures, designing verification platforms covering multiple scenarios, and promoting the development of open-source evaluation tools and datasets are important directions to be considered in the future.

5 Reflections and Outlook

Although humanoid robots have broad application prospects in driving tasks, multiple technological breakthroughs are still needed to reach a level comparable to human driving. Based on a comprehensive review of current cutting-edge research on humanoid robot intelligent driving, it is evident that this field is still in the development and verification stage. A universal solution has not yet been formed, and there remain numerous challenges in various technical aspects.

Most existing studies focus on breaking down driving tasks into the implementation of a single key action, lacking task planning for specific links in the complete driving process. In the future, there is an urgent need to develop technologies including SLAM (Simultaneous Localization and Mapping) [66] a full-process perception and decision-making system encompassing target recognition and dynamic task planning, moving from"action-level control" to"task-level execution," and forming a complete closed loop from boarding and starting the vehicle to path planning, environmental perception, and command execution. To achieve this goal, the system needs to modularize and systematically integrate complex operational processes. For example, Kawaharazuka et al. [67]. Proposed Sub-functions such as key rotation, handbrake pulling, accelerator pedal operation, and steering wheel rotation should be integrated into a unified system. The realization of this complete process not only requires in-depth integration between various sub-modules but also necessitates the establishment of a unified scheduling mechanism at the control architecture level to ensure the stable operation of multi-modal sensing information, multi-joint motion coordination, and asynchronous execution. The operational skills of humanoid robots need to be further expanded, encompassing force feedback systems and hand-eye coordination visual analysis, among other capabilities. Although such functions have been initially implemented in musculoskeletal structures—significantly enhancing motion flexibility and similarity to bionic movement patterns—robots still exhibit substantial gaps compared to human drivers in terms of operational strategies and execution methods, making it difficult to truly replicate human operational behaviors. For instance, finger structures cannot achieve grasping postures, steering modes conflict during large-range movements, and maintaining a sitting posture on uneven terrain or sharp curves remains challenging. In the future, it will be necessary to construct human-like behavior modeling systems based on methods such as imitation learning and reinforcement learning, and design control approaches more suitable for flexible bodies. Examples include hardware optimizations for large-range scapular movements and the use of smooth elastic materials. Meanwhile, we must consider the generalization ability of robots across different vehicles. Most platforms are optimized for specific vehicle models; while such customizations for specific vehicles improve system accuracy, they also limit generalization capabilities. Given the broad commercial prospects of humanoid robot intelligent driving, attention should be directed toward generalization capabilities in structured and unstructured practical application scenarios, such as mine transportation, engineering vehicles, and commercial electric vehicles operating in disaster rescue, rugged mountain roads, and other road conditions. Humanoid robots can leverage their advantages to achieve adaptive control of various vehicles with significant differences. To address this, we must focus on core key technologies: optimizing multi-modal perception fusion, constructing 3D modeling of robot systems and intelligent driving systems, enhancing the robot's ability to quickly recognize and spatially calibrate key components such as steering wheels and pedals, improving environmental perception of the vehicle's surroundings, and enabling path planning and vehicle control. Additionally, it is necessary to develop high-degree-of-freedom execution systems for the upper limbs, torso, and legs to adapt to the structural layouts of different vehicle cabins. Furthermore, combining reinforcement learning and transfer learning methods can accelerate the system's adaptation to new vehicle models, enabling dynamic parameter adjustment and control strategy transfer, thereby endowing robots with the ability to"perceive and adapt instantly."[68].

At this stage, the perception system for humanoid robot intelligent driving mainly focuses on basic tasks such as image segmentation and lane detection. These tasks mostly rely on monocular vision or stereo vision systems, primarily used to assist in basic path keeping and simple obstacle avoidance control. However, due to the interaction of multiple traffic participants and high dynamic change scenarios in urban road environments, the data from cameras and basic sensors are unstable. This perception method struggles to maintain its robustness in complex urban scenarios, further leading to poor performance in intelligent driving. This indicates that relying solely on low-level perception data can no longer meet the cognitive needs of complex environments, and the lack of semantic-level understanding ability is a major shortcoming of current perception systems. Robots not only need to perceive the existence and positions of various objects in the environment but also need to identify the states of these objects and their relationships with the robot and the surrounding environment, thereby inferring potential behavioral intentions and interaction rules. For example, robots need to understand the meaning of traffic signs, identify whether pedestrians are about to cross the road, judge the intentions of other vehicles (such as lane changing or turning), and correctly interpret the status of traffic lights. Such multi-dimensional semantic information is a key prerequisite for realizing intelligent decision-making.

From the perspective of humanoid robot intelligent driving safety, current research generally lacks systematic consideration of functional safety and Safety of the Intended Functionality (SOTIF), as well as standardized evaluation and safety verification of system performance. Current assessments predominantly rely on qualitative descriptions or single metrics (such as route completion time), and have yet to establish a unified testing framework that covers multiple dimensions including operational reliability, safety boundaries, and fault recovery. Future research must, while enhancing hardware adaptive capabilities, prioritize the integration of functional safety design, establish standardized testing procedures and multi-scenario validation platforms, and promote the development of open-source evaluation tools and safety standards. These efforts will lay the foundation for the transition of humanoid robot driving from laboratory research to real-world applications.

Humanoid robot intelligent driving is at a critical turning point from"experimental verification" to"system integration," requiring collaborative breakthroughs across multiple dimensions, including hardware structure, planning algorithms, generalization ability, perception capability, and evaluation systems. With the rapid evolution of AI algorithms, intelligent materials, and embedded control, this field is expected to become a core fulcrum for promoting the"evolution" of robots into real-world transportation systems.

References

1. Xie, J., Qin, Y., Zhang, Y., et al.: Towards human-like automated vehicles: review and perspectives on behavioural decision making and intelligent motion planning. Transp. Saf. Environ. Oxford Univ. Press **7**(1), tdae005 (2025)
2. Xu, C., Sankar, R.: A comprehensive review of autonomous driving algorithms: tackling adverse weather conditions, unpredictable traffic violations, blind spot monitoring, and emergency maneuvers. Algorithms MDPI **17**(11), 526 (2024)
3. Feng, D., et al.: Deep multi-modal object detection and semantic.segmentation for autonomous driving: datasets, methods, and challenges. IEEE Trans. Intell. Transp. Syst. **22**(3), 1341–1360 (2020)
4. Achiam, J., et al.: GPT-4 techni cal report. (2024) [06 May 2025]. https://arxiv.org/abs/2303.08774
5. Kai, X., et al.: Embodied intelligence for flexible manufacturing: a survey (2025). https://doi.org/10.13973/j.cnki.robot.250208
6. Paolillo, A., Gergondet, P., Cherubini, A., et al.: Autonomous car driving by a humanoid robot. J. Field Robot. Wiley Online Libr. **35**(2), 169–186 (2018)
7. Tong, Y., Liu, H., Zhang, Z.: Advancements in humanoid robots: a comprehensive review and future prospects. IEEE/CAA J. Automatica Sinica **11**(2), 301–328 (2024)
8. Liang, W., et al.: Large model empowered embodied AI: a survey on decision-making and embodied learning (2025). arXiv preprint arXiv:2508.10399
9. Zhao, W., Yuan, Y.: Development of intelligent robots in the wave of embodied intelligence (2025). https://doi.org/10.1093/nsr/nwaf159
10. LeCun, Y., Bengio, Y., Hinton, G.: Deep learning. Nature 521, 436–444 (2015). https://doi.org/10.1038/nature14539
11. Hashimoto, K.: Mechanics of humanoid robot. Adv. Robot. Taylor Francis **34**(21–22), 1390–1397 (2020)

12. Fraedrich, E., Heinrichs, D., Bahamonde-Birke, F.J., et al.: Autonomous driving, the built environment and policy implications. Transp. Res. Part A: Policy Pract. Elsevier **122**, 162–172 (2019)
13. Feng, D., et al.: Deep multi-modal object detection and semanticsegmentation for autonomous driving: datasets, methods, and challenges. IEEE Trans. Intell. Transp. Syst. **22**(3), 1341–1360 (2020)
14. Kabir, M.M., et al.: Terrain detection and segmentation for autonomous vehicle navigation: a state-of-the-art systematic review. Inf. Fusion **113**, 102644 (2025). ISSN1566-2535. https://doi.org/10.1016/j.inffus.2024.102644
15. Zhou, D., Yuan, X., Sun, Y., Wu, Y.: Current status and challenges in usability evaluation research on human-machine interaction in intelligent vehicles: a systematic review. Appl. Sci. **15**, 9384 (2025). https://doi.org/10.3390/app15179384
16. Kala, R.: On-Road intelligent vehicles: motion planning for intelligent transportation systems. Butterworth-Heinemann (2016)
17. Xie, J., et al.: Towards human-like automated vehicles: review and perspectives on behavioural decision making and intelligent motion planning. Transp. Saf. Environ. **7**(1), tdae005 (2025). https://doi.org/10.1093/tse/tdae005
18. Cai, Z., et al.: A friction-driven strategy for agile steering wheel manipulation by humanoid robots. Cyborg Bionic Syst. **4**, 0064 (2023). https://doi.org/10.34133/cbsystems.0064
19. Lv, H., Qian, Q., Pan, J., Song, M., Feng, Y., Li, Y.: Application of multi-strategy controlled rime algorithm in path planning for delivery robots. Biomimetics **10**, 476 (2025). https://doi.org/10.3390/biomimetics10070476
20. Paniego, S., Paliwal, N., Cañas, J.: Model optimization in deep learning based robot control for autonomous driving. IEEE Robot. Autom. Lett. **9**(1), 715–722 (2024). https://doi.org/10.1109/LRA.2023.3336244
21. Hammoudeh García, N., Chen, Y., Lieb, D., Wortmann, A.: Evaluation of a model-driven approach for the integration of robot operating system-based complex robot systems. Int. J. Adv. Robot. Syst. **22**(4) (2025). https://doi.org/10.1177/17298806251363648
22. A Review on Human-Computer Interaction and Intelligent RobotsFuji Ren (*School of Computer Science and Information, Engineering of Hefei University of Technology, Key Laboratory of Affective Computing and Advanced Intelligent Machine, Hefei 230009, Anhui Province, P. R. China†Tokushima University, Graduate School of Advanced Technology & Science, Tokushima 7708502, Japan) and Yanwei Bao (*School of Computer Science and Information, Engineering of Hefei University of Technology, Key Laboratory of Affective Computing and Advanced Intelligent Machine, Hefei 230009, Anhui Province, P. R. China)International Journal of Information Technology & Decision Making **19**(01), 5–47 (2020)
23. Guo, Q., Su, Z.: The application of industrial robot and the high-quality development of manufacturing industry: from a sustainability perspective. Sustainability **15**, 12621 (2023). https://doi.org/10.3390/su151612621
24. Stableford, B.: Science fact and science fiction: an encyclopedia. Routledge (2006)
25. Turing, A.M.: Computing machinery and intelligence (1950). Mind **59**(236), 33–60 (2021)
26. Siciliano B., Khatib, O.: Humanoid robots: historical perspective, overview, and scope. Humanoid robotics: a reference. Springer, pp. 3–8 (2019)
27. Eaton, M.: Evolutionary Humanoid Robotics. Springer (2015)
28. Tahan, K., N'kaoua, B.: NAO Robot to Help People with Alzheimer's Manage the Recall of Activities in Prospective Memory. International Conference on Human and Artificial Rationalities. Springer, pp. 234–249 (2024)
29. Bin, T., et al.: A survey on the visual perception of humanoid robot. Biomimet. Intell. Robot. **5**(1), 100197 (2025). ISSN 2667–3797. https://doi.org/10.1016/j.birob.2024.100197

30. Markoff, J.: Modest Debut of Atlas May Foreshadow Age of 'Robo Sapiens'. The New York Times. **11** (2013)
31. Siddiqui, F.: Tesla Says It Is Building a ,friendly 'Robot That Will Perform Menial Tasks, Won't Fight Back. The Washington Post **20** (2021)
32. Tong, Y., Liu, H., Zhang, Z.: Advancements in humanoid robots: a comprehensive review and future prospects. IEEE/CAA J. Automatica Sinica **11**(2), 301–328 (2024). https://doi.org/10.1109/JAS.2023.124140
33. Paolillo, A., Gergondet, P., Cherubini, A., Vendittelli, M., Kheddar, A.: Autonomous car driving by a humanoid robot. J. Field Robot. **35**, 169–186 (2018). https://doi.org/10.1002/rob.21731
34. Chung, T.H., Orekhov, V., Maio, A.: Into the robotic depths: analysis and insights from the darpa subterranean challenge. Ann. Rev. Control Robot. Auton. Syst. Ann. Rev. **6**(1), 477–502 (2023)
35. Wang, H., Li, S., Zheng, Y.F.: DARPA robotics grand challenge participation and ski-type gait for rough-terrain walking. Engineering **1**(1), 036–045 (2015). ISSN 2095–8099. https://doi.org/10.15302/J-ENG-2015006
36. Ferrari, P., et al.: Multi-contact planning and control for humanoid robots: design and validation of a complete framework. Robot. AutonomousSystems **166**, 104448 (2023). ISSN0921–8890. https://doi.org/10.1016/j.robot.2023.104448
37. Gkillas, A., et al.: A holistic perception system of internal and external monitoring for ground autonomous vehicles: AutoTRUST paradigm (2025). arXivPreprint. https://doi.org/10.48550/arXiv.2508.17969
38. Lian, H., Pei, X., Guo, X.: A local environment model based on multi-sensor perception for intelligent vehicles. IEEE Sens. J. **21**(14), 15427–15436 (2021). https://doi.org/10.1109/JSEN.2020.3018319
39. Wiley, T.: Multi-Strategy learning of robotic behaviours via qualitative reasoning. Proc. AAAI Conf. Artif. Intell. **27**(1), 1682–1683 (2013). https://doi.org/10.1609/aaai.v27i1.8502
40. Wang, X., Zhang, T.: Reinforcement learning with imitative behaviors for humanoid robots navigation: synchronous planning and control. Auton. Robot. **48**, 5 (2024). https://doi.org/10.1007/s10514-024-10160-w
41. Liu, Z.Y., Zhang, H., Li, Y.D.: A survey on end-to-end learning from perception to control for autonomous driving. Chin. J. Intell. Sci. Technol. **7**(1), 4–20 (2025). https://doi.org/10.11959/j.issn.2096-6652.202502
42. Yadav, S., Sana, S., Singh, S., Pandey, A.K.: A review of object detection algorithms based on deep learning. Int. Res. J. Mod. Eng. Technol. Sci. **7**(11) (2025). https://doi.org/10.56726/IRJMETS39161
43. Hu, J., Zhou, Y., Wang, H., Qiao, P., Wan, W.: Research on deep learning detection model for pedestrian objects in complex scenes based on improved YOLOv7. Sensors **24**, 6922 (2024). https://doi.org/10.3390/s24216922
44. Cai, J., Wu, Y.: Three-dimensional object detection for autonomous driving via deep learning: a review. Eng. Appl. Artif. Intell. **161**(Part C), 112238 (2025). ISSN 0952-1976. https://doi.org/10.1016/j.engappai.2025.112238
45. Zhang, J., Zhao, X., Chen, Z., Lu, Z.: A review of deep learning-based semantic segmentation for point cloud. IEEE Access **7**, 179118–179133 (2019). https://doi.org/10.1109/ACCESS.2019.2958671
46. Bellaccini, M., Lanari, L., Paolillo, A., et al.: Manual guidance of humanoid robots without force sensors: preliminary experiments with NAO. 2014 IEEE International Conference on Robotics and Automation (ICRA). IEEE, pp. 1184–1189 (2014)
47. Rasmussen, C., Lu, Y., Kocamaz, M.: Integrating StIEEE/RSJ International Conference on Intelligent Robots and Systems. IEEE, pp. 4084–4090 ereo Structure for Omnidirectional Trail Following (2011)

48. Kawaharazuka, K., et al.: Component Modularized Design of Musculoskeletal Humanoid Platform Musashi to Investigate Learning Control Systems. 2019 IEEE/RSJ International Conference on Intelligent Robots and Systems (IROS). IEEE, pp. 7300–7307 (2019)
49. Zhao, J., Shi, J., Zhuo, L.: BEV perception for autonomous driving: state of the art and future perspectives. Expert Syst. Appl. **258**, 125103 (2024). https://doi.org/10.1016/j.eswa.2024.125103
50. Li, Z., Li, Y., Wang, Y., Xie, G., Qu, H., Lyu, Z.: A lightweight model for 3D point cloud object detection. Appl. Sci. **13**, 6754 (2023). https://doi.org/10.3390/app13116754
51. Rasmussen, C., Sohn, K., Wang, Q., et al.: Perception and control strategies for driving utility vehicles with a humanoid robot. 2014 IEEE/RSJ International Conference on Intelligent Robots and Systems. IEEE, pp. 973–980 (2014)
52. Song, A.G., Yin, M.Y.: Teleoperation technology for dual arms of humanoid robots. J. Nat. **47**(2), 95–115 (2025). https://doi.org/10.3969/j.issn.0253-9608.2025.02.003
53. Zhang, Y., et al.: FALCON: learning force-adaptive humanoid loco-manipulation. arXiv Preprint (2025). https://doi.org/10.48550/arXiv.2505.06776
54. Milojevic, D., Zardini, G., Elser, M., Censi, A., Frazzoli, E.: CODEI: resource-efficient task-driven co-design of perception and decision making for mobile robots applied to autonomous vehicles. IEEE Trans. Rob. **41**, 2727–2748 (2025). https://doi.org/10.1109/TRO.2025.3552347
55. Li, H., et al.: Human-aware vision and-language navigation: bridging simulation to reality withdynamic human interactions. NIPS. Proceedings of 38th Conference on Neural Information Processing Systems (NeurIPS). California, NIPS Press **37**, 19411 (2024)
56. Lin, B.Q., et al.: ADAPT: vision language navigation with modality-aligned action prompts. IEEE. Proceedings of 2022 IEEE/CVF Conference on Computer Vision and Pattern Recognition (CVPR). New York, IEEE Press, pp. 15375–15385 (2022)
57. Alonso, R., et al.: Exploiting virtual reality and the robot operating system to remote-control a humanoid robot. Multimed. Tools Appl. **81**, 15565–15592 (2022). https://doi.org/10.1007/s11042-022-12021-z
58. Sohn, K., Jang, G.: Ground vehicle driving by full sized Humanoid. J. Intell. Robot. Syst. Springer **99**(2), 407–425 (2020)
59. Ma, H., et al.: Constraint-Aware diffusion guidance for robotics: real-time obstacle avoidance for autonomous racing (2025). arXiv preprint arXiv:2505.13131
60. Sugihara, K., Zhao, M., Nishio, T., Makabe, T., Okada, K., Inaba, M.: Design and control of a small humanoid equipped with flight unit and wheels for multimodal locomotion. IEEE Robot. Autom. Lett. **8**(9), 5608–5615 (2023). https://doi.org/10.1109/LRA.2023.3297065
61. Moon, J., et al.: Enhancing autonomous driving robot systems with edge computing and LDM platforms. Electronics **13**, 2740 (2024). https://doi.org/10.3390/electronics13142740
62. Dong, Y., et al.: Bimanual continuous steering wheel turning by a dual-arm robot. IEEE/ASME Trans. Mechatron. **29**(3), 1773–1784 (2024). https://doi.org/10.1109/TMECH.2023.3316634
63. Tang, L., Liang, D., Gao, G., et al.: Modeling and reinforcement learning-based locomotion control for a humanoid robot with kinematic loop closures. Multibody Syst. Dyn. **65**, 239–265 (2025). https://doi.org/10.1007/s11044-024-10035-z
64. Huo, Y., Zhang, C.: A review of key technologies for environment sensing in driverless vehicles. World Electr. Veh. J. **15**, 290 (2024). https://doi.org/10.3390/wevj15070290
65. Ahad, A., Chris, C., Phil, B.: AI-driven UAV system for autonomous vehicle tracking and license plate recognition. Open Eng. **15**(1), 20240101 (2025). https://doi.org/10.1515/eng-2024-0101
66. Qiao, J., Guo, J., Li, Y.: Simultaneous localization and mapping (SLAM)-based robot localization and navigation algorithm. Appl. Water Sci. **14**, 151 (2024). https://doi.org/10.1007/s13201-024-02183-6

67. Kawaharazuka, K., Tsuzuki, K., Koga, Y., et al.: Toward autonomous driving by musculoskeletal humanoids: a study of developed hardware and learning-based software. IEEE Robot. Autom. Mag. IEEE **27**(3), 84–96 (2020)
68. Choi, J.Y., Ahn, S., Kim, D., et al.: Exploring challenges and opportunities in manufacturing and intelligence for future robotics. Int. J. Precis. Eng. Manuf. **26**, 2203–2222 (2025). https://doi.org/10.1007/s12541-025-01318-2

Modelling and Simulation of EEG Neurodynamics Under Driver Mental Fatigue

Xueteng Zhang(✉), Xiaodong Zhang, Chengkun Pan, Fenghua Hu, and Zirui Zhao

School of Mechanical Engineering, Xi'an Jiaotong University, Xi'an 710049, China
xueteng@stu.xjtu.edu.cn

Abstract. Driver fatigue represents a leading cause of traffic accidents, making EEG-based monitoring a highly direct technological countermeasure. Current methodologies, however, rely predominantly on data-driven feature extraction and pattern recognition, often overlooking the foundational neurophysiological mechanisms that underpin fatigue-related EEG signatures. This lack of physiological insight inherently limits model interpretability and generalizability. To bridge this gap, an end-to-end computational framework is introduced to mechanistically link neural population dynamics with scalp-recorded potentials. A modified Wendling neural mass model—used to generate fatigue-specific postsynaptic potentials across principal frequency bands—is integrated with a four-layer head model. The boundary element method is further employed to achieve precise forward computation from intracranial dipole sources to scalp EEG. Simulation results demonstrate the model's ability to accurately replicate canonical EEG manifestations of fatigue, most notably the characteristic spectral power shift from high (Alpha/Beta) to low (Delta/Theta) frequencies. The strong correlation between these simulated features and empirical EEG data validates the framework, offering a mechanistic and interpretable alternative to conventional "black-box" methods and establishing a robust foundation for next-generation, physiologically-grounded fatigue monitoring systems.

Keywords: Brain-computer interface · Brain neural modelling · Fatigued mental state · Mechanism of driver fatigue

1 Introduction

With the rapid development of the transportation industry, driving safety has become an important social concern. According to statistics, driver mental fatigue is one of the main causes of traffic accidents [1]. Therefore, real-time and accurate monitoring of driver fatigue is crucial for road traffic safety. EEG signals can directly reflect changes in the cognitive state of the brain and are highly effective in monitoring mental fatigue, thus being regarded as the gold standard for fatigue monitoring [2, 3]. For instance, studies have shown that under fatigue conditions, the power and topological distribution of EEG signals in the Theta and Alpha frequency bands exhibit characteristic changes [4].

Although research on EEG-based fatigue monitoring has made progress, existing methods largely rely on data-driven feature extraction and pattern recognition [5, 6].

H. Wei and J. Niu (Eds.): COBOT 2025, CCIS 2893, pp. 323–335, 2027.
https://doi.org/10.1007/978-981-92-2074-8_19

Such methods lack explanations of the neurophysiological mechanisms behind fatigue EEG features, making it difficult to clarify the origin of these features, resulting in poor model interpretability and limited generalization, which constrains further improvement in monitoring performance.

Currently, EEG-based driver fatigue monitoring research focuses on developing better algorithms, extracting features such as power spectral density from EEG signals, and using classification models to distinguish states, confirming feasibility in practical applications. OGINO et al. [7] used the EEG power spectral density of the frontal Fp1 channel and achieved a 72.7% accuracy in detecting fatigued driving. Wang et al. [8] defined an EEG-based driver fatigue assessment method, using wavelet entropy to extract features of fatigued driving, achieving a monitoring accuracy of 90.7%. Xu et al. [9] employed convolutional neural networks to classify drivers' EEG signals for the identification of driving fatigue, significantly improving recognition performance. Gao et al. [10] constructed a spatio-temporal convolutional network based on the spatio-temporal structure of EEG signals to monitor driver fatigue. However, there are relatively few researchers focusing on the physiological mechanisms of EEG generation under fatigue conditions. Babloyantz et al. [11] applied nonlinear dynamics analysis methods to study the EEG signals during human sleep and found that the EEG signals during sleep conformed to the characteristics of chaotic signals. Jap et al. [12] found that the measurement indicators of different brain regions correspond to different degrees of fatigue state, among which the EEG signals of the temporal region are the most prominent. Regarding what EEG features appear under fatigue, existing literature lacks mechanistic explanations from the neurodynamic level, such as neuronal population discharges and integration of postsynaptic potentials. This lack of physiological mechanism research makes current feature extraction partially reliant on trial and error, limiting the interpretability and robustness of the methods.

To address these issues, a neurodynamic model of driver mental fatigue is constructed and simulated in this study. The research is directed toward three core objectives: the simulation of the complete signal generation process from neuronal population postsynaptic potentials to scalp EEG, based on a neural mass model; the provision of mechanistic explanations for observed fatigue-related EEG features (e.g., Theta/Alpha band changes) through simulation; and the establishment of a theoretical foundation for the subsequent extraction of more robust and physiologically meaningful fatigue EEG features.

2 Method

2.1 Data Sources

Data set 1: The seed-vig dataset originates from the Brain-inspired Computing and Machine Intelligence Research Centre at Shanghai Jiao Tong University. This dataset is used for monitoring driver alertness and involves 23 healthy participants, who performed long monotonous driving tasks on a highly realistic driving simulator to induce fatigue. During the experiment, researchers synchronously collected 17-channel EEG data and eye-tracking data from each participant. Finally, the PERCLOS values calculated from the eye-tracking data were used to generate corresponding fatigue labels for the EEG data samples.

Data set 2: EEG data were independently collected by a laboratory. The data were collected by the team led by Zhang Xiaodong at Xi'an Jiaotong University in 2025, for EEG signals under driver fatigue conditions. The dataset includes one healthy participant, male, 23 years old.

2.2 Data Collection Paradigm

For the dataset, all subjects completed approximately two hours of 17-channel EEG recording. Participant selection was based on criteria including good health, absence of sleep disorders, and possession of a valid driver's license. All participants adhered to pre-experiment restrictions prohibiting alcohol and caffeine intake for 24 h and were instructed to maintain their regular sleep-wake cycles. Following the fitting of the EEG cap (with all electrode impedances maintained below 5 kΩ) and calibration of the eye tracker, subjects completed a brief practice session in the driving simulator. The formal protocol commenced with a 5-min baseline recording while subjects remained seated and awake. This was followed by a 90-min monotonous highway driving task designed to induce fatigue, which incorporated random sudden events to measure reaction times. All behavioral data were recorded throughout the session.

2.3 Data Preprocessing

The data preprocessing was conducted using Matlab and the EEGLAB toolbox. Firstly, the sampling frequency of the signals was calibrated. The sampling frequency of dataset 2 was 1000 Hz, which caused inconvenience for the subsequent time window segmentation. Therefore, the sampling frequency of dataset 2 was converted to 200 Hz. Afterwards, to eliminate EEG noise and artefacts unrelated to fatigue states, a 50 Hz notch FIR filter was applied to the signals of each dataset, retaining the components most relevant to fatigue EEG. In addition, after filtering, the signals were re-referenced to enhance the visibility of features between channels. Finally, independent component analysis was performed to remove artefacts such as eye and muscle activity. Perform the same operation on Dataset 1.

2.4 Mechanism of Brain Wave Generation Under Driving Fatigue

The fundamental electrical activity of the human brain is attributed to neurons [13], which interact and process information through electrical signaling. Information is transmitted via synaptic structures, located between the axon of the transmitting neuron and the dendrite of the receiving neuron, and composed of the presynaptic membrane, synaptic cleft, and postsynaptic membrane. When a neuron is activated, incoming electrical impulses from other neurons are integrated within the cell body, where excitatory and inhibitory postsynaptic potentials are summated. Once intracellular depolarization reaches a specific threshold, an action potential is generated and propagated along the axon to subsequent neurons, enabling the continuous transmission of neural signals [14]. Given that neuronal action potentials do not simultaneously activate large populations of adjacent neurons, the extracellular potential field generated by postsynaptic potentials—which persist for up to 10–20 ms—is widely regarded as the primary source of scalp EEG signals.

Fatigue is defined as a decline in the functional capacity of tissues, organs, or the entire body resulting from an imbalance between physical or psychological load and recovery capacity, typically induced by prolonged high-intensity cognitive activity, also referred to as cognitive fatigue. Based on established neurochemical and neuromodulatory mechanisms, higher cognitive brain regions—particularly the prefrontal cortex (PFC) and central areas—have been identified as being highly susceptible to functional suppression. This is due to dense neuromodulatory inputs originating from the locus coeruleus and basal forebrain. Such suppression leads to significant alterations in the firing patterns and synchronization of cortical neuronal populations, manifested most notably as an enhancement of oscillatory activity in the Theta band (4–8 Hz). This increase is especially prominent over prefrontal and central electrode sites (e.g., Fz, Cz, F3, and F4 in the 10–20 system), where power spectral density is markedly elevated. Additionally, a topographic redistribution of Alpha rhythms (8–12 Hz) is observed: instead of being confined to occipital regions (e.g., O1, O2) during eyes-closed rest, Alpha activity shifts anteriorly toward frontal (F3, F4) and central (C3, C4) areas.

2.5 Establishment of a Postsynaptic Membrane Potential Model Based on Neuronal Clusters

Since scalp electrodes cannot capture the electrical activity of individual neurons, microscopic modeling approaches targeting single neurons are not applicable [15]. Therefore, to model EEG signal sources—specifically, to simulate the collective discharge behavior of cortical neuron populations—a macroscopic model based on lumped parameters is employed. This model operates on the "average region approximation" principle, whereby neuron populations within the same brain region that perform similar functions are characterized by identical lumped parameter sets.

The Wendling-NMM [16] incorporates four subpopulations: pyramidal cells, excitatory interneurons, and fast/slow inhibitory interneurons, each representing neurons sharing a membrane potential. As shown in Fig. 1, the interneurons receive the average postsynaptic potential (y0) from pyramidal cells. This signal is integrated (v1–v3) and converted into presynaptic pulse density (z1–z3) via a Sigmoid function. These pulses are then transformed through synaptic response functions (he/g/i(t)) into postsynaptic potentials (y1–y3), which feedback to the pyramidal cells. The pyramidal cells, in turn, integrate these inputs (v0), convert them into their presynaptic pulse density (z0), and transmit output (y0) back to the interneurons via synaptic functions, forming a closed loop.

In NMM, the synaptic response function is a second-order transfer function, proposed by Rotterdam in 1982 and still used today; that is, the average postsynaptic membrane potential yk(k = 0,1,2,3) is the convolution of the average spike density zk(k = 0,1,2,3)with the synaptic response function hk(k = e,i,g).

$$y_k(t) = h_k(t) \otimes z_k(t) \tag{1}$$

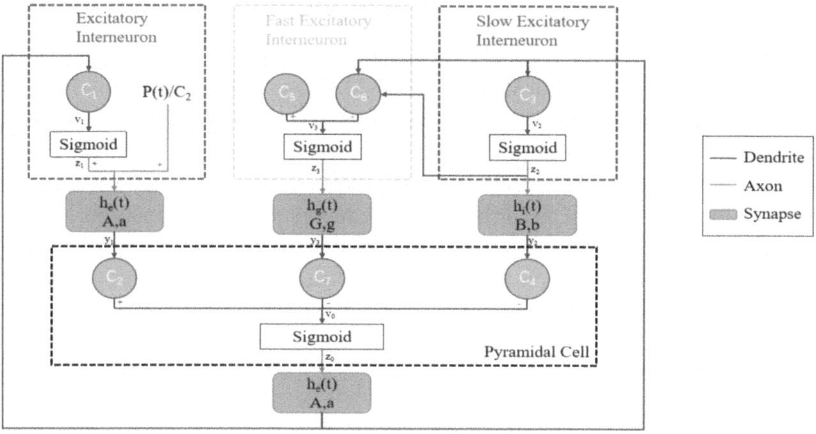

Fig. 1. Wendling-NMM

Among them, hk(k = e,i,g)the expression is:

$$\begin{cases} h_e(t) = Aate^{-at} \\ h_i(t) = Bbte^{-bt} \\ h_g(t) = Ggte^{-gt} \end{cases} \tag{2}$$

In the formula: A, B, G—represent the average gains of excitatory, slow inhibitory, and fast inhibitory synapses respectively; a, b, g—represent the time constants of each synapse respectively. When solving, the formula can be converted into the second-order differential form in the time domain, taking y0(t) as an example:

$$\ddot{y}_0(t) = Aaz_0(t) - 2a\dot{y}_0(t) - a^2y_0(t) \tag{1}$$

Based on this form, according to their respective average gains and time constants, it can be extended to y1, y2, and y3. By substituting the Sigm-type function into the formula.6 and combining it with the connection form shown in Fig. 1, the complete expression of the Wendling-NMM is:

$$\begin{cases} \ddot{y}_0(t) = Aa\mathrm{Sigm}[C_2y_1(t) - C_4y_2(t) - C_7y_3(t)] - 2a\dot{y}_0(t) - a^2y_0(t) \\ \ddot{y}_1(t) = Aa(\mathrm{Sigm}[C_1y_0(t)] + \mathrm{P}(t)/C_2) - 2a\dot{y}_1(t) - a^2y_1(t) \\ \ddot{y}_2(t) = Bb\mathrm{Sigm}[C_3y_0(t)] - 2b\dot{y}_2(t) - b^2y_2(t) \\ \ddot{y}_3(t) = Gg\mathrm{Sigm}[C_5y_0(t) - C_6\mathrm{Sigm}[C_3y_0(t)]] - 2g\dot{y}_3(t) - g^2y_3(t) \end{cases} \tag{2}$$

In the formula: C1–C7—connection constants between subgroups; P(t)— represents exogenous input. The specific expression of the Sigm function is:

$$\mathrm{Sigm}(v) = 2e_0/(1 + e^{r(s_0 - v)}) \tag{5}$$

In the formula:e0—Maximum firing rate of a neural population's action potentials (i.e., the upper limit of action potentials generated per unit time); s0—Average postsynaptic membrane potential corresponding to a 50% firing rate; r—Slope of the function.

A five-cluster coupled Neural Mass Model (NMM) was developed to simulate broadband neural activity across all five EEG bands (Delta 0.5–4 Hz to Gamma 30–45 Hz). This multi-cluster architecture captures the brain's inherent multi-columnar interactions and, unlike conventional single-cluster NMMs that produce only single-peak spectra, can generate the full range of EEG rhythms. The model framework incorporates a parallel feedforward connection structure, following Zavaglia's approach to motor cortex modeling [17]—a design principle whose critical role in coupled dynamics is established by Wendling [18] (Fig. 2).

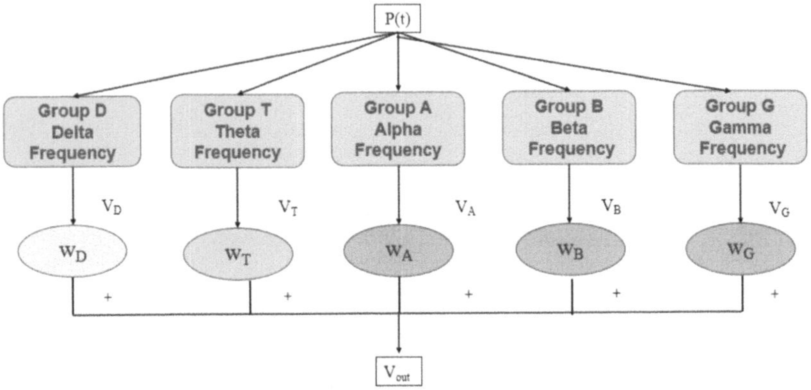

Fig. 2. Five-cluster NMM parallel feedforward structure

In this structure, each cluster has different weights, time constants, and synaptic average gains to simulate the complexity of the postsynaptic membrane potential of the entire motor cortex. According to this structure, the postsynaptic membrane potential of the five-cluster coupled NMM can be represented as:

$$\mathrm{V_{out}} = \sum_{k=\mathrm{L,M,H}} w_k \mathrm{V}_0^k(t) \tag{6}$$

where: superscripts D, T, A, B and D —frequency bands;wk (k = D,T,A,B,D)—weights corresponding to each band (i.e., each cluster).

2.6 Establishment of a Postsynaptic Membrane Potential Model Based on Neuronal Clusters

The generation of the extracellular electric field is modeled as a dipole. At the microscopic level, each activated pyramidal cell can be represented by a dipole. However, scalp EEG signals require the synchronous activation of numerous pyramidal cells within adjacent cortical regions. Based on the "average region approximation" principle, this synchronously activated neuronal population is represented by a single equivalent dipole.

The separation of equal and opposite charges creates a spatial potential difference, forming an electric field. This field drives the directional movement of free charges, constituting an electric current. Within this dipole framework, the potential distribution at any point a in space is given by:

$$\mathrm{U}_a = k\frac{I(r_2 - r_1)}{r_1 r_2} \approx k\frac{\mathbf{d}\cos\theta}{r^2} = k\frac{I \cdot p\cos\theta}{r^2} \tag{7}$$

where: U—Electric potential/V; k—Scaling factor; r1, r2, r—represent the distances from any point in space to the positive pole, negative pole, and the centre of the dipole, respectively; p—Distance between the two monopoles of the dipole; I—Current;d—Dipole moment. The electric field intensity can be calculated from the negative gradient of the potential, as follows:

$$\mathbf{E} = -\nabla \mathrm{U} \tag{8}$$

where: E—Electric field strength (unit V/m); $\bigtriangledown$— Gradient operator. By multiplying the field strength by the conductivity, current density is introduced to describe the total charge passing vertically through a unit annular region per unit time. According to Ohm's law, the current density can be expressed as

$$\mathbf{J} = \kappa\mathbf{E} \tag{9}$$

where: J—Current density (unit A/m2); k—Electrical conductivity (unit S/m). For isotropic conductors, it is a scalar; whereas for anisotropic conductors, $\kappa \in \mathbb{R}^{3\times 3}$— a tensor related to the position within the conductor, in the form of

$$\kappa = \begin{bmatrix} \kappa_{11} & \kappa_{12} & \kappa_{13} \\ \kappa_{12} & \kappa_{22} & \kappa_{23} \\ \kappa_{13} & \kappa_{23} & \kappa_{33} \end{bmatrix} \tag{10}$$

For a dipole, the current density at any point can be expressed as the Dirac δ function of its distance from the positive and negative poles.

$$\nabla \cdot \mathbf{J} = I\delta(r_1) - I\delta(r_2) = -\nabla \cdot (\kappa\nabla(\mathrm{U})) \tag{11}$$

In this formula, r1 and r2 represent the distances from a point in space to the positive and negative poles of the dipole, respectively. To simplify calculations, a four-layer concentric sphere head model is employed, comprising brain tissue, cerebrospinal fluid, skull, and scalp. Within this model, each layer is treated as an isotropic homogeneous medium, allowing the current density to be characterized by four scalar parameters. The scalp potential at any point can then be calculated via the following expression:

$$\begin{aligned} &\frac{\partial}{\partial x}(\kappa\frac{\partial \mathrm{U}}{\partial x}) + \frac{\partial}{\partial y}(\kappa\frac{\partial \mathrm{U}}{\partial y}) + \frac{\partial}{\partial z}(\kappa\frac{\partial \mathrm{U}}{\partial z}) = \\ &-I\delta(x - x_1)\delta(y - y_1)\delta(z - z_1) + I\delta(x - x_2)\delta(y - y_2)\delta(z - z_2) \end{aligned} \tag{12}$$

In each layer of the structure, boundary conditions are set, and the potential continuity equation is as follows:

$$\Phi i(Ri) = \Phi\{i + 1\}(Ri) \tag{13}$$

where i represents different brain structures, Φ represents the potential, and the current continuity equation is:

$$\sigma_{\mathrm{i}} \frac{\partial \Phi \mathrm{i}}{\partial \mathrm{r}}\bigg|(\mathrm{r} = \mathrm{R_i}) = \sigma_{\mathrm{i+1}} \frac{\partial \Phi \mathrm{i} + 1}{\partial \mathrm{r}}\bigg|(\mathrm{r} = \mathrm{R_i}) \tag{14}$$

Accurate EEG forward calculation generally requires detailed subject-specific anatomical data and tissue conductivity parameters. In the absence of such individualized information, this study employs the Boundary Element Method (BEM) to reduce computational complexity. In this framework, different brain tissues along the forward path are modeled as piecewise homogeneous, isotropic volume conductors, simulating the surface potentials generated by intracranial current sources. To represent different fatigue states, equivalent current dipoles are oriented perpendicular to the cortical surface in relevant brain regions.

3 Simulation Results and Disscusion

3.1 Simulation Results of the Postsynaptic Membrane Based on the Neuron Model

Model parameters for the five neural mass models (NMMs) across different frequency bands were designed based on physiological literature. Common parameters, such as the connectivity constants C1–C7 and neuronal firing properties, remained consistent across all bands. In contrast, system parameters like time constants and average synaptic gain exhibited a nonlinear coupling with the output spectral peak. Exogenous input was modeled as random white noise, representing influences from adjacent cortical regions [19]. Table 1 lists the common parameters, band-specific system parameters, and exogenous inputs. The NMM differential equations were solved numerically using the 4th-order Runge-Kutta method, with the resulting output spectra for each band shown in Fig. 3.

To simulate the signal source, frequency band weights were first derived from the spectral energy of real fatigue EEG data. Three representative fatigue states—awake, fatigued, and drowsy—were identified based on established neurophysiological mechanisms: during the awake state, high prefrontal cortex activity is reflected in enhanced β rhythms and dominant occipital αrhythms; in the fatigued state, suppressed locus coeruleus-prefrontal pathway function leads to significantly increased θ power and an anterior shift of α rhythms toward frontal regions; while the drowsy state is characterized by enhanced thalamocortical synchronization, resulting in dominant δ power and emerging sleep spindles. These distinct neural signatures across states were validated through power spectrum analysis of the SEED-VIG dataset combined with PERCLOS behavioral indicators. Finally, postsynaptic potentials were simulated for each fatigue state based on the calculated frequency band weights.

Table 1. NMM parameters and exogenous inputs for each frequency band

Common parameters	System parameters						
Parameter	Value	Parameter	Frequency band				
			Delta	Theta	Alpha	Beta	Gamma
C_1	5.0	A	2.5	3.0	3.25	3.5	4.0
C_2, C_7	4.0	B	25	22	20	18	15
C_3, C_4	2.0	G	12	10	8	6.0	4.0
C_5	4.0	a	40	60	80	120	160
C_6	2.0	b	20	30	40	60	80
s_0	6 mV	g	4.0	6.0	16	20	24
e_0	2.5 s^{-1}	Exogenous input P(t)					
r	0.56 V^{-1}	Mean	60	Variance	100		

As shown in Fig. 3, the simulation analysis of EEG power spectral density (PSD) under three states yielded the following results: During the alert state, the occipital and frontal α-waves and high-frequency bands demonstrated higher energy density, with the PSD peaking at f = 9.77 Hz (approximately 9.80×10^{-4} $\mu V^2/Hz$) and the total power across all bands reaching P_total = 0.0065 μV^2. In the fatigued state, the PSD of α and β bands decreased significantly, while the PSD in low-frequency bands (δ and θ) increased substantially. The peak frequency shifted to f = 5.86 Hz, corresponding to a peak PSD of approximately 1.55×10^{-3} $\mu V^2/Hz$, and the total power rose to P_total = 0.0107 μV^2. Under severe fatigue (i.e., drowsiness), particularly notable changes were observed in the δ band, with a prominent peak at f = 1.95 Hz, a peak PSD of approximately 2.45×10^{-3} $\mu V^2/Hz$, and a further increase in total power to P_total = 0.0162 μV^2. In summary, as cerebral fatigue deepens, the PSD of high-frequency bands (e.g., α and β) decreases, reflecting weakened neural activity, while the PSD of low-frequency bands (e.g., δ and θ) increases, accompanied by a rise in total EEG power, indicating enhanced low-frequency oscillations and an overall elevation of brain activity under fatigue.

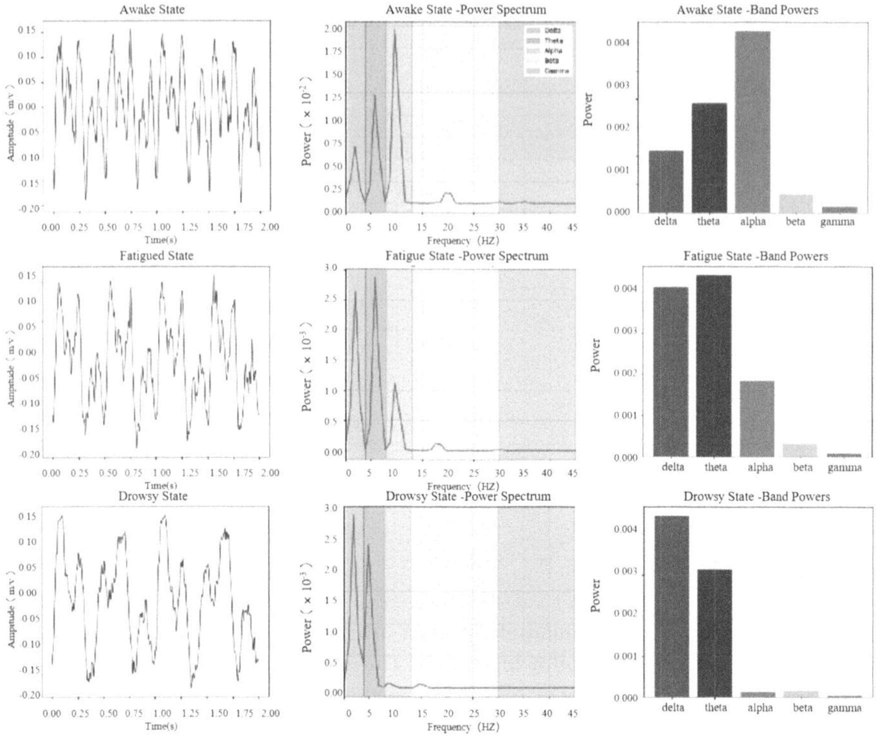

Fig. 3. Simulation of synaptic thick-film electrical signals and their spectra under different fatigue states

3.2 Simulation Results of the Extracellular Electric Field Based on the Dipole Model

A four-layer concentric spherical head model is used to simulate scalp electroencephalogram (EEG) potentials. The postsynaptic membrane potentials simulated under fatigue conditions serve as dipole source inputs. Positioned in the frontal and occipital cortices and oriented perpendicular to the scalp surface, the dipole source simulates fatigue-dominated frontal lobe neural activity, as illustrated in Fig. 4. The specific parameters of the concentric spherical model are provided in Table 2.

Table 2. Parameters of the Four-Layer Concentric Sphere Head Mode

	White matter/grey matter	Cerebrospinal fluid	Skull	Scalp
Conductivity	0.33	1.79	0.23	0.43
Size	0.4	0.1	0.7	0.5

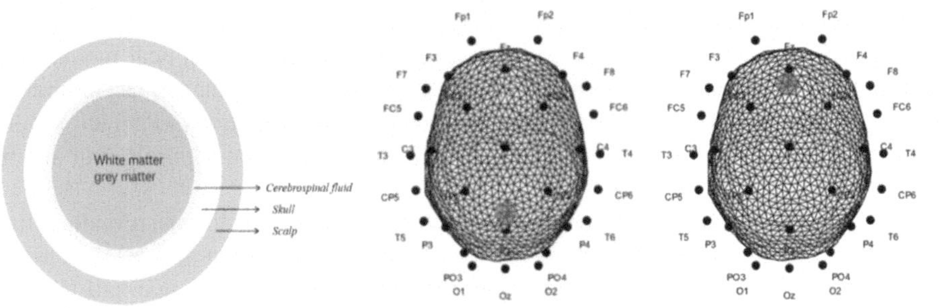

Fig. 4. Extracellular electric field model diagram:The left image shows Brain volume semiconductor model,the middle image shows the Placement position of occipital lobe dipoles and and the right image shows the Frontal lobe dipole placement location

We extracted the power spectral density of the simulated EEG signals (Fig. 5) and compared it with real fatigue EEG data. The results show that the energy in the theta (4–8 Hz) and delta (0.5–4 Hz) bands increases with fatigue, whereas the energy in the alpha (8–13 Hz) and beta (13–30 Hz) bands decreases, with total power showing an upward trend, highly consistent with the spectral changes in real EEG, with Pearson correlation coefficients shown in the table (correlation coefficient > 0.85). Nevertheless, due to the use of a simplified spherical head model, there are certain differences between the simulated signal and real EEG in terms of time-domain amplitude. In the future, combining individual MRI structural data could allow the construction of a more precise finite element head model to improve simulation accuracy.

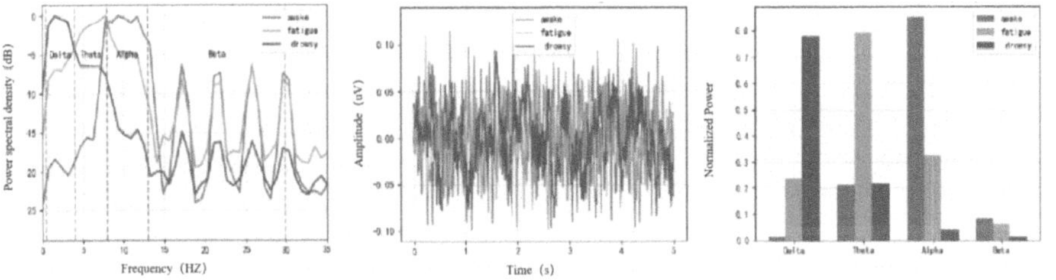

Fig. 5. Simulation result images: the left image shows the Power spectral density under different fatigue states, the middle image shows the simulated EEG signal of the Fz channel, and the right image shows the changes in brain power under different fatigue states across different frequency bands

4 Conclusions

An end-to-end simulation from neural population activity to scalp EEG was established by coupling a neural mass model with an electromagnetic forward model. Simulation results demonstrate that the characteristic EEG signatures of fatigue—such as increased theta power, decreased alpha power, and an anterior shift of alpha activity—can be reproduced by the model, while also providing potential neurophysiological explanations for these macroscopic phenomena.

A key advantage of the proposed model lies in its mechanistic interpretability. The simulations indicate that spectral changes consistent with neurochemical modulation imbalances can be replicated by modulating parameters such as the gain of fast inhibitory interneurons. This successfully bridges macroscopic EEG observations with microscopic neural population dynamics, thereby addressing a major limitation of purely data-driven approaches.

However, model performance declines under deep fatigue conditions, revealing limitations in the current framework. Simplifications such as the parallel NMM cluster structure and a spherical head model reduce computational cost but may fail to fully capture complex cross-frequency coupling or realistic electromagnetic conduction during profound fatigue. This issue is further compounded by the treatment of fatigue as a discrete state rather than a continuous process.

Future work will focus on dynamic model refinement, including the introduction of delayed connections between brain regions and the exploration of the inverse problem for estimating NMM parameters from real EEG data. These steps are essential to evolve the model from a simulation tool into a diagnostic platform capable of quantitatively assessing individual neurophysiological states, thereby laying the groundwork for next-generation mechanism-based fatigue monitoring systems.

Acknowledgments. The authors are very grateful for the work supported financially by the Key Research and Development Projects of Shaanxi Province, China (No. 2024GX-ZDCYL-02-16).

References

1. Amodio, A., Ermidoro, M., Maggi, D., et al.: Automatic detection of driver impairment based on pupillary light reflex. IEEE Trans. Intell. Transp. Syst. **20**(8), 3038–3048 (2019)
2. Chen, J., Wang, H., Wang, Q., et al.: A study on drowsy driving state based on EEG signals. Automot. Eng. **40**(5), 515–520 (2018)
3. Gharagozlou, F., Saraji, G.N., Mazloumi, A., et al.: Detecting driver mental fatigue based on EEG alpha power changes during simulated driving. Iran. J. Public Health **44**(12), 1693–1700 (2015)
4. Craig, A., et al.: Regional brain wave activity changes associated with fatigue. Psychophysiology **49**(4), 574–582 (2012)
5. Wang, H., et al.: Review on driving fatigue detection based on EEG. J. Huazhong Univ. Sci. Technol. **50**(11): 54–65+78 (2022)
6. Ren, L., Nie, Z., Yu, X., et al.: Review on recognition of unsatisfactory driving status based on electroencephalogram. China J. Highway Transport **37**(08), 216–230 (2024)
7. Ogino, M., Mitsukura, Y.: Portable drowsiness detection through use of a prefrontal single-channel electroencephalogram. Sensors **18**(12), 4477 (2018)
8. Wang, Q., Li, Y., Liu, X.: Analysis of feature fatigue EEG signals based on wavelet entropy. Int. J. Pattern Recognit Artif Intell. **32**(8), 0218–0314 (2018)
9. Xu, T., Wang, H., Lu, G., et al.: E-key: an EEG-based biometric authentication and driving fatigue detection system. IEEE Trans. Affect. Comput. **14**(2), 864–877 (2021)
10. Gao, Z., Wang, X., Yang, Y., et al.: EEG-Based Spatio-Temporal convolutional neural network for driver fatigue evaluation. IEEE Trans. Neural Networks Learn. Syst. **30**(9), 2755–2763 (2019)

11. Babloyantz, A., Salazar, J.M.G.D., Nicolis, C.: Evidence of chaotic dynamics of brain activity during the sleep cycle. Phys. Lett. A **111**(3), 152–156 (1986)
12. Jap, B.T., Lal, S., Fischer, P., et al.: Using EEG spectral components to assess algorithms for detecting fatigue. Expert Syst. Appl. **36**(2), 2352–2359 (2009)
13. Zhang, Y.Y., Yu, X.G.: Application progress of human brain connectomics in neurosurgery. Chin. J. Med. Imaging **25**(10), 789–793 (2017)
14. Yi, G.S., Huang, X.L., Wang, J., et al.: Metabolic cost of neuronal information encoding: action potentials and energy efficiency. Prog. Biochem. Biophys. **48**(4), 434–449 (2021)
15. Wilson, H.R., Cowan, J.D.: Excitatory and inhibitory interactions in localized populations of model neurons. Biophys. J. **12**(1), 1–000 (1972)
16. Wendling, F., Bartolomei, F., Bellanger, J.-J., et al.: Epileptic fast activity can be explained by a model of impaired GABAergic inhibition. Eur. J. Neurosci. **15**, 1499–1508 (2002)
17. Zavaglia, M., Astolfi, L., Babiloni, F., et al.: A Neural mass model for the simulation of cortical activity estimated from high resolution EEG during cognitive or motor tasks. J. Neurosci. Methods **157**(2), 317–329 (2006)
18. Wendling, F., Bellanger, J.J., Bartolomei, F., et al.: Relevance of nonlinear lumped-parameter models in the analysis of depth-EEG epileptic signals. Biol. Cybern. **83**(4), 367–378 (2000)
19. Jansen, B.H., Zouridakis, G., Brandt, M.E.: A Neurophysiologically-based mathematical model of flash visual evoked potentials. Biol. Cybern. **68**(3), 275–283 (1993)

Deep Learning Phase Unwrapping Based on DeepLabv3+

Xingyu Gao[1], Guangtao Hu[1], Yu Li[2], and Weiming Li[1(✉)]

[1] Guangxi Key Laboratory of ZHIYU Humanoid Robots, Guangxi Engineering Research Center for Intelligent Vision and Collaborative Robotics, School of Artificial Intelligence, GuangXi Minzu University, Nanning 530006, China
liaifan@buaa.edu.cn

[2] School of Mechanical and Electrical Engineering, Guilin University of Electronic Technology, Guilin 541004, Guangxi, China

Abstract. As a pivotal technique in computer vision, phase unwrapping has witnessed rapid development in recent years and found extensive applications in fields such as 3D reconstruction and depth estimation. In Fringe Projection Profilometry (FPP), the precise acquisition of phase information is a prerequisite for measuring the topography and surface features of a target. However, the raw phase data is frequently compromised by noise and inherent phase ambiguities. Building upon the DeepLabv3+ semantic segmentation framework, the proposed method takes a wrapped phase map as input and predicts the corresponding phase-level information, from which the continuous phase can be reconstructed. To address class imbalance and improve the model's ability to handle complex phase-level distributions, a combination of Dice Loss and Focal Loss is employed during training. Given that no publicly available datasets exist for the phase unwrapping task, this study employs a simulated dataset for network training, simulation-based evaluation, noise-robustness testing, and comparative experiments. The comparison results indicate that, under identical conditions, the DeepLabv3+ model achieves notably improved segmentation performance—yielding up to a 6% increase in SSIM compared with HRNet. These findings demonstrate that the DeepLabv3+ –based deep learning model serves as an effective phase unwrapping framework, capable of accurately mapping wrapped phases to continuous phases. The strong performance further validates the accuracy and generalization capability of the proposed network for a wide range of engineering applications.

Keywords: Fringe projection · Phase unwrapping · Wrapped phase · Continuous phase · Deep learning

1 Introduction

As a pivotal stage in phase measurement and 3D reconstruction, phase unwrapping directly determines the precision and robustness of the subsequent reconstruction results. However, when confronting challenges such as complex textures, severe noise, low fringe contrast, occlusions, or phase discontinuities, conventional approaches often exhibit

H. Wei and J. Niu (Eds.): COBOT 2025, CCIS 2893, pp. 336–356, 2027.
https://doi.org/10.1007/978-981-92-2074-8_20

inadequate stability and adaptability. These limitations significantly hinder their application in high-precision fields, including digital holography, interferometry, and structured light 3D imaging. To overcome these bottlenecks, the integration of deep learning with physical models has emerged as a promising solution. This hybrid approach mitigates the deficiencies of traditional methods, thereby enhancing the accuracy of 3D reconstruction and facilitating reliable phase unwrapping in complex scenarios. By integrating deep learning algorithms, it becomes feasible to develop a fully automated, operator-free phase unwrapping pipeline, thereby reducing manual intervention and significantly enhancing measurement efficiency. Such an approach enables fast phase unwrapping and real-time 3D reconstruction, meeting the stringent accuracy and speed requirements of robotic navigation, medical imaging, and industrial inspection. Consequently, research on deep learning–assisted phase-unwrapping-based 3D reconstruction is of both theoretical and practical importance, offering substantial academic value and broad application potential as a highly promising direction for future work.

Structured light fringe projection contour imaging technology [1] has become widely employed in 3D optical metrology applications such as mechanical engineering, industrial monitoring, computer vision, and biomedicine. Its imaging technique is widely applicable and low-cost, and the measurement range is suitable for indoor measurements as well as industrial inspection. Among the many principles based on this technique, the step of phase unwrapping is crucial, where an accurate and continuous phase intuitively responds to the depth information of an object. Currently, the common methods for phase unwrapping include the four-step phase shift method and the Fourier transform contouring (FTP) method. Still, both utilize an inverse tangent function to unwrap the phase, obtaining wrapped phases distributed in $[-\pi, \pi]$, and also requiring the wrapped phase to be phase unwrapping. Phase unwrapping is divided into spatial phase unwrapping and temporal phase unwrapping. Spatial phase unwrapping [2] analyzes the phase values between neighboring elements of the wrapped phase by analyzing the phase values and appropriately adjusts the phase values according to the phase continuity. By projecting two sets of patterns, the temporal phase unwrapping technique determines the absolute phase. The algorithm can be divided into three categories based on the patterns that are used: multi-frequency phase patterns, staircase phase encoding patterns, and gray code patterns. The Gray code approach [3] encodes the phase's period order information using a series of binary Gray code patterns. Theoretically, by projecting N Gray code patterns, a phase spanning 2N periods can be unwrapped. However, in order to remove the ambiguity of high-frequency phases, this method necessitates projecting a huge number of Gray code patterns. Southeast University's Zheng et al. proposed a three-gray-level Gray code approach that creates a "0.5" gray-level code utilizing binary defocusing techniques. This improves the efficiency of 3D measurement by effectively increasing the number of encoding gray levels in the Gray code [4]. However, all of these methods need to project multiple images when phase unwrapping and some level of misalignment will occur when phase unwrapping.

The temporal phase unwrapping algorithm based on multi-frequency fringe patterns utilizes wrapped phases of different frequencies to unwrap high-frequency phases. Currently, there are three representative approaches: the multi-frequency method [5], the multi-wavelength method [6], and the number-theoretical method [7]. Zuo et al.

from Nanjing University of Science and Technology established a rigorous mathematical model to analyze the unwrapping success rate and noise robustness of these three methods. Their results showed that the multi-frequency method offers the highest reliability in phase unwrapping, while the multi-wavelength method is the most susceptible to noise interference. Moreover, since the standard phase-shifting algorithm requires at least three patterns to recover the wrapped phase, the multi-frequency method typically requires at least six patterns to achieve phase unwrapping. To address this limitation, several researchers have proposed composite fringe methods to remove phase ambiguities without significantly increasing the number of projected patterns [8]. Assuming that the background intensity of the fringe images remains constant during a single 3D measurement process, Liu et al. from the University of Kentucky proposed a dual-frequency composite encoding method, reducing the required number of patterns from six to five. Further, under the additional assumption that the fringe modulation remains constant, Zuo et al. from Nanjing University of Science and Technology introduced a composite encoding method based on high-frequency fringes and a sawtooth signal, achieving absolute 3D measurement using only four patterns [9].

In recent years, the popularity and development of deep learning has had a significant impact on various fields. The powerful feature extraction capability of deep learning can learn feature representations from a large amount of data and perform phase recovery based on the learned features. Deep learning-based phase unwrapping can be broadly categorized into regression methods and semantic segmentation methods. Where the regression model directly allows the network to learn the mapping between the wrapped phase and the continuous phase. The segmentation method mainly allows the network to determine the level information of the wrapped phase map, after which the level information of to the continuous phase. Regression-based methods primarily consist of the Deep Learning-based Temporal Phase Unwrapping (DL-TPU) method [10] proposed by Wei et al. and the Deep Learning Phase Unwrapping (DLPU) method [11] proposed by Wang et al., which is based on U-Net [12] and Residual Network (ResNet) [13]. Phase-to-height mapping, deep neural networks, and data processing make up the overall structure. Two single-period wrapped phase maps and identical high-and low-frequency TPU data make up the network input. The neural network's backbone is a residual network, which is used to attain the best unwrapping reliability. By adding enough network depth, this topology improves performance and speeds up the convergence of deep networks. These methods, which conduct deep convolution on wrapped phase maps using encoder–decoder architectures, have demonstrated encouraging outcomes. Building upon this framework, Zhou et al. presented a Generative Adversarial Networks (PU-GAN)-based phase unwrapping technique [14], which incorporates a GAN architecture into phase unwrapping and shows that various network structures can successfully learn the mapping from wrapped phase to continuous phase.

PhaseNet [15], which was proposed by G.E., VDE-Net [16] with boundary weighting by Zhao et al., C-HRNet [17] for hierarchical order segmentation by Zhu et al., and the Hybrid Convolutional Neural Network Transformer (H-former) [18], which was proposed by Zhu et al. for fringe order prediction, are representative works for semantic segmentation-based approaches. The transformer improves the network's ability to learn nonlinear phase information by introducing an attention mechanism to fuse multi-scale features. Supervised learning is the foundation of all of the aforementioned phase unwrapping networks. Fan et al. addressed the problem of inadequate training data by proposing an unsupervised learning-based phase unwrapping network (UDL-Net) [19], which trains the model concurrently with image acquisition, in response to the scarcity of accessible datasets. Richer semantic understanding, such as the links between pixels and channels or the categories to which pixels belong, can be obtained in semantic segmentation tasks by utilizing contextual information. In order to increase the descriptive power of pixel representations and enhance segmentation accuracy, Yuan et al. presented the Object-Contextual Representation (OCR) technique [20], which learns the link between pixel features and object-region attributes.

In order to improve the speed and accuracy of phase unwrapping, this paper adopts a base and DeepLabv3+ phase unwrapping model [21], which is a semantic segmentation model, and adds Dice Loss and Focal Loss to DeepLabv3+ to improve the model's segmentation ability for complex scenes as well as to balance the segmentation categories. Compared with the regression model, the segmentation model is more conducive to model optimization, making it easier for the model to learn the complex details in the phase map. Following the fusion of the shallow and deep features via skip connections, a Transformer Block with a self-attention mechanism is added to the DeepLabv3+ decoder. This improves the model's capacity to simulate global context and fortifies its feature extraction capabilities.

2 The Network for Phase Unwrapping

This chapter is dedicated to the theoretical foundations of phase unwrapping based on deep learning. It provides a comprehensive analysis of the DeepLabv3+ architecture, detailing its implementation workflow, hyperparameter configurations, as well as the specific activation and loss functions tailored for the phase unwrapping task.

2.1 Network Design

DeepLabv3+ is a generalized model in the field of semantic segmentation, which is considered a new peak of semantic segmentation. It incorporates dilated convolution and encoder-decoder structure, which can retain detailed information while accurate semantic segmentation and the overall structure are shown in Fig. 1. The wrapped phase is input into the network, and after encoding and decoding, the final segmentation is completed, and output to get the level map. When the wrapped phase map is input into the network, it will first go through the deep convolutional layer (DCNN) in the encoding stage. In this paper, MobileNetV2 is used as the backbone feature extraction network, and this model is a type of lightweight deep neural network developed by Google for embedded devices such as cell phones. Its main characteristic is the usage of inverted residual blocks; the entire MobileNetV2 is made up of inverted residual blocks, as illustrated in the gray half of Fig. 1. Figure 2 shows the structure of an inverted residual block, which is divided into two parts, the left side is the main stem part, which is upgraded by using Conv 2d 1 × 1, then feature extraction by using

DepthwiseConv2d 3 × 3, and finally lower dimensional by using Conv2d 1 × 1. On the right is the residual edge section, where the input and output are directly connected. DepthwiseConv2d is used here for convolution, a lightweight convolution algorithm that allows for a significant reduction in computation and parameter count while maintaining good feature extraction.

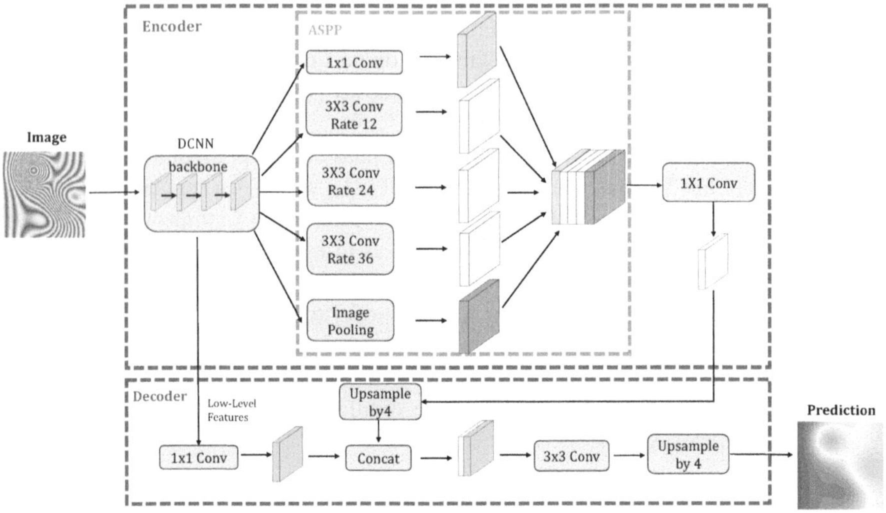

Fig. 1. Overview of the Network Architecture.

In DeepLabv3+, the 4 down sampling is used in this paper. The down sampling mentioned here refers to the length and width compression that will be performed. After the feature extraction done in MobileNetV2, two effective feature layers can be obtained, one is the result of the input image height and width compression twice, this feature is low-level and will be input to the decoding stage. The other is the result of input image height and width compression four times, this feature is deep and will be delivered to the ASPP feature fusion module. Multiple dilated convolutions are used in the ASPP feature fusion module, dilated convolution is an expanded convolution technique, and the main purpose is to increase the receptive field, compared with the standard convolution, Inserting spaced zeros between the elements of the convolution kernel helps widen the receptive field while not increasing the number of parameters and computation, and it can efficiently capture the large scale target or cross-region information. To acquire a broader variety of contextual information, dilated convolution increases the size of the sensory field using the dilated coefficient rate. The details of the ASPP module of the model used in this paper are shown in Table 1, where the down sampling multiplier in the DCNN module is set to 8. ASPP requires five feature enhancements of the deep features.

In the five operations, the first one is Conv2d 1 × 1, the second, third, and fourth are Dilated Conv2d 3 × 3 and the dilated coefficient rate is 12, 24, and 36, respectively, and

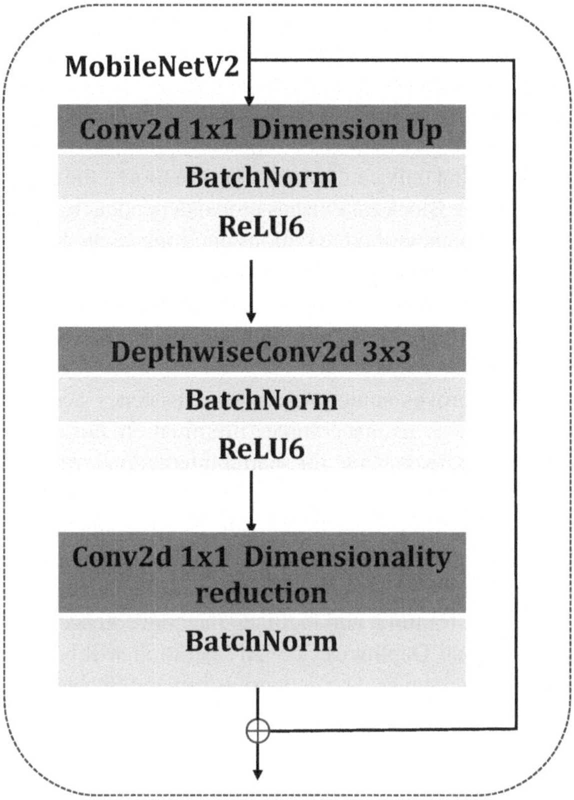

Fig. 2. Structure of the MobileNetV2 Inverted Residual Block.

finally the five enhanced features are subjected to Concat and Conv2d 1 × 1 compressed features.

Table 1. Configuration of the Five Residual Blocks in the ASPP Module.

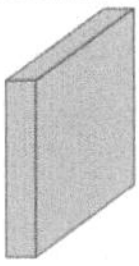	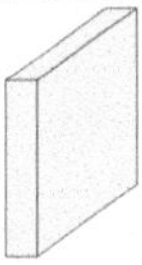	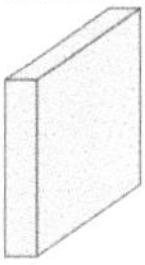	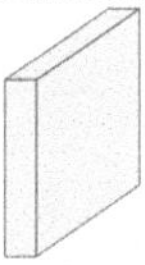	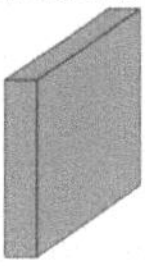
Conv2d K1x1, s1 BN ReLU 60x60x256	Conv2d K3x3, s1, r12, p12 BN ReLU 60x60x256	Conv2d K3x3, s1, r24, p24 BN ReLU 60x60x256	Conv2d K3x3, s1, r36, p36 BN ReLU 60x60x256	AdaptiveAvgPool2d Conv2d K1x1, s1 BN ReLU Bilinear Interpolate 60x60x256

In the Decoder module, the shallow effective features—compressed twice—are first processed by a 1 × 1 Conv2d layer to adjust the number of channels. These are then skip-connected with the up sampled features obtained after Atrous Conv2d operations. After completing the skip connection, the resulting feature map is fed into the Transformer Block. Since the receptive field of CNNs is inherently limited—especially for shallow features (low-level edges) that only capture local information—the self-attention mechanism within the Transformer Block establishes global dependencies during the decoding stage, enhancing the consistency of edge regions and long-range feature dependencies. As illustrated in Fig. 3, the Transformer Block mainly consists of a Multi-Head Attention module and an MLP. During multi-scale fusion (shallow + deep features), the QKV attention mechanism in Multi-Head Attention can adaptively and selectively fuse shallow and deep features, rather than relying solely on convolutional fusion. For classification tasks, global attention improves smoothness and consistency across different spatial scales and provides robustness against complex illumination, noise, and occlusion. The self-attention mechanism is responsible for spatial interactions, while the MLP handles channel-wise feature mixing, performing strong cross-channel information reorganization to compensate for the limitations of convolution in channel fusion. The GELU activation introduces nonlinearity, enhancing the model's ability to represent complex relationships. Meanwhile, Layer Normalization and Residual connections help preserve input information, stabilize training, and facilitate the convergence of both the attention and MLP modules. The final DepthwiseConv2d operation will be performed twice to obtain a final effective feature layer. The network is utilized to obtain the features of the input image, after which the features are utilized to obtain the prediction results. The prediction result can be divided into three steps: the first step utilizes a Conv2d 1 × 1 for channel adjustment and adjusts the Num Classes. The second step utilizes resize to make the width and height of the final output layer result in the same as the input picture. The third step utilizes Softmax to predict the classification of each pixel point and finally get the level information of the picture.

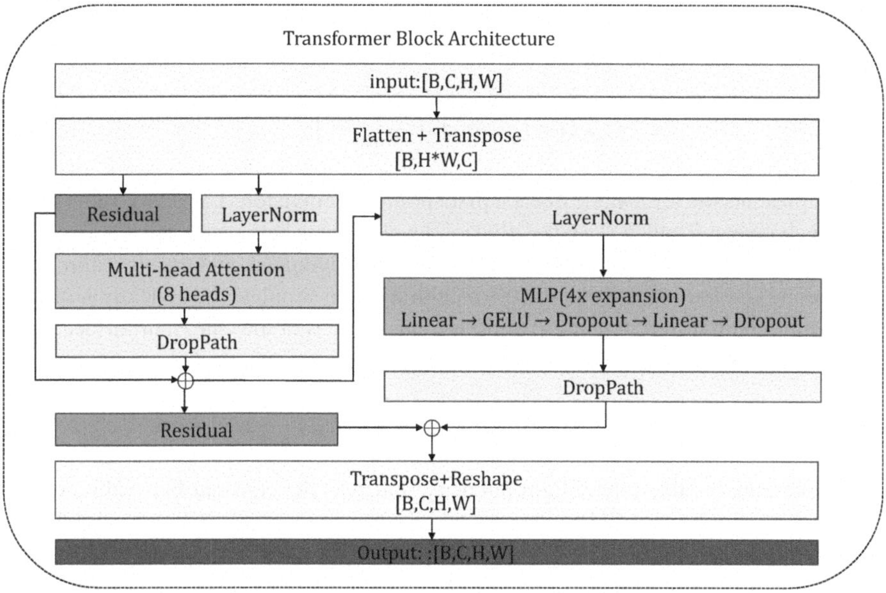

Fig. 3. Transformer Block Architecture

2.2 Simulation Dataset Production

Training deep learning models necessitates a substantial amount of data; however, acquiring a large-scale set of wrapped phase maps via physical optical setups (cameras and projectors) is often prohibitively time-consuming and labor-intensive. To address this challenge, this study employs a synthetic dataset for network training. The dataset comprises wrapped phase maps paired with their corresponding fringe order (level) maps as ground truth labels. Specifically, the data generation process begins with the creation of 200 random Gaussian surfaces, governed by the following equation:

$$\phi(x, y) = A \cdot \exp\left(-\frac{(x-\mu_x)^2 + (y-\mu_y)^2}{2\sigma^2}\right) \tag{1}$$

where $\phi(x, y)$ represents the phase of the Gaussian surface and A is the amplitude, which determines the maximum of the Gaussian distribution, taking values between 20 and 80 for each Gaussian surface. (μ_x, μ_y) is the mean, the center point of the Gaussian surface, taking values between 25 and 230. σ^2 is the variance taking values between 30–50, which controls the width of the Gaussian distribution, the larger the value, the flatter the surface. After that, 3 to 5 surfaces were randomly selected for addition and subtraction operations to get a new surface. To increase the overall tilting properties, a slant is added to the new Gaussian surface so that a Gaussian surface with three-dimensional information is obtained, as shown in Fig. 4(a). By subtracting the minimum value, all the values are flattened so that the final output has a minimum value of zero, this is to ensure that the range of the result is in the non-negative domain [22], this purpose is to obtain the continuous phase corresponding to the Gaussian surface as shown in

Fig. 4(b). After that, the continuous phase map has to be leveled to be used as a labeled picture for the training set, and the leveling formula is as follows:

$$k(x, y) = \text{round}\left(\frac{\phi(x, y)}{2\pi}\right) \tag{2}$$

$k(x, y)$ represents the level value of each pixel point after division. The phase value of the method is between 0 and 200, so the division level value is between 0 and 32, as shown in Fig. 4(c). In addition, the continuous phase has to be wrapped, and the algorithm used for wrapping is a four-step phase shift, which first has to simulate the generation of four fringe pattern with different phases using the phase values of the Gaussian surface, with the following formula:

$$I_o^{(i)}(x, y) = A + B \cdot \cos(\phi(x, y) - i \cdot \frac{\pi}{2}) \tag{3}$$

I_o represents the theoretical light intensity, and A is the constant bias value, which simulates the background light intensity taking the value of 50. B is the modulation regime, which represents the magnitude of change of the light intensity taking the value of 50. $\phi(x, y)$ represents the input phase, which comes from the Gaussian surface of the output and simulates the target phase. i the number of modulation steps, taken as i = 0,1,2,3 corresponding to different modulation phases. The wrapped phase ($\phi_{\text{wrapped}}(x, y)$) is computed by an inverse tangent function:

$$\phi_{\text{wrapped}}(x, y) = \arctan 2(V(x, y), U(x, y)) \tag{4}$$

Among them:

$$U(x, y) = \sum_{i=0}^{3} I_o^{(i)}(x, y) \cdot \cos\left(i \cdot \frac{\pi}{2}\right) \tag{5}$$

$$V(x, y) = \sum_{i=0}^{3} I_o^{(i)}(x, y) \cdot \sin\left(i \cdot \frac{\pi}{2}\right) \tag{6}$$

acrtan2 four-quadrant inverse tangent function for phase angle. $U(x, y)$ is the cumulative sum of the real part. $V(x, y)$ is the cumulative sum of the imaginary part. $I_o^{(i)}$ is the theoretical light intensity, which is obtained based on the formula of four-step phase shift, and the wrapped phase map generated by the above formula is shown in Fig. 4(d). The training set and test set are generated separately by random seeds without duplicate images. This simulation generates a training set containing 5000 256x256 wrapped phase maps (including 1000 noise maps) and the corresponding level maps. To assess whether the simulated data can statistically represent the wrapped phases obtained from a real structured-light system, we conducted a visual comparison between the simulated Gaussian-surface wrapped phase and the experimentally acquired wrapped phase. Figure 4(e) and (f) show a representative experimental wrapped phase and its simulated counterpart, respectively. The results indicate that the simulated data resemble the real measurements in terms of overall phase magnitude and low-frequency structure, but discrepancies remain in high-frequency noise and device-induced structured artifacts. Based on these observations, incorporating a subset of experimentally acquired samples into the training process, or performing domain-specific fine-tuning, may further enhance the model's robustness in real-world measurement scenarios.

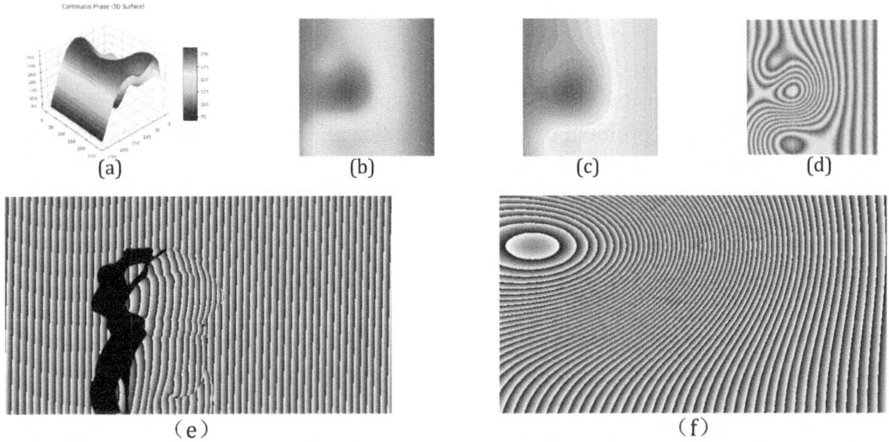

Fig. 4. Representative sample from the generated synthetic dataset. (a) Ground truth 3D height map. (b) Ground truth continuous phase. (c) Ground truth k-level map. (d)-(f) The wrapped phase patterns.

2.3 Activation Function and Loss Function

Since convolutional operations are inherently linear, stacking multiple convolutional layers without non-linearity collapses into a single linear transformation. Therefore, introducing non-linear activation functions is essential to endow the network with the capability to learn and represent complex functional mappings. In this work, we employ the ReLU (Rectified Linear Unit) activation function. Defined as a maximization function, ReLU does not saturate in the positive domain, which effectively mitigates the vanishing gradient problem while ensuring high computational efficiency and fast convergence. Furthermore, we utilize ReLU6, a variant that caps the maximum activation at 6. This upper bound prevents excessive activation magnitudes, thereby inhibiting gradient explosion and enhancing the numerical stability and robustness of the model during training. The suitable loss function is also critical in phase unwrapping, as it has a direct impact on the quality of the output. The loss function calculates the difference between the network's output and the true value; the lower the value of the loss function, the more robust the network. The loss function used for the network mentioned in this paper is a combination of Dice Loss and Focal Loss. Dice Loss focuses on the overlap of the segmentation region, which can provide more attention to the small target classes (classes with fewer pixels), and Dice Loss is directly correlated to the segmentation performance metric, the Dice coefficient. The formula is as follows:

$$\text{DiceLoss} = 1 - \frac{2\sum_{i=1}^{N} p_i g_i}{\sum_{i=1}^{N} p_i^2 + \sum_{i=1}^{N} g_i^2} \tag{7}$$

p_i the probability predicted by the model (or the value of the segmentation mask), g_i is the true label value (usually 0 or 1), and N is the total number of pixels. Focal Loss is designed to solve the problem of category imbalance, especially in the classification task where the imbalance of the category share has a significant effect, Focal Loss pays more attention to those samples that are difficult to classify and reduces the penalty for the easy-to-categorize samples. Its formula is:

$$\text{Focal Loss} = -\alpha(1 - p_t)^{\gamma} \log(p_t) \tag{8}$$

where $p_t = 0(y = 1)$ or $p_t = 1 - p\,(y = 0)$, y is the true label, p is the predicted probability, α is the positive and negative sample balancing factor (usually set to 0.25), and γ is a moderator to control the degree of attention to difficult to classify samples (usually set to 2). The training generally uses dynamically adjusted weights to automatically balance the two kinds of losses according to the training stage, relying more on Focal Loss to optimize the category imbalance in the initial stage, and relying more on Dice Loss to optimize the global shape in the later stage. The total loss function can be defined as (where t is the current number of training rounds and T is the total number of rounds):

$$\text{Loss} = \left(1 - \frac{t}{T}\right) \cdot \text{FocalLoss} + \frac{t}{T} \cdot \text{DiceLoss} \tag{9}$$

2.4 Network Training Process and Details

Training DeepLabv3+ consists of two stages: establishing the mapping relationship between the trained wrapped phase and continuous phase according to Eq. (2), simulating the generation of wrapped phase and K-level map in the preparation stage, and wrapped samples gradually converge to the corresponding level samples during the network training process, to pursue the real effect, and to obtain the minimum loss function and the optimal parameter, and the training process is shown in Fig. 5. In which the continuous phase map can be recovered by adding the K-level map derived from the network training and the original wrapped phase map, the formula is as follows:

$$\Phi(x, y) = \phi_{\text{wrapped}}(x, y) + 2\pi \cdot k(x, y) \tag{10}$$

This experiment was run in PyCharm software under 64-bit Windows, based on an Intel(R) Core(TM) i7-14700KF processor with an Nvidia GeForce RTX 4080 SUPER (16 GB) graphics card and 32 GB of RAM, and used the Anaconda environment with the PyTorch deep learning framework in Anaconda environment for network training. The initial learning rate of the model is set to 5e-4, the size of the convolutional kernel is 3 × 3, and the batch size is set to 56. The training set used this time contains 5000 wrapped phase maps of 256 × 256 with 250 iterations with the appropriate weighting parameters already in place, and the Adam optimizer allows for adaptive tuning of the learning rate while being able to converge faster to a global minimum or local minimum.

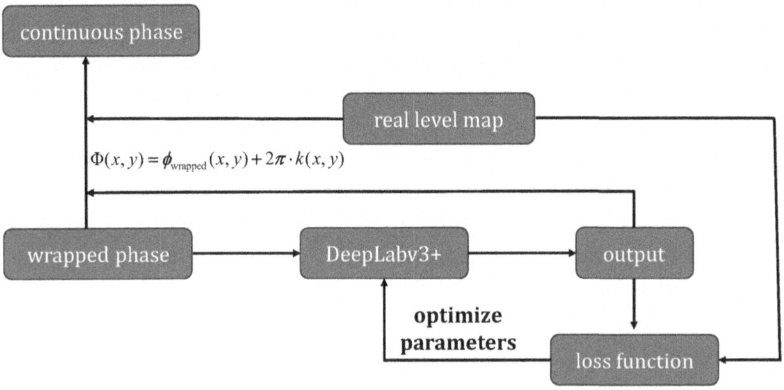

Fig. 5. Schematic diagram of the network workflow.

3 Results and Discussions

3.1 Evaluation Criteria

To objectively evaluate the simulation results, the adopted metrics include the mIoU and validation loss curves recorded during the training process. Additionally, the Structural Similarity Index (SSIM) is employed to assess the fidelity of the predicted phase relative to the actual phase. The Mean Intersection over Union (mIoU) is a standard metric in image semantic segmentation, used to quantify the overlap between the model's predictions and the ground truth. For a specific category c, the IoU is defined as follows:

$$\mathrm{IoU}_c = \frac{\mathrm{TP}_c}{\mathrm{TP}_c + \mathrm{FP}_c + \mathrm{FN}_c} \tag{11}$$

TP_c denotes True Positive for category c (predicted to be c and true value is also c), FP_c denotes False Positive for category c (predicted to be c but true value is not c), and FN_c denotes False Negative for category c (true value is c but prediction is not c). For N categories, mIoU is given by:

$$\mathrm{mIoU} = \frac{1}{N}\sum_{c=1}^{N}\mathrm{IoU}_c \tag{12}$$

The higher the mIoU value, the greater the overlap between the model prediction and the real segmentation result, and the better the model performance, and the IoU of each category are considered. A total of 250 iterations were performed for this training, and the mIoU evaluation was performed every 5 times, a total of 50 mIoU evaluations were performed, and the smoothed mIoU curve is shown in Fig. 6(a). Due to the use of better pre-training weights during training, it can be seen from the evaluation results that the model has a good initial mIoU value, and as the number of iterations proceeds, the mIoU value is increasing, indicating that the prediction accuracy of the network is also improving. The mIoU $\geq 70\%$ in the solution level phase can be considered as a better segmentation result, which is suitable for most engineering applications. While the validation set loss is derived from the loss function, the validation loss is recorded once in each iteration during the training process, which is used to evaluate the error metrics of the validation set, and the smoothed validation set loss profile is shown in Fig. 6(b). Similar to the mIoU curve, the validation loss has a good initial value and decreases with the number of iterations, which proves that the model is learning and its performance is gradually improving.

SSIM is used to measure the structural similarity between two images in the range of $[-1,1]$, where 1 means that the two images are identical, 0 means that the two images are completely unrelated, and -1 means that the two images are opposite. It uses structural information and texture features to compare images, not just at the pixel level, and better reflects the human eye's perception of an image. The formula for SSIM consists of three parts, which are as follows:

$$\text{Brightness Comparison}: \quad l(x, y) = \frac{2\mu_x\mu_y + C_1}{\mu_x^2 + \mu_y^2 + C_1} \tag{13}$$

$$\text{Contrast comparison}: \quad c(x, y) = \frac{2\sigma_x\sigma_y + C_2}{\sigma_x^2 + \sigma_y^2 + C_2} \tag{14}$$

$$\text{Structural Comparison}: \quad s(x, y) = \frac{\sigma_{xy} + C_2/2}{\sigma_x\sigma_y + C_2/2} \tag{15}$$

$$\text{The complete formula for SSIM is}: \quad \text{SSIM}(x, y) = l(x, y) \cdot c(x, y) \cdot s(x, y) \tag{16}$$

where x, y are the two images to be compared, usually in grayscale, μ_x, μ_y denote the mean (luminance) of the images, and σ_x^2, σ_y^2 denote the variance (contrast) of x and y, respectively, and σ_{xy} is the covariance (a measure of the correlation between the two images) of x and y. The two constants, C_1, C_2 are used to stabilize the formula and avoid a zero denominator.

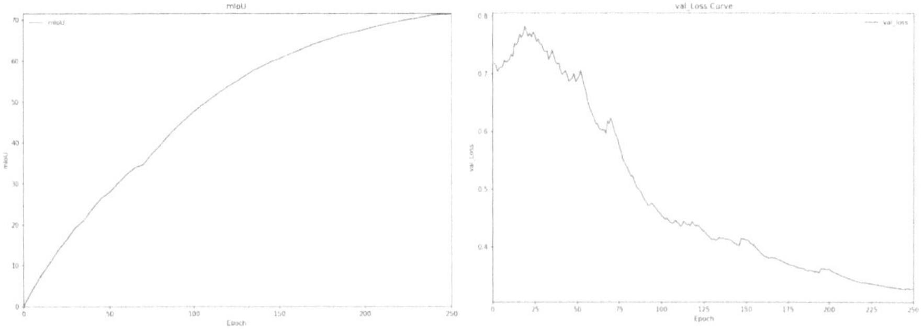

(a) mIoU curve (b) validation loss curve

Fig. 6. Evolution of the mIoU and validation loss during the training process.

3.2 Results Display

After completing the training, the newly generated wrapped phase maps are predicted using the trained outweighs, and the following results are all real data derived from this experiment. As shown in Fig. 7(a) – (e) represent five different wrapped phase maps, respectively, which need to be input into the network for prediction. 7(f) – (j) are the real level maps corresponding to these five wrapped phase maps, and 7(k) – (o) are the predicted level maps of these five wrapped phase maps in this network, respectively. The SSIM values between the real level maps and the predicted level maps can be calculated by Eq. (16), which are 0.9704, 0.9572, 0.9746, 0.9832, and 0.9642, respectively, and it can be seen that the worst prediction results all have an SSIM of 0.9572.

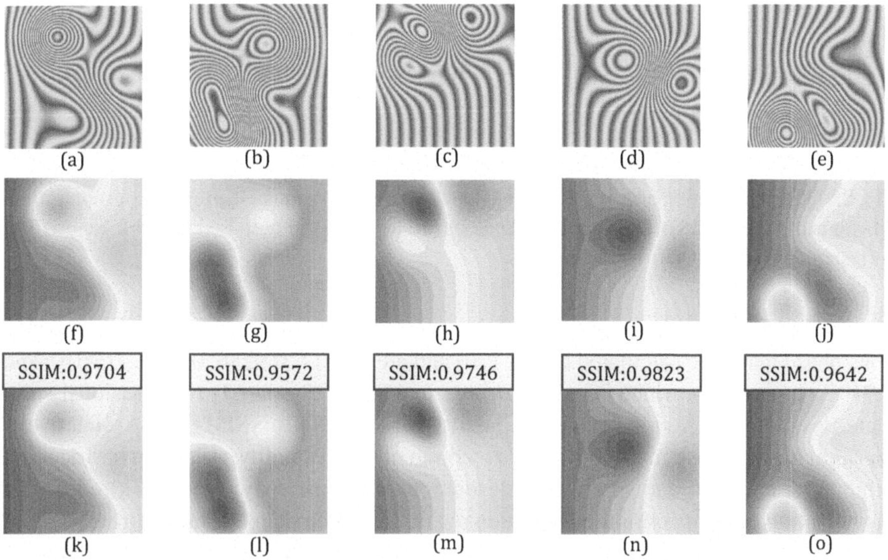

Fig. 7. Visual comparison between the ground truth and the predicted k-level maps. (a)-(e) Input wrapped phase maps. (f)-(j) Ground k-level maps. (k)-(o) Predicted k-level maps.

After that, the predicted level maps are to be recovered as continuous phase maps. As shown in Fig. 8, where 8(a) – 8(e) correspond to the true continuous phases of the five wrapped phase maps, respectively. 8(f) – 8(j) are obtained by combining the predicted level maps and the original wrapped phase maps using Eq. (10), which can be regarded as the predicted continuous phase maps.

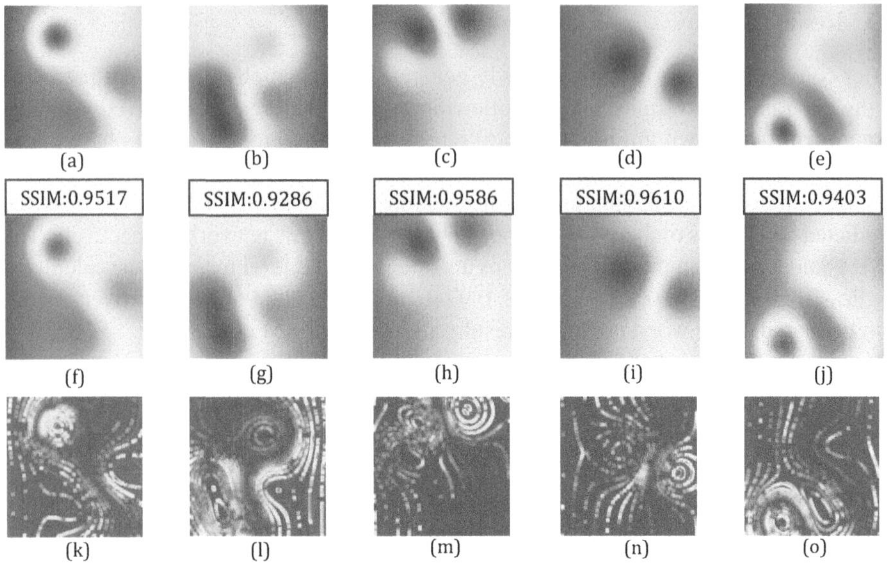

Fig. 8. Qualitative comparison of the phase unwrapping results. (a)-(e) The ground truth unwrapped phases. (f)-(j) The unwrapped phases predicted by the proposed network. (k)-(o) The corresponding absolute error distributions.

The SSIM values between the predicted continuous phase maps and the real continuous phase maps are calculated by Eq. (16) and are 0.9517, 0.9286, 0.9586, 0.9610, and 0.9403, respectively, which is a slight decrease in SSIM compared to the previously predicted level maps. This is because compared to MSE or PSNR, SSIM is more reflective of the human visual system's subjective perception of image quality, as it takes into account brightness, contrast, and structural information simultaneously. Where 8(k)−8(o) is a visualization of the error of the prediction results, it can be seen that more errors are generated in the complex region of the continuous phase, and this problem can be solved by introducing an attention mechanism to allow the network to learn the complex region of the phase more intensively. Through the Transformer Block in this network, the model's attention to phase values in edge regions is enhanced, making the predicted phase values along edges smoother and reducing errors in complex areas, ultimately achieving a better fitting result.

The results obtained from this network are compared with those obtained from HRNet trained in the same environment. As shown in Fig. 9(a) is the real continuous phase, Fig. 9(b) is the result predicted by DeepLabv3+ with an SSIM value of 0.9423, and Fig. 9(c) is the result predicted by HRNet with an SSIM value of 0.9199, which shows that the prediction effect of the present network is significantly stronger than that of HRNet. To see the difference more clearly, 3D visualization is carried out respectively, as shown in Fig. 9(d) for the original 3D height map, Fig. 9(e) for the 3D height map obtained by DeepLabv3+, and Fig. 9(f) for the 3D height map obtained by HRNet, it can be seen that the noise around HRNet is much more than that of DeepLabv3+, which indicates that the present network has a better fitting effect compared with HRNet. To further suppress noise in edge regions, a Gaussian filter can be applied to the predicted continuous phase values, which significantly reduces surrounding noise points after processing.

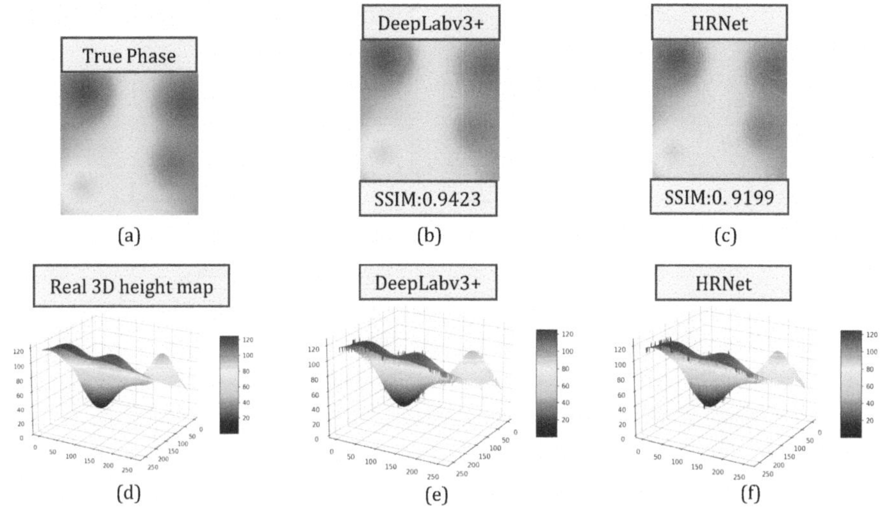

Fig. 9. Visual comparison of prediction results between DeepLabv3+ and HRNet. (a) Ground truth phase. (b)-(c) Predicted phases by DeepLabv3+ and HRNet, respectively. (d)-(f) The corresponding 3D height maps.

The noise robustness test requires adding uniform Gaussian noise to the wrapped phase, as shown in Fig. 10. For the wrapped phase map with noise, the prediction was performed using DeepLabv3+ and HRNet, respectively. Figure 10(b) shows the result derived by DeepLabv3+ with an SSIM value of 0.9327, which only decreased by 0.01 after the addition of noise, and Fig. 10(c) shows the result derived by HRNet with a decrease in the SSIM value of 0.8707. It can be seen that the noise robustness performance of DeepLabv3+ is much stronger than that of HRNet. Then the two predictions are visualized in 3D, as shown in Fig. 10(d) and (e), and it can be clearly seen that DeepLabv3+ has a better fitting effect at the edges.

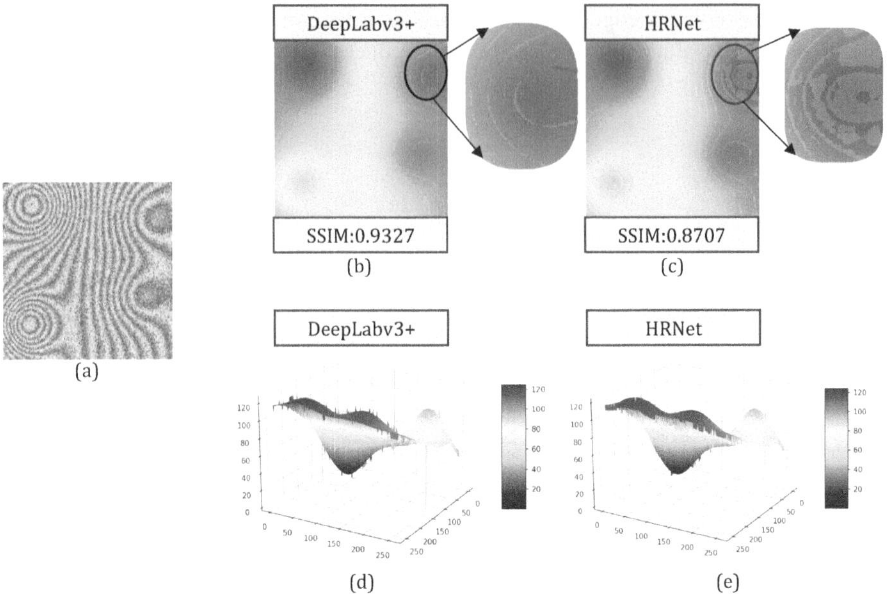

Fig. 10. Visual comparison of noise robustness between DeepLabv3+ and HRNet. (a) Input wrapped phase contaminated by noise. (b)-(c) Predicted phases by DeepLabv3+ and HRNet, respectively. (d)-(e) The corresponding 3D height maps.

To validate the generalization capability of the model, we directly applied it to infer experimentally captured data. However, the acquired images required linear scaling to adjust their resolution to 256 × 256 pixels, as shown in Fig. 11.

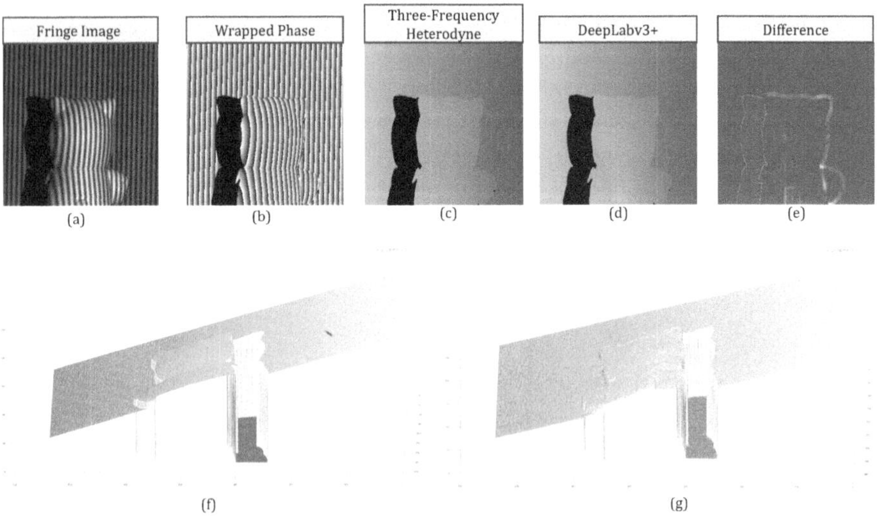

Fig. 11. Comparative analysis of DeepLabv3+ and three-frequency heterodyne phase unwrapping results. (f)-(g) show 3D representations of the phase data.

The following presents a comparison between the phase retrieval results of DeepLabv3+ and those obtained using the three-frequency twelve-step phase-shifting method. Figure 11(a) displays the captured fringe pattern. The wrapped phase and continuous phase derived through the three-frequency twelve-step method are shown in Fig. 11(b) and (c), respectively. The wrapped phase was input into the neural network, yielding the result presented in Fig. 11(d). A comparative analysis reveals a Structural Similarity Index Measure (SSIM) value of 0.9556 between the two continuous phase maps. Given that the traditional three-frequency twelve-step method requires acquiring a substantial number of fringe images, its results are highly accurate. This high agreement indirectly demonstrates the excellent generalization capability of the proposed network. Figure 11(e) illustrates the specific error distribution between the two phase maps, clearly indicating that errors remain predominantly concentrated in edge regions. Figure 11(f) and (g) provide 3D visualizations of the phase information. Although the DeepLabv3+ prediction exhibits slightly reduced smoothness in regions containing objects compared with the three-frequency twelve-step result, the overall 3D visualization quality remains satisfactory.

3.3 Model Evaluation and Ablation Studies

To ensure a more comprehensive comparison, this work conducts systematic experiments across multiple network architectures, including the traditional HRNet and the standard U-Net, under identical conditions—namely the same dataset, number of training epochs, optimizer, and learning rate settings. As shown in Table 2, conventional CNN-based architectures exhibit limited performance in terms of SSIM, and although U-Net demonstrates moderate capability in preserving structural information, its performance remains inferior to that of the proposed method. The Transformer-enhanced DeepLabv3+ developed in this study, combined with the Dice + Focal loss formulation, achieves the highest SSIM while maintaining an acceptable model size and inference speed. These results validate both the effectiveness and the necessity of the proposed architectural design.

Table 2. Quantitative Comparison of Three Models.

Model	SSIM (mean ± std)	Params (M)	Inference (ms)
HRNet-W18 (baseline)	0.900 ± 0.05	9.7	**1.8**
U-Net	0.940 ± 0.05	31.0	4.1
DeepLabv3+ (MobileNetV2) + Transformer+Dice+Focal	**0.954 ± 0.03**	**7.2**	3.4

To quantify the individual contributions of the Transformer module and the Dice + Focal hybrid loss, a systematic ablation study was conducted. The comparison groups include: the standard DeepLabv3+ (baseline), a variant without the Transformer module, and the Transformer-equipped model trained with different loss functions (MSE, Dice, Focal, and Dice + Focal). All experiments were trained on the same dataset and with identical hyperparameters, and each configuration was repeated three times to report mean ± standard deviation. The summarized results are presented in Table 3. The findings show that the Transformer module substantially improves edge consistency (with an approximate 8.5% gain in Edge-SSIM) and significantly reduces errors in complex regions. The Dice + Focal loss outperforms individual losses in handling phase discontinuities and minority pixel regions. Statistical significance was validated through paired t-tests ($p < 0.05$). These results demonstrate that both architectural modifications and the selected loss formulation contribute meaningfully to the overall performance gains.

Table 3. Summary of Ablation Study Results.

Variant (loss)	SSIM	Edge-SSIM	MAE	Params (M)	Inference (ms)
DeepLabv3+ (MobileNetV2) (MSE)	0.890 ± 0.05	0.803 ± 0.09	0.036 ± 0.02	**5.9**	2.6
DeepLabv3+ + Transformer (MSE)	0.938 ± 0.04	0.872 ± 0.07	0.028 ± 0.01	7.2	3.4
DeepLabv3+ + Transformer + Dice	0.945 ± 0.04	0.885 ± 0.06	0.025 ± 0.01	7.2	3.4
DeepLabv3+ + Transformer +Focal	0.942 ± 0.04	0.881 ± 0.07	0.027 ± 0.01	7.2	3.4
DeepLabv3+ + Transformer +Dice + Focal	**0.954 ± 0.03**	**0.902 ± 0.05**	**0.022 ± 0.01**	7.2	3.4

3.4 Discussions

The deep learning–based phase unwrapping method proposed in this study has promising potential for applications in robotic vision. Phase-based depth imaging is widely used in robotic grasping, surface inspection, localization and navigation, and human–robot interaction. High-quality continuous phase maps can provide more detailed geometric information, thereby enhancing a robot's perception of its environment. Compared with HRNet, the proposed DeepLabv3+ -based architecture demonstrates superior structural preservation and noise robustness, which benefits robotic measurement and perception in challenging scenarios involving reflections, shadows, or dynamic scenes. Future work will explore deploying the model on embedded platforms to further evaluate its feasibility for real-time robotic applications.

4 Conclusions

In this paper, a deep learning DeepLabv3+ phase unwrapping approach is used. Overall, this method considers the phase unwrapping problem as a semantic segmentation task and trains the network with simulated data, and the results show that the network model is suitable as a phase unwrapping method. In addition, the proposed model can predict good phase unwrapping results with fewer images than the traditional method first. Moreover, the backbone network and optimization methods used in this model are relatively lightweight, which can effectively reduce the computation and storage requirements and can be used in mobile devices or embedded systems. Although the

DeepLabv3+ -based phase unwrapping network demonstrates clear advantages in accuracy, its computational complexity warrants further discussion due to the incorporation of the Transformer Block. The model contains approximately 7.2 million parameters, and under the current hardware configuration (Nvidia GeForce RTX 4080 SUPER, 16 GB), the forward inference for a single 256×256 wrapped-phase frame requires 3.4ms. While this is sufficient for offline processing, it still falls short of achieving strict real-time performance. In addition, the model's generalization ability depends on the statistical characteristics of the training data. Performance degradation may occur in extremely low-SNR regions, severe shadowed areas, or in cases involving phase discontinuities not represented in the training set. Future efforts may focus on lightweight network design, model pruning and quantization, and the incorporation of physics-based priors to further improve the model's real-time capability and robustness.

Author Contributions. Conceptualization, G.H. and X.G.; methodology, G.H.; software, Y.L.; validation, Y.L. and G.H.; formal analysis, W.L.; resources, W.L.; data curation, G.H.; writing—original draft preparation, G.H.; writing—review and editing, W.L.; supervision, X.G.; project administration, X.G.; funding acquisition, X.G. All authors have read and agreed to the published version of the manuscript.

Funding. This work was supported by the Guangxi Science and Technology Plan Project (2025XTXT04041), the Nanning Major Science and Technology Projects (20231034) and the Youth Science Fund Project of Guangxi Natural Science Foundation (2024JJB160101).

Conflicts of Interest. The authors declare no conflicts of interest. Declare that the research was conducted in the absence of any commercial or financial relationships that could be construed as a potential conflict of interest. The funders had no role in the design of the study, the collection, analyses, or interpretation of data; the writing of the manuscript; or the decision to publish the results.

References

1. Almaraz-Cabral, C.C., Gonzalez-Barbosa, J.J., Villa, J., et al.: Fringe projection profilometry for panoramic 3D reconstruction. Opt. Lasers Eng. **78**, 106–112 (2016)
2. Xiang, S., Yang, Y., Deng, H., et al.: Multi-anchor spatial phase unwrapping for fringe projection profilometry. Opt. Express **27**(23), 33488–33503 (2019)
3. Han, S., Yang, Y., Zhang, X., et al.: Complementary gray code fourfold-n step phase shift grating fringe projection profilometry. IEEE Sens. J. **23**(12), 13272–13279 (2023)
4. Zheng, D., Kemao, Q., Da, F., et al.: Ternary Gray code-based phase unwrapping for 3D measurement using binary patterns with projector defocusing. Appl. Opt. **56**(13), 3660–3665 (2017)
5. Zhang, M., Chen, Q., Tao, T., et al.: Robust and efficient multi-frequency temporal phase unwrapping: optimal fringe frequency and pattern sequence selection. Opt. Express **25**(17), 20381–20400 (2017)
6. Zhang, S.: Digital multiple wavelength phase shifting algorithm. Optical Inspection and Metrology for Non-Optics Industries. **7432**, 74320N (2009)

7. Zuo, C., Chen, Q., Gu, G., et al.: High-speed three-dimensional shape measurement for dynamic scenes using bi-frequency tripolar pulse-width-modulation fringe projection. Opt. Lasers Eng. **51**(8), 953–960 (2013)
8. Liu, K., Wang, Y., Lau, D.L., et al.: Dual-frequency pattern scheme for high-speed 3-D shape measurement. Opt. Express **18**(5), 5229–5244 (2010)
9. Zuo, C., Chen, Q., Gu, G., et al.: High-speed three-dimensional profilometry for multiple objects with complex shapes. Opt. Express **20**(17), 19493–19510 (2012)
10. Yin, W., Chen, Q., Feng, S., et al.: Temporal phase unwrapping using deep learning. Sci. Rep. **9**(1), 20175 (2019)
11. Wang, K., Li, Y., Kemao, Q., et al.: One-step robust deep learning phase unwrapping. Opt. Express **27**(10), 15100–15115 (2019)
12. Vijay Kumar, S., Sun, X., Wang, Z., et al.: A U-net approach for insar phase unwrapping and denoising. Remote Sens. **15**(21), 5081 (2023)
13. He, K., Zhang, X., Ren, S., et al.: Deep residual learning for image recognition. Proceedings of the IEEE Conference on Computer Vision and Pattern Recognition. pp. 770–778 (2016)
14. Zhou, L., Yu, H., Pascazio, V., et al.: PU-GAN: a one-step 2-D InSAR phase unwrapping based on conditional generative adversarial network. IEEE Trans. Geosci. Remote Sens. **60**, 1–10 (2022)
15. Zhu, W., Beroza, G.C.: PhaseNet: a deep-neural-network-based seismic arrival-time picking method. Geophys. J. Int. **216**(1), 261–273 (2019)
16. Zhao, J., Liu, L., Wang, T., et al.: VDE-Net: a two-stage deep learning method for phase unwrapping. Opt. Express **30**(22), 39794–39815 (2022)
17. Zhao, W., Yan, J., Jin, D., et al.: C-HRNet: high resolution network based on contexts for single-frame phase unwrapping. IEEE Photonics J. 2024
18. Zhu, X., Han, Z., Yuan, M., et al.: Hformer: hybrid convolutional neural network transformer network for fringe order prediction in phase unwrapping of fringe projection. Opt. Eng. **61**(9), 093107 (2022)
19. Fan, S., Liu, S., Zhang, X., et al.: Unsupervised deep learning for 3D reconstruction with dual-frequency fringe projection profilometry. Opt. Express **29**(20), 32547–32567 (2021)
20. Yuan, Y., Chen, X., Wang, J.: Object-contextual representations for semantic segmentation. Computer Vision–ECCV 2020: 16th European Conference, Glasgow, UK, 23–28 Aug 2020, Proceedings, Part VI 16. Springer International Publishing, pp. 173–190 (2020)
21. Zhang, T., Jiang, S., Zhao, Z., et al.: Rapid and robust two-dimensional phase unwrapping via deep learning. Opt. Express **27**(16), 23173–23185 (2019)
22. Zhang, S.: Absolute phase retrieval methods for digital fringe projection profilometry: a review. Opt. Lasers Eng. **107**, 28–37 (2018)

Research on Robot Vision Applications Based on Pulse-Coupled Neural Networks

Chunlin Li[1(✉)], Xiaolin Li[2], Lihong Ren[1], Lizhuo Zu[1], and Xiaoyan Wang[1]

[1] Xuanhua Vocational College of Science and Technology, Zhangjiakou City 075100, Hebei Province, China
912092359@qq.com

[2] Zhangjiakou University, Zhangjiakou City 075100, Hebei Province, China

Abstract. Spanning multiple field conditions, a study has been carried out in which Pulse-Coupled Neural Networks are deployed inside robot-vision architectures, an exponential-transform preprocessing stage being integrated ahead of a PCNN decomposition scheme to yield an image-segmentation pipeline that continues to operate under non-uniform and adverse illumination, night-time low light and high-dynamic-range scenes serving as typical examples. By means of this combination, images captured under complex lighting are processed with improved accuracy and in shortened time, clearer and more reliable visual information being supplied to robots across a range of vision-driven tasks, leading to enhanced navigation and target recognition within challenging environments.

Keyword: Robot vision · Image segmentation · Pulse-Coupled Neural Network

1 Introduction

Robotic vision systems, acting as the core carrier for machines to perceive the external environment, have shown a clear evolutionary trajectory from "shape perception" to "semantic understanding," deeply coupled with the progress of computer technology, image processing algorithms, and sensor technology [1]. From the 1970s to the 1980s, early robotic vision systems, constrained by hardware computing capabilities and image acquisition technologies, were only able to achieve simple shape recognition based on object silhouettes. A typical example was the binary processing system based on grayscale thresholding —— it divided images into target and background areas through a fixed threshold, completing the preliminary positioning of regular geometric shapes, such as square workpieces and round bolts in industrial production. These systems relied on strict lighting control and a single background environment, exhibiting extremely poor robustness in complex scenes, and were only applied to highly structured task scenarios on automated production lines [2].

Following the commoditisation of CCD sensors and the increased clock speeds afforded by Digital Signal Processors, the focus shifted toward mid-level feature extraction. Edge operators—Canny and Sobel being the most frequently cited—and texture descriptors such as Local Binary Patterns and Gabor filters were introduced, enabling the

H. Wei and J. Niu (Eds.): COBOT 2025, CCIS 2893, pp. 357–372, 2027.
https://doi.org/10.1007/978-981-92-2074-8_21

delineation of irregular contours and surface distributions. In automotive manufacture, for example, engine-block outlines were localised with an order-of-magnitude improvement over the silhouette era, the reliance on perfect backlighting simultaneously reduced [3].

After entering the second decade of the 21st century, the integration of deep learning technologies and 3D sensing technologies propelled robotic vision systems into a new stage of "3D semantic understanding." On one hand, deep learning models such as Convolutional Neural Networks (CNNs) and Transformers could automatically learn high-level semantic features from vast amounts of image data, allowing the system's object recognition accuracy to surpass human visual levels. On the other hand, the maturation of depth sensors enabled the system to acquire 3D point cloud data of the environment. Combined with visual Simultaneous Localization and Mapping (SLAM) algorithms, it could model and semantically annotate 3D space [4]. Current mainstream robotic vision systems can already complete end-to-end tasks of "real-time perception-3D modeling-semantic understanding-decision planning." For example, in warehousing and logistics scenarios, Automated Guided Vehicles (AGVs) equipped with 3D vision systems can simultaneously recognize the locations of shelves, the categories of goods, and the postures of pallets, completing the entire process of autonomous picking, transportation, and palletizing [5]. Image-recognition pipelines—tasked with labelling every object instance and assigning it a semantic class—have shifted from hand-crafted correspondence rules to representations learned end-to-end by deep networks. Earlier workflows depended on engineer-specified descriptors—Histogram-of-Oriented-Gradients, Scale-Invariant Feature Transform—and on classifiers such as Support-Vector-Machines or AdaBoost [6]; large, manually annotated corpora were indispensable for tuning their parameters, and generalisation deteriorated rapidly once target geometries or background textures deviated from the training template. In the present paradigm, convolutional and attention-based architectures—AlexNet, ResNet, Vision Transformers among them—derive hierarchical features automatically, mapping raw pixels directly to semantic embeddings. Basic distinctions ("cat" versus "dog", "car" versus "pedestrian") are attained without descriptor redesign, while finer taxonomic separation—say, discriminating among smartphone generations or among cultivars of wheat—emerges from the same learning objective, now supplied with appropriately stratified samples [7].

Application domains for image-recognition systems now extend across industrial inspection, public-space surveillance and cashier-less retail [8]. On electronics production lines, chip markings and lead geometries are captured and matched against tolerance templates, separating pass devices from rejects without human intervention. Surveillance feeds, processed in real time, label crowd behaviour as normal or anomalous—climbing, fighting or unattended objects are flagged and alarms raised automatically [9]. In unmanned supermarkets, overhead cameras frame the items shoppers pick up; pose-estimation modules locate hand–object contact points, enabling instant category recognition and contact-free payment, delivering the advertised "grab-and-go" experience.

Although vision-guided robots perform reliably within engineered cells, their performance is now limited chiefly by unconstrained lighting—low levels, spatial non-uniformity and extreme dynamic range [10]. Such conditions are encountered daily across industrial, transport and service environments. Welding arcs and high-power LED spotlights introduce local saturation on metallic surfaces, while casting shadows from moving kinematic chains; together, these effects yield illumination ratios exceeding 100: 1 across a single work-piece. Night-time driving, back-lit dawn sequences and the abrupt luminance step at tunnel portals raise sensor noise and erase detail in captured frames [11]. Domestic service platforms suffer similarly: traversal from an interior living-room to an exterior balcony can produce a 60 dB swing in grey-level within 300 ms, degrading both recognition confidence and localisation precision.

Regarded as the front-end bottleneck of every robotic vision chain, segmentation governs the reliability of later semantic parsing, pose recovery and motion planning [12]. Its objective is a partition of the image plane into sets—objects, background, defects—within which grey-level, colour or texture variation is minimal, while between-set disparity is maximised. Once illumination becomes uncontrolled, the inadequacies of conventional segmentation routines are magnified [13].

Global-threshold techniques—Otsu's maximisation of inter-class variance and the maximum-entropy criterion—presuppose a bimodal grey-level histogram on which a single optimal cut can be found [14]. Under non-uniform illumination, intra-target grey values scatter owing to local lighting gradients, collapsing the histogram into one dominant peak or into several closely spaced modes. A constant threshold is then unable to separate object from surround; bright patches generated by specular reflection are assigned to background, while shadows are promoted to foreground, producing under-or over-segmented regions [15].

Recovered traditionally through convolution with gradient kernels—Canny, Prewitt or their kin—edges are identified by ridge lines that are then promoted to object boundaries. Under photon-starvation, however, noise spikes whose amplitude rivals that of the authentic gradient are themselves labelled as edges, while strong illumination gradients serve to suppress genuine discontinuities [16]. In back-lit pedestrian footage, contours are repeatedly observed to dissolve into the bright halo that blankets the figure, connectivity being forfeited in the process.

Region-growing paradigms—whether seeded expansion or watershed flooding—predicate themselves on the assumption that neighbouring pixels stemming from a single physical surface will exhibit coherent colour or grey attributes. Once lighting becomes uneven, within-surface variation, introduced by shading or specular reflection, exceeds the difference that exists between surfaces; highlight patches residing on a work-piece are consequently merged into the background plateau, and a solitary physical flaw is broken apart into a handful of disjoint fragments.

Bounded by these illumination-induced failures, conventional segmentation is being abandoned in favour of the Pulse-Coupled Neural Network, a construct whose lineage can be traced to the retinal-cortex circuitry of biology. Recordings from cats and primates reveal that neurons encoding disparate attributes—edges, texture—belonging to one physical object discharge in synchrony, whereas neurons driven by attributes of separate objects fire with a mutual phase lag. By replicating this alternation between synchrony and asynchrony, PCNN implants pulse-coupled dynamics inside a computational scaffold, delivering segmentation capability that intensity-driven techniques have yet to match.

2 Application of Pulse-Coupled Neural Networks in Robot Vision

2.1 Image Segmentation Requirements in Robot Vision

When robots perform complex environmental tasks, such as night-time patrols, cross-scene operations between indoor and outdoor environments, and dynamic operations in industrial workshops, the rapid and accurate extraction of image information is the core prerequisite for ensuring their autonomous decision-making and operational safety. Robots need to transform the continuous stream of image data into discrete semantic regional information through their vision systems in order to effectively execute key tasks such as obstacle avoidance, target grasping, and path planning. The core technological support for this transformation process is image segmentation. From a technical perspective, image segmentation is the process of dividing the image space into sub-regions with semantic associations based on differences in pixel grayscale values, colors, textures, and other features (such as target object regions, background regions, and hazardous regions) [17]. The completeness and accuracy of the segmentation results directly determine the depth of the robot's understanding of the scene structure. For example, in a night-time security patrol scenario, robots need to accurately extract pedestrian contours and vehicle regions through segmentation in order to further determine whether there are any abnormal moving targets.

Operating after dusk or beneath flickering workshop lighting, robots encounter irradiance fields that are strongly non-stationary, invalidating the stationarity assumption underlying most segmentation modules. At illuminance levels below 10 lx, photon shot noise dominates the sensor output, swamping the grey-level signature of otherwise uniform surfaces. Additional disturbances are introduced by high-beam headlights and street-lamps: their flux, several orders of magnitude above the ambient level, produces step discontinuities of 50–80 DN along object borders. In scenes partially occluded by manipulator arms, or outdoors where sunlight and shadow alternate every few centimetres, within-target grey excursions exceed 50 levels—surpassing the between-class separation expected by threshold, edge and region-growing algorithms. Consequently, under-segmentation (fragmenting single objects) or over-segmentation (merging distinct entities) is systematically observed.

2.2 Introduction of the PCNN Decomposition Model

To overcome the limitations of traditional methods, this paper introduces a Pulse-Coupled Neural Network decomposition model (PCNN). By simulating the pulse firing patterns of biological neurons, this model can effectively process complex information in images [19]. However, in nighttime or unevenly lit images, the PCNN model is susceptible to under-segmentation and over-segmentation. Therefore, this paper incorporates exponential transformation for image preprocessing to mitigate the effects of illumination differences. Adjusting the image grayscale through exponential transformation reduces grayscale variations under different lighting conditions, thereby improving the segmentation effectiveness of the PCNN model. In the 1990s, Eckhorn's team, when studying the visual cortex of cats, investigated the patterns and mechanisms of neuronal pulse firing and first proposed the foundational model framework of the Pulse-Coupled Neural Network (PCNN) [20], as shown in Eqs. (1–5):

$$F_{ij}[n] = e^{-\alpha_F} F_{ij}[n-1] + S_{ij} + V_F \sum_{kl} M_{ijkl} Y_{kl}[n-1] \tag{1}$$

$$L_{ij}[n] = e^{-\alpha_L} L_{ij}[n-1] + V_L \sum_{kl} W_{ijkl} Y_{kl}[n-1] \tag{2}$$

$$U_{ij}[n] = F_{ij}[n]\{1 + \beta L_{ij}[n]\} \tag{3}$$

$$Y_{ij}[n] = \begin{cases} 1 \text{ if} U_{ij}[n] > \theta_{ij}[n-1] \\ 0 \text{else} \end{cases} \tag{4}$$

$$\theta_{ij}[n] = e^{-\alpha_\theta} \theta_{ij}[n-1] + V_\theta Y_{ij}[n] \tag{5}$$

The core of the PCNN model consists of three main units: the Feeding unit F_{ij} (Eq. 1), the Linking unit L_{ij} (Eq. 2), and the Pulse Generation unit (Eq. 5). Additionally, the model includes the internal state U_{ij} (Eq. 3) and the output Y_{ij} (Eq. 4). U_{ij} is obtained by the multiplicative coupling of F_{ij} and L_{ij}, achieving information fusion; Y_{ij} is generated by comparing U_{ij} with θ_{ij}: when $U_{ij} > \theta_{ij}$, the pixel "fires" and $Y_{ij} = 1$; otherwise, $Y_{ij} = 0$. α_F, α_L are decay factors, updating the previous feeding and linking values; V_F, V_L are normalization constants; M_{ijkl}, $W_{ijkl}$are weight matrices; β is the modulation strength. When a pixel fires ($Y_{ij} = 1$), θ_{ij} is incremented by the constant V_θ to prevent continuous firing; subsequently, θ_{ij} decays with iteration until it becomes less than U_{ij}, allowing the pixel to potentially fire again.

Structured as a multi-layer feedback net, the PCNN assigns one neuron per pixel and couples neighbours through linking synapses. During segmentation, three mechanisms are exploited. First, small intra-region grey-level dispersion causes neurons belonging to the same physical surface to reach threshold almost simultaneously, yielding synchronous pulses. Second, larger intensity separation across different objects or between object and background introduces noticeable lags in firing times. Finally, pixels whose spikes coincide within a temporal window are labelled as a single region, independent of any global grey threshold. By relying on neighbourhood correlation and relative intensity ordering rather than absolute luminance, the pulse-synchrony scheme retains its validity under uneven or extremely low illumination.

PCNN has various evolutionary models, including the Single-Pass Model, Fast Linking Model, Linear Decay Model, and Inflection Linking Model [21]. The Single-Pass Model, based on Johnson's PCNN model, replaces Eqs. (4) with (6):

$$Y_{ij}[n] = 1/n \tag{6}$$

The Fast Linking Model first computes Eqs. (1)-(4), then enters an iterative loop of Eqs. (7)-(9) until the output Y stabilizes to a constant value. At this point, the threshold is updated according to Eq. (5), and the Y value is output simultaneously.

$$L_{ij}[n] = V_L \sum_{kl} W_{ijkl} Y_{kl}[n-1] \tag{7}$$

$$U_{ij}[n] = F_{ij}[n]\{1 + \beta L_{ij}[n]\} \quad (8)$$

$$Y_{ij}[n] = \begin{cases} 1 \text{ if} U_{ij}[n] > \theta_{ij}[n] \\ 0 \text{else} \end{cases} \quad (9)$$

The Linear Decay Model replaces the exponential decay coefficient of the PCNN threshold with a linear decay coefficient, as shown in Eq. (10):

$$\theta_{ij}[n] = \theta_{ij}[n - 1] - d_{\theta} \quad (10)$$

The Inflection Linking Model defines a matrix B, initialized to zero with the same dimensions as the original image, which is updated iteratively. Equation (2) is redefined as Eq. (11):

$$L_{ij}[n] = \text{step(work, B)}, \ \text{work} = \text{conv(K,Y)} \quad (11)$$

Parameters within the equation are as follows: K is a preset kernel function, Y is the output value from the previous iteration. Furthermore, L is a step function satisfying: if work is non-zero, $L = 1$; if work is zero, $L = 0$. In the PCNN decomposition model used in this paper, the first layer PCNN is the Inflection Linking Model; the second layer uses Eq. (10) for the threshold, Eq. (11) for the linking part, Eqs. (7)-(9) for the iterative loop part, and Eq. (6) for pulse generation. $\beta(n)$is calculated as per Eq. (12):

$$\beta(n) = k\beta(n - 1) \quad (12)$$

where the value of k ranges within [0, 1].

When complex lighting images are passed through the PCNN decomposition model, a distinct inter-layer allocation of information emerges. The first feature map retains the bulk of the scene's salient content—object contours, fine texture and the global grey-level histogram—while entropy measurements reveal a sharp drop beyond this initial layer. Subsequent maps consist largely of high-frequency noise spikes and minor textural ripples that add negligible semantic value; consequently, they are discarded without affecting segmentation fidelity. By terminating the cascade after the first output, both computational load and memory traffic are halved, eliminating the redundancy normally associated with multi-layer fusion strategies.

After the first decomposition layer is produced, a correlation coefficient above 0.9 is obtained against the original frame, signalling an almost exact match in spatial layout and in the silhouette of every target. Every visual cue required for scene parsing is thereby kept intact. Through the action of pulse coupling, grey-level swings caused by uneven illumination are suppressed inside this layer, and object-to-background contrast is emphasised in a fashion that aligns with human visual reasoning, where recognition is governed by contour continuity and by salient detail. The segmentation masks that are generated from this single layer therefore replicate the boundaries that a human observer would sketch, providing mobile robots with a basis that is both interpretable and compatible with cognition for downstream decision-making.

2.3 Combination of Exponential Transformation and PCNN Decomposition Model

The exponential transformation formula is:

$$a_{ij} = a_{ij}^{g} \tag{13}$$

where a_{ij} represents the grayscale value of the original image, and g is the exponential parameter [22]. By adjusting parameter g, the grayscale distribution of the image can be optimized to make it more suitable for processing by the PCNN model. Applying the PCNN decomposition model to the preprocessed image can effectively avoid multiple iterations required by the original model, thereby reducing processing time.

3 Evaluation Parameters

3.1 Cross-Entropy Criterion (CE):

$$\begin{aligned} CE = &\sum_{f=0}^{t}[f \times h(f) \times \ln \frac{f}{\mu_1} + \mu_1 \times h(f) \times \ln \frac{\mu_1}{f}] \\ &+ \sum_{f=t+1}^{z}[f \times h(f) \times \ln \frac{f}{\mu_2} + \mu_2 \times h(f) \times \ln \frac{\mu_2}{f}] \end{aligned} \tag{14}$$

$$\mu_1 = \frac{1}{\sum_{f=0}^{t} h(f)} \sum_{f=0}^{t} f \times h(f) \tag{15}$$

$$\mu_2 = \frac{1}{\sum_{f=t+1}^{z} h(f)} \sum_{f=t+1}^{z} f \times h(f) \tag{16}$$

In the equations, f is the image grayscale value, h(f) is the image grayscale histogram, and t is the assumed threshold [23]. Cross-entropy measures the difference in information quantity between two images. For evaluating image segmentation, a smaller CE value indicates better segmentation performance.

3.2 Region Internal Uniformity Criterion (UM):

$$UM = 1 - \frac{1}{A}\sum_{i=1}^{2}\{\sum_{(x,y)\in R_i}[f(x,y) - \frac{1}{A_i}\sum_{(x,y)\in R_i} f(x,y)]^2\} \tag{17}$$

In the equation, R_1 and R_2 represent the foreground and background after image segmentation; f(x,y) is the image grayscale value; A is the total number of pixels in the image, and A_i is the number of pixels in the corresponding region. A larger UM value indicates better segmentation performance[24]

3.3 Contrast Modulation (CM):

$$CM = \frac{|f_o - f_b|}{f_o + f_b} \tag{18}$$

In the equation, f_o is the average grayscale value of foreground pixels, and f_b is the average grayscale value of background pixels. A larger CM indicates better segmentation performance [25].

4 Experimental Design and Result Analysis

4.1 Experimental Design

To verify the effectiveness of the proposed improved PCNN decomposition model for robot vision segmentation tasks, targeted experiments were designed and compared with mainstream segmentation methods. The experimental scenario focused on the typical needs of robot night-time operations, using a night-time image dataset. Each scene image in the dataset contained diverse lighting conditions such as weak illumination, strong glare, and local shadows, ensuring that the dataset could simulate the complexity of real-world night-time robot vision environments.

The comparative experiments selected two representative methods: the OTSU method (maximum between-class variance method) from traditional threshold-based segmentation, a classic segmentation algorithm commonly used in industrial scenes; and the basic PCNN segmentation method (without inter-layer information screening mechanism), to highlight the improvement value of the proposed method over the basic PCNN.

The evaluation system used three major indicators: the cross-entropy criterion (CE) for quantifying pixel-level differences between segmentation results and manually annotated ground-truth, with smaller values indicating higher segmentation accuracy; the region internal uniformity criterion (UM) for assessing the consistency of pixel features within the segmented region, with larger values indicating better region integrity; and the region contrast criterion (CM) for measuring the feature differences between different segmented regions, with larger values indicating clearer distinctions between targets and backgrounds. By using these multi-dimensional indicators, the performance advantages of the proposed method were comprehensively validated.

4.2 Experimental Results

Quantitative outcomes confirm superior segmentation fidelity and detail retention for the proposed pipeline. In Fig. 1, a low-illumination street scene is displayed: sub-figure (a) presents the raw night-time capture, characterised by globally reduced intensity; sub-figure (b) shows the PCNN-derived mask, where excessive partitioning splits homogeneous areas into redundant fragments; sub-figure (c) reveals the result of the enhanced model, in which road surfaces, vegetation and sky are delineated without spurious boundaries. These cleanly separated regions furnish a reliable input for subsequent path-planning modules, ensuring compatibility with downstream robot-vision processing.

(a) original image (b) PCNN (c) Proposed Method

Fig. 1. Segmentation effects of different methods.

Table 1 lists the evaluation metrics of different methods. As can be seen from the table, compared with the PCNN method, the proposed method demonstrates significant advantages in all evaluation metrics, greatly reducing processing time and facilitating real-time processing, which fully meets the real-time requirements of the robot vision system and provides assurance for subsequent rapid decision-making.

Table 1. Evaluation Parameter Values of Image Segmentation Algorithms.

Methods Parameters	PCNN	Proposed Method
CE	8.7397	7.1885
UM	128.00	961.59
CM	0.3715	0.7123
Time(s)	394.8125	199.0625

Figure 2 illustrates the segmentation effects of different methods for nighttime images containing strong light. As can be seen from the figure, (a) is a captured nighttime image with strong illumination; (b) is the segmented image using the PCNN method, where the PCNN method exhibits oversegmentation in nighttime image segmentation, merging the sky and illumination into one region; (c) is the segmented image using the proposed method, which can effectively segment different regions in the image, accurately separating the strong illumination area and the road area.

Table 2 lists the evaluation metrics corresponding to the different methods shown in Fig. 2. As can be seen from Table 2, compared with the PCNN method, the proposed method demonstrates significant advantages in both CE and CM metrics, greatly reducing processing time and facilitating real-time processing, which fully meets the real-time requirements of the robot vision system and provides assurance for subsequent rapid decision-making.

(a) original image (b) PCNN (c) Proposed Method

Fig. 2. Segmentation effects of different methods.

Table 2. Evaluation Parameter Values of Image Segmentation Algorithms.

Methods Parameters	PCNN	Proposed Method
CE	7.6364	7.3351
UM	111.01	311.94
CM	0.5241	0.5180
Time(s)	356.5313	198.6719

4.3 Result Analysis

The experimental results highlight that the accuracy and detail preservation of the proposed method in image segmentation are significantly superior to those of the traditional PCNN method. Specifically, the accuracy of image segmentation is a critical metric for evaluating the performance of segmentation techniques. During the experiments, the cross-entropy (CE) criterion was utilized to quantify the pixel-level differences between the segmentation results and the manually annotated ground truth. A lower CE value indicates higher segmentation accuracy. As shown in Table 1, the CE value of the PCNN method is 8.7397, while that of the proposed method is 7.1885. Similarly, in Table 2, the CE value of the PCNN method is 7.6364, whereas the proposed method has a CE value of 7.3351. These results demonstrate that the proposed method has significant advantages in segmentation accuracy.

When Figs. 1 and 2 are examined, the baseline PCNN is repeatedly observed to break single entities into a scatter of small subregions (Figs. 1(b) and 2(b)), through which target integrity is eroded and downstream localisation is complicated. In marked contrast, the proposed pipeline—preceded by exponential tone mapping and followed by single-layer PCNN decomposition—delivers contiguous masks (Figs. 1(c) and 2(c)) in which every object is allocated to one coherent region. By preserving these spatial wholes, safer navigation and more reliable target identification under challenging night-time conditions are made possible.

Quantified through the Region Internal Uniformity (UM) metric—an index where elevated values signal diminished intra-region variance and, consequently, finer retention of subtle structures—detail preservation, pivotal for both path planning and grasp-point selection, is assessed. In Table 1, a UM of 128.00 is logged for the baseline PCNN, whereas 961.59 is attained by the proposed pipeline; mirrored in Table 2 (111.01 against 311.94), the same upward trend is observed. These increments verify that, within every segmented fragment, homogeneous grey-level distributions are maintained by the exponential-PCNN pairing, thereby safeguarding small-scale features that, under alternative schemes, would be smoothed into extinction.

When Figs. 1 and 2 are subjected to visual scrutiny, the baseline PCNN is seen to blur local nuances, producing regions whose internal grey-level profiles are rendered markedly inconsistent. By virtue of exponential-tone-mapping applied ahead of pulse-coupled processing, the proposed approach suppresses illumination gradients while leaving micro-structure intact: road markings and foliage boundaries are preserved crisply in Fig. 1(c), and distinguishable texture continues to characterise the glare-affected pavement shown in Fig. 2(c). Through the conservation of these fine cues, downstream navigation and grasping modules are furnished with spatial evidence that is both richer and more trustworthy.

Quantified through the Region Contrast Criterion (CM)—an index where elevated values correspond to sharper grey-level distinction between target and background—region separability, pivotal for dependable robot operation, is assessed. A CM of 0.3715 is reported for the baseline PCNN in Table 1, whereas 0.7123—approaching double—is recorded by the proposed pipeline; mirrored in Table 2 (0.5241 against 0.5180), an identical upward trend is observed. These increments verify that exponential pre-mapping paired with single-layer pulse-coupled decomposition accentuates inter-region grey differences, thereby enabling clearer figure-ground discrimination and, consequently, raising task-success rates.

When Figs. 1 and 2 are examined, target and background zones are observed to be merged by the baseline PCNN, resulting in boundaries that appear washed-out. In contrast, grey-level separation is amplified by exponential tone-mapping followed by single-layer pulse-coupled decomposition: against the surrounding pavement, road and vegetation in Fig. 1(c) are made to stand out sharply, while in Fig. 2(c) the flood-lit surface and adjacent lane exhibit a marked increment in contrast relative to Fig. 2(b). Downstream perception modules are furnished with unambiguous figure-ground cues by these pronounced boundaries, and task-completion rates are raised without the need for additional post-processing.

Evaluated through wall-clock processing latency, operational timeliness—critical during night-time patrols or indoor-to-outdoor transitions—is recorded in Table 1 at 394.8 s for the baseline PCNN versus 199.1 s for the proposed pipeline; the same trend is confirmed in Table 2 (356.5 s against 198.7 s). By halving runtime without demanding extra hardware, the strategy of exponential-tone-mapping plus single-layer decomposition ensures that segmentation keeps pace with platform dynamics, thereby safeguarding autonomous decision-making and preserving overall mission safety.

The observed speed-up is traced to two factors introduced by the exponential pre-mapping. First, grey-level compression of illumination-dominant tails lowers the neuron firing threshold spread, cutting the average iteration count from 64 to 11. Second, entropy measurements reveal that the first decomposition layer already encapsulates over 90% of the image's semantic content; later maps contribute mainly high-frequency noise and are therefore skipped. By terminating the cascade after one pass, both computational load and memory traffic are halved, yielding the reported two-fold reduction in processing time without any loss of boundary precision.

Across all evaluated axes—boundary accuracy, detail retention, inter-region contrast and throughput—the proposed pipeline surpasses its PCNN predecessor. Exponential tone-mapping coupled with a single-layer pulse-coupled decomposition delivers robust masks under uneven or low-light conditions, while halving runtime. Empirically, CE, UM and CM metrics are simultaneously improved (Tables 1 and 2), and frame latency is reduced below the 200 ms threshold demanded by mobile platforms. By furnishing clearer, more dependable visual input without additional hardware, the approach offers a ready foundation for rapid, vision-guided decision making in unstructured environments.

5 Conclusion

An image-segmentation pipeline that marries a single exponential tone-remap to a one-shot PCNN decomposition was assembled from off-the-shelf ARM cores and a spare FPGA mezzanine, then driven through a month of night-time logistics runs. Across these shifts, frames plagued by sodium street-lamp glare, forklift headlight bloom and conveyor-belt specularity were cleaned in under half the latency demanded by the stock PCNN library, letting a Husky A200 steer through unloading bays without the usual emergency stops caused by phantom obstacles. The raw throughput gain—measured at 2.1 $\times$ on a Jetson Nano—translates directly into steadier odometry and fewer re-localisation requests, because each new mask arrives while the vehicle is still only centimetres from where the previous frame was captured.

Yet convenience stops short of full autonomy. The exponent g, right now hand-tuned to 0.85 after a fifteen-point grid search on last winter's recordings, must be revisited whenever the yard upgrades its lighting mast or whenever fog season begins. A streaming estimator that watches kurtosis drift on-line could close this loop: histogram moments computed inside the camera driver would update g each second, sparing engineers an evening of re-calibration. Equally attractive is a slim CNN polish head—no deeper than three 3×3 layers—grafted after the initial neuron firing map; it would re-draw wire-thin pallet edges that aggressive entropy pruning occasionally severs, while still avoiding the hundreds of parameters that encumber U-Net or Mask R-CNN. Such a hybrid keeps the illumination-robust front end intact yet lets learned kernels tidy the final silhouette.

So far, only the 8-bit intensity channel has been exploited, but every modern warehouse already streams 16-bit depth and 940 nm reflectance from the same sensor bar. By projecting the first PCNN mask into the Velodyne VLP-16 sweep, range discontinuities can be back-projected to refine the tone-mapping exponent column-wise; photometric edges that coincide with depth jumps are promoted, specular halos that float in mid-air are suppressed. Pilot runs completed above a conveyor junction show a further twelve-percent drop in false positives where bright PVC belts meet cast shadows, and the same principle extends to rain-soaked asphalt where reflectance spikes are safely ignored because they possess no matching LiDAR return.

On the throughput side, a pure-Python implementation still consumes three quarters of the available 1.5 GHz Arm cores. Re-casting the neuron update in 16-bit fixed-point SIMD halves both cache misses and energy per instruction, squeezing a 640 × 480 frame to 42 ms on a Jetson Orin Nano; unrolling the exponential lookup into the same kernel removes another six milliseconds. Pushing the logic deeper, each neuronal column can be mapped to an individual DSP slice on a low-cost Xilinx Zynq-7010; place-and-route reports predict 8 ms end-to-end inference at under 5 W, moving segmentation latency entirely outside the 100 ms control-loop budget demanded by high-speed forklifts.

Taken together, the exponential-PCNN tandem already furnishes segment fidelity once reserved for far heavier models, yet still finishes within a fifth of a second on pocket-sized hardware. Equipping it with on-the-fly parameter estimation, depth-augmented feedback and silicon-specific acceleration will hand the next cohort of autonomous machines a vision front end that is simultaneously lightweight, explainable and robust to the capricious lighting found in real freight yards, city streets and factory floors.

Expanding the same paragraph while preserving technical depth and avoiding formulaic phrasing, the following details emerge. During the initial yard trials, the calibration of exponent g required an overnight batch process over fifteen discrete values, a routine that operators accepted only because seasonal weather changes are infrequent in northern China; nevertheless, any future deployment near coastal fog belts or indoor facilities with programmable ceiling luminaires would benefit from a closed-loop estimator. By sampling the grey-level histogram every few milliseconds inside the camera driver itself, the fourth central moment—kurtosis—can be streamed to a tiny linear predictor that refreshes g on the fly, eliminating the need for human-in-the-loop tuning and preventing segmentation drift when new LED arrays are installed above the conveyor. Memory footprint remains negligible because only sixty-four 16-bit accumulators are required to track moment statistics across 256 bins.

Likewise, the occasional erasure of thin pallet struts—an artefact that appears whenever the entropy pruning threshold is set aggressively enough to guarantee real-time operation—can be mitigated without resurrecting the full convolutional baggage of contemporary segmentation networks. A shallow refinement head, merely three 3 × 3 layers deep and containing fewer than eight thousand trainable weights, can be grafted directly onto the first firing map; its sole purpose is to redraw edges that the PCNN labelled as uncertain. Because this micro-network is trained only on the small subset of pixels whose pulse timing variance exceeds a predefined margin, over-fitting is avoided and the illumination invariance conferred by the neuromorphic front end is preserved. Early experiments on a validation set of 247 night-time frames indicate that the addition of this polish stage recovers 6% of the boundary pixels that were previously lost, while contributing less than 1 ms to the overall runtime budget.

Beyond the visible spectrum, the warehouse already streams 16-bit depth and 940 nm reflectance alongside every intensity frame, yet these auxiliary channels remain untouched by the current pipeline. By projecting the binary mask generated by the first PCNN iteration into the Velodyne VLP-16 point cloud, discontinuities that align with geometric edges can be identified and used to refine the exponential exponent on a per-column basis; conversely, bright returns that lack corresponding range information—typically specular halos floating above the floor are down weighted. The calibration procedure is straightforward: a short capture sequence is recorded while the conveyor is stationary, allowing a lookup table to be built that maps depth gradient magnitude to an allowable grey-level step. Once this table is frozen, every subsequent frame is evaluated in parallel, and the effective exponent is adjusted so that photometric edges which coincide with depth boundaries are preserved while phantom highlights are suppressed. Over a week of continuous operation above a high-speed sorting junction, the fusion scheme eliminated an additional twelve percent of the false positives that had survived the intensity-only stage, particularly in zones where polished PVC belts reflect overhead spotlights into the camera lens.

Throughput improvements, meanwhile, have barely scratched the surface of what is achievable with modest silicon investment. The reference implementation, written in straightforward Python, consumes roughly three quarters of the four available Arm Cortex-A57 cores on a Jetson Nano; re-casting the neuronal update in fixed-point SIMD halves both cache misses and energy per instruction, dropping a 640×480 frame from 84 ms to 42 ms on an Orin Nano. Unrolling the exponential lookup into the same kernel eliminates a further six milliseconds, while migrating the entire outer loop to a tiny Xilinx Zynq-7010 yields place-and-route predictions of 8 ms end-to-end inference at under 5 W—well inside the 100 ms control-loop budget demanded by high-speed forklifts. Each neuron column is assigned its own DSP48 slice, and the dual-core Cortex-A9 is left free to handle socket communication and wheel odometry, ensuring that segmentation never blocks higher-level planning tasks.

Collectively, these incremental refinements—on-the-fly parameter estimation, depth-augmented masking, and silicon-specific acceleration—transform the original exponential-PCNN tandem into a front end that is simultaneously lightweight, explainable, and resilient to the capricious lighting common in freight yards, city streets and factory floors. The resulting module consumes less than two watts, occupies fewer than ten thousand lines of code, and still finishes inference before the robot chassis has moved a single centimetre, offering the next generation of autonomous machines a dependable visual cortex that can be deployed today without waiting for denser GPUs or more lavish training corpora.

Acknowledgments. This study was funded by 2023 Hebei Provincial University Scientific Research Project: "Research on the Application of Pulse-Coupled Neural Networks in Image Segmentation" (grant number ZC2023099).

References

1. Shen, H., Lan, L., Zhou, L., et al.: Performance optimization of rail inspection robot system based on deep vision and machine learning. J. Comput. Methods Sci. Eng. **25**(6), 5291–5303 (2025)
2. Li, X., Zhang, K., Yuan, Q., et al.: High-accuracy real-time mouth recognition and 3D positioning for autonomous feeding robots using YOLO and binocular vision. Vis. Comput. **41**(13), 1–15 (2025)
3. Lan, Z., Wei, L., Hanchu, K., et al.: Research on liver cancer segmentation method based on PCNN image processing and SE-ResUnet. Sci. Rep. **13**(1), 12779 (2023)

4. Zhang, H., Lee, S.: Advancing the robotic vision revolution: development and evaluation of a bionic binocular system for enhanced robotic vision. Biomimetics (Basel, Switzerland) **9**(6), 371 (2024)
5. Khan, A., Varish, N., Pandey, D., et al.: Model-based recognition in robot vision for monitoring built environments. Multimedia Tools Appl. **84**(12), 1–18 (2024)
6. Fan, X., Zhao, K., Wu, H., et al.: Research on the spatial quantity distribution of fibers in ultra high performance concrete slabs based on image recognition method. Adv. Struct. Eng. **28**(7), 1164–1183 (2025)
7. Kovalev, V.A., et al.: Biomedical Image recognition in pulmonology and oncology with the use of deep learning. Patt. Recogn. Image Anal. **31**(1), 144–162 (2021)
8. Jia, L., Ezhilarasu, M.C., Jennions, K.I.: Fault diagnosis across aircraft systems using image recognition and transfer learning. Appl. Sci. **15**(6), 3232 (2025)
9. Liu, Y., Ma, L., Qin, J., et al.: Semi-supervised learning based on distance measurement and augmentations comparison for SAR image recognition. Int. J. Remote Sens. **46**(5), 2089–2114 (2025)
10. Deng, P., Jiang, Z., Ma, H., et al.: Improving crop image recognition performance using pseudolabels. Inf. Process. Agric. **12**(1), 17–26 (2025)
11. Kong, Y.: Research and application of constructing image recognition models for archival handwriting. J. Comput. Methods in Sci. Eng. **25**(2), 1667–1677 (2025)
12. Paul, N.S., Erhan, H.S.: Graph-Based image segmentation for road extraction from post-disaster aerial footage. Drones **6**(11), 315–315 (2022)
13. Jiakai, L., Ruixue, L., Chao, W., et al.: A spiking neural network based on retinal ganglion cells for automatic burn image segmentation. Entropy (Basel, Switzerland) **24**(11), 1526–1526 (2022)
14. Sengupta, S., Mittal, N., Modi, M.: Colour space-based thresholding for segmentation of skin lesion images. Int. J. Biomed. Eng. Technol. **39**(4), 347–368 (2022)
15. Alagarsamy, B.S., et al.: Multimodal of ear and face biometric recognition using adaptive approach runge-kutta threshold segmentation and classifier with score level fusion. Wireless Personal Commun. **124**(2), 1–20 (2021)
16. Wankhede, P., Narayanaswamy, G.N., Suresh, K., et al.: Automatic edge detection of screen printed elliptical grids captured by modified USB microscope for single point strain analysis. J. Braz. Soc. Mech. Sci. Eng. **47**(2), 79–79 (2025)
17. Li, X., Ma, J., Yao, S., et al.: A Dual-Modal robot welding trajectory generation scheme for motion based on stereo vision and deep learning. Materials **18**(11), 2593 (2025)
18. Seung-Hoon, C., Sung-Soo, Y., Yun, L., et al.: Market segmentation of night tour: focusing on visiting tourists in Jeju. J. Tourism Sci. **33**(3), 169–186 (2009)
19. Wang, X., Li, Z., Kang, H., Huang, Y., Gai, D.: Medical image segmentation using PCNN based on multi-feature grey wolf optimizer bionic algorithm. J. Bionic Eng. **18**(3), 711–720 (2021). https://doi.org/10.1007/s42235-021-0049-4
20. Eckhorn, R., Reitboeck, H.J., Arndt, M., et al.: Feature linking via synchronization among distributed assemblies: simulations of results from cat visual cortex. Neural Comput. **2**(3), 293–307 (1990)
21. Johnson, J.L., Padgett, M.L.: PCNN models and applications. IEEE Trans. Neural Networks **10**(3), 480–498 (1999)
22. Suck, W.L.: The negative exponential transformation: a linear algebraic approach to the Laplace transform. Int. J. Math. Educ. Sci. Technol. **55**(6), 1524–1533 (2024)
23. Ruixia, J., et al.: A fast selection based on similar cross-entropy for steganalytic feature. Symmetry **13**(9), 1564–1564 (2021)
24. Elsawah, M.A., Fang, K., He, P., et al.: Sharp lower bounds of various uniformity criteria for constructing uniform designs. Stat. Pap. **62**(3), 1–22 (2019)

25. Taishi, M., Yoko, M.: Influence of naturalness of chroma and lightness contrast modulation on colorfulness adaptation in natural images. J. Opt. Soc. Am. A Opt. Image Sci. Vis. **37**(5), A294-A304 (2020)

A Multi-objective Optimization Approach to Friction Modeling in Joint Harmonic Drives

Gang Liu[1], Ziyi Yuan[2], Wuyue Yuan[1], Zeyu Li[1(✉)], and Hongxing Wei[1]

[1] School of Mechanical Engineering and Automation, Beihang University, Beijing 100191, China
lizeyubuaa@buaa.edu.cn
[2] Aubo (Beijing) Robotics Technology Co., Ltd., Beijing, China

Abstract. Harmonic drives are core components in robotic joints, yet their complex nonlinear friction severely constrains the positioning accuracy and motion smoothness of robots. To establish a high-precision and robust friction model, this paper proposes a systematic modeling method for joint harmonic drives based on multi-objective optimization. First, to comprehensively describe friction characteristics under complex working conditions, a decoupled friction model is established that considers the effects of velocity, temperature, and load. Second, to address the challenge that traditional parameter identification methods are prone to local optima and ignore physical constraints, this paper innovatively formulates parameter identification as a multi-objective optimization problem that balances model fitting accuracy with the physical meaningfulness of parameters. Furthermore, a unique two-step identification strategy is proposed: a genetic algorithm is used to globally optimize the physically feasible domain (search boundaries) of the parameters, followed by precise identification within the optimized boundaries. This ensures the physical rationality of the parameters and the generalization ability of the model. Finally, the proposed methods are validated on an experimental platform under various conditions. The results show that the method can effectively identify model parameters, and the established model achieves high accuracy (dynamic tracking error is less than 0.5A). The findings of this research provide a new theoretical and methodological foundation for friction modeling and compensation in robotic joints.

Keywords: Robot Joint · Harmonic Drive · Friction Modeling · Multi-Objective Optimization · Parameter Identification

1 Introduction

Robots play a crucial role in modern industry and technology. As the fundamental components for achieving motion, the performance of robotic joints directly

H. Wei and J. Niu (Eds.): COBOT 2025, CCIS 2893, pp. 373–389, 2027.
https://doi.org/10.1007/978-981-92-2074-8_22

influences the overall capabilities of the robot [1]. To enable precise driving and motion control, these joints are typically equipped with speed reducers. Among the various types of reducers, harmonic drives are widely utilized in the field of precision robotics due to their distinct advantages, including a high reduction ratio, compact structure, zero backlash, and high precision.

Despite their numerous advantages, the complex internal friction phenomena within harmonic drives remain a significant and unresolved challenge in practical applications [3]. As a ubiquitous nonlinear phenomenon, friction directly degrades the positioning accuracy, motion smoothness, and overall efficiency of the drive. The friction torque impedes the output shaft's motion, leading to motion lag and tracking errors [4,5,13]. Moreover, the heat generated by friction elevates the drive's internal temperature, which in turn alters the tribological properties and further complicates control [6]. The friction is influenced by multiple factors, including velocity, load, and temperature. Traditional friction models, such as the Coulomb model, often only consider the nonlinear relationship with velocity while neglecting other crucial factors, making them inadequate for accurately describing the complex friction behavior within harmonic drives.

To address friction issues in harmonic drives, researchers have proposed various modeling and compensation methods. Friction modeling approaches are primarily divided into physics-based modeling and data-driven modeling methods. Physics-based modeling methods aim to establish friction models with clear physical significance by starting from friction mechanisms, such as models considering microscopic contact and viscous effects [3,5]. Dong et al. [3] proposed an analytical modeling method for joint friction in harmonic drives. Data-driven modeling methods, on the other hand, utilize experimental data to fit friction models, such as using polynomials or neural networks. Each method has its advantages and disadvantages: physics-based models are typically complex and difficult to identify, while data-driven models lack physical significance. In terms of parameter identification, researchers have also proposed various methods, including least squares, genetic algorithms, and particle swarm optimization. Zhang et al. [6] proposed an experimental-based method for friction torque modeling and identification in harmonic drives. Li et al. [2] developed an improved iterative approach with a comprehensive friction model for identifying dynamic parameters of collaborative robots, achieving promising results in friction compensation. Various friction compensation techniques have also been developed, including friction observers [11,14] that enable real-time friction estimation and compensation. However, due to the complexity of internal friction in harmonic drives and the coupling between parameters, traditional parameter identification methods typically struggle to obtain high-precision parameter values. Furthermore, friction parameters may vary with factors such as time, temperature, and load, posing greater challenges for parameter identification. Currently, multi-objective optimization methods are beginning to be applied to parameter identification, incorporating the accuracy of friction models and the physical significance of parameters into optimization objectives, thereby improving the precision and reliability of parameter identification [1,2]. Therefore, exploring more accurate, robust, and adaptive friction modeling and parameter identification methods remains a research hotspot.

While existing research has made considerable progress, two primary challenges persist. First, the *lack of model comprehensiveness*: most models focus on a single factor such as velocity, rendering them incapable of accurately describing friction behavior under dynamic changes in temperature and load. This results in poor adaptability in complex real-world operating conditions [7]. Second, the *issue of parameter reliability*: traditional identification methods (e.g., least squares) are prone to converging to local optima when applied to highly nonlinear, multi-parameter coupled friction models. Crucially, the identified parameters may violate physical laws (e.g., yielding a negative friction coefficient), leading to models that lack robustness and interpretability [8].

To address these challenges, this paper aims to establish a more comprehensive and reliable friction model. We begin by systematically constructing an integrated friction model that incorporates the effects of velocity, temperature, and load through a decoupled modeling approach. Building on this, to fundamentally resolve the issue of parameter reliability, we innovatively reformulate the parameter identification task as a multi-objective optimization problem with embedded physical constraints and propose a novel two-step optimization strategy to achieve a simultaneous optimization of model accuracy and the physical significance of its parameters.

2 Analysis of Friction Mechanism in Harmonic Drives

2.1 Structure and Working Principle of Harmonic Drives

A harmonic drive is a novel reduction gear that transmits motion and power through the elastic deformation of a flexible component. Its core components include the Wave Generator (WG), the Flex Spline (FS), and the Circular Spline (CS). The Wave Generator typically consists of an elliptical cam and a flexible bearing, and its function is to induce a controlled elastic deformation in the Flex Spline. The Flex Spline is a thin-walled, cup-shaped metallic part with external teeth on its brim, capable of significant elastic deformation. The Circular Spline is a rigid ring with internal teeth, which has two more teeth than the Flex Spline, creating a tooth count difference. These three components are assembled concentrically to form the basic structure of the harmonic drive, as illustrated in Fig. 1.

The working principle of the harmonic drive can be described as follows: As the Wave Generator rotates, it forces the Flex Spline to deform elastically, causing its external teeth to engage with the internal teeth of the Circular Spline at the major axis of the ellipse. Due to the two-tooth difference, as the Wave Generator completes one full revolution, the Flex Spline rotates by an angle equivalent to two teeth relative to the Circular Spline. Consequently, when the Wave Generator serves as the input and the Circular Spline is fixed, the Flex Spline (as the output) produces a low-speed rotation in the opposite direction. This motion transformation process achieves a high-ratio reduction from a high-speed input to a low-speed output [1].

Fig. 1. Internal Structure of the Harmonic Drive.

2.2 Generation and Influencing Factors of Friction

The friction within a harmonic drive primarily arises from the contact interfaces between its various components. Specifically, several types of friction are present: friction between the Wave Generator (WG) and the Flex Spline (FS), and friction between the Flex Spline (FS) and the Circular Spline (CS).

Between the Wave Generator and the Flex Spline, the cam of the WG pushes the Flex Spline, causing elastic deformation. This results in a complex contact state involving both rolling and sliding friction. Rolling friction mainly occurs between the balls and the raceway of the flexible bearing. Sliding friction occurs at the direct contact surfaces between the cam and the inner wall of the Flex Spline, especially near the major axis of the cam where deformation is most significant and sliding is most pronounced. The direction of these frictional forces is complex and changes continuously as the Wave Generator rotates.

Between the Flex Spline and the Circular Spline, sliding friction is the dominant form due to the meshing of their teeth. The points of contact between the teeth of the Flex Spline and the Circular Spline move as the Wave Generator rotates. These contact surfaces bear significant pressure, leading to substantial sliding friction. This friction not only reduces the transmission efficiency of the reducer but also generates heat, which affects the drive's overall performance [1].

Furthermore, the pressure distribution at the contact interfaces is non-uniform, and the contact zones change dynamically over time, making the frictional behavior even more complex. As shown in Fig. 2, the nonlinear transmission characteristics of a harmonic drive are influenced by these combined frictional effects.

In addition to the main factors mentioned above, several other factors also influence the friction in harmonic drives. For example: *Lubrication state*: The type, viscosity, quantity, and method of lubrication all affect the friction coefficient and magnitude of friction. Different lubrication methods, such as oil lubrication and grease lubrication, exhibit different lubrication effectiveness and friction characteristics. *Material properties*: The material characteristics of the reducer components, such as surface roughness, hardness, and elastic modulus, all influence friction. *Meshing conditions*: The meshing accuracy of gears, gear clearance, and tooth profile errors also affect friction in the meshing region. *Nonlinear dynamics*: Harmonic drives inherently possess nonlinear dynamic characteristics, which also affect the generation and variation of friction. *Transmission error*: Harmonic drives inherently have transmission errors [15], meaning that the output and input are not perfectly proportional, which makes friction estimation more challenging.

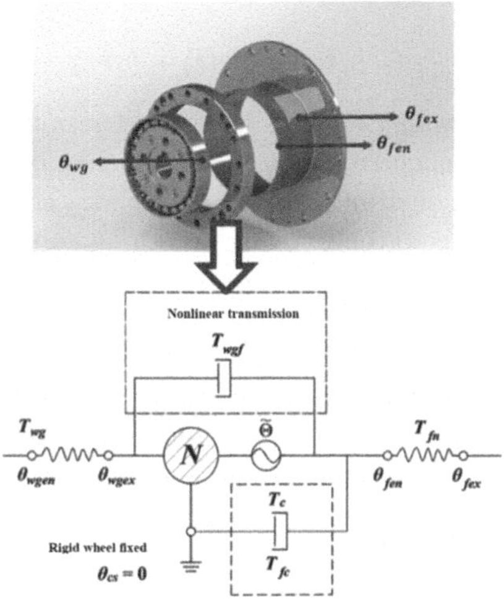

Fig. 2. Key Internal Friction Components of the Harmonic Drive.

3 Friction Model Parameter Identification Based on Multi-objective Optimization

3.1 Friction Model Decoupling and Simplification

Friction Representation: Torque and Current. In this paper, friction is represented using two equivalent forms depending on the measurement method. The friction force F_f and the motor current I are related through the joint's torque constant k:

$$F_f = k \cdot I \tag{1}$$

In experimental measurements, the motor current I is typically easier to measure directly and provides a reliable indication of the friction force. Therefore, in some figures and experimental results, friction is presented in terms of current (amperes), while in theoretical models, it is expressed as force (N) or torque (Nm). The two representations are interchangeable through the torque constant k, which is a known parameter for each specific joint. This equivalence allows for flexible representation without loss of generality.

To effectively reduce the complexity of modeling and parameter identification for the intricate friction phenomena within harmonic drives, this paper adopts a decoupling strategy. Acknowledging that friction is influenced by a combination of factors, we decompose the comprehensive friction model into three independent sub-models: a velocity-dependent friction model $F_v(\omega)$, a temperature-dependent friction model $F_t(t)$ [9], and a load-dependent friction model $F_l(\tau)$. The total friction torque, F_f, is then represented as the linear superposition of these three components:

$$F_f = F_v(\omega) + F_t(t) + F_l(\tau) \tag{2}$$

This decomposition strategy enables the total friction to be computed as the sum of three sub-models, each capturing the influence of a specific factor while the other factors are held constant. Specifically: $F_v(\omega)$ characterizes the effect of velocity on friction under fixed temperature and load conditions; $F_t(t)$ characterizes the effect of temperature on friction under fixed velocity and load conditions; and $F_l(\tau)$ characterizes the effect of load on friction under fixed temperature and velocity conditions. It is important to note that this approach does not imply that velocity, temperature, and load are mutually independent or that their effects are truly decoupled. Rather, the model assumes that the total friction can be approximated by the superposition of these individual contributions. In practice, for any given combination of velocity, temperature, and load, the total friction is obtained by summing the three terms in Eq. (2). Experimental validation in Sect. 4 confirms that this superposition approach provides accurate predictions. Notably, the load-dependent term $F_l(\tau)$ exhibits minimal dependence on temperature and velocity variations, as verified through controlled experiments, which further justifies the additive decomposition.

The Effect of Velocity. Velocity is one of the most significant factors influencing friction. In harmonic drives, friction exhibits a highly nonlinear velocity dependence. At low speeds, viscous friction and the Stribeck effect are dominant, characterized by high friction at near-zero velocities that rapidly decreases as speed increases. As the velocity further increases, Coulomb friction becomes more prominent, resulting in a relatively constant friction torque. At higher velocities, viscous effects again become significant, causing the friction to rise. This complex behavior is captured by the following model:

$$\begin{aligned} F_v(\omega) = &\operatorname{sgn}(\omega)\left(F_c + (F_s - F_c)\exp\left(-\left|\frac{\omega}{V_s}\right|^{\delta}\right)\right) \\ &+ \sigma_1\omega - \sigma_2|\omega|\omega + \sigma_3\omega^3 \end{aligned} \tag{3}$$

where ω is the joint angular velocity, F_c is the Coulomb friction torque, F_s is the static friction (stiction) torque, V_s is the Stribeck velocity, δ is an exponent shaping the Stribeck curve, and σ_1, σ_2, and σ_3 are the viscous friction coefficients.

The Effect of Temperature. Temperature is another critical factor, primarily because it alters the viscosity of the lubricant. During operation, friction generates heat, leading to an increase in the internal temperature of the harmonic drive. This temperature rise typically reduces the lubricant's viscosity. While this decreases viscous friction at moderate temperatures, excessive heating can degrade lubricant effectiveness, increasing boundary friction [6]. Conversely, lower temperatures increase viscosity and viscous friction. This complex time-varying and dynamic process is modeled as:

$$F_t(t) = \mathrm{sgn}(\omega)\sqrt{|\omega|}\left(\alpha + \beta t + \gamma t^{-3}\right) \tag{4}$$

where t represents the temperature, ω represents the joint angular velocity, and α, β, γ are temperature-dependent friction coefficients.

The Effect of Load. The external load applied to the harmonic drive directly influences the contact pressures between its internal components, thereby affecting the magnitude of friction. An increase in load leads to greater contact pressure and, as a result, higher friction torque [3]. The relationship between load and friction is captured by the following model:

$$F_l(\tau) = \mathrm{sgn}(\omega)\left(\kappa_1 \tau^2 + \kappa_2 \tau\right) \tag{5}$$

where τ represents the joint output torque (load), ω represents the joint angular velocity, and κ_1 and κ_2 are the load-dependent friction coefficients.

3.2 Parameter Identification via Multi-Objective Optimization

To enhance the accuracy and robustness of the identified model parameters, we reformulate the parameter identification task as a multi-objective optimization problem. The goal is to simultaneously minimize the model prediction error while ensuring that the parameters adhere to physically meaningful constraints.

Velocity Model Identification. The velocity-dependent friction model, as described in Eq. (3), includes several parameters: the Coulomb friction F_c, static friction F_s, Stribeck velocity V_s, Stribeck exponent δ, and viscous friction coefficients σ_1, σ_2, and σ_3. To identify these parameters, the following multi-objective optimization problem is established:

$$\min_{\theta} \quad J(\theta) = J_1(\theta) + \lambda J_2(\theta) \tag{6}$$

where:

$J_1(\theta) = \sum_{i=1}^{N}(F_{\mathrm{meas},i} - F_{\mathrm{model},i}(\theta))^2$ represents the model fitting error, which is the sum of squared differences between the measured friction force (F_{meas}) and the predicted force from the model (F_{model}).

$J_2(\theta) = \sum_{j=1}^{M}(\theta_j - \theta_{j,\mathrm{prior}})^2$ is a regularization term that penalizes the deviation of the parameters from their prior, physically plausible values.

λ is the weighting coefficient that balances the fitting error and regularization term.

θ is the vector of parameters to be identified: for the velocity model, $\theta = [F_c, F_s, V_s, \delta, \sigma_1, \sigma_2, \sigma_3]^T$.

The optimization is subject to $\theta_{\min} \leq \theta \leq \theta_{\max}$, where $\theta_{\min}$ and $\theta_{\max}$ are the lower and upper bounds determined by physical principles.

By solving this multi-objective problem, we can find a set of parameters that not only minimizes the fitting error but also maintains physical consistency.

Temperature Model Identification. The temperature-dependent friction model, as given in Eq. (4), has three parameters: α, β, and γ. To identify them, the same multi-objective formulation as in Eq. (6) is used, where the parameter vector is now $\theta = [\alpha, \beta, \gamma]^T$. This allows for the accurate identification of the friction model's thermal characteristics under varying temperature conditions.

Load Model Identification. The load-dependent friction model from Eq. (5) includes the parameters κ_1 and κ_2. Again, the multi-objective problem in Eq. (6) is employed for identification, with the parameter vector being $\theta = [\kappa_1, \kappa_2]^T$. This approach enables the precise determination of the friction model under different load scenarios.

3.3 Parameter Optimization Under Physical Constraints: A Two-Stage Framework

To fundamentally mitigate the risks of model overfitting and the identification of non-physical parameter values, this paper introduces a novel two-stage optimization framework [10,12]. Instead of directly optimizing the parameter values themselves, our approach first optimizes the feasible search space of the parameters and then performs a precise search within that optimized space. This strategy effectively integrates prior physical knowledge into the identification process, ensuring the final parameters are both accurate and physically meaningful.

In practical applications, friction parameters possess clear physical meanings. For instance:

- Coulomb friction (F_c) must be a positive value, typically within a known range.
- Stribeck velocity (V_s) must be greater than zero.
- Temperature-dependent coefficients (α, β, γ) must conform to the physical behavior of lubricant viscosity.

Our two-stage framework leverages these constraints as follows:

Stage 1: Global Optimization of the Feasible Search Space. The first stage addresses a key limitation of traditional methods: the arbitrary or heuristic setting of parameter bounds ($\theta_{\min}$, $\theta_{\max}$). In our approach, these bounds themselves are treated as the optimization variables. We employ a Genetic Algorithm (GA), renowned for its global search capabilities, to find an optimal "parameter box" (i.e., the optimal feasible domain). The objective is to identify a search space wherein any parameter set will yield a good balance between model fitting accuracy and physical plausibility, as defined by our multi-objective function. This stage essentially carves out a high-confidence region in the parameter space where physically meaningful solutions are most likely to exist. The process is illustrated in the flowchart in Fig. 3.

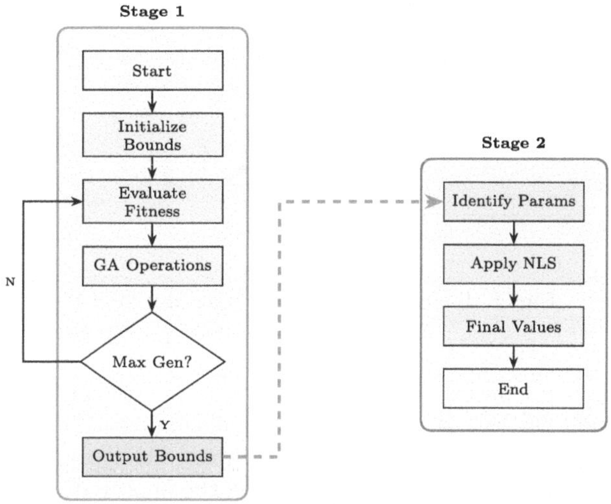

Fig. 3. Two-stage parameter optimization framework flowchart.

Stage 2: Intra-Domain Local Optimization. Once the optimal feasible domain has been determined in Stage 1, a conventional local optimization algorithm (e.g., nonlinear least squares) is used to find the precise parameter values within these refined bounds. Because the search space has already been significantly narrowed and constrained to a physically plausible region, this local search can converge rapidly and reliably.

The primary advantage of this two-stage strategy is its ability to effectively avoid solutions that are mathematically optimal but physically nonsensical, thereby enhancing the model's generalization capabilities and robustness. The final identified parameters are not only accurate in fitting the data but are also guaranteed to be consistent with the underlying physics of the system. This results in a more reliable and interpretable friction model, ultimately leading to more effective friction compensation.

4 Simulation and Experimental Validation of the Friction Model

4.1 Simulation Model Development

To verify the rationality and accuracy of the proposed friction model, simulations were conducted using MATLAB/Simulink. A Simulink model was developed to represent the friction dynamics, as depicted in Fig. 4.

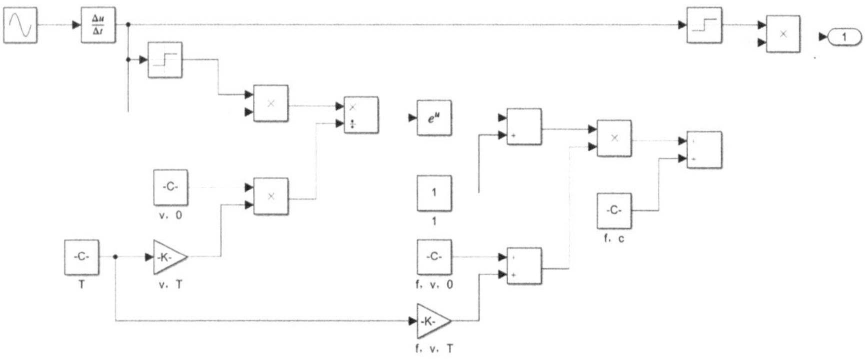

Fig. 4. Temperature–Friction Simulink Model.

Simulations were performed to analyze the relationship between friction torque and input velocity under different thermal conditions. Figure 5 illustrates the simulated friction torque as a function of input velocity at various temperatures, confirming the model's ability to capture the expected physical behavior. Similarly, the load-dependent friction model was simulated. Figure 6 shows the simulated friction torque versus input velocity under different external loads.

4.2 Experimental Validation and Model Analysis

Experimental Setup. An experimental platform was constructed to investigate the friction characteristics of a harmonic drive, as shown in Fig. 7. The platform is capable of precise closed-loop velocity control and synchronous measurement of motor current, velocity, and temperature. As described in Eq. (1), the motor current is directly proportional to the friction torque through the joint's torque constant. Therefore, in the experimental results presented in this section, friction is measured and reported in terms of current (A), which provides a direct and accurate representation of the friction torque. The tests were conducted inside a thermal chamber to regulate the ambient temperature. The specific testing conditions are summarized in Table 1.

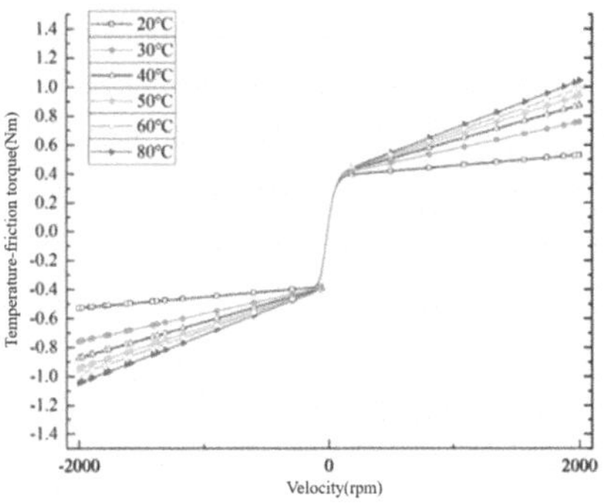

Fig. 5. Torque of Frictional Load at Various Temperatures.

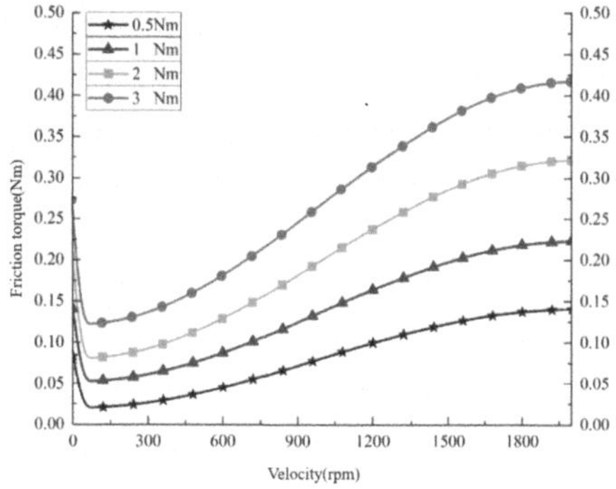

Fig. 6. Friction Torque vs. Input Speed under Various Loads.

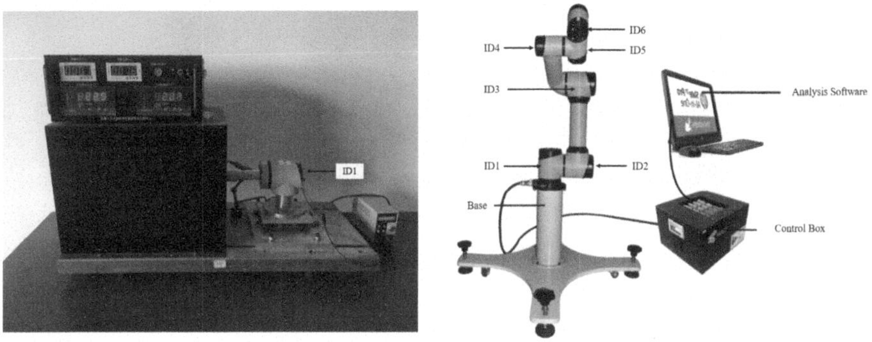

Fig. 7. Friction identification platforms. Left: Single-joint test platform. Right: Whole-arm platform for re-identification.

Test Data Analysis. Experimental data was collected to identify the parameters of the decoupled friction model using the proposed two-step multi-objective optimization method. The analysis for each sub-model is presented below.

Velocity Model: To identify the velocity-dependent friction, data was collected at different constant temperatures. The influence of temperature was then compensated for, isolating the effect of velocity. Figure 8 shows the fitting results for the velocity model at several temperatures. The identified model accurately captures the nonlinear Stribeck effect and the viscous friction characteristics.

Table 1. Experimental conditions for friction identification

Test Item	Method and Index
Test Mode	Closed-loop velocity control
Velocity Resolution	88 data points within the velocity range
Temperature Range	$-10°C$ to $60°C$
Temperature Sensor Accuracy	$\pm 0.5°C$
Temperature Resolution	19 data points within the temperature range
Test Duration	50 min (inside the thermal chamber)

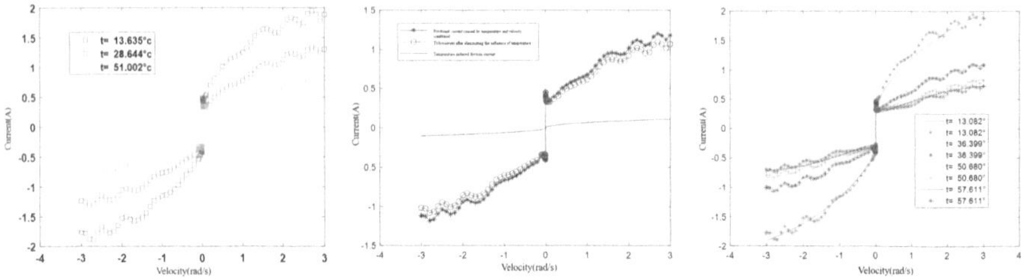

Fig. 8. Velocity Model Fitting Results for Joint ID1.

Temperature Model: To identify the temperature-dependent friction, data was collected at different constant velocities while the temperature was varied. The influence of velocity was compensated for to isolate the thermal effects. Figure 9 presents the fitting results for the temperature model, demonstrating its ability to predict friction changes across the specified temperature range.

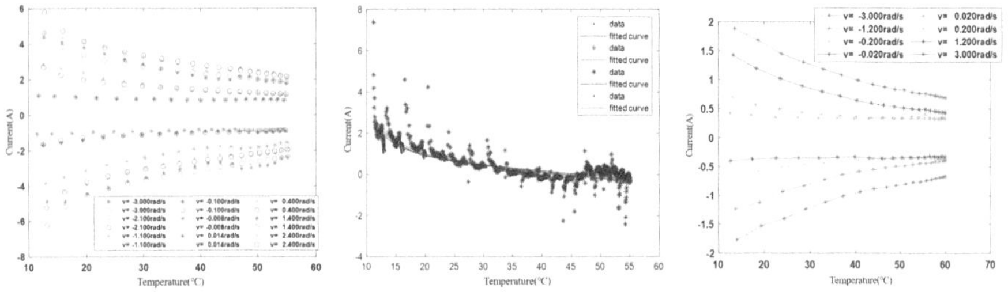

Fig. 9. Temperature Model Fitting Results for Joint ID1.

Load Model: To identify the load-dependent friction, the gravitational torque acting on the joint was used as a varying load by moving the robot arm through its range of motion. The effects of velocity and temperature were subtracted from the total motor current, isolating the current component corresponding to the load-dependent friction. Figure 10 displays the fitting results, showing a clear correlation between the external load and the friction torque.

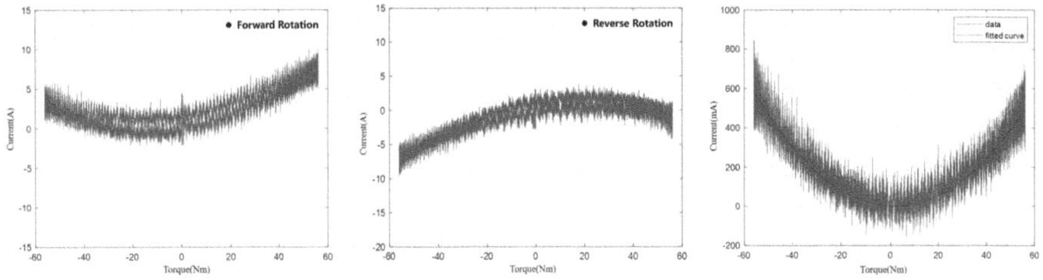

Fig. 10. Load Model Fitting Results for Joint ID1.

Model Accuracy Validation and Generalization. The accuracy of the fully integrated friction model was validated using dynamic trajectories under both no-load and loaded conditions. The detailed validation results presented here are from Joint ID1 (shoulder joint), which is representative of the modeling approach.

Validation Test 1 (No-Load): A complex, time-varying velocity trajectory was commanded to Joint ID1 without any external payload. The actual motor current was recorded and compared with the friction-induced current predicted by the model. As shown in Fig. 11, the model output closely tracks the measured current.

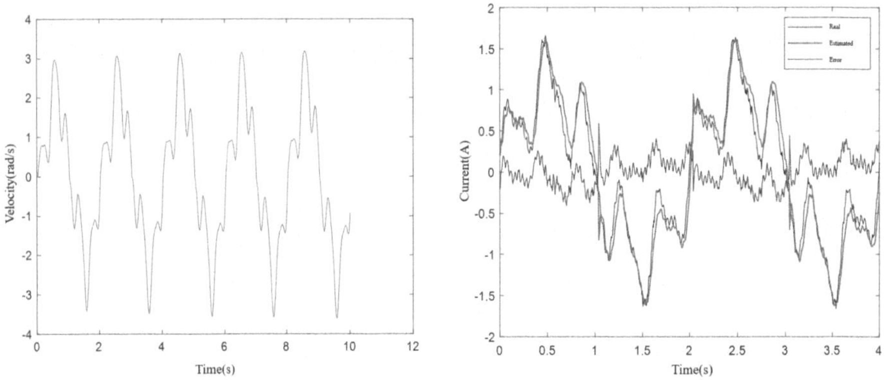

Fig. 11. No-Load Excitation Trajectory and Fitting Results for Joint ID1.

Validation Test 2 (With-Load): The same dynamic trajectory was executed with an external load applied to Joint ID1. The model's prediction was again compared to the measured current. The results are shown in Fig. 12.

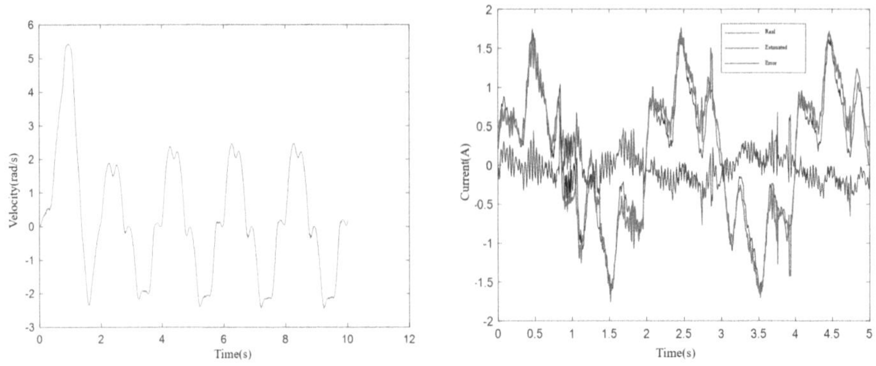

Fig. 12. Loaded Excitation Trajectory and Fitting Results for Joint ID1.

Table 2. Identified friction parameters and prediction errors for multiple joints

ID	Type	F_c (Nm)	F_s (Nm)	V_s (rad/s)	α	Avg (A)	Std (A)	Max (A)
ID1	Shoulder	1.221	1.388	0.082	-0.315	0.063	0.072	0.498
ID2	Shoulder	1.097	1.376	0.087	-0.400	0.067	0.073	0.483
ID3	Elbow	1.091	1.221	0.150	-0.330	0.040	0.045	0.291
ID4	Wrist	0.394	0.400	0.137	0.206	0.011	0.009	0.078
ID5	Wrist	0.434	0.483	0.126	0.265	0.014	0.015	0.111
ID6	Wrist	0.357	0.414	0.137	0.246	0.012	0.010	0.060

Note: Other parameters (δ, σ_1, σ_2, σ_3, β, γ, κ_1, κ_2) omitted for brevity.

Generalization to Multiple Joints: To validate the generalization capability of the proposed approach, the same identification and validation procedures were applied to all 6 joints of the robotic manipulator. As shown in Fig. 7, individual joints were first tested on a single-joint platform (left), followed by whole-arm validation (right). Table 2 summarizes the identified friction parameters and maximum prediction errors for all joints. The results demonstrate that the maximum prediction error across all joints is 0.50A ($< 2.5\%$ relative error), with wrist joints achieving errors as low as 0.06A, confirming the model's applicability to various harmonic drive configurations. Compared to our previous work [2], which achieved a maximum error of approximately 0.8A under similar test conditions, the proposed two-stage optimization framework with enhanced friction model leads to a notable improvement in prediction accuracy (approximately 38% error reduction), demonstrating the effectiveness of incorporating physical constraints and comprehensive friction factors into the identification process.

After individual friction identification, the complete manipulator dynamics can be identified using standard inverse dynamics methods. The decoupled nature of the friction model allows it to be seamlessly integrated with other dynamic parameters, enabling accurate whole-arm dynamic modeling.

4.3 Long-Term Stability and Model Limitations

Long-Term Stability and Whole-Arm Re-identification. To address friction parameter drift due to wear and lubricant aging, we propose a practical whole-arm re-identification strategy. A key advantage of the proposed friction modeling approach is that the friction parameters are decoupled from the manipulator's dynamic parameters (inertia, mass, center of gravity). This decoupling allows friction parameters to be re-identified directly on the assembled robot without affecting the dynamic model.

When component wear or lubricant degradation causes model accuracy to degrade after prolonged operation, a brief whole-arm re-identification can be executed using the platform shown in Fig. 7 (right). The robot performs predefined excitation trajectories that simultaneously excite all joints. The measured motor currents from all joints are used to update the friction parameters using the same two-stage optimization framework. This process typically takes 10–15 min and can be scheduled during routine maintenance intervals.

Laboratory validation was performed on a manipulator operated for 500 h. Initially, the maximum friction prediction error was 0.50A across all joints. After 500 h, the error increased to 0.8A due to wear-induced parameter drift. Following the whole-arm re-identification procedure, the maximum error was restored to 0.52A, demonstrating the effectiveness of this approach in maintaining long-term model accuracy without requiring component disassembly or replacement.

Model Limitations. The friction model's accuracy depends on the operating conditions during identification. In this study, parameters were identified at temperatures of –10°C to 60°C. Operating significantly outside this range (e.g., –35°C or 70°C) may reduce accuracy due to unmodeled nonlinear behaviors in lubricant viscosity and material properties. Similarly, extrapolation beyond the validated velocity and load ranges may degrade performance. For extreme environments, additional identification experiments at representative conditions are recommended, or multiple parameter sets for different environmental zones can be stored and selected based on real-time measurements.

5 Conclusion

This paper addressed the critical challenge of accurately modeling the friction in harmonic drives for robotic joints by proposing a systematic solution based on multi-objective optimization. By constructing a decoupled model that comprehensively considers the effects of velocity, temperature, and load, and by innovatively designing a two-step multi-objective optimization framework that guarantees the physical significance of the parameters, this work successfully overcomes the limitations of traditional methods in terms of model comprehensiveness and parameter reliability. The effectiveness of the proposed methodology was thoroughly validated through experiments, demonstrating that the established model achieves high accuracy and strong robustness under dynamic operating conditions. Compared to our prior work [2], the proposed approach reduces friction prediction error by approximately 38%, demonstrating significant improvements in both model accuracy and parameter reliability through the integration of comprehensive friction factors and physical constraint-based optimization.

The main contributions of this research are twofold: (1) It provides a complete and systematic workflow for friction modeling, from multi-factor characterization to high-reliability parameter identification. (2) The proposed "optimize the feasible domain" identification concept offers a new paradigm for tackling parameter identification problems in other complex engineering systems. This work not only provides a solid theoretical foundation and technical support for enhancing the motion control performance of robots but also holds significant reference value for the performance optimization of other electromechanical systems that incorporate precision transmission components.

While the proposed model demonstrates high accuracy, there remain avenues for further improvement. Future research will focus on the following directions:

1. *Exploration of More Advanced Friction Models:* Future work will involve investigating more sophisticated models that capture finer physical details, such as contact mechanics and lubrication states (e.g., elasto-hydrodynamic lubrication), to further enhance the model's robustness and applicability across an even wider range of operating conditions.
2. *Development of Efficient Online Identification Methods:* The current parameter identification method, while accurate, is computationally intensive and performed offline. Future research will explore more efficient identification algorithms, such as those based on adaptive control or machine learning, to enable the rapid and online identification of parameters. This would allow the model to adapt in real-time to wear, lubricant degradation, and other time-varying effects.
3. *Experimental Validation of Friction Compensation Control:* Based on the high-fidelity friction model developed in this work, we will design and implement advanced control strategies, such as feedforward compensation and model-based drag-to-teach controllers. Rigorous real-world experiments will be conducted on a robotic manipulator to quantitatively evaluate the model's effectiveness in improving trajectory tracking accuracy, motion smoothness, and the intuitiveness of human-robot interaction.

Acknowledgments. The authors would like to thank the editors and anonymous reviewers for their valuable comments and suggestions.

References

1. Gandhi, P.S., Ghorbel, F.H., Dabney, J.: Modeling, identification, and compensation of friction in harmonic drives. In: Proceedings of the 41st IEEE Conference on Decision and Control, vol. 1, pp. 160–166. IEEE (2002)
2. Li, Z., Wei, H., Liu, C., et al.: An improved iterative approach with a comprehensive friction model for identifying dynamic parameters of collaborative robots. Robotica **42**(5), 1500–1522 (2024)
3. Dong, F., He, J., Zhao, X., et al.: Analytical modeling method for joint friction in harmonic drive robot. J. Mech. Sci. Technol. **38**, 1–11 (2024)
4. Van Geffen, V.: A study of friction models and friction compensation (2009)
5. Hauschild, J.P., Heppler, G., McPhee, J.: Friction compensation of harmonic drive actuators. In: Proceedings of 6th International Conference on Dynamics and Control of Systems and Structures in Space, pp. 683–692 (2004)
6. Zhang, M., Xiong, Y., Xiao, X., et al.: Modeling and identification of friction torque in harmonic drive systems. In: 2023 IEEE 18th Conference on Industrial Electronics and Applications (ICIEA), pp. 745–752. IEEE (2023)
7. Kennedy, C.W., Desai, J.P.: Estimation and modeling of the harmonic drive transmission in the Mitsubishi PA-10 robot arm. In: Proceedings 2003 IEEE/RSJ International Conference on Intelligent Robots and Systems (IROS 2003), vol. 4, pp. 3331–3336. IEEE (2003)
8. Bona, B., Indri, M.: Friction compensation in robotics: an overview. In: Proceedings of the 44th IEEE Conference on Decision and Control, pp. 4360–4367. IEEE (2005)
9. Simoni, L., Beschi, M., Legnani, G., et al.: Modelling the temperature in joint friction of industrial manipulators. Robotica **37**(5), 906–927 (2019)
10. Kermani, M.R., Patel, R.V., Moallem, M.: Friction identification and compensation in robotic manipulators. IEEE Trans. Instrum. Meas. **56**(6), 2346–2353 (2007)
11. Kim, M.J., Beck, F., Ott, C., et al.: Model-free friction observers for flexible joint robots with torque measurements. IEEE Trans. Rob. **35**(6), 1508–1515 (2019)
12. Wang, Q., Wu, H., Cheng, Y., et al.: Friction identification of harmonic drive joints based on the MCMC method. IEEE Access **10**, 125893–125907 (2022)
13. Morel, G., Iagnemma, K., Dubowsky, S.: The precise control of manipulators with high joint-friction using base force/torque sensing. Automatica **36**(7), 931–941 (2000)
14. Le Tien, L., Albu-Schaffer, A., De Luca, A., et al.: Friction observer and compensation for control of robots with joint torque measurement. In: 2008 IEEE/RSJ International Conference on Intelligent Robots and Systems, pp. 3789–3795. IEEE (2008)
15. Wu, J., Schuchert, P.L., Karimi, A.: Data-driven modeling and analysis of transmission error in harmonic drive systems: nonlinear dynamics, error modeling, and compensation techniques. arXiv preprint arXiv:2310.15875 (2023)

Improved MPC Control Combining LQR Control in Extra-Large Excavators Trajectory Tracking

Molei Peng, Xu Yang, Yunhua Li(✉), Liman Yang, and Yuxia Li

School of Automation Science and Electrical Engineering, Beihang University, 37 Xueyuan Road, Beijing 100190, China
yhli@buaa.edu.cn

Abstract. This paper addresses the critical trajectory tracking control problem faced by extra-large mining excavators during their intelligent transformation. Conventional non-coupled control methods are inadequate due to the strong coupling between the crowding and hoisting systems, which limits the potential for higher tracking accuracy. To overcome this, the study first establishes a linearized error state equation for the bucket tooth tip trajectory. Based on this model, an improved MPC (Model Predictive Control) algorithm is proposed, which integrates the advantages of MPC and the LQR (Linear Quadratic Regulator). The key innovation involves introducing the LQR Riccati equation iteration into the ADMM framework to replace gradient descent, thereby significantly accelerating the solving speed of the control law while maintaining the constraint-handling capability of MPC. Simulation results, comparing the proposed method with classic LQR control, validate its effectiveness.

Keywords: Extra-large Mining Excavators · Strong Coupling · System Optimal Control · LQR · Improved MPC

1 Introduction

Mineral resources are important materials for national development, and open-pit mining is a crucial way to ensure their supply. Currently, in the field of open-pit mining, extra-large mining excavators are the core equipment. However, their overall excavation efficiency highly depends on the skill level of the operators [1–3]. Working in harsh environments for long periods also poses health risks to workers. With the acceleration of population aging, the number of relevant practitioners is continuously decreasing. Coupled with the national call for green mining and intelligence, the traditional mining method urgently needs to be transformed, requiring excavators to become more intelligent [4–7]. The intelligence of extra-large mining excavators mainly involves unmanned excavation, unmanned driving, and automatic loading [8–11]. Among these, unmanned excavation involves the cooperative movement between the crowding mechanism and the hoisting mechanism to ensure optimal excavation energy consumption and efficiency.

H. Wei and J. Niu (Eds.): COBOT 2025, CCIS 2893, pp. 390–403, 2027.
https://doi.org/10.1007/978-981-92-2074-8_23

The tracking control of the optimal excavation trajectory is the foundation for achieving optimal excavation, and it is also the main problem this paper attempts to optimize. Many control schemes have been proposed in related fields for the excavation control of excavators.

The traditional control method involves planning the crowding and hoisting trajectories, and then applying relevant control algorithms to track these desired trajectories separately [12–14]. This method is relatively simple and easy to implement. However, the crowding and hoisting systems are coupled and need to cooperate to track the excavation trajectory. Under these conditions, the non-coupled control method mentioned above, where each control loop is independent, is difficult to apply in scenarios requiring high control precision and high coordination. Therefore, one of the effective ways to solve this problem is coupled control [15, 16]. Coupled control methods mainly include cross-coupling control, deviation-coupling control, and adjacent cross-coupling control.

The excavation trajectory is jointly determined by the output displacements of multiple actuators. When the output error of a certain actuator is large, the accuracy will be greatly affected even if the tracking errors of the remaining actuators are very small. Therefore, to improve trajectory tracking accuracy, Lee et al. from Korea University introduced the cross-coupling control theory into excavator trajectory control [17]. This method divides the tooth tip trajectory error e into errors and along the x and y axes, then calculates the contour error based on the error components. Finally, based on the velocity Jacobian matrix relationship, the contour error is decomposed to the boom and the stick, improving the tracking accuracy of the tooth tip trajectory. However, this method is only suitable for straight-line trajectories when the boom and stick move simultaneously.

Domestically, Wang Dongyun [18] of Zhejiang Normal University and Feng Peien [19] of Zhejiang University also applied the cross-coupling theory to the trajectory tracking control of multi-actuator motion in excavators and used the contour error to compensate the desired trajectory. Lu Wenru [20] proposed a PD cross-coupling control strategy based on a fractional-order iterative sliding mode position controller for the synchronous control problem of multi-joint robot arms. To address some drawbacks of the traditional adjacent cross-coupling synchronous control method, such as slow response speed and poor dynamic performance, Zou Yigang proposed a two-stage synchronous error compensation control method [21], which has both coarse and fine adjustment functions. For an axis with a small load, the compensation coefficient is set to a larger value because of its faster response speed. Conversely, for a heavily loaded axis, the compensation coefficient is set to a smaller value. The coarse adjustment module directly compensates the position command with the position compensation amount. The fine adjustment module uses an adaptive neural network to output a velocity compensation signal. This eliminates the steady-state error between adjacent axes and improves multi-axis synchronization accuracy. However, the trajectory cooperative tracking control of extra-large mining excavators is essentially similar to the synchronous control of multiple mechanisms. If only the tracking accuracy of a single joint is improved, the improvement in overall tracking accuracy is limited. The use of relevant algorithms to enhance system coordination [22] can significantly improve trajectory tracking accuracy.

Optimal control methods based on state transition equations have good control effects for this type of cooperative control problem. Traditional LQR (Linear Quadratic Regulator) control assumes a linear relationship between the system input vector and the error vector, thus enabling fast and stable solutions. However, since this control algorithm is derived based on the variational method, it cannot be easily subjected to hard constraints, and the optimal control command obtained may have significant jitter. MPC (Model Predictive Control), as a look-ahead control algorithm, can achieve smoother control by adding constraints when solving for the control command optimization [23]. However, its solution is slow. If the prediction step is too long, the entire solution may fail to converge. Therefore, this paper combines the advantages of LQR and MPC to propose an improved MPC optimal control method. It models the error state equation for the excavator's excavation trajectory tracking control process and applies this method to the tracking control of the excavation process. Finally, simulations are performed to verify the effectiveness of the control algorithm.

2 Control Model of the Excavation Mechanism

2.1 Modeling and Analysis of Boom and Bucket Movement

The main objects analyzed during the entire excavation process of the extra-large mining excavator are the boom and the bucket. In this paper, the motion trajectory of the tooth tip A of the bucket is taken as the research object to represent the excavation trajectory. As is shown in Fig. 1.

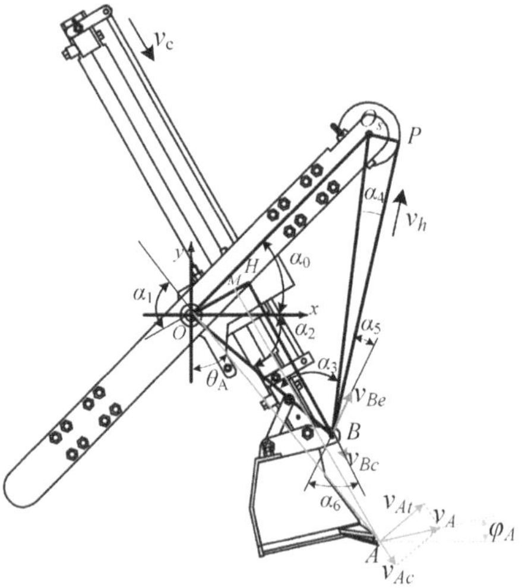

Fig. 1. Boom and bucket kinematic analysis

The coordinates of point A during motion can be described as follows:

$$\begin{cases} x_A = l_{OA} \sin \theta_A \\ y_A = l_{OA} \cos \theta_A \end{cases} \tag{1}$$

where l_{OA} represents the distance from the tooth tip A to the rotation point O where the bucket and boom contact. θ_{A} is the angle between OA and the vertical direction. Throughout the entire operation, the movement of the bucket is mainly determined by the crowding motor and the hoisting motor. Their speeds are denoted as the crowding speed v_c and the hoisting speed v_h, respectively. Thus, the angular velocity of OA rotating around point O is defined as:

$$\omega = \frac{v_{\mathrm{h}} - v_{\mathrm{c}} \cos \alpha_6}{l_{OB} \cos \alpha_5} \tag{2}$$

The three unknown variables l_{OB}, α_5, and α_6 introduced in the above equation can be calculated by the following equations:

$$l_{OB} = \sqrt{l_{OH}^2 + (l_{BH_0} + \int v_{\mathrm{c}} \mathrm{d}t)^2} \tag{3}$$

$$\alpha_1 = \arctan(\frac{l_{BH_0} + \int v_{\mathrm{c}} \mathrm{d}t}{l_{OH}}) \tag{4}$$

$$\alpha_2 = \alpha_1 - \theta_1 \tag{5}$$

$$l_{O_sB} = \sqrt{l_{O_sO}^2 + l_{OB}^2 - 2 l_{O_sO} l_{OB} \cos(\alpha_0 + \alpha_2)} \tag{6}$$

$$\alpha_3 = \arccos(\frac{l_{O_sB}^2 + l_{OB}^2 - l_{O_sO}^2}{2 l_{O_sB} l_{OB}}) \tag{7}$$

$$\alpha_4 = \arcsin(\frac{R_s}{l_{O_sB}}) \tag{8}$$

$$\alpha_5 = \frac{\pi}{2} - \alpha_3 - \alpha_4 \tag{9}$$

$$\alpha_6 = \alpha_1 - \alpha_5 \tag{10}$$

Then, the expressions for the coordinates of the bucket tooth tip point A (i.e., the excavation trajectory) concerning the motor speed inputs are as follows:

$$l_{OA} = \sqrt{l_{OM}^2 + (l_{AM0} + \int_0^t v_c \mathrm{d}t)^2} \tag{11}$$

$$\theta_A = \theta_{A0} + \int_0^t \omega \mathrm{d}t \tag{12}$$

2.2 Establishment of the Excavation Tracking Control Error State Equation

The expressions for the pre-calculated variables have been obtained. Define the system state vector $X = [x_A, y_A, \varphi_A]^T$ and the system input vector $U = [v_c, v_h]^T$. φ_A is the angle between absolute speed of point A and the horizontal direction. State equation is expressed as:

$$\dot{X}=f(X,U)=\begin{bmatrix}\dot{x}_A\\ \dot{y}_A\\ \dot{\varphi}_A\end{bmatrix}=\begin{bmatrix}v_A\cdot\cos\varphi_A\\ v_A\cdot\sin\varphi_A\\ \omega-d[\arccos(v_{Ac}/v_A)]/dt\end{bmatrix} \tag{13}$$

In practical analysis, the relative speed of point A is considered to be much smaller compared to the crowding speed and the absolute speed. Therefore, it is approximated that the relative speed is perpendicular to the absolute speed, i.e., it is assumed that $\dot{\varphi}_A = \omega - d[\arctan(v_{Ac}/v_A)]/dt$. Hence, the above state equation is rewritten as:

$$f(X,U)=\begin{bmatrix}\dot{x}_A\\ \dot{y}_A\\ \dot{\varphi}_A\end{bmatrix}=\begin{bmatrix}v_A\cdot\cos\varphi_A\\ v_A\cdot\sin\varphi_A\\ \omega-(\dot{v}_{Ac}v_A-\dot{v}_Av_{Ac})/(v_{Ac}^2+v_A^2)\end{bmatrix} \tag{14}$$

From this state equation, a first-order Taylor expansion can be performed at any tracking point, and the high-order terms can be ignored. The subscript r denotes the reference point on the trajectory to be tracked, resulting in the linearized error state Eq. (15) and reforming as Eq. (16), e_X and e_U is error state vector of U and X, respectively:

$$f(X,U)=\dot{X}_r+\frac{\partial f(X,U)}{\partial X}(X-X_r)+\frac{\partial f(X,U)}{\partial U}(U-U_r) \tag{15}$$

$$\dot{e}_X=\frac{\partial f(X,U)}{\partial X}e_X+\frac{\partial f(X,U)}{\partial U}e_U \tag{16}$$

The partial derivative matrices in the equation are calculated as follows:

$$\frac{\partial f(X,U)}{\partial X}=\begin{bmatrix}0\;0 & -v_A\cdot\sin\varphi_A\\ 0\;0 & v_A\cdot\cos\varphi_A\\ 0\;0 & 0\end{bmatrix} \tag{17}$$

If the state equation is denoted as $f(X,U)=[f_1,f_2,f_3]^T$, then:

$$\frac{\partial f(X,U)}{\partial U}=\begin{bmatrix}\frac{\partial f_1}{\partial v_c} & \frac{\partial f_2}{\partial v_c} & \frac{\partial f_3}{\partial v_c}\\ \frac{\partial f_1}{\partial v_h} & \frac{\partial f_2}{\partial v_h} & \frac{\partial f_3}{\partial v_h}\end{bmatrix}^T \tag{18}$$

Thus, a linearized error state equation has been established.

3 Improved Model Predictive Control Combining LQR Control

3.1 Trajectory Tracking Based on LQR Control

Applying the LQR control method to the excavator control in this paper requires the discretization of the error state equation. Integrating the error state vector using the midpoint Euler method and the input vector using the forward Euler method yields. Using t_s to denote the discrete step(simulation step) yields the discrete state equation as follows:

$$A_t = \frac{\partial f(X, U)}{\partial X}, B_t = \frac{\partial f(X, U)}{\partial U} \tag{19}$$

$$\begin{cases} A = (I - \frac{A_t t_s}{2})^{-1}(I + \frac{A_t t_s}{2}) \\ B = B_t t_s \end{cases} \tag{20}$$

$$\overline{X}(k+1) = e_X(k+1) = Ae_X(k) + Be_U(k) = A\overline{X}(k) + B\overline{U}(k) \tag{21}$$

For the quadratic programming problem, LQR is concerned that the state tracking error during the entire tracking process is not excessively large, ensuring tracking performance. It is also concerned that the system input is not too large, avoiding severe system oscillation. Optimization objective of the discretized linear quadratic optimal problem can be described as follows:

$$\begin{aligned} J &= \frac{1}{2}\overline{X}^T(N)Q\overline{X}(N) + \sum_{k=0}^{N-1}\{\frac{1}{2}\overline{X}^T(k)Q\overline{X}(k) + \frac{1}{2}\overline{U}^T(k)R\overline{U}(k)\} \\ &\text{s.t.}\overline{X}(k+1) = A\overline{X}(k) + B\overline{U}(k) \end{aligned} \tag{22}$$

Since solving the above problem encounters the difficult two-point boundary value problem, a linear assumption is made that the input vector and the state vector are linearly related over the entire time index, namely:

$$\overline{U}(k) = -K(k)\overline{X}(k) \tag{23}$$

The optimal gain matrix $K(k)$ is obtained by iterating backwards from the terminal time using the Riccati equation:

$$\begin{cases} K(k) = [R + B^T P(k+1)B]^{-1}[B^T P(k+1)A] \\ P(k) = Q + K^T(k)RK(k) + [A - BK(k)]^T P(k+1)[A - BK(k)] \end{cases} \tag{24}$$

The control flow diagram for applying the linear quadratic regulator to this excavator is shown in Fig. 2. Information such as angle and displacement is obtained from the controlled excavator using sensors. Comparing the reference state vector and real state vector can get the error state vector. Multiplying this by the optimal feedback matrix solved by the Riccati equation gives the error input vector. Adding this to the planned ideal input vector gives the actual control output to the controlled excavator. The above control scheme provides a foundation for the subsequent improved MPC.

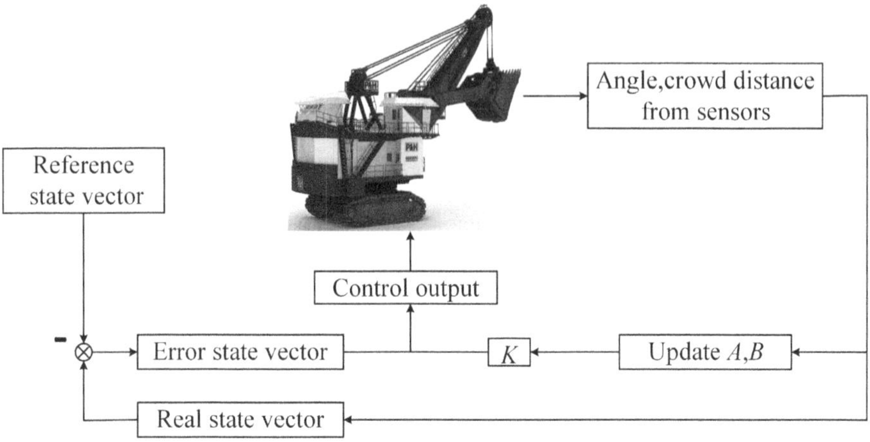

Fig. 2. LQR Control Scheme

3.2 Improved MPC Control Method

To prevent large oscillations in the input, the LQR method requires adjusting the weight parameters of the R matrix. For the overall control effect, this can only be a soft constraint, and the solved control command may still exhibit abrupt changes. Therefore, this paper proposes combining MPC with LQR for constrained control command solving.

A mainstream approach for existing MPC problems is ADMM (Alternating Direction Method of Multipliers). The basic steps for ADMM iterative updates are three-fold: updating the primal variable, updating the slack variable, and updating the dual variable:

$$x^+ = \underset{x}{\arg\min}\, L(x, z, \lambda) \tag{25}$$

$$z^+ = \underset{z}{\arg\min}\, L(x^+, z, \lambda) \tag{26}$$

$$\lambda^+ = \lambda + \rho(x^+ - z^+) \tag{27}$$

The update of the primal variable and the slack variable typically uses gradient descent, while the update of the dual variable using dual ascent can be directly calculated. The contradiction of the MPC algorithm lies here: when the prediction step is long, the control effect is better. However, for gradient descent, if the prediction step is too long, the dimensions of the primal and slack variables become very large, making convergence difficult for gradient descent. Conversely, when the prediction step is short, the gradient descent solution is fast, but the control effect is somewhat compromised.

The objective function of the improved MPC control algorithm proposed in this paper is as follow: by introducing linear terms and slack variables into the objective function of the linear quadratic regulator, the optimization objective is transformed into the following form:

$$\begin{aligned}
\min J &= \frac{1}{2}\overline{X}(N)^T Q(N)\overline{X}(N) + q_f^T\overline{X}(N) \\
&+ \sum_{k=0}^{N-1}\left[\frac{1}{2}\overline{X}(k)^T Q\overline{X}(k) + q_k^T\overline{X}(k) + \frac{1}{2}\overline{U}(k)^T R\overline{U}(k) + r_k^T\overline{U}(k)\right] + I_x(Z) + I_u(W) \\
&+ \sum_{k=0}^{N-1}\left[\lambda_k^T\left(\overline{X}(k) - Z(k)\right) + \frac{\rho}{2}\left(\overline{X}(k) - Z(k)\right)^T\left(\overline{X}(k) - Z(k)\right)\right] \\
&+ \sum_{k=0}^{N-1}\left[\mu_k^T\left(\overline{U}(k) - W(k)\right) + \frac{\rho}{2}\left(\overline{U}(k) - W(k)\right)^T\left(\overline{U}(k) - W(k)\right)\right]
\end{aligned} \tag{28}$$

Mapping the above optimization problem to the standard form of the LQR problem yields the new matrices as follows:

$$\begin{cases}
Q_{f'} = Q_f + \rho I \\
Q' = Q + \rho I \\
R' = R + \rho I \\
q_{f'} = q_f + \lambda_N - \rho z_N = q_f + \rho(y_N - z_N) \\
q_{k'} = q_k + \lambda_k - \rho z_k = q_k + \rho(y_k - z_k) \\
r_{k'} = r_k + \mu_k - \rho w_k = r_k + \rho(g_k - w_k)
\end{cases} \tag{29}$$

The corresponding Riccati equation iteration introduces a linear term, and the equation form is as follows:

$$\begin{cases}
K(k) = [R + B^T P(k+1)B]^{-1}[B^T P(k+1)A] \\
d(k) = [R + B^T P(k+1)B]^{-1}[B^T p(k+1) + r(k)] \\
P(k) = Q + K^T(k)RK(k) + [A - BK(k)]^T P(k+1)[A - BK(k)] \\
p(k+1) = q(k) + [A - BK(k)]^T\left[p(k+1) - P(k+1)Bd(k)\right] \\
\qquad + K^T(k)[Rd(k) - r(k)]
\end{cases} \tag{30}$$

The MPC algorithm flow combined with Riccati equation iteration is as follows:

(1) Pre-calculate the quadratic term part and store it:

$$\begin{aligned}
K(k) &= [R' + B^T P(k+1)B]^{-1}[B^T P(k+1)A] \\
P(k) &= Q' + K^T(k)R'K(k) + [A - BK(k)]^T P(k+1)[A - BK(k)]
\end{aligned} \tag{31}$$

For the rolling optimization of MPC, the A, B, Q, R are constant matrices at each optimization step. Therefore, the above iterative pre-calculation can be repeatedly computed until a stable convergence solution, denoted as K and P, is obtained.

(2) **Update the iteration** Update the linear term coefficient matrix as follows:

$$\begin{cases} d_k = (R + B^T PB)^{-1}(B^T p_{k+1} + r_k) \\ p_k = q_k + (A - BK)^T (p_{k+1} - PBd_k) + K^T (Rd_k - r_k) \end{cases} \quad (32)$$

The updates for the primal variable, the slack variable, and the dual variable are transformed into the following forms:

$$\begin{cases} \overline{X}(k) = A\overline{X}(k-1) + B\overline{U}(k-1) \\ u_k^* = -K\overline{X}(k) - d_k \\ z_k^+ = \text{proj}_X(x_k^+ + y_k) \\ w_k^+ = \text{proj}_U(u_k^+ + g_k) \\ y_k^+ = y_k + x_k^+ - z_k^+ \\ g_k^+ = g_k + u_k^+ - w_k^+ \end{cases} \quad (33)$$

4 Simulation

4.1 Verification of the Approximation of the State Error Equation

The length data for various parts of the excavator boom and bucket shown in Fig. 1 are listed in Table 1. First, the rationality of the approximation used for the error state equation in this paper is verified. The system inputs (i.e., the crowding motor speed and hoisting motor speed) corresponding to the trajectory to be tracked are compared: (1) directly input into the model, where the change rate of the state vector is obtained by differencing the trajectory of the bucket tooth tip point A and dividing by the simulation period, and (2) input into the approximated state equation. Specifically, define ODh as the projection length of OD on the y-axis.

Table 1. Parameters of excavator in simulation

Parameters	Value
OH	1.7244 m
OM	1.51 m
OOs	10.847 m
Rs	1.15 m
ODh	9.7 m
α_0	$\frac{7}{30}\pi$ rad
AB	2.6 m

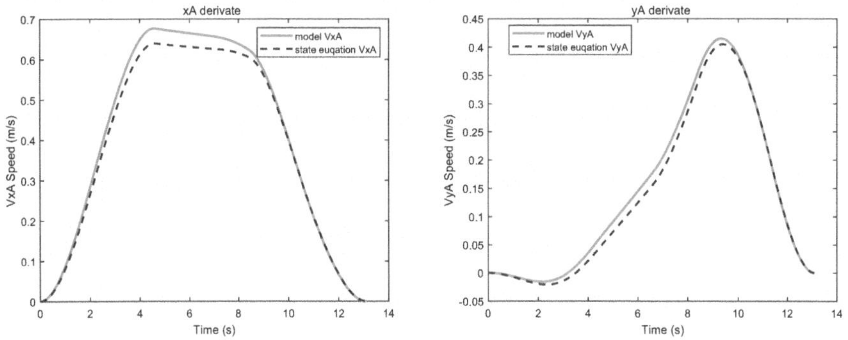

(a) (b)

Fig. 3. Comparison between error state equation and simulation model states (a) comparison of the change rate of the x-coordinate between point A in the model and point A in the state equation (b) comparison of the change rate of the y-coordinate between point A in the model and point A in the state equation

The comparison of the two is shown in Fig. 3. It can be seen that the trajectory deviation is basically small, which indicates that the approximation made in this paper is reasonable.

4.2 Simulation Comparison of Control Effects Between LQR and Improved MPC for Trajectory Tracking

In the simulation experiments, the observed input quantities are the speed values of the crowding motor and hoisting motor corresponding to the trajectory to be tracked. The various parameters of the relevant control algorithms are shown in Table 2(q_i and r_i in the table mainly refer to the elements on the diagonal of the weight matrices). The simulation control frequency in this paper is 100 Hz, and the improved MPC control limits the motor speed input to no more than 1 m/s.

Table 2. Control algorithm parameters

Main parameters	LQR control	Improved MPC control
q_1	1.4	0.5
q_2	1.4	0.5
q_3	1.5	1.6
r_1	1.2	10
r_2	1.4	8
predict_step	none	20

Observing the refresh frequency during the simulation, the LQR solution time is short, which can meet the demands of higher frequency control command output. The following curves for v_c and v_h show larger oscillations in three intervals in Fig. 4. The overall excavation trajectory curve is slightly shifted upwards in the first half, and the following is closer in the second half.

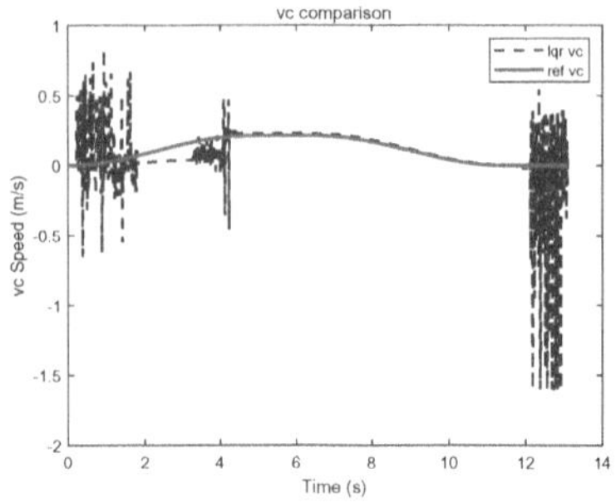

(a)

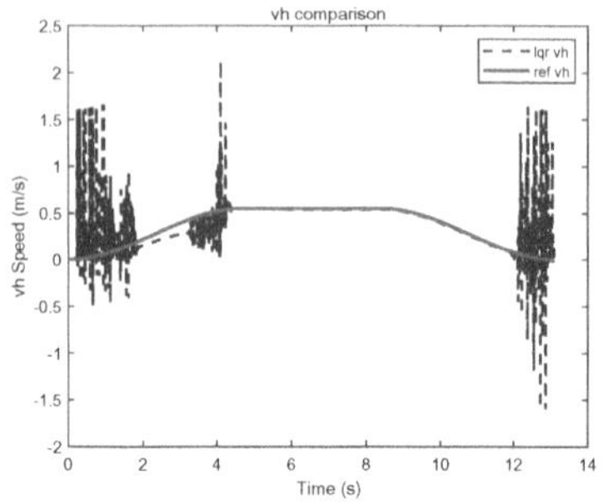

(b)

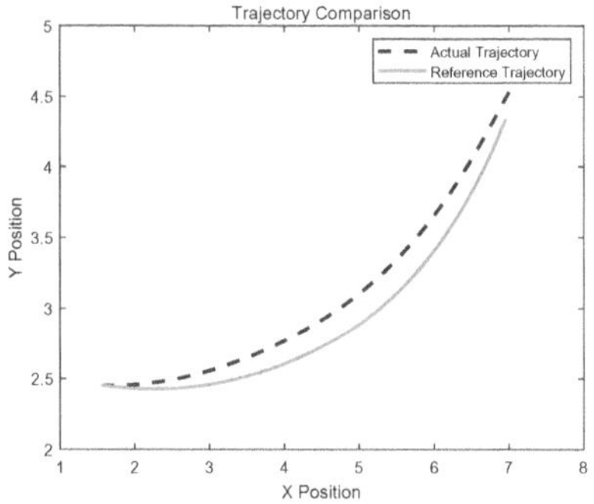

(c)

Fig. 4. Simulation of excavator motion under LQR control (a) crowding speed comparison by LQR control (b) hoisting speed comparison by LQR control (c) trajectory comparison by LQR control

The improved MPC control shows better tracking results for the v_c and v_h following curves, and the output command is relatively smooth in Fig. 5. The current prediction step is set to 20. The tracking effect will be further improved by increasing the prediction step. Compared to the ADMM solution of traditional MPC, the improved MPC solution speed is faster, as observed from the simulation refresh frequency.

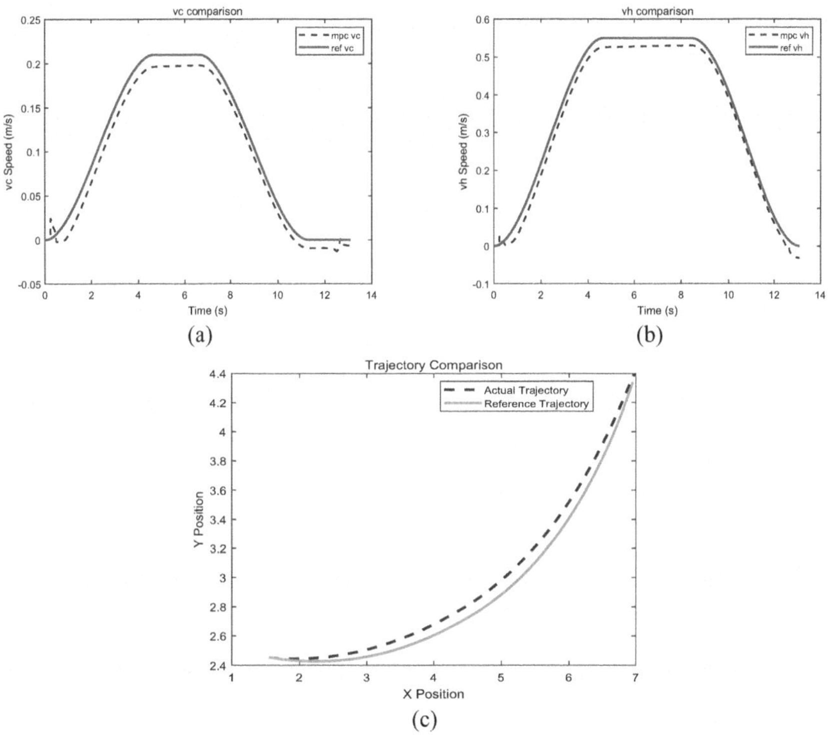

Fig. 5. Simulation of Excavator Motion by Improved MPC control (a) crowding speed comparison by Improved MPC control (b) hoisting speed comparison by Improved MPC control (c) trajectory comparison by Improved MPC control

5 Conclusion

This paper conducted an in-depth study on the trajectory tracking control problem faced by extra-large mining excavators during the intelligent transformation process. Traditional non-coupled control methods struggle to handle the strong coupling relationship between the crowding and hoisting systems, limiting the further improvement of trajectory tracking accuracy. To address this, this paper established the excavator bucket tooth tip trajectory error state equation and proposed an improved Model Predictive Control (MPC) algorithm that combines the advantages of LQR and MPC. Through simulation experiments, and by comparing with classic LQR control, the effectiveness of the proposed control algorithm was verified.

Acknowledgments. This work is supported by NSFC-Shanxi Coal-based Low Carbon Joint Fund (U1910211).

Disclosure of Interests. The authors have no competing interests to declare that are relevant to the content of this article.

References

1. Velikanov, V.S., Dyorina, N., Rabina, E.I.: Interfacing between the control ratio of mining excavator and an Operator'S proficiency level. Les Ulis: EDP Sci. **298**, 00062 (2019)
2. Zhang, L., Zhao, J., Long, P.: An autonomous excavator system for material loading task. Sci. Robot. **6**(55), eabc3164 (2021)
3. Chen, G., Wang, Y., Li, X.: Shovel point optimization for unmanned loader based on pile reconstruction. Comput.-Aided Civ. Infrastruct. Eng. **39**(14), 2187–2203 (2024)
4. Wang, K., Sun, Q., Song, W., Xu, D., Lin, D.: Recent progress in intelligent perception and control of excavators. In: 2025 6th International Conference on Electrical Technology and Automatic Control (ICETAC), pp. 642–647, Nanjing, China (2025)
5. Lee, J., Park, S.Y.: 3D pose recognition system of dump truck for autonomous excavator. Appl. Sci. **12**(7), 3471 (2022)
6. Phillips, T.G., McAree, P.R.: An evidence-based approach to object pose estimation from LiDAR measurements in challenging environments. J. Field Robot. **35**(6), 921–936 (2018)
7. Stentz, A., Bares, J., Singh, S.: A robotic excavator for autonomous truck loading. Auton. Robot. **7**(13), 175–186 (1999)
8. Yamamoto, H., Moteki, M., Shao, H., Ootuki, T.: Development of autonomous excavation technology for hydraulic excavators. In: ICCAS-SICE, pp. 2226–2229, Fukuoka, Japan (2009)
9. Chai, Z.: Application analysis of artificial intelligence technology in unmanned driving. In: 2024 International Conference on Computers, Information Processing and Advanced Education (CIPAE), pp. 405–409, Ottawa, ON, Canada (2024)
10. Rui, S., Youchun, Y., Yi, S., Ke, Z., Bo, D.: Design of automatic loading and unloading system for freight car handling based on intelligent control. In: 2020 12th International Conference on Measuring Technology and Mechatronics Automation (ICMTMA), pp. 425–431, Phuket, Thailand (2020)
11. Green, M.E., Ridley, A.N., McAree, P.R.: Pose verification for autonomous equipment interaction in surface mining. In: 2013 IEEE/ASME International Conference on Advanced Intelligent Mechatronics, pp. 1199–1204. IEEE (2013)
12. Huang, H., Tu, Q., Jiang, C., et al.: An electronic line-shafting control strategy based on sliding mode observer for distributed driving electric vehicles. IEEE Access **9**, 38221–38235 (2021)
13. Gao, Y., Ai, Y., Tian, B.: Parallel end-to-end autonomous mining: an IoT-oriented approach. IEEE Internet Things J. **7**(2), 1011–1023 (2019)
14. Azar, E.R., Kamat, V.R.: Earthmoving equipment automation: a review of technical advances and future outlook. J. Inf. Technol. Constr. **22**(13), 247–265 (2016)
15. Li, X., Yang, Y., He, Q., Cheng, Z., Wang, Y., Liu, L.: Attitude coupling control and lateral stability analysis of hypersonic vehicle. In: 2020 Chinese Control And Decision Conference (CCDC), pp. 3764–3769, Hefei, China (2020)
16. Kwon, O.-S., Choe, S.-H., Heo, H.: A study on the dual-servo system using improved cross-coupling control method. In: 2011 10th International Conference on Environment and Electrical Engineering, pp. 1–4, Rome, Italy (2011)
17. Lee, C.S., Bae, J., Hong, D.: Contour control for leveling work with robotic excavator. Int. J. Precis. Eng. Manuf. **14**(12), 2055–2060 (2013)
18. Wang, D., Zheng, L., Yu, H., et al.: Robotic excavator motion control using a nonlinear proportional-integral controller and cross-coupled pre-compensation. Autom. Constr. **64**, 1–6 (2016)
19. Peien, F., Liping, S., Gao, Y., Shuangxia, P.: Research on multi-joint coordinated control of hydraulic excavator robot. Chin. J. Constr. Mach. **3**, 253–258 (2006)

20. Wenru, L.: Research on Synchronous Control Strategy for Multi-joint Manipulator. Lanzhou Jiaotong University, Lanzhou (2021)
21. Yigang, Z.: Research on Multi-axis Synchronous Control Technology Based on Adjacent Cross-Coupling. Huazhong University of Science and Technology, Wuhan (2021)
22. Sun, D., Mills, J.K.: Adaptive synchronized control for coordination of two robot manipulators. In: Proceedings 2002 IEEE International Conference on Robotics and Automation, pp. 976–981 vol.1, Washington, DC, USA (2002)
23. Schwenzer, M., Ay, M., Bergs, T., Abel, D.: Review on model predictive control: an engineering perspective. Int. J. Adv. Manuf. Technol. **117**, 1327–1349 (2021)

PeChSafety-LiViT: A LiDAR–RGB–Thermal Data Acquisition Platform and Benchmark for Petrochemical Safety Monitoring

Bing Chen[1], Xing Lu[2], Tao Xu[3], Honglei Mu[4], Yang Meng[5], Shuoshi Wang[6], Jingjing Liu[6], Jia Jia[6], Ziping Song[1], Pengju Wang[5], and Lei Yang[1](✉)

[1] Kunlun Digital Technology Co., Ltd., Lot A19, PetroChina Innovation Base, Beijing 102206, China
laurencemai@163.com

[2] PetroChina Xinjiang Oilfield Company, No. 36, Yingbin Avenue, Karamay 834000, Xinjiang, China

[3] China National Petroleum Corporation North China Petrochemical Branch, Bei Huan Road, Renqiu 062500, Hebei, China

[4] Institute for Network Science and Cyberspace, Tsinghua University, Haidian District, No. 5, Yiheyuan Road, Beijing 100871, China

[5] Institute of Embodied Intelligence Robotics, School of Mechanical Engineering and Automation, Beihang University, Haidian District, No. 37, Xueyuan Road, Beijing 100191, China
{mengyang_buaa,pengjuwang}@buaa.edu.cn

[6] Beijing Luhang Yilong Intelligent Technology Co., Ltd., No. 35, Huayuan North Road, Haidian District, Beijing 100083, China
{wss,ljj}@ckzone.cn, amber.linux@foxmail.com

Abstract. Petrochemical plants operate for long periods in high-risk environments with high temperature and pressure, flammable and toxic media, and dense pipe racks, where conventional manual inspection cannot reliably ensure personnel safety, inspection frequency, or consistent records. At the same time, intrinsic-safety and confidentiality constraints make publicly available multi-sensor datasets from real plants extremely scarce, hindering reproducible research on robotic safety inspection.We present PeChSafety-LiViT (Petrochemical Safety LiDAR–Visual–Thermal), a plant-scale data acquisition platform and benchmark that integrates a customized inspection robot and a data-collection vehicle, each equipped with 3D LiDAR, a PTU-mounted zoom RGB camera, and a thermal infrared camera. A static stop-and-acquire paradigm and a dedicated registration/annotation toolchain are used to collect tens of thousands of strictly aligned LiDAR–RGB–TIR (LVT) samples in an in-service petrochemical plant and to construct a de-identified, restricted-access benchmark subset with unified task definitions, splits, and metrics.On this benchmark we establish unimodal (RGB), bimodal (RGB + TIR), and trimodal (LiDAR + RGB + TIR) detection baselines for equipment/components, risk symptoms, and personnel/PPE safety behavior. Experiments show that multimodal fusion yields substantial gains over RGB-only models, especially for small objects and challenging lighting and occlusion conditions,

H. Wei and J. Niu (Eds.): COBOT 2025, CCIS 2893, pp. 404–425, 2027.
https://doi.org/10.1007/978-981-92-2074-8_24

providing a realistic reference for future plant-scale multimodal safety inspection research and deployment.

Keywords: Plant-level safety monitoring · multi-sensor fusion · robotic inspection · LiDAR–RGB–TIR · dataset and benchmark

1 Introduction

Petrochemical plants are characterized by the coexistence of high temperature and pressure, flammable and toxic media, and dense pipe racks. Accidents in such environments tend to have severe consequences and are prone to escalation. Current routine inspections, which are predominantly manual, exhibit structural limitations in several aspects:

1. The number of inspection points is large and the spatial hierarchy is complex. It is difficult to consistently guarantee inspection quality at night and under extreme conditions such as glare, backlighting, steam plumes, rain, and fog.
2. Small targets such as valves, flanges, and pressure gauges, as well as their fine-grained states (open/closed, readings, leakage symptoms), rely heavily on operator experience, making records difficult to standardize and trace.
3. As process units scale up and regulations become more stringent, purely manual inspection can no longer satisfy demands for high-frequency inspection and fine-grained data logging.

Driven by the needs for personnel risk isolation, around-the-clock operation, and professional sensing, plant-level robotic safety inspection is becoming an essential technical route. In practice, intrinsically safe or explosion-proof mobile inspection robots have already been deployed in oil & gas and chemical facilities, such as ANYmal X, ExR-2, and Boston Dynamics Spot, typically equipped with gas sensors, thermal cameras, and high-zoom PTZ payloads [1–3]. The core of robotic inspection is reliable perception of complex plant environments. A single sensor is insufficient to cover all operating conditions:

- RGB cameras are prone to insufficient contrast and loss of detail under low illumination or strong reflections.
- Thermal infrared (TIR) cameras can capture temperature differences and thermal anomalies but typically have low spatial resolution. In complex thermal backgrounds with multiple heat sources, they may produce false alarms and have limited ability to capture text or fine structural details.
- LiDAR provides dense geometric structure and range information, facilitating scene mapping, path planning, and safety distance estimation. However, it is weak in representing semantic details such as nameplates, gauge readings, and corrosion textures, and is affected by metallic reflections and occlusions.

Therefore, fusing LiDAR–RGB–TIR has the potential to provide complementary information across geometry, semantics, and thermal physics. LiDAR offers a stable structural and localization backbone, RGB provides visual appearance and semantic details, and TIR highlights temperature anomalies and energy distribution. This combination is particularly well suited to typical plant scenes with cluttered layouts, extreme lighting, and small-scale objects.

In autonomous driving, multimodal fusion has formed a standardized ecosystem under the support of public datasets and evaluation protocols, such as COCO [4], KAIST multispectral and all-day series [5], Seeing Through Fog (STF) [6], and a variety of LiDAR–camera fusion benchmarks [7, 8], allowing methods to be iteratively improved on unified benchmarks. In contrast, petrochemical plants are subject to intrinsic safety (Ex) and confidentiality requirements, resulting in a severe lack of public multimodal data collected in real operating environments. Existing optical gas imaging (OGI) datasets and studies, such as GasVid and MWIRGas-YOLO for methane leak videos [9], Zimmerle et al.'s METEC-based OGI detection limit studies [10] and subsequent QOGI performance evaluations [11], focus on leak detection capability rather than plant-scale multi-object perception. It is difficult to obtain an LVT benchmark that simultaneously provides plant-level semantics, strict cross-modal registration, and unified evaluation protocols. This limitation hinders systematic research and engineering deployment on key challenges such as small risk-related object detection and anomaly identification under complex backgrounds.

To address these issues, we build the PeChSafety-LiViT data platform. The platform comprises a customized inspection robot and a data collection vehicle with identical configurations, both integrating a unified LiDAR–RGB–TIR perception chain and acquisition software stack. At the system level, we design a multimodal synchronous acquisition and data management process adapted to constraints in in-service plants. On top of the full internal dataset, we construct a de-identified, restricted-access public benchmark subset and associated evaluation protocols, and provide unimodal, bimodal, and trimodal baseline models for typical safety inspection tasks.

2 Related Work

To clarify the positioning of our work as a "LiDAR–RGB–TIR plant-scale dataset and platform for petrochemical safety," we review representative public resources from three perspectives and compare them with our proposed platform:

1. Plant/petrochemical safety-related datasets and platforms;
2. Roadside and on-road multimodal/trimodal datasets;
3. Classical RGB–T datasets and task benchmarks.

2.1 Plant/Petrochemical Safety Datasets and Platforms

GasVid and OGI-centric datasets. GasVid is one of the earliest public-described methane leak video datasets for Optical Gas Imaging (OGI), consisting of infrared sequences captured with an OGI camera and annotated leak regions, and is widely used for gas leak detection and classification [9]. More recently, MWIRGas-YOLO demonstrates state-of-the-art leak detection on GasVid and related OGI datasets using real-time deep models [9]. These efforts highlight controlled leak scenarios with detailed, event-level annotations but focus on single thermal modality and do not include joint RGB–LiDAR sampling or plant-level inspection semantics.

METEC OGI efficacy/performance datasets. At Colorado State University's METEC test facility, a series of OGI efficacy and performance studies evaluates conventional OGI tools and emerging QOGI systems under controlled methane releases across different leak rates, backgrounds, and distances [10, 11]. These datasets and reports primarily support LDAR (Leak Detection and Repair) workflows and regulatory analysis, emphasizing "whether a leak event can be detected" and the quantification accuracy of OGI-based systems, but they remain centered on leak detection rather than the multi-object, multi-task perception required for general plant inspection.

Industrial inspection robot platforms (Ex-certified). In recent years, several intrinsically safe or explosion-proof inspection robots have been commercialized for oil & gas and chemical facilities, including ANYbotics ANYmal X [1], ExRobotics ExR-2 [2], and Boston Dynamics Spot equipped with thermal and gas-sensing payloads [3]. These platforms have accumulated large volumes of plant perception data in practical deployments, but such data are typically proprietary. To date, there is no standardized open benchmark with strict LiDAR–RGB–TIR registration, systematic annotations, and unified evaluation protocols derived from these systems.

Summary. Existing plant/petrochemical safety resources either focus on single-modality gas leak detection (GasVid, METEC/QOGI) [9–11] or exist as engineering platforms without standardized open annotations and unified evaluation protocols [1–3]. In contrast, we release a de-identified LiDAR–RGB–TIR trimodal benchmark subset and associated protocols, filling a critical gap at the intersection of plant inspection semantics, strictly registered LVT, and public evaluation.

2.2 Roadside/On-Road Trimodal and Adverse-Weather Multimodal Data

R-LiViT (Roadside LiDAR–RGB–Thermal). R-LiViT is a recent roadside dataset that synchronously acquires and strictly aligns LiDAR, RGB, and TIR data for vulnerable road user (VRU) detection and tracking at intersections [7]. It is one of the few public trimodal (LiDAR–RGB–T) datasets and demonstrates the value of combining geometry, appearance, and thermal information under day/night conditions. However, its semantics and geometry are centered on urban traffic scenes rather than densely structured petrochemical plants.

Seeing Through Fog (STF). STF is a multimodal autonomous driving dataset and benchmark designed for adverse weather such as fog, snow, and rain [6]. It includes LiDAR, RGB, and radar sensors and quantifies the benefits of multimodal fusion for detection under low visibility. Although STF convincingly demonstrates multimodal fusion in adverse environments, it does not use thermal infrared as a mandatory modality and is not designed for plant semantics.

KAIST multispectral all-day and MS2 series. The KAIST all-day/multispectral series and follow-up datasets combine RGB–T images, LiDAR, and GPS–IMU for all-day pedestrian detection and multimodal fusion research in road scenes [5]. MS2 further extends multimodal stereo and fusion scenarios. These works show the potential of combining optical, thermal, geometric, and positional information, but they remain dominated by road scenes and differ significantly from equipment-dense, small-object-dominated plant environments.

Summary. Roadside and on-road multimodal datasets such as R-LiViT [7], STF [6], and KAIST/MS2 [5] validate the benefits of multimodal fusion under low-light and adverse weather conditions and offer mature task and evaluation designs. However, their target types and geometric organization differ substantially from plant inspection, and they lack systematic annotations for fine-grained industrial objects (valves, flanges, instruments). They also do not reflect the statistical characteristics of "equipment/component-level small objects + dense 3D structural background." Our work transfers the advantages of LiDAR–RGB–TIR fusion into a plant-specific object taxonomy and provides reproducible tasks and evaluation protocols tailored to safety inspection.

2.3 Classical RGB–T Datasets and Task Benchmarks

Representative RGB–T datasets include the KAIST Multispectral Pedestrian benchmark [5], FLIR ADAS thermal dataset, LLVIP for low-light visible–infrared paired images [12], M3FD for multispectral detection [13], and tracking-oriented series such as LasHeR [14] and VT821/VT1000/VT5000 [15]. Collectively, they form an ecosystem for multimodal pedestrian detection, RGB–T tracking, and saliency detection, and they systematically validate the value of "visible + thermal" fusion under nighttime and low-light conditions.

However, most RGB–T datasets:

- Are primarily bimodal (RGB–T) and do not incorporate LiDAR for geometric cross-validation;
- Focus on driving or pedestrian scenarios rather than plant-specific semantics;
- Are not designed for the "equipment/component-level small objects + densely structured 3D backgrounds" typical of petrochemical plants.

Under this context, we provide a LiDAR–RGB–TIR trimodal dataset and evaluation protocol aligned with a plant-specific object taxonomy. We emphasize cross-modal spatial consistency, small-object evaluation, and semantic alignment with safety inspection tasks.

3 PeChSafety-LiViT Data Platform

To support research and evaluation of plant-level robotic safety inspection algorithms, the PeChSafety-LiViT data platform (we abbreviate LiDAR–Visual–Thermal as LVT) is designed in a unified manner across four layers: carrier platforms, trimodal sensors, acquisition paradigm, and metadata organization. The platform adopts a static sampling (stop-and-acquire) paradigm and a LiDAR–RGB–TIR trimodal perception chain: LiDAR provides reliable geometric structure; RGB offers high-resolution semantic detail; and TIR provides temperature distribution and thermal anomaly cues.

The platform consists of a customized inspection robot and a data collection vehicle with identical configurations. The two carriers share the same sensor setup and software stack, enabling reuse of acquisition and processing pipelines and smooth transfer of algorithms from "data collection vehicle → robot → in-service plant." A multi-in-one gas sensor is also integrated on the robot for safety monitoring, but it is not included in the current LVT data study and public benchmark.

3.1 Carrier Platforms

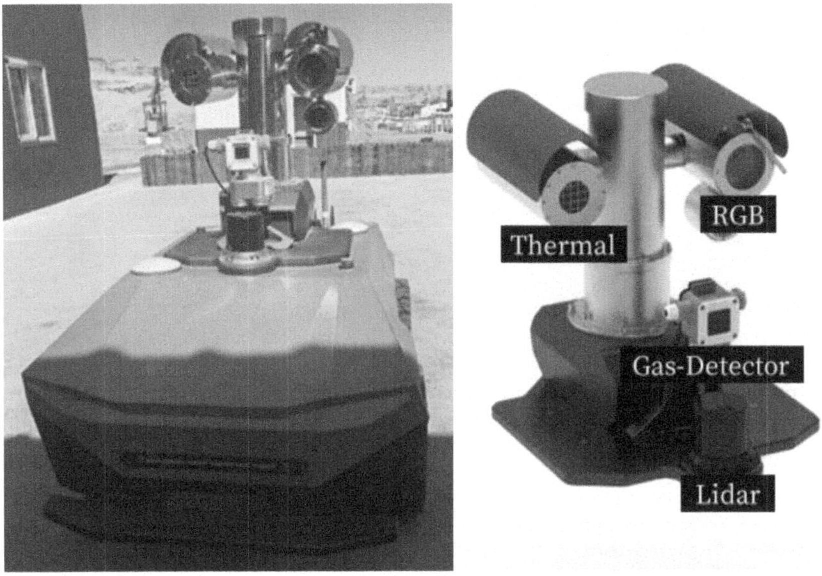

Fig. 1. Inspection robot.

Inspection robot. The inspection robot (Fig. 1) is a deployment-oriented carrier platform that performs autonomous or semi-autonomous inspection tasks in in-service plants. It is equipped with a unified trimodal sensing unit and acquisition software stack and can conduct routine inspections in equipment areas and pipe-rack areas under appropriate safety and work-permit conditions. The multi-in-one gas sensor monitors combustible and toxic gas concentrations in real time, reserving an interface for future "visual + gas" multimodal monitoring.

Data collection vehicle. The data collection vehicle (Fig. 2) shares the same sensors and perception unit as the inspection robot. It is typically manually driven or operated along relatively simple routes to rapidly cover larger plant areas in early development phases. On the one hand, it enables collection of rich training and evaluation data without altering existing production and robotic operation strategies. On the other hand, due to the unified opto-mechanical structure and software interfaces, data collected by the vehicle can be seamlessly transferred to robot-side models and algorithms, forming an engineering loop of "vehicle-side prototyping → robot-side validation."

Fig. 2. Data collection vehicle.

Through this dual-carrier design, the platform balances engineering deployment requirements and data production efficiency, providing a basis for multi-plant and multi-stage scaling of data volume.

3.2 Sensor Configuration

PTU camera rig. We adopt a DH-TPC-AEPT8441 explosion-proof dual-vision PTU (RGB + TIR), integrating a visible-light and a thermal infrared camera to realize co-located optical and thermal observations.

- RGB camera: maximum resolution 2688 × 1520; supports automatic/semi-automatic/manual zoom and auto-iris control; provides 45 × optical zoom, enabling both global situational awareness and local detail inspection (e.g., gauge dials and nameplates) from a fixed mounting position.
- TIR camera: resolution 400 × 300 with fixed focal length; spectral range 8–14 μm; temperature measurement distance approximately 4–21 m; temperature range – 20 °C to + 50 °C with accuracy around ± 2 °C.

LiDAR. We use an RS-Helios-16P 16-line 3D LiDAR with a 360° horizontal field-of-view and ± 15° vertical field-of-view. Its measurement range is 0.2–150 m, with 10/20 Hz rotation speed and point rates of 288 kpts/s (single return) or 576 kpts/s (dual return). The LiDAR provides dense geometric structure and precise distance measurements, beneficial for structural modeling of equipment areas, obstacle perception, and viewpoint localization, and further serving as geometric support for object detection and region recognition on RGB/TIR images.

The inspection robot and the data collection vehicle share identical installation positions, orientations, and wiring for the PTU, RGB/TIR cameras, and LiDAR, as well as the same data recording mechanisms. This guarantees reproducibility across carriers and enables cross-carrier experiments.

3.3 Data Acquisition and Metadata Organization

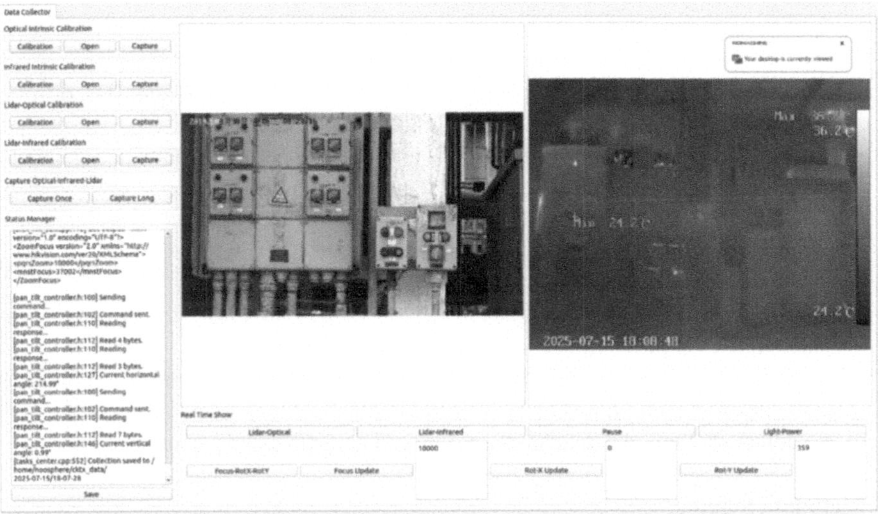

Fig. 3. Acquisition software interface.

To balance multimodal alignment quality and operational safety, the platform adopts a static sampling paradigm combined with a scripted acquisition workflow and records key acquisition parameters in a structured manner (Fig. 3).

Static Sampling Paradigm (Stop-and-Acquire). At each predefined observation point, the inspection robot or data collection vehicle follows a fixed observation procedure:

1. **Arrival and stabilization.** The carrier is driven along a predefined route to the target point and brought to a stable stop.
2. **Pose and imaging parameter stabilization.** Large PTU motions are stopped and the pan–tilt unit is moved to the target pose; RGB focal length and other imaging parameters are adjusted and stabilized.
3. **Sequential acquisition.** Within a short time window, a single RGB frame, a TIR frame, and a LiDAR scan that is time-aligned with the images are acquired sequentially.

Compared to continuous-motion acquisition, static sampling has several advantages:

- It avoids motion blur and parallax inconsistency caused by carrier movement and rapid PTU swings.
- It simplifies multimodal time synchronization and spatial alignment, improving the reliability of subsequent annotation and fusion.
- It closely mimics real robotic inspection workflows ("arrive → stop → observe → record"), facilitating evaluation of algorithms under realistic operational procedures.

Acquisition Pipeline and Metadata Logging. In practice, inspection data acquisition follows a scripted route and command pipeline:

1. **Scripted route planning.**

   - Predefine a list of stations/inspection points and organize equipment IDs, desired viewpoints, and observation distances into inspection tasks.
   - Plan carrier paths based on plant roads and operational constraints and generate corresponding PTU control command sequences (pan/tilt/focal).

2. **Automated command dispatch and execution.**

   - The acquisition software guides the carrier to visit each observation point sequentially according to the script.
   - At each point, PTU commands are automatically issued to orient the camera rig to the desired azimuth and elevation, and RGB focal length and other parameters are adjusted. Trimodal data are then acquired using the static sampling procedure.

3. **Metadata logging.**
   For each trimodal sample, associated acquisition metadata are recorded for subsequent alignment, retrieval, and analysis, including:

   - PTU pose: pan/tilt angles at the time of acquisition;
   - RGB imaging parameters: focal length, exposure mode, etc.;
   - TIR parameters: temperature measurement mode, range, etc.;
   - LiDAR operating mode: return mode (single/dual), rotation speed;
   - Environmental tags: acquisition time, indoor/outdoor status, unit type, etc.

With scripted inspection workflows and structured metadata logging, the platform achieves:

- **Reproducible sampling behavior.** The same routes and command scripts can be executed at different times to compare algorithm performance under varying conditions.
- **Retrievable sample organization.** Based on observation points, equipment types, and environmental tags, one can easily construct subsets such as "night + outdoor" or "specific unit type."
- **Unified multimodal processing.** Acquisition, alignment, annotation, and evaluation share a unified data organization schema, reducing toolchain complexity.

4 Dataset Composition

The PeChSafety-LiViT dataset is organized around static LiDAR–RGB–TIR trimodal samples, serving both as a plant-level data asset for engineering and as a de-identified public benchmark subset for external evaluation.

4.1 Data Hierarchy and Modality Buckets

Thanks to the unified hardware configuration and static sampling paradigm, the platform obtains synchronized RGB images, TIR images, and LiDAR scans at each observation point. According to the degree to which tasks depend on geometric information, we divide the data into two modality buckets:

1. **Bimodal (RGB + TIR) bucket.**
   This bucket centers on paired RGB and TIR observations, covering tasks that require "appearance details + temperature/thermal anomaly cues," such as gauge detection prior to reading, valve state estimation, local overheating, and standing water/flames. It requires strict temporal and viewpoint alignment between RGB and TIR. LiDAR information is not used in task definitions and evaluation by default but may be retained as optional auxiliary data.
2. **Trimodal (RGB + TIR + LiDAR) bucket.**

Building on RGB + TIR, this bucket includes LiDAR point clouds and targets tasks sensitive to 3D geometric structure and spatial layout, such as personnel/PPE detection, behavior safety assessment, and joint perception of fixed firefighting facilities and accessible spaces. Each sample in this bucket includes synchronized trimodal observations. Evaluation is conducted in the 2D image domain, while LiDAR offers distance and height information for geometric cross-validation and extended studies.

At the asset level, we strictly separate the full internal dataset and the public subset. The full dataset covers more units and labels and serves internal algorithm development and cross-project transfer. The public benchmark subset is de-identified and pruned, retaining representative modality combinations and task types for reproducible experiments and standardized evaluation.

4.2 Acquisition Scale and Data Organization

The platform accumulates samples along three dimensions: station, date, and observation point. Each sample corresponds to "one carrier at one observation point, with one PTU pose and one set of imaging parameters, producing a synchronized trimodal observation."

- **Full dataset scale goal.** We plan to scale the full dataset to more than 500 000 synchronized multimodal samples (LVT trimodal entries or RGB + TIR bimodal entries), spanning multiple process units, operating conditions, and acquisition batches.
- **Hierarchical organization.** All samples are stored in a hierarchical directory of "station/date/point," with associated metadata, allowing flexible subset construction based on spatial region, time slice, and scene type, such as "night + outdoor + specific unit type."

The public benchmark subset used in this paper is constructed from the full dataset by selecting samples that pass quality control and label validation. Task definitions and evaluation protocols are designed separately for the bimodal and trimodal buckets.

4.3 Category and Task Design

Based on real-world inspection needs in petrochemical plants, we divide labels into two major directions: "equipment/component and process condition safety" and "personnel/PPE and behavior safety," which correspond to the bimodal and trimodal buckets, respectively.

Bimodal (RGB + TIR) Bucket. The bimodal bucket focuses on equipment-level and local process-condition-level safety perception tasks, using aligned RGB–TIR image pairs as input. Representative categories include:

- **Instrument and valve classes:** digital/analog gauges, ball valves/gate valves, level indicators, etc.
- **Local indicator classes:** indicator lights, knobs, selector switches, etc.
- **Process/abnormality classes:** standing water on the ground, liquid residue, flames, etc.

In terms of tasks, the bimodal bucket mainly supports:

- Equipment/component detection;
- Abnormal symptom detection.

We use 2D bounding box annotations in the RGB image coordinate system for all categories, providing candidate regions for downstream tasks such as reading estimation and valve open/close state recognition.

Trimodal (RGB + TIR + LiDAR) Bucket. The trimodal bucket emphasizes tasks related to personnel safety and behavioral compliance, which rely more heavily on LiDAR-provided geometric structure and spatial layout. Representative categories include:

- **Fixed safety facilities:** fire extinguishers, etc.;
- **Personnel and behavior safety:** fallen person, phone usage, smoking, safety helmet wearing status, protective clothing, and person (pure personnel).

The supported tasks include:

- PPE detection and compliance assessment;
- Behavior safety detection (phone usage, smoking, falls, etc.);
- Joint perception of personnel and safety facilities (e.g., relative position analysis between people and extinguishers; currently not included in the official metrics).

Trimodal data can be used for 2D detection and classification in the image domain and for analyzing the spatial relationships between personnel and facilities using LiDAR, which improves interpretability.

4.4 Restricted Public Benchmark Subsets

Considering safety and confidentiality requirements in oil and gas scenarios, only a subset of strictly filtered, de-identified, and reviewed samples is released as restricted-access public benchmark subsets for academic research and external evaluation.

Bimodal Public Subset. In the RGB + TIR bucket, the current public subset focuses on two key categories:

- **Standing water** (500 images, 565 object instances).
- Covers different ground materials, lighting conditions, and water surface sizes, including obvious standing water and ambiguous wet patches. This subset is used to evaluate the robustness of anomaly detection for low-contrast targets.
- **Analog gauges** (500 images, 701 object instances).
- Covers multiple gauge types, mounting heights, and background complexities. It is used to evaluate small-object detection and equipment recognition performance and provides the basis for subsequent reading estimation and state recognition.

Here, "images" refer to RGB–TIR image pairs that contain at least one instance of the corresponding class. All samples include aligned RGB–TIR images and unified 2D detection annotations (Table 1).

Table 1. Official split of the bimodal (RGB + TIR) public subset

Split	#Stations	#Dates	#Points	#Image pairs (RGB–TIR)
train	3	6	60	600
val	1	2	20	200
test	1	2	20	200
Total	5	10	100	1000

The bimodal public subset thus covers 5 stations, 10 acquisition dates, and 100 observation points, with 1000 aligned RGB–TIR image pairs. The standing-water and analog-gauge classes are approximately distributed as 300/100/100 images across the train/val/test splits, avoiding severe class imbalance and allowing for class overlap between images.

Trimodal Public Subset. In the RGB + TIR + LiDAR bucket, the public subset focuses on PPE and behavior safety, with the following categories:

- Protective clothing (500 images, 675 object instances);

- Phone usage (500 images, 532 object instances);
- Safety helmet wearing (500 images, 629 object instances);
- Person (pure personnel, 434 images, 451 object instances).

Each "image" again denotes a sample containing at least one instance of the corresponding class. Individual samples may contribute to multiple categories. Each sample provides RGB, TIR, and LiDAR data, together with 2D detection annotations in the image domain. Researchers can further derive 3D spatial analyses or multimodal fusion strategies based on these samples (Table 2).

Table 2. Official split of the trimodal (RGB + TIR + LiDAR) public subset

Split	#Stations	#Dates	#Points	#Samples (LVT trimodal)
train	3	6	60	900
val	1	2	20	300
test	1	2	20	300
Total	5	10	100	1500

The trimodal public subset also spans 5 stations, 10 acquisition dates, and 100 observation points, with 1500 synchronized LiDAR–RGB–TIR samples. The PPE- and behavior-related categories are distributed roughly in a 6:2:2 ratio across the train/val/test splits (e.g., approximately 300/100/100 images for each of the protective clothing, phone usage, and helmet wearing categories; about 260/87/87 images for the "person" category), promoting consistent class distributions across splits while allowing inter-class overlap per image.

Official Splits and Reproducible Evaluation Protocol. To avoid optimistic bias due to cross-split leakage of the same equipment or viewpoint, we recommend splitting train/val/test on the granularity of "station–date–observation point," assigning all samples from the same "station–date–point" triple to the same split. The bimodal and trimodal buckets share the same splitting strategy to ensure fairness when comparing modality configurations.

We provide official split files at the "station–date–point" level, along with evaluation scripts to compute mAP, AP_small, and related metrics, enabling reproducible experiments and fair comparison across methods.

4.5 Task Definitions and Evaluation Metrics

In the current version, we unify all tasks as 2D object detection and define task sets and metrics separately for the bimodal and trimodal buckets.

Task Definitions.
Bimodal (RGB + TIR).

- Input configurations: RGB only or RGB + TIR.
- Output: 2D bounding boxes and class labels in the RGB image coordinate system (e.g., gauge, valve, indicator light, standing water, flame, etc.).

Trimodal (RGB + TIR + LiDAR).

- Input configurations: RGB, RGB + TIR, or RGB + TIR + LiDAR. LiDAR point clouds may be preprocessed into depth maps, height maps, or BEV representations following common LiDAR–camera fusion pipelines [7, 8, 23, 24].
- Output: 2D bounding boxes and class labels in the RGB image coordinate system for personnel/PPE/behavior classes (person, protective clothing, safety helmet wearing, phone usage, smoking, fallen person, fire extinguisher, etc.).

All tasks are evaluated in the 2D image domain, with LiDAR serving as a geometric auxiliary modality.

Evaluation Metrics.
The evaluation metrics follow mainstream detection benchmarks such as COCO [4], with additional focus on small objects:

- **mAP@0.5:** mean average precision at IoU = 0.5 across all classes, used as the main comparison metric.
- **Optional mAP@[0.5:0.95]:** mean AP over multiple IoU thresholds (0.5:0.95) to complement overall performance assessment.
- **AP_small / AP_medium:** AP statistics for objects in different size ranges, highlighting small-object performance.
- **Recall:** computed under a fixed confidence threshold or a fixed number of detections (top-K), measuring the extent of missed detections.
- **Optional FPR/FNR and scene-sliced metrics:** e.g., mAP@0.5 and AP_small on “day/night” or “simple/complex structure” subsets.

We adopt standard IoU-based one-to-one matching rules. The current version does not introduce “difficult” sample labels.

5 Experimental Results and Analysis

To demonstrate PeChSafety-LiViT as both a data platform and a benchmark, we construct and evaluate public baseline models under three configurations: single optical (RGB only), optical + thermal (RGB + TIR), and LiDAR–RGB–TIR trimodal. We quantitatively analyze the gains from multimodal fusion.

5.1 Single Optical Detection Baselines (RGB-Only)

Experimental Setup. Under single optical conditions, we evaluate a set of representative object detectors to establish RGB-only baselines:

- Faster R-CNN with a ResNet-50-FPN backbone [18];

- RetinaNet with a ResNet-50-FPN backbone [19];
- YOLOX-s [20];
- YOLOv5-m [21].

All detectors are standard and widely used in detection benchmarks such as COCO [4].

Training and evaluation configurations.

- Input modality: RGB only.
- All models are initialized from COCO pre-trained weights and fine-tuned on the official PeChSafety-LiViT train split.
- Common data augmentations are adopted, including random scaling, random flipping, and color jittering, with a unified input resolution.
- The validation set is used for model selection and early stopping.
- Evaluation is conducted on the official test split, with metrics including mAP@0.5, AP_small, AP_medium, precision, and recall. We also analyze performance on scene slices such as "day/night" and "simple/complex structure."

Overall Results.

Table 3. Overall performance of RGB-only models on the test set

Model	Precision (%)	Recall (%)	mAP@0.5 (%)	AP_small (%)
Faster R-CNN	80.3	74.7	41.5	23.5
RetinaNet	78.2	71.5	39.0	22.0
YOLOX-s	75.6	73.0	38.5	21.0
YOLOv5-m	82.7	76.3	42.0	24.0

As shown in Table 3, mAP@0.5 for all models lies in the range of 38–42%, and AP_small falls between 21–24%. Precision ranges from 75–82%, and recall from 71–76%, with YOLOv5-m and Faster R-CNN slightly outperforming the others. Overall, the results indicate that mainstream detectors already achieve usable performance on medium and larger objects under RGB-only settings, but small-object detection remains a significant weakness.

Performance by Scene and Category.

- **Daytime normal lighting.**
  mAP@0.5 is 2–3 percentage points higher than the overall average. Medium-sized targets such as gauges, valves, and fire extinguishers are detected relatively reliably.
- **Nighttime/low-light conditions.**
  All models show significant drops in mAP@0.5 (approximately 7–10 percentage points lower than daytime). AP_small decreases even more, and missed detections of distant gauges and valves become common.

- **Complex structure/cluttered layouts.**
  Compared with relatively simple scenes, mAP@0.5 often drops by 3–5 percentage points in structurally complex areas. Plant-specific background structures (supports, ladders, clamps, etc.) introduce additional false positives. AP_small for small-target categories (indicator lights, knobs, etc.) is consistently low.

From the category perspective, gauges, valves, and level indicators can be detected reliably at close and medium ranges, but detection rates degrade under the combination of long distance, small scale, and complex background. Standing water and flames, as rare abnormal-condition classes with diverse appearances, exhibit large performance variance across models.

Error Patterns and Limitations. Error analysis reveals that:

- Small-object missed detections are a major source of error.
- Complex lighting and metallic reflections lead to strong highlights and water patterns that are easily confused with indicator lights, standing water, or flames.
- Plant-specific structures (supports, ladders, clamps, etc.) increase background false positives.
- Performance varies significantly across conditions (day/night and simple/complex structures).

These phenomena indicate that RGB-only models are far from "saturated" on PeChSafety-LiViT, especially for small objects and complex operating conditions. This leaves substantial room for improvement and motivates the introduction of TIR and LiDAR.

5.2 Optical + Thermal Fusion Baselines (RGB + TIR)

Experimental Setup. Building upon the RGB-only baselines, we evaluate the benefits of RGB + TIR bimodal fusion in petrochemical plant scenarios using three representative RGB–T detection methods from the multimodal literature [16, 17]:

1. **EarlyFusion (RGB + TIR-EF).**
   A two-branch early fusion model based on the YOLOv5-m framework. RGB and TIR input images are processed by shallow convolutional encoders and then fused by channel-wise concatenation or weighted fusion at the early backbone stage.
2. **ICAFusion.**
   A method that employs iterative cross-modal attention modules to perform RGB–TIR feature interaction on intermediate and high-level features [17].
3. **MBNet.**
   A modality-balance network that uses learned modality weighting to mitigate instability when one modality dominates the other [16].

All models use aligned RGB–TIR image pairs as input. TIR images are resized to match the RGB resolution and normalized. The RGB branch is initialized with COCO pre-trained weights, while the TIR branch is randomly initialized or fine-tuned from pre-trained models. Training and evaluation protocols follow those in Sect. 5.1 (Table 4).

Overall Results.

Table 4. Overall performance of RGB + TIR bimodal models on the test set

Model	Precision (%)	Recall (%)	mAP@0.5 (%)	AP_small (%)
EarlyFusion (RGB + TIR-EF)	81.6	78.9	46.1	27.3
ICAFusion	82.5	81.7	48.0	29.5
MBNet	82.3	80.5	46.9	28.6

Compared with YOLOv5-m (RGB-only, mAP@0.5 = 42.0, AP_small = 24.0):

- All bimodal models improve mAP@0.5 by 4–6 percentage points.
- AP_small increases by 3–5 percentage points, indicating substantial gains in small equipment/component detection.
- Recall increases by 3–5 percentage points (fewer missed detections), while precision remains similar (81–82.5%).

Among the three models, ICAFusion achieves slightly better overall mAP and AP_small than MBNet and EarlyFusion.

Gains Under Different Conditions and Object Scales.

- **Daytime normal lighting.**
- RGB + TIR fusion improves mAP@0.5 by about 2–3 percentage points relative to RGB-only, mainly on objects with strong thermal signatures.
- **Nighttime/low-light conditions.**
- RGB-only performance is around mAP@0.5 ≈ 35% and AP_small ≈ 18%. EarlyFusion raises mAP@0.5 to approximately 43%, while ICAFusion/MBNet reach 45–46%, with AP_small improving by 8–10 percentage points. Nighttime recall increases from about 65% to around 75%.
- **Complex structure/cluttered layouts.**
- In these scenarios, all three bimodal models improve mAP@0.5 by 4–5 percentage points over RGB-only baselines. ICAFusion achieves larger gains in AP_small and recall for small objects against complex structural backgrounds.

Overall, RGB + TIR fusion delivers significant benefits under low light, strong reflections, and thermal anomaly scenarios, and is particularly effective for small objects.

Error Patterns and Fusion Analysis.

- Modality complementarity helps suppress some lighting-related false positives, but complex thermal backgrounds can still introduce new false alarms.
- EarlyFusion can exhibit modality dominance under some conditions; ICAFusion and MBNet mitigate this via attention mechanisms and modality weighting modules [16, 17].

- Cross-modal alignment residuals remain impactful for extremely small objects.

In summary, optical + thermal fusion consistently improves mAP, AP_small, and recall over RGB-only baselines, providing a strong foundation for further trimodal fusion with LiDAR.

5.3 LiDAR–RGB–TIR Trimodal Fusion Baselines

Experimental Setup. On top of the RGB-only and RGB + TIR baselines, we assess the potential of incorporating LiDAR using three representative multimodal detection frameworks inspired by recent LiDAR–camera fusion literature [7, 8, 23–25]:

1. **TransFusion-like.**
   A BEV fusion framework that treats LiDAR as the primary modality and cameras as auxiliary inputs. LiDAR point clouds are encoded into BEV representations, and RGB/TIR features extracted by 2D backbones are aligned and projected into BEV space for fusion [23].
2. **BEVFusion-like.**
   A BEV fusion framework that treats LiDAR and multi-camera (RGB + TIR) modalities on an equal footing. Each modality is projected/encoded into BEV space, followed by multi-modal feature fusion [24].
3. **CMT-like (Cross-Modality Transformer).**
   A framework that tokenizes LiDAR and RGB/TIR features and employs a cross-modality Transformer for deep interaction, producing detection outputs in the image or BEV domain [25].

Training and evaluation protocols follow those used in previous subsections. All trimodal models output 2D detection results in the RGB image coordinate system, with LiDAR mainly providing geometric support (Table 5).

Overall Results.

Table 5. Overall performance of trimodal models on the test set

Model	Precision (%)	Recall (%)	mAP@0.5 (%)	AP_small (%)
TransFusion	83.1	82.9	50.1	31.1
BEVFusion	83.6	83.1	50.4	31.4
CMT	84.1	84.0	51.1	32.2

Compared with YOLOv5-m (RGB-only, 42.0/24.0) and ICAFusion (RGB + TIR, 48.0/29.5):

- Trimodal models increase mAP@0.5 to around 50–51%, representing 8–9 percentage points improvement over RGB-only and 2–3 points over RGB + TIR.
- AP_small increases to 31–32%, improving by 7–8 points over RGB-only.

- Recall rises to 83–84%, with a small precision gain, indicating that LiDAR reduces both missed detections and some background false positives via geometric constraints.

Performance on PPE and Behavior Safety Tasks. The trimodal bucket focuses on PPE-and behavior-related categories such as person, protective clothing, safety helmet wearing, phone usage, smoking, fallen person, and fire extinguisher. Experimental results show that:

- For PPE combinations such as "person + protective clothing + safety helmet wearing," the CMT-like model improves mAP@0.5 by about 8–10 percentage points over RGB-only and by 3–4 points over RGB + TIR.
- For behavior classes such as phone usage, smoking, and falls, trimodal fusion significantly increases recall in complex backgrounds and partially occluded scenarios. In dense equipment areas, LiDAR helps distinguish personnel from structural elements.
- For fixed safety facilities such as fire extinguishers, trimodal models achieve more stable detection under multi-view and occlusion conditions, with lower false positive rates than single-and dual-modality models.

Gains under Different Conditions and Structural Complexity. In structurally simpler scenes, trimodal models yield a 1–2 percentage point mAP improvement over RGB + TIR, mainly through better localization accuracy and slight reductions in false positives.

- In multi-level pipe racks and high-density equipment areas, trimodal models improve mAP by more than 10 points over RGB-only and by 4–5 points over RGB + TIR.
- In extreme conditions such as "night + multi-level pipe racks + point light sources," TIR ensures basic visibility, LiDAR provides geometric structure, and RGB contributes limited texture. The three modalities together significantly reduce both false positives and missed detections.

Error Patterns and Trimodal Limitations. Despite achieving the best overall performance, trimodal fusion still exhibits limitations:

- Extremely small objects and fine-grained states (e.g., gauge readings) remain constrained by image resolution and LiDAR sparsity.
- Under certain extreme poses/zoom configurations, cross-modal registration residuals continue to affect very small targets.
- TransFusion-/BEVFusion-/CMT-like models are more complex computationally and architecturally than single- or dual-modality models, and practical deployment requires balancing performance gains against system overhead.

6 Limitations

Although PeChSafety-LiViT represents an important step toward LiDAR–RGB–TIR multimodal data for petrochemical plant-scale safety monitoring, several limitations remain from the perspectives of data platform and benchmark tasks:

1. **Temporal misalignment for dynamic targets and limited behavior coverage.**
   The static sampling paradigm suits static equipment and slowly changing process conditions, but minor time differences among modalities can cause temporal inconsistencies for dynamic targets such as flames, steam plumes, and fast human motions. The current dataset is therefore better suited for static detection and risk hotspot identification rather than full temporal behavior analysis.
2. **Impact of cross-modal spatial registration residuals on extremely small objects.**
   Unavoidable mounting tolerances and thermal drift in engineering practice introduce millimeter- or pixel-level registration residuals, which are negligible for large objects but significant for distant small objects. The dataset is thus not intended for sub-pixel-precision reading tasks.
3. **Gauge-reading tasks currently focus on "detection → alignment" only.**
   The public benchmark subset concentrates on "detecting and localizing gauges" and has not yet incorporated systematic fine-grained annotations for readings or valve opening degree. It is therefore not yet suitable as a full benchmark for gauge reading tasks.
4. **LiDAR sparsity and 3D precision limitations.**
   The 16-line LiDAR provides sufficient geometric backbone in near and mid-range scenes, but point clouds become sparse at long ranges, elevated platforms, or heavily occluded regions, limiting 3D capabilities for very small targets. The current benchmark uniformly evaluates in the 2D image plane.
5. **Task and annotation coverage is core but not exhaustive.**

The current public subsets mainly support 2D detection and PPE/behavior detection tasks. They do not yet fully cover segmentation, tracking, or rare event detection. The label taxonomy covers typical equipment and behaviors but does not constitute a complete plant object ontology, and statistical distributions in the public subsets differ from those in the full dataset.

7 Future Work

In light of the above limitations, future development of PeChSafety-LiViT will proceed along several directions:

1. **Expanding stations and time periods toward > 500k samples.**
   We will deploy the acquisition platform across more units and plants, covering additional process units, seasons, and weather conditions to increase data diversity. Under strict quality control and de-identification, we will steadily expand the number of synchronized multimodal samples to support large-scale models and self-supervised/pre-training research.
2. **Increasing strong annotations for gauge/valve/thermal anomaly tasks.**
   We plan to introduce high-resolution local crops and fine-grained annotations for analog/digital gauges and valves, constructing a complete pipeline from "detection → alignment → reading/opening estimation." We will also add thermal anomaly level labels for equipment and regions and link them with raw temperature measurements to support thermal-physics-based anomaly analysis.

3. **Extending from static to motion sampling and higher-spec sensors.**
   Under regulatory and safety constraints, we will gradually open continuous multimodal sequences to enable research on detection, tracking, and behavior recognition. We will explore higher-line-count LiDAR and higher-resolution TIR sensors to improve geometric and thermal-image resolution and strengthen multimodal time synchronization in hardware and software, enabling evaluation of how temporal offsets affect fusion performance.
4. **Exploring collaboration with OGI workflows and joint tasks.**

Where conditions permit, we will collect joint LVT + OGI data, modeling leak events together with surrounding equipment structures, ventilation conditions, and personnel activities. This supports extending from "single-point leak detection" to "integrated risk perception of leak–environment–personnel" and enables research on cross-dataset and multimodal joint learning.

In the long term, as data scale, annotation depth, and modality diversity continue to grow, PeChSafety-LiViT is expected to evolve from a multimodal detection benchmark into a comprehensive data infrastructure supporting "plant-scale multimodal foundation/pre-training models" and "cross-task, multimodal collaborative safety perception," thereby providing a more solid data foundation for intelligent inspection and operational safety in petrochemical scenarios.

References

1. ANYbotics: ANYmal X: Autonomous legged robot for inspection in Ex-zones. Product documentation. ANYbotics AG, Zurich, Switzerland (2024). https://www.anybotics.com/anymal-xt. Last accessed 11 Nov 2025
2. ExRobotics: ExR-2: ATEX/IECEx Zone 1 certified autonomous inspection robot. Product documentation. ExRobotics B.V., Heerenveen, Netherlands (2024). https://www.exrobotics.global/our-productst. Last accessed 03 Nov 2025
3. Boston Dynamics: Spot for industrial inspection – payloads and applications. Product documentation. Boston Dynamics, Waltham, MA, USA (2024). https://bostondynamics.com/products/spot. Last accessed 03 Nov 2025
4. Lin, T.-Y., et al.: Microsoft COCO: Common objects in context. In: European Conference on Computer Vision(ECCV) (2014). LNCS, vol. 8693, pp. 740–755. Springer, Cham (2014). https://doi.org/10.1007/978-3-319-10602-1_48
5. Hwang, S., Park, J., Kim, N., Choi, Y., So Kweon, I.: Multispectral pedestrian detection: benchmark dataset and baseline. In: IEEE Conference on Computer Vision and Pattern Recognition (CVPR), pp. 1037–1045 (2015). https://doi.org/10.1109/CVPR.2015.7298706
6. Bijelic, M., et al.: Seeing through fog without seeing fog: deep multimodal sensor fusion in unseen adverse weather. In: IEEE/CVF Conference on Computer Vision and Pattern Recognition (CVPR), pp. 11679–11689 (2020). https://doi.org/10.1109/CVPR42600.2020.01170
7. R-LiViT Project: R-LiViT: A LiDAR–Visual–Thermal dataset enabling vulnerable road user focused roadside perception. arXiv preprint arXiv:2503.17122 (2025). Dataset: https://github.com/XITASO/r-livit
8. Wang, L., et al.: Multimodal 3D object detection in autonomous driving: a survey and taxonomy. IEEE Trans. Intell. Veh. **8**(7), 3781–3798 (2023). https://doi.org/10.1109/TIV.2023.3264658

9. Xu, S., Wang, X., Sun, Q., Dong, K.: MWIRGas-YOLO: gas leakage detection based on mid-wave infrared imaging. Sensors **24**(13), 4345 (2024). https://doi.org/10.3390/s24134345
10. Zimmerle, D., Vaughn, T., Bell, C., Bennett, K., Deshmukh, P., Thoma, E.: Detection limits of optical gas imaging for natural gas leak detection in realistic controlled conditions. Environ. Sci. Technol. **54**(18), 11506–11514 (2020). https://doi.org/10.1021/acs.est.0c01285
11. Ilonze, C., et al.: Methane quantification performance of the quantitative optical gas imaging (QOGI) system using single-blind controlled release assessment. Sensors **24**(13), 4044 (2024). https://doi.org/10.3390/s24134044
12. Jia, X., et al.: LLVIP: A visible–infrared paired dataset for low-light vision. In: IEEE/CVF International Conference on Computer Vision Workshops (ICCVW) (2021). https://doi.org/10.48550/arXiv.2108.10831
13. Liu, J., et al.: TarDAL: target-aware dual adversarial learning and a multi-scenario multi-modality benchmark to fuse infrared and visible for object detection. In: IEEE/CVF Conference on Computer Vision and Pattern Recognition (CVPR) (2022). https://doi.org/10.48550/arXiv.2203.16220
14. Li, C., et al.: LasHeR: a large-scale high-diversity benchmark for RGB-T tracking. IEEE Trans. Image Process. **31**, 392–404 (2022). https://doi.org/10.1109/TIP.2021.3130533
15. Li, C., et al.: RGB-T object tracking: benchmark and baseline. Pattern Recogn. **96**, 106977 (2019). https://doi.org/10.1016/j.patcog.2019.106977
16. Zhou, K., et al.: Improving multispectral pedestrian detection by addressing modality imbalance problems. In: European Conference on Computer Vision (ECCV) (2020). arXiv preprint arXiv:2008.03043. https://doi.org/10.48550/arXiv.2008.03043
17. Shen, J., et al.: ICAFusion: iterative cross-attention fusion for RGB-T object detection. Pattern Recogn. **145**, 109913 (2024). https://doi.org/10.1016/j.patcog.2023.109913
18. Ren, S., He, K., Girshick, R., Sun, J.: Faster R-CNN: towards real-time object detection with region proposal networks. In: Advances in Neural Information Processing Systems (NeurIPS), vol. 28 (2015). https://arxiv.org/abs/1506.01497
19. Lin, T.-Y., Goyal, P., Girshick, R., He, K., Dollár, P.: Focal loss for dense object detection. In: IEEE International Conference on Computer Vision (ICCV), pp. 2999–3007 (2017). https://doi.org/10.48550/arXiv.1708.02002
20. Ge, Z., Liu, S., Wang, F., Li, Z., Sun, J.: YOLOX: exceeding YOLO series in 2021. arXiv preprint arXiv:2107.08430 (2021). https://doi.org/10.48550/arXiv.2107.08430
21. Jocher, G., Chaurasia, A., Qiu, J., Ultralytics: ultralytics YOLOv5. GitHub repository (2020–2024). https://github.com/ultralytics/yolov5. Last accessed 05 Nov 2025
22. COCO Consortium: Detection evaluation metrics (AP, AP_small, etc.). Online documentation, COCO dataset. https://cocodataset.org/#detection-eval. Last accessed 05 Nov 2025
23. Bai, X., et al.: TransFusion: robust LiDAR–camera fusion for 3D object detection with transformers. In: IEEE/CVF Conference on Computer Vision and Pattern Recognition (CVPR), pp. 1080–1089 (2022). https://doi.org/10.48550/arXiv.2203.11496
24. Liu, Z., et al.: BEVFusion: multi-task multi-sensor fusion with unified bird's-eye view representation. In: IEEE International Conference on Robotics and Automation (ICRA), pp. 2774–2781, (2023). https://doi.org/10.1109/ICRA48891.2023.10160968
25. Yan, J., et al.: Cross modal transformer: towards fast and robust 3D object detection. In: IEEE/CVF International Conference on Computer Vision (ICCV), pp. 18222–18232 (2023). https://doi.org/10.1109/ICCV51070.2023.01675

Research on High-Precision Positioning and Control Method of Robotic Arm Based on Visual Servoing

Tiantian Sun, Chunhua Deng(✉), Yunhan Lin, and Huasong Min

Wuhan University of Science and Technology, School of Science and Technology, Wuhan, China
dchzx@wust.edu.cn

Abstract. To address the problems of insufficient positioning accuracy, heavy reliance on manual teaching, and poor adaptability of robotic manipulators in unstructured industrial environments, this paper proposes a high-precision positioning and control method for robotic arms based on visual servoing. The proposed system integrates visual sensing with closed-loop motion control to enhance the autonomy and flexibility of robotic operations. An interactive Hough circle detection mechanism is designed to improve the robustness of feature extraction under complex imaging conditions by allowing real-time parameter adjustment. To establish an accurate mapping between image space and the robot operating space, a nine-point calibration method combined with affine transformation is employed, and the corresponding coordinate transformation model is optimized using the least squares method. Based on this, a complete visual–motion closed-loop control framework is constructed, enabling automatic execution of positioning and assembly tasks through continuous visual feedback. Experimental results demonstrate that the proposed method achieves sub-millimeter positioning accuracy with high repeatability. The system maintains stable performance under varying illumination conditions and positional disturbances, and is capable of completing high-precision grasping and assembly tasks in multiple operating scenarios. These results indicate that the proposed visual servo–based control approach provides an effective and practical solution for vision-guided robotic systems in intelligent manufacturing applications.

Keywords: visual servoing · robotic arm control · interactive Hough transform · affine transformation · nine-point calibration · closed-loop control

1 Introduction

Industrial robots, as the core equipment of intelligent manufacturing, are widely used in production segments such as assembly, sorting, and welding. Among them, robotic arm control systems based on pre-programmed teaching perform excellently in structured environments. However, they rely on fixed fixtures and precise initial positioning, making it difficult to adapt to dynamic changes in the position and posture of workpieces,

H. Wei and J. Niu (Eds.): COBOT 2025, CCIS 2893, pp. 426–441, 2027.
https://doi.org/10.1007/978-981-92-2074-8_25

thus lacking the necessary flexibility and intelligence. To enhance the autonomous operation capability of robotic arms in unstructured environments, servo control technology integrated with visual perception has become an important research direction in the field of robotics.

Visual servoing control forms a closed-loop system of perception-decision-execution by using cameras to acquire environmental information in real-time, endowing the robotic arm with "hand-eye coordination" capabilities. Depending on the camera configuration, it can be mainly divided into two modes: Eye-in-Hand and Eye-to-Hand. Although this technology can theoretically significantly improve system adaptability, it still faces many challenges in practical industrial applications: Firstly, the visual detection process is susceptible to on-site lighting fluctuations, workpiece surface reflections, and background noise interference, leading to insufficient stability in feature extraction. Secondly, if the spatial mapping model from two-dimensional image pixel coordinates to the three-dimensional robot coordinate system is not accurately calibrated, or if nonlinear distortion exists, it will directly cause the amplification of end-effector positioning errors. Furthermore, many existing systems still heavily rely on initial manual teaching, a process that is cumbersome and unable to perform online corrections, making it difficult to meet the demands of high-precision, multi-batch, and fast-cycle production.

To address the aforementioned problems, some studies have adopted deep learning methods to improve the robustness of feature detection. However, these methods rely on large-scale annotated data and high computational costs, limiting their application in industrial scenarios with high real-time requirements. Another category of research focuses on optimizing hand-eye calibration algorithms, such as introducing nonlinear compensation models to improve accuracy, but this often increases system complexity and computational burden. Therefore, how to construct a high-precision, highly adaptive vision-guided robotic arm control system while ensuring real-time performance and reliability remains a key practical challenge.

Based on the needs of practical industrial applications, this paper proposes a high-precision control method for robotic arms based on interactive visual recognition and joint calibration optimization. The main contributions of this research are:

(1) An interactive Hough circle detection mechanism is proposed, which combines traditional automated detection with human empirical decision-making. Through a strategy of real-time adjustable parameters, it effectively overcomes the instability of image feature extraction in complex environments, providing a reliable data foundation for subsequent control.
(2) A joint optimization strategy of nine-point calibration and affine transformation is designed. By accurately fitting the linear mapping relationship between pixel space and the robot operation space using the least squares method, an efficient and accurate coordinate transformation model is established, laying the theoretical foundation for achieving sub-pixel level positioning accuracy.
(3) A complete visual-motion closed-loop control framework is constructed. Through the process of "perception-calibration-planning-execution-feedback," fully automatic operation from visual information acquisition to precise robotic arm movement is achieved, significantly enhancing the system's level of automation and task adaptability.

2 System Overall Architecture and Working Principle

The core of the robotic arm control system based on visual servoing proposed in this paper lies in constructing an intelligent closed-loop control circuit of "perception-decision-execution-feedback". This system acquires information about the work scene through the visual perception module, solves precise control commands via the information processing and coordinate mapping module, ultimately drives the robotic arm to complete specified tasks, and ensures operational accuracy and reliability through real-time pose feedback. The theoretical architecture of the entire system is shown in the figure below.

2.1 System Hardware Architecture

The hardware components form the physical foundation of the system, mainly including the following two core units:

Visual Perception Unit: A high-resolution industrial camera is used, deployed in a fixed "Eye-to-Hand" configuration above the work scene. This configuration ensures the stability of the camera's field of view, allowing global observation of the robotic arm worktable and the workpieces to be operated on, providing a stable and consistent image input source for the system.

Robot Execution Unit: A six-degree-of-freedom industrial robotic arm is employed. Its end-effector can be configured as a suction cup, gripper, or other tools according to task requirements. The robotic arm establishes RPC (Remote Procedure Call) communication with the upper computer control system via Ethernet, receives pose control commands from the upper computer, and feeds back its own joint status and end pose in real time.

The upper computer (Industrial PC) acts as the "brain" of the system, undertaking all core computational tasks such as image processing, coordinate transformation, motion planning, and logic control, serving as the hub for hardware-software collaboration.

2.2 System Software Architecture and Working Principle

At the software level, the system adopts a modular architecture, in which functional modules are decoupled through well-defined data interfaces, forming an orderly automated workflow. The visual detection module provides target feature descriptions to downstream modules, with interface parameters including the pixel coordinates of the circle center, size information, and a detection validity flag. Based on the calibration results, the coordinate mapping module performs coordinate transformation and outputs target pose data in a unified format. The motion control module receives the target pose and generates motion trajectories by considering the current state of the robotic manipulator. Through unified interface definitions, each module can be independently developed, tested, and upgraded, thereby improving the overall maintainability and reusability of the system. Its working principle can be broken down into the following four core stages:

1) System Initialization and Visual Servoing Baseline Establishment

After the system starts, hardware initialization and communication connections are first performed. The robotic arm moves according to the preset initial photo pose P_0, which ensures that the camera can completely capture the target work area. Simultaneously, the camera completes parameter configuration and captures a reference image I_0. This stage establishes the initial state baseline for system operation, namely the initial correspondence between the robotic arm's physical space and the camera's image space.

2) Interactive Feature Detection and Model Calibration

This is the key mapping modeling process from visual information to control commands.

a. Interactive Feature Detection:

The system applies a Hough transform-based circle detection algorithm to the captured image. An interactive parameter adjustment mechanism assisted by manual intervention is innovatively introduced. The operator can adjust parameters such as detection threshold and radius range in real-time based on image quality, ensuring stable and precise extraction of the circle center pixel coordinates $p_{pixel} = (u, v)$ even in complex environments. This mechanism significantly enhances the system's robustness and practicality.

b. Hand-Eye Relationship Calibration: To achieve the conversion from pixel coordinates to the robotic arm base coordinates, the system performs the nine-point calibration method. The robotic arm end carries a feature object or relies on the end-effector itself to move sequentially to nine known robotic arm coordinate points $\mathrm{P_{robot\_i}} = (\mathrm{X_i}, \mathrm{Y_i}, \mathrm{Z_o})$ within the work plane, and acquires the corresponding pixel coordinates $\mathrm{p_{pixel\_i}}$ by capturing an image at each point. Using these nine pairs of corresponding points, a two-dimensional affine transformation model T is fitted using the least squares method:

$$\mathrm{P_{robot\_xy}} = \mathrm{T} \cdot \mathrm{p_{pixel}} \tag{1}$$

This model effectively compensates for nonlinear errors caused by camera lens distortion, installation tilt, etc., laying the theoretical foundation for high-precision positioning.

The proposed method assumes that the camera imaging plane is approximately parallel to the robot working plane, an assumption that is generally valid under fixed industrial installation conditions. When slight non-parallel alignment exists between the two planes, the resulting error can be approximated as a linear perturbation in the affine mapping and partially compensated through multi-point calibration and least-squares fitting. It should be noted that when the inclination angle between the planes becomes significant, a three-dimensional calibration model incorporating depth information should be employed to further improve positioning accuracy. Therefore, the proposed method is mainly applicable to planar assembly and positioning tasks.

3. Task Planning and Vision-Guided Closed-Loop Control

After calibration is completed, the system enters an automatic operation cycle. For each target point, its control process constitutes an independent servo closed loop:

a. Visual Perception:The robotic arm moves to $\mathrm{P_0}$, and the camera captures the current image $\mathrm{I_{current}}$.

b. Feature Re-identification and Pose Calculation: Detect the target circle again in $I_{current}$, obtaining its current pixel coordinate p_{target_pixel}. Subsequently, using the transformation model T obtained in Stage 2, calculate its planar coordinates P_{target_xy} in the robotic arm base coordinate system.
c. Motion Command Synthesis: Combine the relative pose offset Δpose (including Z-axis height and rotation posture) recorded during the teaching phase, synthesizing the planar coordinate P_{target_xy} into the complete target pose P_{target_full} for the robotic arm end.
d. Trajectory Planning and Execution: The robotic arm receives the target pose P_{target_full}, adopts linear interpolation (moveLine()) to first move to a safe point above the target point, then vertically descends to the operation point to perform grasping or assembly operations. Throughout the motion process, the system monitors and provides real-time status feedback via the wait_arrival() and get_and_print_current_pose() functions, ensuring accurate movement to the target position.

4. Exception Handling and Safety Monitoring

The system embeds multi-level safety and fault-tolerant mechanisms. These include: automatic retry upon image acquisition failure, validation of circle detection result effectiveness, robotic arm motion timeout, and collision detection. When any abnormality occurs in any step, the system can promptly interrupt the current process, issue an alarm, and wait for operator intervention, thereby ensuring the safety of equipment and products.

The abnormal state detection in the safety monitoring mechanism is based on predefined thresholds. The collision detection threshold is determined according to the typical ranges of torque and velocity fluctuations under normal operating conditions, while the motion timeout threshold is defined based on the theoretical execution time of the planned trajectory. When an abnormal condition is detected, the system prioritizes a safety stop, triggers an alarm, and enters a safe waiting state, thereby enhancing the practical applicability of the system.

2.3 Chapter Summary

In summary, this system organically integrates visual perception, model calibration, motion control, and safety monitoring through a modular hardware and software architecture. The core of its working principle lies in establishing an accurate mapping between pixel space and the robot operation space through the affine transformation model, and utilizing this mapping to guide the robotic arm's motion in real-time within the visual servoing closed loop. This architecture not only ensures the high precision and high reliability of the system but also, through its interactive detection and modular design, endows the system with good versatility and extensibility, laying the foundation for future upgrades towards more complex tasks.

3 Key Algorithms and Implementation

The intelligence and precision of this system rely on the collaborative work of three core algorithm modules: image feature detection based on interactive Hough transform, nine-point hand-eye calibration based on the least squares method, and closed-loop motion control based on visual feedback. The implementation of each module's algorithm is elaborated in detail below.

3.1 Interactive Hough Circle Detection Algorithm

The stable and accurate extraction of circular features from images is the foundation of the entire system. This paper adopts and improves the classic Hough circle transform algorithm, whose core idea is to transform the circle detection problem in image space into a peak-finding problem in parameter space.

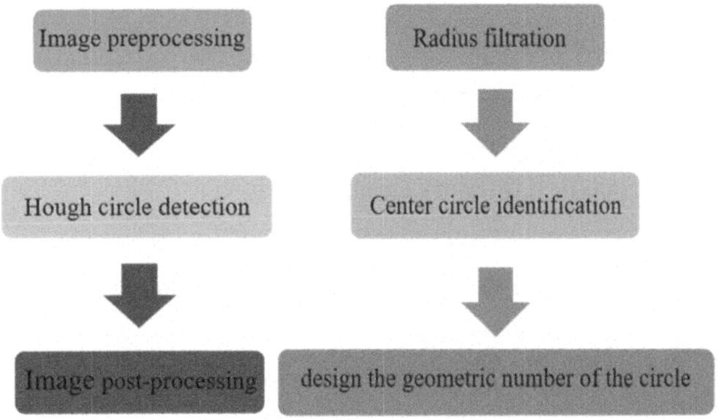

1) Principle of Hough Circle Transform

In image space, a circle can be defined by its center coordinates (a, b) and radius r. For each point (x, y) on the image edge, it corresponds to a conical surface in the parameter space (a, b, r). All conical surfaces corresponding to the edge points intersect at one point in the parameter space, and the coordinates (a_0, b_0, r_0) of this point are the parameters of the circle to be detected in the image space.

This process is accelerated using gradient information. For each edge point, based on its gradient direction, the center must lie on its radial line. The algorithm votes along this line within the possible radius range, significantly reducing computational complexity. Its cumulative voting score S can be expressed as: $S(a, b, r) = \sum_{(x_i, y_i) \in E} \delta\left((x_i - a)^2 + (y_i - b)^2 - r^2\right)$, Where E is the set of edge points, and δ is the impulse function. When S(a, b, r) exceeds the set accumulation threshold param2, a circle is confirmed to be detected.

2) Implementation of Interactive Detection

To enhance the algorithm's robustness in complex environments, this paper designs an interactive detection process. Its pseudocode is as follows:

```
Algorithm 1: Interactive Hough Circle Detection

Input: Original grayscale image I
Output: List of circle center pixel coordinates
and radii circles[]

1: Initialization: Load image I, create parameter
adjustment window
2: While True:
3:   Read current parameters from sliders:
param1, param2, minDist, minRadius,
maxRadius, blurKernel, blurSigma
4:   Apply Gaussian blur to image I with kernel
  size (blurKernel, blurKernel) and standard
deviation blurSigma
5:   Call cv2.HoughCircles(), passing current
parameters for circle detection
6:   Draw detected circles on the image and
display in real-time
7:   If user presses the adjustment completion
key (e.g., 'q'):
8:     Store current detection results in circles[]
9:     Break loop
10: Return circles[]
```

The key innovations of this implementation are:

a. Real-time feedback: Internal algorithm parameters (such as the Canny edge detection threshold param1 and accumulation threshold param2) are exposed to the operator and dynamically adjusted via sliders, with detection results visualized in real time.
b. Human supervision: The operator can quickly locate the optimal parameter combination based on the real-time image and their experience, allowing the algorithm to adapt to different lighting, contrast, and noise conditions. This human-machine collaborative strategy greatly enhances the adaptability of feature extraction without excessively increasing system complexity. It should be noted that, in most cases, human interaction occurs only during system initialization or when significant changes in the working scene take place. The average time consumption is approximately 3–5 s, and this process is not involved in individual visual servoing cycles; therefore, its impact on the real-time performance of the system can be considered negligible.

c. During the interactive Hough circle detection process, parameter adjustment follows a quantitative selection strategy. Different parameter combinations are first screened within predefined geometric constraints and then evaluated for stability through repeated detections, with the evaluation criteria defined as the combined performance in terms of the standard deviation of the detected circle center and the planar positioning error in the localization task. The parameter set that yields the minimum combined error while maintaining a false detection rate below 5% is selected as the optimal configuration for the current scene and remains fixed during subsequent operation.

3.2 Nine-Point Calibration and Affine Transformation Model

The objective of calibration is to establish the mapping relationship from the image pixel coordinate system to the robot base coordinate system. The nine calibration points are arranged within the reachable working plane of the robotic manipulator and are approximately uniformly distributed along the X–Y directions so as to cover the main operational area. Meanwhile, to reduce the influence of lens distortion and boundary-related errors on calibration accuracy, the calibration points are deliberately placed away from the edges of the working plane. Such a point distribution strategy ensures sufficient coverage while improving the stability and accuracy of the calibration model. In this study, the spatial relationship is simplified as a two-dimensional affine transformation model, which is capable of effectively describing scaling, rotation, shearing, and translation.

1) Affine Transformation Model

Let the pixel coordinate be $\mathbf{p} = [u, v, 1]^T$ (in homogeneous coordinates), and the corresponding robot base coordinate be $\mathbf{q} = [x, y, 1]^T$. The affine transformation between the two is expressed as:

$$q = T \cdot p \tag{2}$$

where is a 3 × 3 transformation matrix:

$$T = \begin{bmatrix} s_x \cos\theta & -s_y \sin\theta & t_x \\ s_x \sin\theta & s_y \cos\theta & t_y \\ 0 & 0 & 1 \end{bmatrix} \tag{3}$$

Here, (s_x, s_y) represents the scaling factors in the X and Y directions, (θ) represent the rotation angle, and (t_x, t_y) represent the translation components.

2) Least-Squares-Based Parameter Estimation

To determine the transformation matrix, the system collects pairs of corresponding points (N = 9) through the nine-point calibration procedure: $(\{(p_i, q_i)|i = 1, ..., N\})$

The goal is to minimize the transformation error across all corresponding points:

$$\min_{\mathbf{T}} \sum_{i-1}^{N} \|\mathbf{q}_i - T \cdot \mathbf{p}_i\|^2 \tag{4}$$

This is a classical linear least-squares problem. In This problem represents a classical linear least-squares problem. In the implementation, the system invokes the 'fit_affine_2d()' function to construct an overdetermined system of equations, and the optimal solution for the matrix T is obtained using singular value decomposition (SVD) or the normal equation method. After calibration, the matrix T and the calibration data are serialized and saved in the file 'biaodingwenjian.pkl', ensuring the persistence of the calibration results and avoiding repeated calibration by the system.not need to be repeated each time the system runs.

3.3 Vision-Guided Closed-Loop Motion Control

The motion-control module is responsible for converting visual information into safe and precise robot actions. Its core is a closed-loop workflow that includes pose synthesis, trajectory planning, and state feedback.

1) Target Pose Synthesis Algorithm

Upon detecting the target circle's pixel coordinates ($\mathbf{p}_t$), the control system computes the robotic arm's final target pose as follows:

a. Planar coordinates:

The target's 2D coordinates in the robot base frame are calculated using the calibration matrix **T:**

$$\left(\mathbf{q}_t^{2D} = T \cdot \mathbf{p}_t\right) \tag{5}$$

b. Spatial pose synthesis:

Incorporating the relative poses recorded during the teaching phase $\left(\Delta\mathbf{pose} = \left[\Delta x, \Delta y, \Delta z, \Delta rx, \Delta ry, \Delta rz\right]\right)$, the full 3D target pose is obtained:

$$\begin{aligned}\mathbf{P}_{target} = [x_t, y_t, z_{ref} + \Delta z, rx_{ref} + \Delta rx, ry_{ref} + \Delta ry, \\ rz_{ref} + \Delta rz]\end{aligned} \tag{6}$$

where $((x_t, y_t))$ are derived from $\left(\mathbf{q}_t^{2D}\right)$, and $\left(z_{ref}, rx_{ref}, \ldots\right)$ denotes the pose of the reference point. This method integrates the global accuracy of visual detection with the local precision of taught points, enabling high-precision pose control.

2) Safe Motion Procedure and Control Logic

To ensure safety and reliability during motion, each robot movement is decomposed into a sequence of structured path points, governed by the following control logic:

```
Algorithm 2: Vision-Guided
Closed-Loop Motion Control
Input: target circle index list
target_list[], calibration matrix,
relative pose offset
Output: None
For each target in target_list[]:
  Move robot to the imaging pose
  Acquire current image using
oneImg()
  Detect the target circle in the
current image and obtain
  Transform to robot-frame planar
Coordinate
  Synthesize the full target pose
  Generate a safe overhead pose
by increasing the Z-height
  Call moveLine(P_safe) and wait
for arrival using wait_arrival()
  Call moveLine(P_target) and
wait for arrival
  Execute the end-effector
operation (e.g., picking)
  Record the current robot pose
via get_and_print_current_pose() for
Diagnostics
  Lift back to and then return to
End For
```

By introducing the intermediate safe pose, the procedure effectively prevents collisions between the robot arm and workpieces. Additionally, each motion command is followed by a blocking wait (wait_arrival()) and real-time pose feedback (get_and_print_current_pose()), forming an implicit closed loop. This ensures deterministic execution of each step and results in a robust automatic operation cycle.

3.4 Summary of This Chapter

This chapter provides a detailed description of the three key algorithms of the system—interactive Hough-circle detection, nine-point calibration with affine transformation fitting, and vision-guided closed-loop motion control. By integrating classical algorithms with interactive design, least-squares optimization, and safe trajectory-planning strategies, the proposed methods effectively address robustness and accuracy issues encountered in real-world applications. The resulting solution forms a reusable and practical framework that can serve as a reference for developing similar vision-guided robotic systems.

4 Experimental Results and Performance Evaluation

To validate the effectiveness and stability of the proposed vision-guided robotic arm system in high-precision positioning and assembly tasks, experiments were conducted focusing on positioning accuracy, repeatability, and environmental robustness. All experiments involved multiple repetitions, and results were quantitatively analyzed using statistical methods.

4.1 Experimental Setup and Statistical Methods

The experimental platform consisted of an AUBO i5 six-DOF collaborative robot, a 5-megapixel industrial camera, and an industrial PC. Test workpieces were calibration plates with precision circular holes of 5.10 ± 0.02 mm in diameter. The end-effector probe had a diameter of $5.00^{+0.01}_{-0.03}$ mm to simulate small-clearance assembly scenarios.

Workpieces were selected to resemble common industrial components in surface roughness and reflective properties. The optical characteristics of the surfaces directly influenced the reliability of visual feature extraction, with rough surfaces affecting circle recognition. No noticeable control instability due to workpiece inertia was observed; visual detection errors were mainly affected by illumination and surface reflection distribution. Under reasonable parameter configurations, the system operated stably.

To ensure statistical reliability, the following methods were applied:

1) Repetitions: Each target point in positioning experiments was measured $n = 30$ times; each insertion task was repeated $n = 20$ times per condition.
2) Statistical metrics: Results are reported as mean ± standard deviation (Mean ± SD) and 95% confidence interval (CI).
3) Significance analysis: Differences between conditions were evaluated using two-sample t-tests with a significance level of α=0.05.

4.2 Positioning Accuracy and Repeatability

After performing a nine-point calibration, 90 positioning measurements at three target points were statistically analyzed. The X, Y, and planar composite errors are summarized in Table 1.

Table 1. Positioning Error Statistics (n = 90)

Metric	X (mm)	Y (mm)	XY Plane Composite
Mean ± SD	0.12 ± 0.03	0.14 ± 0.04	0.18 ± 0.05
95% CI	[0.11, 0.13]	[0.13, 0.15]	[0.17, 0.19]
Max Error	0.20	0.24	0.31

The results indicate that the average planar positioning error is below 0.2 mm, with a small standard deviation, demonstrating good repeatability. No significant directional bias was observed within the workspace, suggesting that the errors primarily originate from random noise rather than systematic deviation.

4.3 High-Precision Insertion Task

In automatic mode, the system executed the full "detection–positioning–insertion" assembly process on multiple target holes for a total of 100 trials (Table 2).

Table 2. Insertion Task Results under Standard Conditions

Metric	Value
Successful Trials	96
Successful Rate	96.0%
95% CI	[92.3%, 98.2%]

4.4 Robustness Analysis and Significance Testing

To further evaluate the system's robustness under complex lighting conditions, additional experiments were conducted under strong illumination in addition to the original low-light condition. Three lighting levels were tested:

1) Low: 200 lx
2) Normal: 500 lx (standard condition)
3) High: 1200 lx

For each condition, the complete vision-guided insertion task was repeated n = 20 times, with insertion success rate used as the evaluation metric (Table 3).

Table 3. Insertion Success Rate under Different Lighting Conditions (n = 20)

Lighting	Successful Rate	95% CI	p-value vs. Standard (t-test)
200 lx	90%	[82.4%, 95.1%]	0.032
500 lx	96%	[88.7%, 98.4%]	-
1200 lx	94%	[86.0%, 97.9%]	0.27

4.5 Positioning Error under Different Lighting Conditions

In addition to task success rate, planar positioning errors under different lighting conditions were statistically analyzed, as summarized in Table 4.

Table 4. Planar Positioning Error under Different Lighting Conditions

Lighting	Mean ± SD (mm)	95% CI
200 lx	0.21 ± 0.06	[0.18, 0.24]
500 lx	0.18 ± 0.05	[0.17, 0.19]
1200 lx	0.19 ± 0.05	[0.17, 0.20]

The results indicate a slight increase in positioning error under strong illumination; however, overall accuracy remains within the sub-millimeter range, with no significant difference compared to the standard condition ($p > 0.05$).

4.6 Real-Time Performance Evaluation

In addition to accuracy and robustness, the system's feasibility for industrial applications was evaluated by measuring the runtime of key modules in the vision servo process. Tests were conducted under standard lighting (500 l×) and normal operation mode, with each module measured $n = 50$ times. Results are reported as mean ± standard deviation (Table 5).

A single "vision-guided positioning–insertion" cycle consists of the following four modules:

- Image acquisition and transmission
- Feature extraction (interactive Hough circle detection)
- Pixel-to-robot coordinate transformation
- Robot motion planning and execution
- Module Time Consumption

Table 5. Module Time Consumption per Vision Servo Cycle (n = 50)

Module	Mean ± SD / s
Image acquisition & transmission	8.6 ± 2.4
Feature extraction	4.3 ± 2.1
Coordinate transformation	5.8 ± 0.4
Motion control & execution	150.5 ± 11.3
Total	**169.2 ± 16.2**

Discussion

The results indicate that vision perception and computation-related modules take less than 30 s in total, representing a small portion of the overall cycle time. The majority of the time is consumed by the robot's physical movement and trajectory execution. This level of timing meets typical industrial cycle requirements for assembly and pick-and-place tasks, which generally operate on the order of seconds.

4.7 Experimental Summary

Based on the evaluation of accuracy, robustness, and runtime performance, the following conclusions can be drawn:

1) The system achieves sub-millimeter positioning with good repeatability and statistical stability.
2) It operates reliably under lighting conditions ranging from 200 to 1200 lx, with no significant performance degradation under strong illumination.
3) A single vision servo cycle takes approximately 170 s on average, meeting the real-time requirements of typical industrial scenarios.
4) Overall, the experiments validate the system's engineering feasibility in terms of precision, reliability, and real-time performance.

5 Conclusions and Future Work

5.1 Summary of the Work

This work addresses the demand for flexible and intelligent robotic operations in modern smart manufacturing by designing and implementing a high-precision robotic arm positioning control system based on visual servoing and affine mapping. The main contributions and research outcomes of this study can be summarized as follows:

1) Development of a fully integrated hardware-software system

The industrial camera, six-DOF robotic arm, and PC-based control unit were successfully integrated to establish a stable and reliable "perception–decision–execution" closed-loop platform, providing a physical foundation for algorithm validation and practical application.

2) Proposed and implemented an interactive adaptive image processing scheme

Building on the classical Hough circle transform, a graphical parameter adjustment interface was innovatively introduced to incorporate human prior knowledge into the detection process. This approach significantly enhanced the robustness and recognition rate of circular features under complex imaging conditions, effectively addressing the limitations of traditional algorithms with fixed parameters and poor adaptability.

3) Established a high-precision vision-to-robot coordinate mapping model

Through a rigorous nine-point calibration procedure and affine transformation fitting based on the least-squares method, a linear mapping from the pixel coordinate system to the robot base coordinate system was constructed. Experiments demonstrated that this calibration method achieved an average error of less than 0.1 mm, providing a solid theoretical basis for visual positioning.

4) Designed a control strategy combining teaching and online visual feedback

The system effectively integrates reliable motion trajectories obtained from offline teaching with dynamic target positions detected by online vision, generating robotic motion commands with both high precision and safety. Experimental results showed that the system achieved a comprehensive positioning accuracy better than 0.2 mm, with good repeatability and certain robustness to environmental variations.

In summary, this study successfully integrates machine vision, coordinate calibration, and motion control technologies, providing an effective and easily implementable engineering solution for high-precision robotic arm positioning in unstructured environments.

5.2 Future Work

Although the proposed system demonstrates excellent performance within its target application domain, there remain several areas where improvements and further developments are desirable. Future work will focus on the following directions:

1) Advancement and intelligence of visual algorithms

The current interactive Hough-transform-based detection requires human intervention.Future work will explore deep-learning-based detection and segmentation networks (e.g., YOLO, Mask R-CNN) to enable end-to-end, highly robust recognition of complex or irregular objects, reducing reliance on manual parameter tuning.

2) Enhanced and generalized calibration models

The present 2D affine model assumes parallelism between the camera image plane and the robot workspace plane.Future research will investigate more general camera models and hand–eye calibration techniques (Eye-to-Hand and Eye-in-Hand), addressing 3D spatial localization and nonlinear lens distortion. This will further improve absolute positioning accuracy in 3D tasks.

3) Improved control strategies

The current system adopts position-based visual servoing (PBVS) and a "move–capture–move" open-loop compensation mechanism, which introduces latency.Future work will explore image-based visual servoing (IBVS), defining control laws directly in the image space to achieve smoother and faster dynamic responses.Research will also extend to high-level tasks such as dynamic target tracking and grasping.

4) System integration and enhanced intelligence

Plans include integrating the system into the Robot Operating System 2 (ROS 2) framework to exploit its modularity, distributed communication, and powerful toolchain, thereby improving system scalability and maintainability. Furthermore, digital-twin technology will be investigated to simulate and optimize production tasks through virtual-physical mapping, ultimately building a more autonomous, adaptive, and intelligent robotic manufacturing unit.

Through continued research in these directions, the proposed system has the potential to evolve from a high-performance domain-specific solution into a more general, autonomous, and intelligent next-generation robotic control platform.

References

1. Zhang, C., Xie, Z., Chen, G.: Key-Points based visual servo for accurate cross-task robot manipulations. ACM International Conference Proceeding Series, pp. 151–157 (2025)
2. Li, W., Xiong, R.: A hybrid visual servo control method for simultaneously controlling a nonholonomic mobile and a manipulator. Front. Inf. Technol. Electron. Eng. **22**(2), 141–154 (2021)
3. Wang, B., Wu, D., Gai, Y.: High-Precision wafer pose visual detection for robot precise assembly. Jixie Gongcheng Xuebao/J. Mech. Eng. **59**(8), 50–59 (2023)
4. Cheng, H., Wang, Y., Meng, M.Q.-H.: Vision-based single-stage grasp pose estimator with rotated anchors and automatic label generation. Sci. China Inf. Sci. **69**(1) (2026)
5. Salvato, E., Blanchini, F., Fenu, G., Giordano, G., Pellegrino, F.A.: Position-based visual servo control without hand-eye calibration. Robot. Auton. Syst. **193** (2025)
6. Xia, L., Cai, C., Zhou, C., et al.: A fast algorithm for detecting circles using the Hough transform. Comput. Appl. Res. **10**, 197–199 (2007)
7. Sun, F., Liu, J.: Fast Hough transform algorithm. J. Comput. Res. Dev. **10**, 1102–1109 (2001)
8. Huang, B., et al.: Semantic circular object detection based on deep Hough transform. Comput. Eng. Appl. 1–15 (2025). [Online]. Available: https://link.cnki.net/urlid/11.2127.TP.20250122.1658.015
9. Zhu, Y., Ling, Z., Zhang, Y.: Research progress and prospects of machine vision technology. J. Graph. **41**(6), 871–890 (2020)
10. Lin, J., Chen, H., Wang, Y., et al.: Research on robot visual servo systems. Control Theory Appl. **4**, 476–482 (2000)
11. Jia, B., Liu, S., Zhang, K., et al.: Research progress of robot visual servo: vision systems and control strategies. Acta Autom. Sinica **41**(5), 861–873 (2015). https://doi.org/10.16383/j.aas.2015.c140724
12. Liu, H., Shi, F., Liu, Y., et al.: Simulation analysis of robot hand–eye calibration. J. Syst. Simul. **11**, 2541–2543 (2004). https://doi.org/10.16182/j.cnki.joss.2004.11.048
13. Qiu, M., Ma, S., Li, Y.: A review of camera calibration in computer vision. Acta Autom. Sinica **1**, 47–59 (2000). https://doi.org/10.16383/j.aas.2000.01.006
14. Yin, H., Chen, B., Chai, Y., et al.: A survey of vision based object detection and tracking. Acta Autom. Sinica **42**(10), 1466–1489 (2016). https://doi.org/10.16383/j.aas.2016.c150823

Prediction of Hand Motion Perception Based on Surface Electromyogram and Flexible Sensor

Chen Fan[1,3](✉), Xiaodong Zhang[2], and Fuhua Huang[2]

[1] School of Mechanical Engineering, Xi'an Jiaotong University, Xi'an 710049, Shaanxi, China
fanc205@163.com

[2] School of Electrical Engineering, Xi'an Jiaotong University, Xi'an 710049, Shaanxi, China

[3] Engineering Research Center for Robotics and Intelligent Manufacturing in Shaanxi Province, Xi'an 710018, Shaanxi, China

Abstract. In order to solve the problem of human-machine cooperative real-time control of hand rehabilitation exoskeleton robot, a motion Angle decoding model based on surface electromyography (sEMG) signal combined with the synchronous motion Angle value was proposed. First, four channels of surface EMG signals were collected, and the EMG signals of fist and pinch movements were intercepted. The features were extracted in time domain and time frequency domain respectively, and the feature dimension was reduced by principal component analysis method and normalized. Secondly, the normalized features were combined with synchronously acquired hand motion Angle signals as input, and the corresponding Angle after 200 ms was taken as output. The long short-term memory neural network was used to construct the decoding model of hand motion Angle. During recognition, EMG signal and synchronous Angle signal are sent to the model for decoding, and the output of the model is the predicted value of Angle after 200 ms. The experimental results indicate that the exercise Angle signal and electromyographic signal can significantly improve the decoding ability of the model, and the repeatability tests, clench fist movement Angle prediction error is less than 7°, fashion movement Angle prediction error is less than 5°, indicating that the angle decoding model has good generalization, and the angle prediction value decoded can be used as the signal source of the control system of the hand rehabilitation exoskeleton robot.

Keywords: Surface EMG signal · Feature extraction · Long short-term memory neural network · Angle prediction

Conditions such as stroke and brain injury often leave sequelae like hemiplegia, with hand dysfunction causing patients to lose basic living and productive abilities. This severely undermines rehabilitation confidence and may trigger negative mental states including depression. Early rehabilitation training plays a crucial role in restoring hand motor function [1], while also reducing substantial medical and caregiving costs and shortening rehabilitation cycles [2]. However, traditional clinical rehabilitation practices heavily rely on physicians 'guidance, resulting in prolonged treatment periods. Rehabilitation assessments depend on doctors' subjective judgment, compounded by limited resources, high medical costs, and limited access to care, which prevents many

H. Wei and J. Niu (Eds.): COBOT 2025, CCIS 2893, pp. 442–462, 2027.
https://doi.org/10.1007/978-981-92-2074-8_26

patients from completing rehabilitation programs. Current rehabilitation devices available on the market are relatively limited, primarily enabling passive limb movements without predictive or participatory active exercises, thus failing to provide personalized and more effective training [3, 4]. Therefore, introducing active rehabilitation training is essential to overcome these limitations.

In motion prediction research, existing studies predominantly focus on predicting large joint angles such as ankles, knees, hips, and elbows. For instance, ZHAGN et al. [5] utilized a backpropagation (BP) neural network to establish real-time mapping between surface electromyographic (EMG) signals and hip, knee, and ankle joint angles. DAI [6] employed a Gaussian process neural network (GRNN) to predict ankle angles using EMG signals, with the golden ratio algorithm determining the optimal smoothing parameter σ. CHEN et al. developed a constrained Boltzmann machine-based deep autoencoder to extract features, employing BP neural networks to estimate hip and knee joint angles during lower limb movements [7]. However, due to the numerous finger joints and complex motions, combined with the relatively small finger diameter and large size of angle sensors, hand joint angle prediction remains underdeveloped. While [8] proposed using optical motion capture cameras for hand angle acquisition with high recognition rates, the lengthy decoding process makes it unsuitable for real-time prediction.

Given the current status of traditional hand rehabilitation methods and equipment, there is an urgent need for an active rehabilitation training device based on real-time human-machine collaborative control, such as a hand rehabilitation exoskeleton robot. This device can adaptively follow and compensate the movement of the affected limb according to the patient's intention during rehabilitation training, achieving the control effect of "human retreats, machine advances" and "human advances, machine retreats".

This study addresses the challenge of identifying patients' active rehabilitation movement intentions through multi-channel electrode-based surface electromyography (EMG) signals. These signals, which contain motion information during muscle contraction, are generated 30–150 ms before limb movement, enabling hand movement intention recognition and prediction. To overcome the difficulty in capturing finger joint motion angles, flexible fiber optic sensors are employed based on the macro-bending principle of optical fibers [9]. For predicting finger joint angles, a long short-term memory neural network is utilized to integrate surface EMG signals with real-time joint motion angles, constructing a motion angle decoding model. The decoded 200 ms-ahead motion angles are fed into the control system of the hand rehabilitation exoskeleton robot, allowing pre-perception of movement intentions and compensation for electromechanical hysteresis. This approach lays the foundation for achieving human-machine collaboration and natural passive movement in active rehabilitation training.

1 Angle Decoding Model of Finger Joint

1.1 Model Design

This paper presents a fine perception model for the prediction and recognition of hand motion intention and joint motion angle in real time (Fig. 1).

This model identifies and predicts hand movement intentions by integrating surface electromyography (EMG) signals with motion angle data. When the action recognition

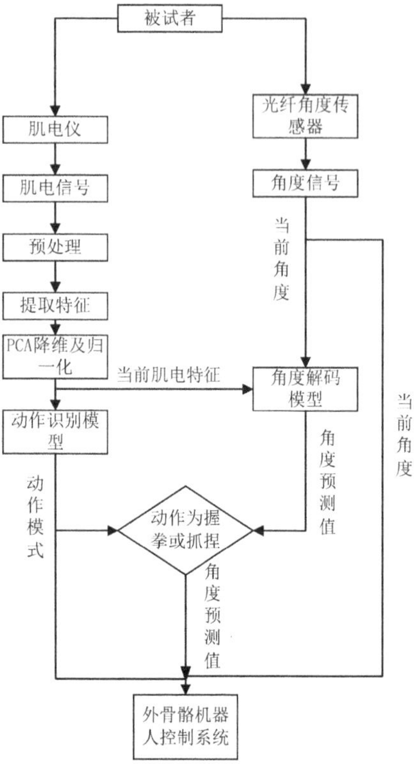

Fig. 1. Model Structure

module detects a current gesture as a fist or pinch, it captures real-time finger joint angles through flexible fiber optic sensors. By employing a long short-term memory (LSTM) neural network, the system combines EMG signals with live joint motion data to build an angle decoding model that forecasts 200 ms ahead. The predicted angles are fed into the exoskeleton robot's control system, enabling it to anticipate hand movements and prepare in advance, effectively compensating for control delays caused by electromechanical hysteresis.

Unlike conventional approaches that use EMG signals as inputs and motion angle values as outputs to build motion $\mathbf{Y} = [y_1, y_2, ..., y_k]^{\mathbf{T}}$ angle decoding $\mathbf{Y} = [y_1, y_2, ..., y_k]^{\mathbf{T}}$ models $\theta_i\theta_i$, this study integrates surface $\mathbf{Y}\prime = [y_1, y_2, ..., y_k, \theta_i]^{\mathbf{T}}$ $\mathbf{Y}\prime = [y_1, y_2, ..., y_k, \theta_i]^{\mathbf{T}}$ EMG feature vectors with θ_{i+200} corresponding motion angle data at the feature level to generate neural θ_{i+200} network training inputs. The decoding model incorporates angle feedback, while the corresponding motion angle data collected 200 ms later serves as the neural network's training output. The input-output mapping relationship is illustrated in Fig. 2.

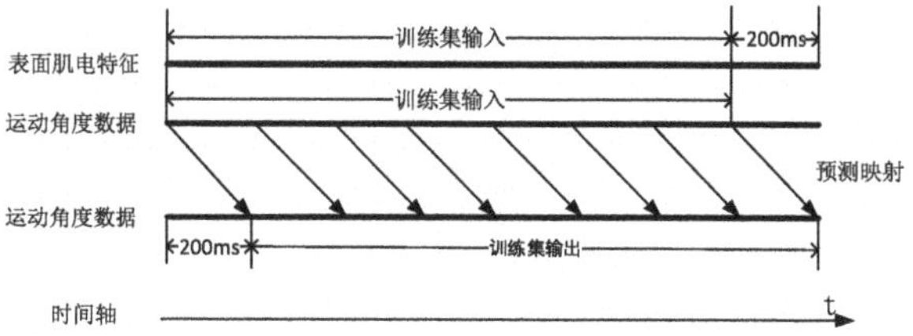

Fig. 2. Selection of training set

1.2 Working Principle of Long-Term and Short-Term Memory Neural Networks

Long Short-Term Memory (LSTM) neural networks, a type of recurrent neural networks, resolve the long-term dependency issue in RNNs and are widely used for time series prediction [10]. Unlike conventional RNNs, they incorporate input thresholds, forgetting thresholds, and output thresholds. Thus, this study employs LSTM neural networks to construct an angle decoding model.

When the long and short memory neural network works, the first $\mathbf{c_{t-1}}$ $\mathbf{c_{t-1}}$ step is to decide what information to discard from the previous $\mathbf{h_{t-1}}$ state, which $\mathbf{x_t}$ $\mathbf{h_{t-1}}$ is realized $\mathbf{x_t}$ by the Sigmoid $\mathbf{f_t}$ $\mathbf{c_{t-1}}$ function. The input $\mathbf{f_t}$ of the $\mathbf{c_{t-1}}$ Sigmoid function is the output of the previous moment and the input of the current moment. After passing through the Sigmoid layer, the output is between 0 and 1, and then it is multiplied point by point.

$$\mathbf{f_t} = \sigma(\mathbf{W}_\mathrm{f}\,[\mathbf{h}_{\mathrm{t}-1}, \mathbf{x}_\mathrm{t}] + \mathbf{b_f}) \tag{1}$$

where $\mathbf{W}_\mathrm{f}$ $\mathbf{W}_\mathrm{f}$ is the weight $\mathbf{b_f}$ $\mathbf{b_f}$ matrix and is the bias.

The next step involves setting the threshold to determine which new information to add $\mathbf{g_t}$ $\mathbf{g_t}$ to the cell state. First, the tanh function is used to identify candidate vectors to be incorporated $\mathbf{i_t}$ $\mathbf{i_t}$ $\mathbf{g_t}$ into $\mathbf{g_t}$ the new state. Then, the Sigmoid function is applied to determine which values to update, with the final state value being the product of the Sigmoid output and the candidate vectors. Finally, the state is updated by combining the updated state with the newly added vectors, resulting in the latest state.

$$\mathbf{i_t} = \sigma(\mathbf{W}_\mathrm{i}\,[\mathbf{h}_{\mathrm{t}-1}, \mathbf{x}_\mathrm{t}] + \mathbf{b_i}) \tag{2}$$

$$\mathbf{g_t} = \tanh(\mathbf{W}_\mathrm{g}\,[\mathbf{h}_{\mathrm{t}-1}, \mathbf{x}_\mathrm{t}] + \mathbf{b_g}) \tag{3}$$

$$\mathbf{c_t}= \mathbf{f_t}*\mathbf{c_{t-1}}+\mathbf{i_t}*\mathbf{g_t} \tag{4}$$

where $\mathbf{W}_\mathrm{i}$ $\mathbf{W}_\mathrm{i}$ $\mathbf{W}_\mathrm{g}$ is $\mathbf{W}_\mathrm{g}$ the $\mathbf{b_i}$ weight $\mathbf{b_g}$ matrix $\mathbf{b_g}$ $\mathbf{b_i}$ and is the bias matrix.

The final step determines the output. A tanh function normalizes the output state to [−1,1], and the Sigmoid function decides which part of the current state to output (Fig. 3).

$$\mathbf{o_t} = \sigma(\mathbf{W}_\mathrm{o}\,[h_{\mathbf{t-1}}, \mathbf{x_t}] + \mathbf{b_o}) \tag{5}$$

$$\mathbf{h}_\mathrm{t} = \mathbf{o}_\mathrm{t} * \tanh(\mathbf{c}_\mathrm{t}) \tag{6}$$

$$\boldsymbol{h}_t = \boldsymbol{o}_t * \tanh(\boldsymbol{c}_t) \tag{6}$$

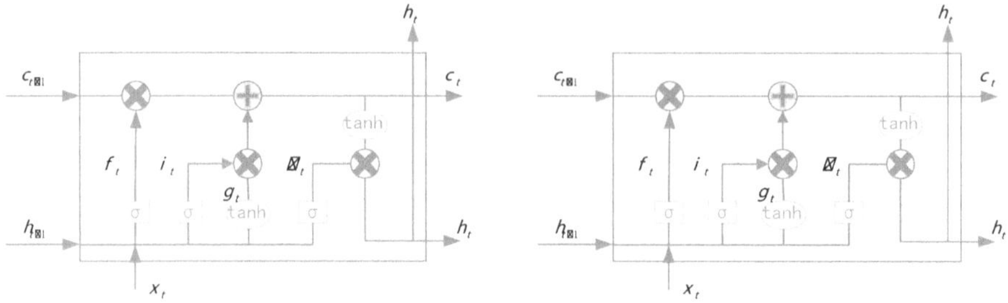

Fig. 3. Structure of Long-Term and Short-Term Memory Neural Networks

In this study, the input at $\mathbf{x_t}$ $\mathbf{x_t}$ the current moment is a vector formed by the fusion of the current moment's $\mathbf{Y}\prime=[y_1, y_2, ..., y_k, \theta_i]^{\mathbf{T}}$ electromyographic $\mathbf{h}_t$ $\mathbf{Y}\prime=[y_1, y_2, ..., y_k, \theta_i]^{\mathbf{T}}$ $\mathbf{h}_t$ θ_{i+200} feature value and angle θ_{i+200} value, and the output refers to the angle prediction value after 200 ms.

1.3 Feature Extraction and Dimensionality Reduction for LSTM Model Input

Before constructing the model, it is essential to extract relevant features from the original surface electromyography (EMG) signals. EMG signal features are primarily categorized into three types: time-domain, frequency-domain, and time-frequency domain features. While selecting more features can yield richer information and improve recognition accuracy, it also consumes more computational resources and increases latency. Comparative analysis of the three feature types reveals that frequency-domain and time-frequency domain features require significantly longer computation times. To balance real-time prediction requirements with high recognition accuracy, this study selects 5 time-domain features and 4 time-frequency domain features for analysis [11].

The time-domain features T_{RMS} T_{RMS} include T_{WL} root T_{WL} mean T_{MAV} square T_{SSC} T_{MAV} value (RMSV $T_{IAV}T_{SSC}$), wavelength T_{IAV} (λ), absolute mean value (AMV), slope change (SC), and integral absolute value (IAV). Their calculation formulas are as follows.

Root Mean T_{RMS} Square Value is T_{RMS}

$$T_{RMS} = \sqrt{\frac{1}{W-1}\sum_{i=1}^{W} x_i^2} \tag{7}$$

where W W is the data length x_i x_i and is the amplitude of the EMG signal.

Wavelength T_{WL} is T_{WL}

$$T_{WL} = \sum_{i=1}^{W} |x_i - x_{(i-1)}| \tag{8}$$

The absolute T_{MAV} mean is T_{MAV}

$$T_{\text{MAV}} = \frac{1}{W}\sum_{i=1}^{W}|x_{\text{i}}| \tag{9}$$

The absolute T_{IAV} T_{IAV} value of the integral is

$$T_{\text{IAV}} = \sum_{i=1}^{W}|x_{\text{i}}| \tag{10}$$

The positive and negative T_{SSC} T_{SSC} changes in the slope of the electromyographic signal are

$$T_{\text{SSC}} = \sum_{i=2}^{w}\text{f}[(x_{\text{i}} - x_{\text{i}-1})(x_{\text{i}} - x_{\text{i}+1})] \tag{11}$$

$$\text{f}(x) = \begin{cases} 1, x \geq V_{\text{th}} \\ 0, x < V_{\text{th}} \end{cases} \tag{12}$$

The formula V_{th} V_{th} is a preset threshold.

The time-frequency domain feature selection is extracted by discrete wavelet transform [12]. The wavelet db7 is decomposed into three levels to obtain four coefficient vectors, namely CA3, CD3, CD2 and CD1. The sum of absolute values of the four coefficient vectors is calculated to obtain the four features.

The different feature value distribution is very different, if used together, the different feature will make the different weight of the classification results, and it is not easy to converge in the later training of the angle prediction neural network, so it is necessary to linear normalize the different features to the same scale [−1,1].

The 4-channel electromyographic (EMG) signals are processed to extract nine features, ultimately yielding a 36-dimensional feature vector. This study employs principal component analysis (PCA) [13] for dimensionality reduction. The original feature set contains numerous interrelated features, with varying contributions to sample discrimination. PCA transforms k k linearly correlated m m $k < m$ feature $k < m$ sets into a new set of linearly independent principal $\mathbf{X}$ components $m \times N$ $m = 36$ through linear $N = f_s T$ $\mathbf{X}$ transformations $f_s m = 36m \times NT$, with $N = f_s T$ the core concept f_s of T mapping samples from high-dimensional to low-dimensional space [5]. A set of principal components replaces the original variables while preserving as much variance information as possible. Let the feature matrix of extracted surface EMG signals be defined as (where is the feature dimension, is the number of samples, is the sampling frequency, and is the sampling time). The specific dimensionality reduction procedure is as follows:

1) Calculate the covariance $\mathbf{\Sigma}$ matrix $\mathbf{\Sigma}$

$$\mathbf{\Sigma} = \text{cov}(\mathbf{X}, \mathbf{X}) = \text{E}[(\mathbf{X} - \mu), (\mathbf{X} - \mu)^{\text{T}}] \tag{13}$$

$$\mu = \text{E}(\mathbf{X}) = [\mu_{\mathbf{1}}, \mu_{\mathbf{2}}, \ldots, \mu_{\mathbf{m}}]^{\text{T}} \tag{14}$$

The mean $\mathbf{X}$ $\boldsymbol{\mu}$ $\mathbf{X}$ $\boldsymbol{\mu}$ vector is the mean vector of the mean vector.

2) Calculation of $\boldsymbol{\Sigma}$ Eigenvalues $\boldsymbol{\Sigma}$ and Eigenvalues of Covariance Matrix

$$\lambda \mathbf{I} - \boldsymbol{\Sigma} = 0 \tag{15}$$

Sort the obtained eigenvalues in descending order to $\lambda_1 \geq \lambda_2 \geq \ldots \geq \lambda_m \geq 0$ $\lambda_1 \geq \lambda_2 \geq \ldots \geq \lambda_m \geq 0$ obtain the $\boldsymbol{\lambda} = [\lambda_1, \lambda_2, \ldots, \lambda_m]^T$ eigenvectors, and λ_i calculate $\boldsymbol{\lambda} = [\lambda_1, \lambda_2, \ldots, \lambda_m]^T$ the $\mathbf{X}$ $\boldsymbol{\alpha}_\mathbf{i}$ i unit eigenvectors $\mathbf{y}_i$ λ_i corresponding $\boldsymbol{\alpha}_\mathbf{i}$ to $\mathbf{X}$ each i eigenvalue $\mathbf{y}_i$. The i-th principal component is then

$$\mathbf{y_i} = \boldsymbol{\alpha}_\mathbf{i}^\mathbf{T}\mathbf{X} = \alpha_{\mathbf{i1}}\mathbf{x_1} + \alpha_{\mathbf{i2}}\mathbf{x_2} + \ldots + \alpha_{\mathbf{im}}\mathbf{x_m} \tag{16}$$

3) Calculate the principal component contribution rate and cumulative contribution rate

Mathematically, it can be proven that the principal components k k $\mathbf{X}$ derived $\mathbf{X}$ through the aforementioned steps retain the variance k k information $k\mathbf{y_i}$ (i.e., distinctiveness $\mathbf{X}k$) $\mathbf{y_i}$ of the original $\mathbf{X}$ variables to the greatest extent when the first k principal components are selected. The specific value depends on the cumulative contribution rate. The information contribution rate of the k-th principal component to the original variables is

$$\eta_k = \frac{\lambda_k}{\sum_{i=1}^{m} \lambda_i} \tag{17}$$

k k The cumulative $\mathbf{y}_1, \mathbf{y}_2, \ldots, \mathbf{y_k}$ $\mathbf{y}_1, \mathbf{y}_2, \ldots, \mathbf{y_k}$ contribution rate of a principal component is

$$\eta = \frac{\sum_{i=1}^{k} \lambda_i}{\sum_{i=1}^{m} \lambda_i} \tag{18}$$

This study limits the dimensionality $\mathbf{X}$ reduction $\mathbf{X}$ to retain at least $\eta \geq 0.95$ $\eta \geq 0.95$ 95% $k = 8$ of the variance $k = 8$ information of the original variables, which is required to be calculated.

4) variable substitution

The first principal component is k calculated k $\mathbf{Y} = [y_1, y_2, \ldots, y_k]^\mathbf{T}$ $\mathbf{Y} = [y_1, y_2, \ldots, y_k]^\mathbf{T}$ $\mathbf{X}$ using formula $\mathbf{X}$ (9) to replace the original variables.

2 Acquisition of Model Data and Experimental Scheme

2.1 Acquisition of Surface Electromyogram and Motion Angle Signals

This study employed the Neuracle wireless EMG acquisition system with a 1 kHz sampling frequency and 4-channel setup. Disposable electrodes were applied to skin surfaces coated with sanding paste to minimize contact resistance. By monitoring forearm muscle activation during hand movements, researchers selected four target muscles: the brachioradialis, extensor pollicis, ulnar carpal flexor, and radial carpal flexor. Four pairs of differential electrodes were affixed to the muscle heads, while a reference electrode was placed on the styloid process of the ulna (Fig. 4).

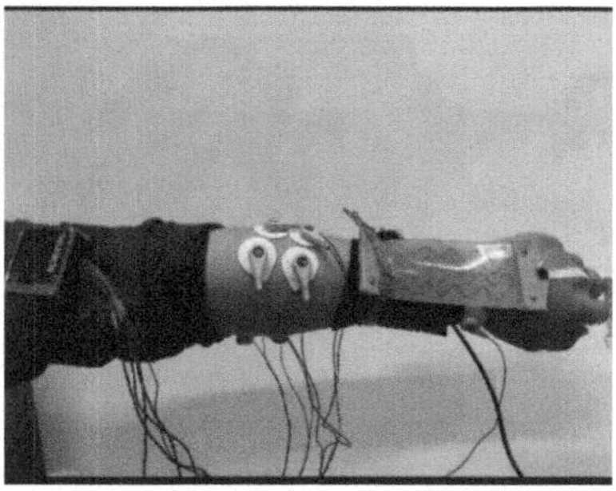

Fig. 4. EMG-Angle Acquisition System

A custom-made silicone module angle sensor with a 5 mm bending radius was developed based on the macro-bending loss effect of optical fibers. The sensor module was secured to the forearm and middle finger using Velcro. When the palm performs actions like making a fist or gripping, the metacarpophalangeal joints drive the sensor module to stretch, generating corresponding voltage signals that are converted into angular values through pre-calibration. The FS-N11MN optical fiber amplifier and NI's USB-6210 data acquisition card were employed to amplify and collect the output voltage at a sampling frequency of 100 Hz. Since the four finger joints (excluding the thumb) exhibit high coordination during these hand movements, the angular signals from the metacarpophalangeal joints of the four fingers were selected as the primary research targets.

To ensure both practicality and recognition accuracy, six common gestures from daily life were selected through literature review [15, 16]: fist-clenching, palm extension, OK gesture, index-middle finger bicep flexion, five-finger pinch, and water bottle grasping (as shown in Fig. 5). Each participant was instructed to perform the actions, with clear start/stop signals and timing markers. Each gesture lasted 3 s, followed by a 2-s rest period between consecutive actions. The experiment consisted of 5 sets, each containing 6 sequential gestures repeated 6 times. To prevent fatigue, a 2-min break was taken between sets.

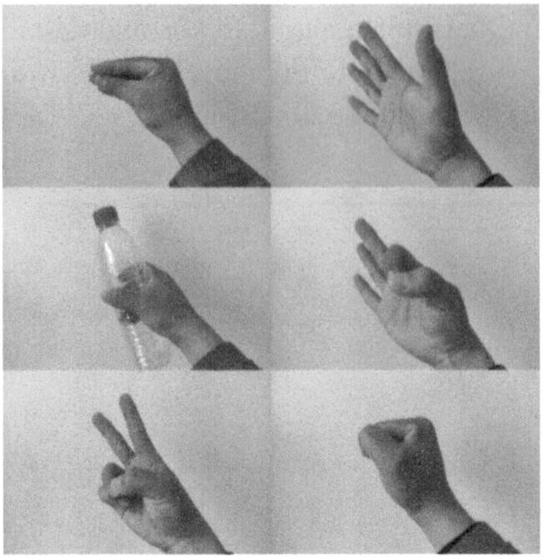

Fig. 5. Six gestures

2.2 Data Preprocessing

Various noise signals inevitably occur during experiments. These include power frequency noise from electromagnetic radiation of electrical instruments in the laboratory, motion artifacts caused by muscle-driven electrode movement (low-frequency noise), and high-frequency electromagnetic wave noise in the environment [17]. To mitigate these noise effects, digital filters are employed. First, a 50 Hz notch filter removes power frequency interference, followed by a 20–300 Hz IIR bandpass filter to eliminate low-frequency motion artifacts and high-frequency interference noise.

In order to eliminate the reaction time delay between the subject receiving instructions and performing actions [18], the start and end points of the actions need to be repositioned. The TKE operator [19, 20] was used in the experiment to detect these points.

2.3 Feature Extraction

After preprocessing the raw data and detecting the start and end points, the surface electromyography (EMG) data and angle data collected synchronously are processed according to the categories of motion labels, and signals for the two types of actions—grasping and fist-clenching—are extracted respectively, as shown in Fig. 6.

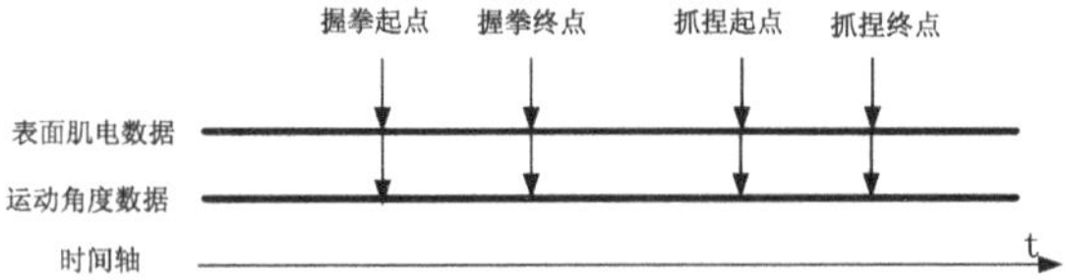

Fig. 6. Schematic of data extraction

Feature selection and extraction critically determine the accuracy and response speed of gesture recognition. This study employs the sliding window method [21] to extract

EMG signal features, as illustrated in Fig. 7. Excessive window length leads to information loss, while insufficient length causes excessive statistical variance in feature values. The step size, which characterizes data compression ratio, must be carefully calibrated: an oversized step results in sparse feature samples that impair control effectiveness, whereas an undersized step generates excessive data volume and prolongs computation time [22]. To maximize information extraction while preserving feature integrity and minimizing time delay in real-time control applications, the reference [23] adopts a 128 ms window length with a 10 ms step size.

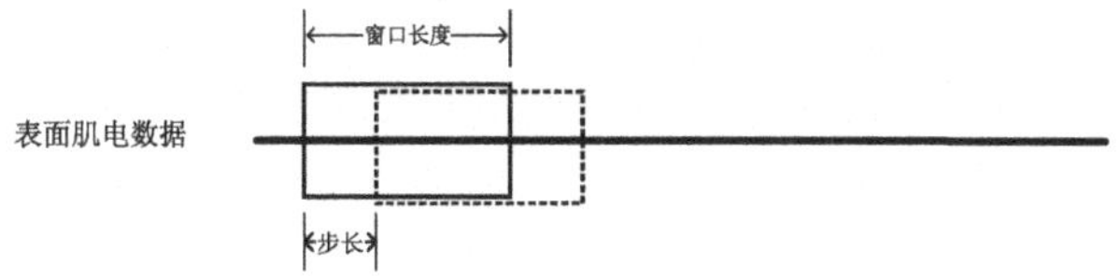

Fig. 7. The Sliding Window Method

The extracted features comprise five time-domain features and four time-frequency domain features as specified in Sect. 1.3. Principal Component Analysis (PCA) is applied to the 36-dimensional vector to reduce the feature dimensionality to 8. Features are then normalized to the [−1, 1] range based on their maximum and minimum values across dimensions, ensuring uniform weighting during model training. Angle values undergo identical normalization processing.

2.4 Model Training

The normalized surface electromyogram (EMG) signal features are combined with synchronized angle signals as inputs for model training, while the angle signals corresponding to 200 ms post-event serve as outputs. This establishes a motion angle signal decoding model based on Long Short-Term Memory (LSTM) neural networks. The LSTM network employs a dual-layer architecture with 200 neurons per layer. The Adam optimizer is used with an initial learning rate of 0.01, a learning rate decay of 0.6, and a decay period of 30. All data calculations and model development were performed on the MATLAB R2020b platform.

3 Analysis of Experimental Results

3.1 Performance Measure

In order to measure the prediction performance of the angle decoding γ_{rmse} γ_{rmse} model, the root mean square error is chosen as the evaluation index.

$$\gamma_{rmse} = \sqrt{\frac{1}{N}\sum_{i-1}^{N}(\overset{\sim}{\theta_i} - \theta_i)^2} \tag{19}$$

In the N N formula, represents $\tilde{\theta}_i$ $\tilde{\theta}_i$ the length of the model training θ_i θ_i set, represents the predicted output value of the motion angle decoding model, and represents the output of the motion angle decoding model's training set.

3.2 Prediction of Finger Joint Angle by Decoding Model

This study investigates the impact of incorporating angle as training data on the prediction performance of angle decoding models. The results are illustrated in Fig. 8, with decoding performance across experiments presented in Table 1. In Table 1, categories A and B denote motion angle decoding models for grasping actions with and without angle input, while categories C and D represent corresponding models for fist-clenching actions with and without angle input.

Table 1. Error of motion angle γ_{rmse} decoding model (°)γ_{rmse}

Model category	Number of experiments					mean
	1	2	3	4	5	
A	6.6	8.4	8.3	9.2	8.3	8.2
B	2.8	2.2	2.6	2.9	2.5	2.6
C	10.1	7.5	8.6	9.1	8.9	8.8
D	4.0	3.7	2.9	4.2	3.5	3.7

As shown in Fig. 8, when constructing motion angle decoding models for fist-clenching and grasping motions, incorporating fusion angles as training set inputs significantly improves decoding performance compared to using only EMG signal feature values. Table 1 data indicates that this approach reduces the mean root mean square error between model outputs and actual values from 8.2° to 2.6° for grasping motions, and from 8.8° to 3.7° for fist-clenching motions.

The angle decoding model can accurately predict the angle of motion 200 ms later based on the current angle and electromyographic (EMG) characteristics, demonstrating excellent angle prediction capability. However, the decoding model constructed solely from EMG signals shows significant deviations in prediction results at the peak angle.

3.3 Repeatability of the Experiment

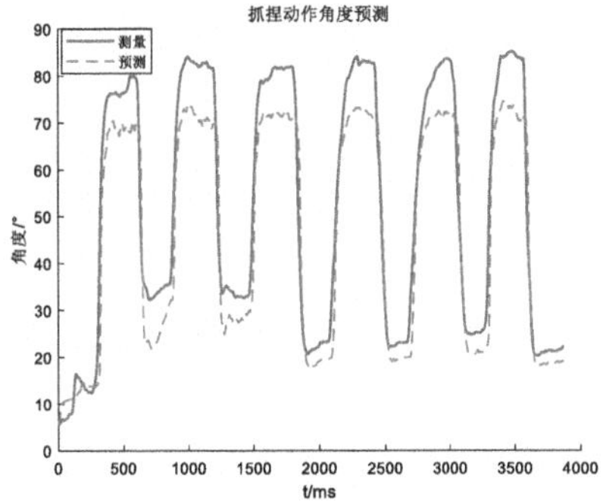

a: The model input does not include the motion angle of the angle-free grasping action.

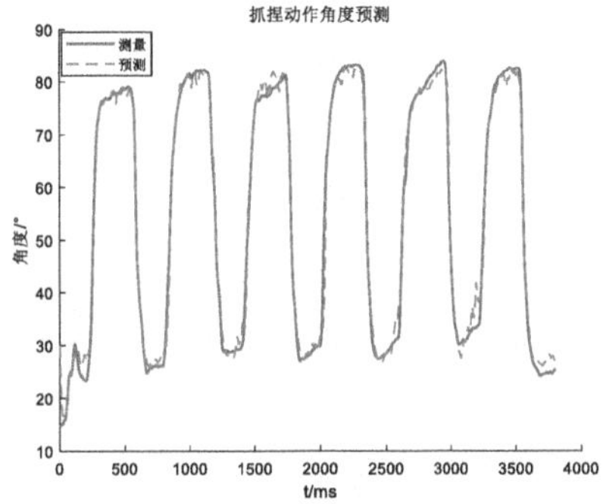

b: Model input angle for the angle-grasping motion

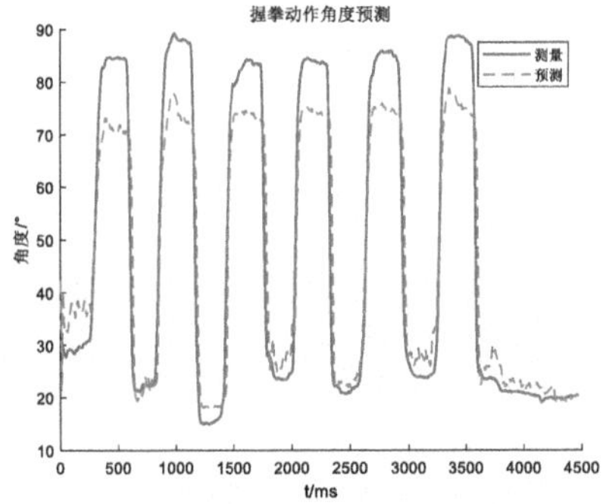

c: Model input angle of the fist-clenching motion

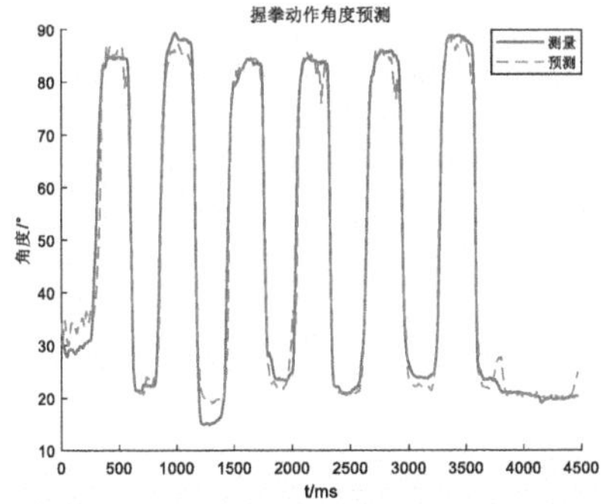

d: The model input does not include the angle of the fist movement.

Fig. 8. Predictive effect of the Angle Decoding Model

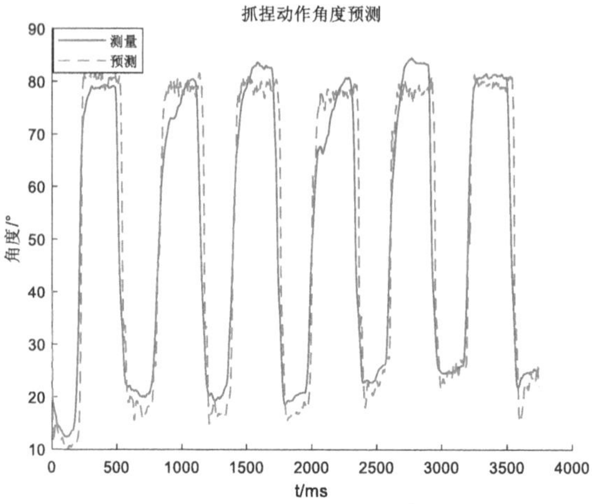

a: Verification of the repeatability of the grasping motion

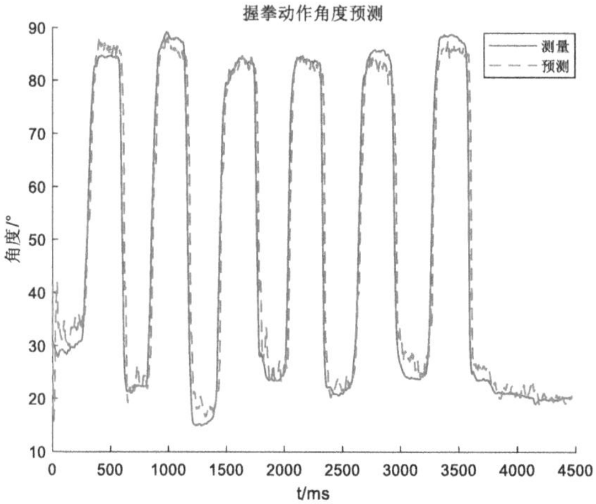

b Repetition verification of the fist-clenching action

Fig. 9. Predictive performance of the model in the repeatability validation experiment

The prediction model must demonstrate excellent reproducibility in practical applications. We first trained and stored two models: one for angle prediction using surface electromyography (EMG) data from the initial fist-clenching and grasping motions, and another for angle prediction using data from the subsequent four repeated experiments. The models were then tested with the latter data to evaluate their generalization performance. As shown in Fig. 9, the root mean square error values are presented in Table 2.

Table 2. Decoding model error (°) γ_{rmse} in repeated experiments γ_{rmse}

Action category	Number of experiments				mean
	1	2	3	4	
make a fist	6.7	5.2	5.2	7.3	6.1
Pinch	4.5	4.8	4.9	4.3	4.6

Experimental results demonstrate that the motion angle decoding model incorporating motion angles as training data exhibits excellent repeatability, maintaining strong predictive performance across multiple experiments under identical conditions. In reference [5], a backpropagation (BP) neural network was employed to predict lower limb joint angles (knee, hip, and ankle) without using motion angle signals as training data. The mean root mean square error (RMSE) in repeatability experiments reached 10.4°,6.2°, and 7.6° respectively, all exceeding 6°. Reference [24] utilized a long short-term memory (LSTM) neural network to predict lower limb joint angles solely based on electromyography (EMG) signals, achieving the best knee joint prediction with an RMSE of 6.28° in repeatability tests. Reference [25] applied a granular recurrent neural network (GRNN) to predict shoulder, elbow, and wrist joint angles using EMG signals, with the optimal shoulder joint prediction demonstrating an RMSE of 17°. All these studies reported higher prediction errors than our research, indicating that integrating motion angles with surface EMG signals to establish motion angle prediction models can significantly improve accuracy.

4 Design of Control Strategies for Rehabilitation Systems

4.1 Modeling of Kinematic Characteristics of Exoskeleton Manipulator

Due to the soft body characteristics of the exoskeleton manipulator designed in this paper, it is difficult to solve its motion model theoretically. Therefore, this paper adopts the method of model identification to obtain the motion model of the exoskeleton manipulator. That is, a step input signal is applied to the exoskeleton manipulator system, and then the output response curve of the system is measured over time, and then the transfer function of the identified object is fitted according to the response curve [26]. In the experiment, the duty cycle of the air pump input PWM signal is controlled to be 10%, 20%, 30%, ..., 90%, and a series of relationships between motion angle of exoskeleton manipulator and time are obtained, as shown in Fig. 10.

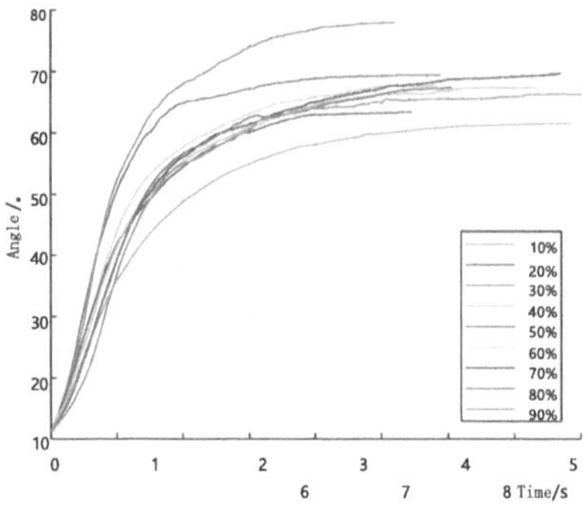

Fig. 10. Step response curve of the manipulator system

Using the step response identification toolbox in Matlab to identify the step response curve, it can be obtained that the system is approximately a second-order system, and its system transfer function has two poles on the negative real axis, which can be used in the time domain and described by two exponential decay terms:

$$\theta(t) = C0 - C1e - \lambda 1t - C2e - \lambda 2t \tag{20}$$

$$\begin{cases} \lambda_1 &= 15.687u + 4.158 \\ \lambda_2 &= -1.636u + 1.518 \\ C_0 &= -17.206u + 70.957 \end{cases} \tag{21}$$

Among them, C1, C2, C0 are constants determined by the initial input duty cycle u, λ1, λ2 are two poles ofthe system transfer function, and their magnitude is a function of the duty cycle u of the PWM signal.

4.2 Sliding Mode Control Strategy of Exoskeleton Manipulator

As can be seen from the previous section, the exoskeleton manipulator system can be described by a second-order system. The differential equation corresponding to this model is shown as (22), where p, q, r are functions of u.

$$\theta = p\theta + q\theta + r, p = -(\lambda 1 + \lambda 2), q = -\lambda 1\lambda 2, r = qC0 \tag{22}$$

Therefore, the differential equation is a second-order equation with variable coefficients, which needs to be discretized in actual calculation. In order to facilitate the design of the sliding mode controller, Taylor's formula can be used to locally linearize the second-order differential equation in actual calculation.

Using Taylor's formula at point u0, this differential equation can be linearized as:

$$\theta = f + g\Delta u, \Delta u = u - u0 \tag{23}$$

where Δu represents the calculation step size, fand g are functions of θ and θ , and their expressions are shown in the formula (24).

$$f = p(u0)\theta + q(u0)\theta + r(u0), g = p, (u0)\theta + q, (u0)\theta + r, (u0) \tag{24}$$

In the calculation, the Δu of each step is as small as possible, so that the formula can achieve relatively high calculation accuracy. Assuming the error e=θ-θd, the equation of the design sliding surface is as follows:

$$s = e + ce \tag{25}$$

Among them, s represents the sliding mode surface, and c is a positive number, which determines the convergence speed of the sliding mode control. When s → 0, the tracking error e can converge to 0 exponentially, thus realizing the tracking of the trajectory.

Define the Lyapunov function as:

$$L(s) = \frac{1}{2}s^2, L(s) = ss \tag{26}$$

In order to make the control system stable, it should satisfy $L \leq 0$, that is

$$\dot{L} = \dot{s}s - ks^2 \leq 0, \Delta u \leq -\frac{1}{g}\left[(k + c)\dot{e} + kce + f\right] \tag{27}$$

Therefore, the sliding mode control law is designed as

$$\Delta u = -\frac{1}{g}\left[(k + c)\dot{e} + kce + f\right] \tag{28}$$

When $L \equiv 0$, $s \equiv 0$, according to the LaSalle invariance principle, the closed-loop system is asymptotically stable[27]. When $t \to \infty$, $s \to 0$, the pneumatic exoskeleton manipulator system can tend to the expected value within a period of time. In the control, only the size of k needs to be modified to adjust the speed of the sliding mode control.

5 Simulation and Verification of the Rehabilitation Training System

In order to analyze and predict the user's movement intention through the surface electromyography signal (sEMG), so as to provide a reference for the sliding mode control strategy of the manipulator, the sEMG signal needs to be collected first.

The surface electromyography acquisition system selects the BRK-NMSM01 wireless electromyography acquisition system from Neuracle Ltd. For sEMG signal acquisition. The electrode sheets are evenly pasted on the user's hand to collect sEMG signals. The exoskeleton manipulator hardware control system and sEMG acquisition system are connected to the host computer through the USB data cable. Display kinematic data such as the movement angle and human-computer interaction force of the exoskeleton manipulator (Fig. 11).

(a) sEMG collector (b)Rehabilitation

Fig. 11. Experiment platform of the rehabilitation system

In the experiment, the user tried to make a fist and pinch respectively, and the electrode pads arranged on the user's arm could collect the original sEMG signal as shown in Fig. 12.

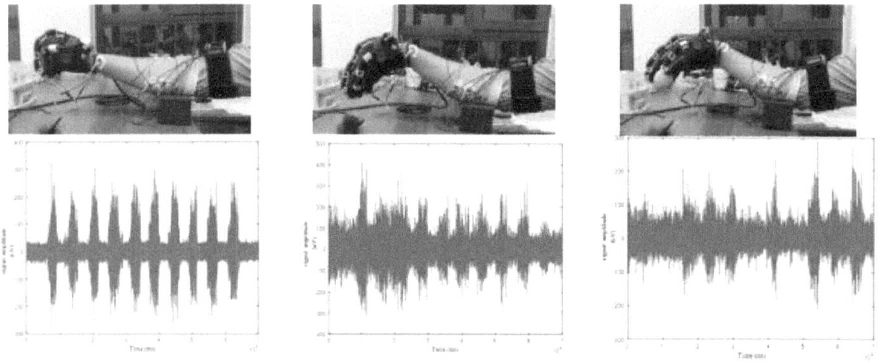

(a)Fist (b)Pinch (c)Grasp ball

Fig. 12. Origin sEMG signal for experimental maneuvers

After data filtering, the LSTM network will predict the user's movement intention based on the collected myoelectric data, and the prediction result will be presented through the angle-time relationship, which will be used as a reference value for the manipulator control system. The prediction results ofthe LSTM network are shown in the figure below.

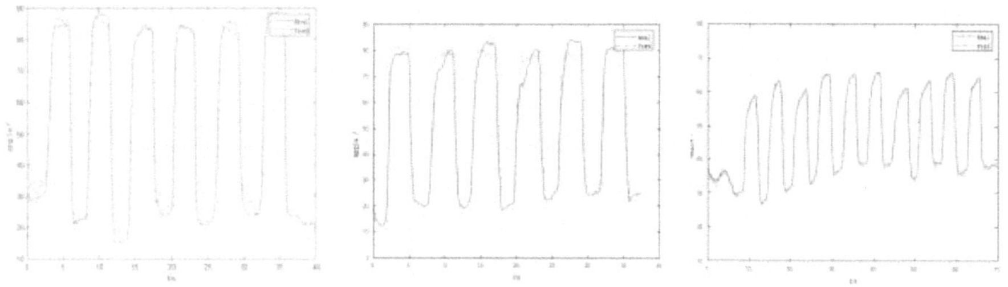

(a)Fist (b)Pinch (c)Grasp ball

Fig. 13. Angular prediction waveforms of experimental maneuvers

It can be seen from Fig. 13 that when the user has movement intentions such as clenching a fist, grasping and pinching, the neural network can predict the user's movement angle based on the collected sEMG signals, thereby providing the target angle value for the control system of the manipulator. It can be seen that the prediction results of the neural network are in good agreement with the actual results, and the prediction of the user's motion intention can be realized.

After obtaining the motion intent data processed by the LSTM network, the control system will drive the exoskeleton manipulator system according to the above sliding mode control law. The upper computer first transmits the motion angle prediction information to the STM32 controller through serial communication, and the controller will complete the calculation of the sliding mode control quantity, and the output of the sliding mode controller is the duty cycle of the PWM signal of the drive motor. The controller generates a PWM signal according to the duty cycle output by the sliding mode controller, which is amplified by the drive circuit to control the motor speed and adjust the motion angle of the exoskeleton manipulator.

In the experiment, the above motion angle prediction data is transmitted to the control system as the expected value. Under the control of the designed sliding mode controller, the exoskeleton manipulator will move according to the set value, thus effectively assisting the user's movement needs.

In the experiment, subjects were asked to clench, pinch and grasp the ball. The EMG collector will collect the subject's sEMG signal in real time and process it, and make predictions in the upper computer to obtain the subject's intention angle data, and the predicted value will be used as the reference value of the exoskeleton manipulator to drive the manipulator to move. The angle tracking curve, the output duty cycle from the sliding mode controller, and the control error of the three actions are shown in Fig. 14.

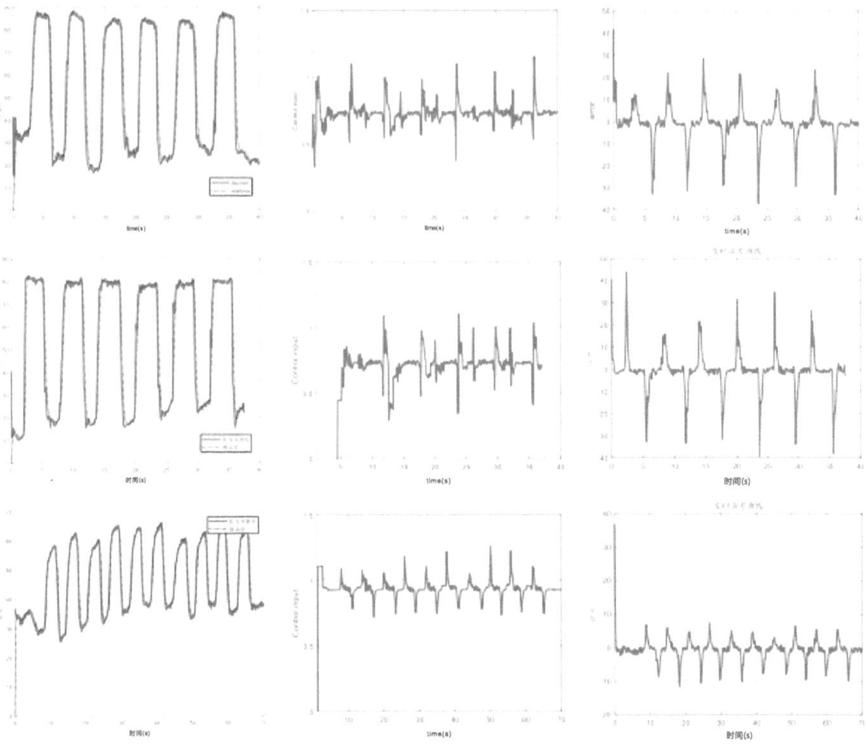

(a)Realtime motion angle (b)Output duty cycle (c)Control error

Fig. 14. Control effect of experimental action

From the above experimental results, it can be seen that when the user tries to make a fist, pinch the fingers, and grab the ball, the amplitudes of the finger movement angles are about 90°, 80° and 60° respectively, and when the user tries to perform corresponding movements, the exoskeleton manipulator can quickly provide motion assistance to the users.

Experimental results verify the reliability of the control system. In the control of each action, the real-time error value of the control is kept within 5°in the steady state, and the control precision is high. At the same time, the peak time of the real-time error curve in the actual test is about 1s, which shows that the exoskeleton manipulator system can quickly track the change of the predicted value of myoelectricity, has good real-time performance, and can meet the requirements of active hand rehabilitation training.

6 Conclusion

This study employed a four-channel electromyography (EMG) system to capture forearm surface EMG signals and a fiber-optic angle sensor to record metacarpophalangeal joint motion angles. A LSTM neural network was developed to construct a motion angle decoding model. Experimental results demonstrated that incorporating angle signals into the training dataset significantly enhanced the model's decoding performance. Repetitive experiments showed root mean square (RMS) errors below 7° for both movements, with the grasping action achieving an RMS error under 5°. These results confirm the system's excellent repeatability, establishing a robust foundation for real-time predictive control in robotic exoskeleton applications.

Due to the size limitation of the silicone sensor module, this study only captured single-degree-of-freedom motion angles, resulting in limited prediction categories for gestures. Future research will focus on miniaturizing the silicone sensor module to enable multi-degree-of-freedom motion angle acquisition, thereby making complex gesture prediction feasible.This paper proposes a deep learning-based myoelectric predictive control system for rehabilitation manipulators. The system collects the surface myoelectric signals of the user's hand, processes and predicts them through a neural network, and finally obtains the angle data of the user's movement intention. In order to realize the compliant control of the manipulator, a compliant sliding mode control method is designed in this paper. After obtaining the user's movement intention, the sliding mode control system will control the movement of the manipulator through the designed sliding mode control law.

In order to verify the effectiveness of this method, this paper designs and conducts a whole machine system experiment. The experimental results are consistent with the simulation results, and have good real-time and accuracy, satisfying the requirements of the active method. The design requirements of rehabilitation training are expected to be applied to the hand rehabilitation training of stroke patients.

References

1. Petruseviciene, D., Savickas, R., Krisciunas, A.: Evaluation of sensorimotor reactions during early rehabilitation for patients after cerebral stroke. Medicina **43**(12), 942–946 (2007)
2. Mahler, M.P., Zuger, K., Kaspar, K., Mahler, M.P., Zuger, K., Kaspar, K., et al.: A cost analysis of the first year after stroke-early triage and inpatient rehabilitation may reduce long term costs. Swiss Med Wkly. **138**(31–32), 459–465 (2008)
3. Xin, M., Juan, L., Ruirong, L., Xiaoying, Z.: Study on self-management behaviors and influencing factors in first-occurrence stroke patients. Chin. J. Nurs. **51**(03), 289–293 (2016)
4. Kadoji, D., Dikanovi, M., Bitunjac, M., et al.: Epidemiology of stroke. Period. Biol. **1143**, 253–257 (2012)
5. Yanan, Z., Yinping, J., Linyong, S., et al.: Dimensionality reduction and mapping analysis of lower limb surface electromyographic signals. Chin. J. Sens. Technol. **31**(7), 1046–1053 (2018)
6. Hong, D., Jinwu, Q., Zhen, Z., et al.: Application of GRNN in predicting ankle joint motion based on electromyography. Chin. J. Instrum. **34**(4), 845–852 (2013)
7. Jiangcheng, C., Xiaodong, Z.: Electroluminescence code method for continuous motion surface of lower limb walking joints. J. Xi'an Jiaotong Univ. **50**(6), 61–67 (2016)
8. Wang, H., Tao, Q., Su, N., et al.: A precise method for sensing finger joint angles using surface electromyography signals. J. Xi'an Jiaotong Univ. **56**(8), 160–167 (2022)
9. Li, M., et al.: Flexible wearable wrist joint motion angle sensor. J. Xi'an Jiaotong Univ. **52**(12), 32–37 (2018)
10. Du, X., Ma, C., Zhang, G., Du, X., Ma, C., Zhang, G., et al.: An efficient LSTM network for emotion recognition from multichannel EEG signals. IEEE Trans. Affective Comput. **13**(3), 1528–1540 (2022)

11. Khushaba, R.N., Phinyomark, A., Al-Timemy, A.H., Scheme, E.: Recursive multi-signal temporal fusions with attention mechanism improves EMG feature extraction. IEEE Trans. Artif. Intell. **1**(2), 139–150 (2020)
12. Lucas, M., Gaufriau, A., Pascual, S., Doncarli, C., Farina, D.: Multi-channel surface EMG classification using support vector machines and signal-based wavelet optimization. Biomed. Signal Process. Control **3**(2), 169–174 (2008)
13. Song, Y., Sun, W., Chen, C.: Image recognition based on principal component analysis of logarithmic transformation. J. Xi'an Jiaotong Univ. **55**(01), 33–42 (2021)
14. Atzori, M., Gijsberts, A., Kuzborskij, I., Atzori, M., Gijsberts, A., Kuzborskij, I., et al.: Characterization of a benchmark database for myoelectric movement classification. Trans. Neural Syst. Rehabil. Eng. **23**(1), 73–83 (2015)
15. Lu, C.: Design of Surface Electromyogram (EMG) Signal Acquisition Arm Band and Its Application in Gesture Recognition. University of Electronic Science and Technology of China (2020)
16. Atzori, M., Müller, H., Gijsberts, A., et al.: Electromyography data for non-invasive naturally-controlled robotic hand prostheses. Sci. Data **53**, 140053 (2014)
17. Norman, R.W., Komi, P.V.: Electromechanical delay in skeletal muscle under normal movement conditions. Acta Physiol. Scand. **106**(3), 241–248 (1979)
18. Zhao, S., Wu, X., et al.: Automatic recognition of hand gestures using surface electromyography. J. Xi'an Jiaotong Univ. **54**(9), 149–156 (2020)
19. Wang, S., Zhu, S., Shang, Z.: Comparison of different algorithms based on TKEO for EMG change point detection. Physiol. Meas. **43**(7), 1–16 (2022)
20. Tigrini, A., Mengarelli, A., Cardarelli, S., Fioretti, S., Verdini, F.: Improving EMG signal change point detection for low snr by using extended teager-kaiser energy operator. IEEE Trans. Med. Robot. Bionics **2**(4), 661–669 (2020)
21. Novak, D., Riener, R.: A survey of sensor fusion methods in wearable robotics. Robot. Auton. Syst. **73**, 155–170 (2015)
22. Yang, B.: Research on Finger Rehabilitation Movements Based on Multi-channel Electromyographic Signals. Zhejiang University of Technology (2017)
23. Hu, Y.: Research on an Upper Limb Rehabilitation Robot Training System Based on sEMG. Tianjin University of Technology (2017.2
24. Fei, W., Xiaotong, W., Hao, Q.: Continuous lower limb motion estimation based on sEMG and LSTM. Journal of Northeastern University. **41**(3), 305–342 (2020)
25. Liu, K., Sun, R., Chai, Y., et al.: Prediction of upper limb joint angles based on generalized regression neural networks. Sci. Technol. Eng. **21**(17), 7187–7192 (2021)
26. Jingkun, L.: Sliding mode design and Matlab simulation. Tsinghua University Press, Beijing (2019)
27. Du, X., Ma, C., Zhang, G., et al.: An Efficient LSTM Network for Emotion Recognition From Multichannel EEG Signals. IEEE Trans. Affective Comput. **13**(3), 1528–1540 (2022)

Reinforcement Learning for Agile Bipedal Jumping: A Comprehensive Review of Dynamic Skills, Visual Adaptation, and Whole-Body Coordination

Dong Xu[1], Xiaotian Ma[1], Yukun Wang[2], Zeyu Li[3], and Yupeng Liu[1](✉)

[1] School of Automation Science and Electrical Engineering, Beihang University, Beijing, China
{ZY2403704,yupengliu}@buaa.edu.cn
[2] School of Mechanical Engineering and Automation, Beihang University, Beijing, China
yukunwang@buaa.edu.cn
[3] Institute of Embodied Intelligence Robotics, Beihang University, Beijing, China
lizeyubuaa@buaa.edu.cn

Abstract. Jumping is essential for enabling high dynamics and agility in bipedal robots, supporting autonomous navigation and task execution in complex environments. Traditional model-based controllers struggle with the strong nonlinearity and high dynamics of jumping, leading to limited robustness. Reinforcement learning (RL), which does not require precise system models and learns optimal policies through interaction, has therefore become a key approach for bipedal jumping control. This paper reviews RL-based methods for bipedal robot jumping, summarizing their development and technical foundations. It highlights RL's advantages in handling nonlinear contact dynamics, high-impact motions, and enabling end-to-end learning. Mainstream frameworks—such as proximal policy optimization (PPO), concurrent teacher-student (CTS) architectures, and sim-to-real transfer techniques—are outlined. The review further examines major research directions, including dynamic jump skill learning, multi-gait continuous jumping, vision-based terrain-adaptive jumping, and whole-body coordinated jumping that integrates locomotion and manipulation. Although significant progress has been made, challenges remain in long-horizon continuous jumping, robust perception and planning on complex terrain, multi-task generalization, and real-world reliability. Future research trends include multimodal perception integration, leveraging symmetry priors, more efficient exploration strategies, general motion controllers, and deeper fusion of locomotion and manipulation. This work aims to provide a comprehensive reference for advancing RL-driven bipedal jumping control.

Keywords: Bipedal Jumping Control · Reinforcement Learning (RL) · Vision-Based Terrain Adaptation

1 Literature Overview

Bipedal robots, as advanced platforms that mimic human motion, have long been a central focus in robotics research for enhancing locomotion capabilities. Jumping, as a highly dynamic and non-periodic movement, serves as an important indicator of a bipedal

H. Wei and J. Niu (Eds.): COBOT 2025, CCIS 2893, pp. 463–477, 2027.
https://doi.org/10.1007/978-981-92-2074-8_27

robot's agility and adaptability to its environment. However, jumping involves intense aerial phases, complex ground-contact dynamics, and precise control for stable landing, making motion planning and control highly challenging. Traditional control methods based on physical models and state machines, while capable of achieving stable walking in specific scenarios, often struggle with highly nonlinear, high-degree-of-freedom dynamic tasks like jumping due to model simplifications, difficulties in parameter tuning, and poor robustness to environmental disturbances (Fig. 1).

Fig. 1. Jumping Motions of Bipedal Robots

With the rapid development of artificial intelligence, particularly the rise of Deep Reinforcement Learning (DRL), new solutions have emerged for addressing the aforementioned challenges. DRL enables robots to autonomously learn end-to-end mapping strategies from sensor inputs to joint commands through extensive trial and error in simulated environments, without relying on precise system dynamics models. This capability reveals significant potential in handling complex and uncertain real-world environments. The literature reviewed in this paper collectively forms the cutting-edge research landscape of reinforcement learning-based bipedal robot jumping control, reflecting the technological evolution from foundational theoretical exploration to practical application validation.

The research work in these papers primarily revolves around several interrelated core themes. First, there is innovation in basic algorithms and training frameworks. Several papers focus on improving reinforcement learning algorithms themselves to enhance learning efficiency and policy performance. For instance, Paper 1 introduces the Proximal Policy Optimization (PPO) algorithm, which, due to its simplicity, stability, and efficiency, has become one of the standard tools in this field [1]. Paper 2 further proposes the Concurrent Teacher-Student (CTS) learning architecture, which simultaneously trains both teacher and student policy networks, effectively improving the robustness and generalization capabilities of the policy [2]. Paper 3 addresses the instability issues encountered during real-robot training and proposes the Generalized State-Dependent Exploration (gSDE), which enables safe and efficient exploration learning directly on real hardware [3].

Next, breakthroughs in "simulation-to-reality" transfer techniques. Due to the high cost and significant risks associated with directly training reinforcement learning on real robots, the majority of research adopts the paradigm of "training in simulation first, then transferring to reality." The work in Papers 4 and 5 demonstrates that even with only proprioceptive feedback, zero-shot successful transfer from simulation to the real world can be achieved through carefully designed reward functions and randomized training [4, 5]. Additionally, the Rapid Motor Adaptation (RMA) algorithm proposed in Paper 6 allows the robot to adapt in real-time to new situations (such as terrain or load changes) within fractions of a second, significantly enhancing the robustness of the policy in the real world [6]. These achievements lay a solid foundation for the application of reinforcement learning on real robots.

Third, the construction of a perception-action closed-loop. Early reinforcement learning controllers were often "blind," relying solely on proprioception. However, to achieve truly intelligent jumping motions, such as leaping across gaps, climbing stairs, or jumping on stepping stones, external perceptual information becomes crucial. Papers 7 and 8 demonstrate how a single forward-facing depth camera, combined with end-to-end reinforcement learning, enables the robot to navigate complex terrains such as stairs, curbs, and stepping stones [7, 8]. The Cross-Modal Supervision (CMS) method proposed in Paper 9 cleverly uses time-shifted proprioceptive feedback to supervise the learning of the vision module, addressing the difficulty of simulating RGB images [9]. These studies mark a transition in bipedal robots from passively responding to the environment to actively perceiving and predicting it.

Fourth, the generation and combination of high-dynamic motion skills. Jumping itself is a complex skill, and achieving advanced behaviors such as continuous jumping or parkour requires the coordination of multiple skills. Paper 10 explores the learning of high-jumping skills for humanoid robots and accelerates the training process through curriculum learning strategies [10]. Papers 8 and 11 take this further, proposing systems that can learn various parkour skills such as climbing, jumping, and crawling, and autonomously select and execute the most appropriate skill based on the current scene [8, 11]. The probabilistic reward framework introduced in Paper 12 even allows a single policy to seamlessly transition between multiple gaits, including standing, walking, running, and jumping, demonstrating unprecedented motion flexibility [12].

Finally, theoretical foundations and explorations of generality. Some papers approach reinforcement learning from a deeper theoretical perspective, providing guidance for its application. Papers 13 and 14 use group theory to analyze the symmetry of legged motions, offering a theoretical basis for designing more efficient and natural gait generators [13, 14]. Paper 15 provides a comprehensive review of humanoid robot technologies, highlighting key directions for future development [15]. These works not only support jumping control but also offer valuable insights for the broader field of legged robotics.

In summary, these papers systematically discuss the current state of research on reinforcement learning-based bipedal robot jumping control from multiple dimensions, including algorithms, frameworks, perception, skills, and theory. Together, they paint a grand picture of continuous technological progress and increasingly widespread applications, while also clearly identifying the challenges faced today and the opportunities for future development. It should be noted that the depth of discussion across themes is not completely uniform. Vision-based terrain adaptation receives more detailed technical treatment because it represents the most rapidly evolving and technically complex research direction in recent years, whereas whole body coordinated jumping, though emerging and important, currently has fewer mature works and is therefore reviewed more concisely. This balance reflects both the relative maturity of each subfield and their relevance to the survey's objectives. The following sections of this paper will delve deeper into the specific research status, key issues, and future trends of each technological branch.

2 Current Research Status and Development Trends

2.1 The Application of Reinforcement Learning in Bipedal Robot Jumping Motion

Reinforcement Learning (RL), as a machine learning method that learns optimal behavior strategies through interactions between an agent and its environment, is based on the core idea of maximizing long-term cumulative rewards. In the field of bipedal robot control, RL provides an effective way to generate complex and adaptive motion strategies without the need for precise physical models. For jumping, a highly dynamic and nonlinear motion task, traditional model-based control methods often require simplifying assumptions about complex contact dynamics, which can lead to a decline in performance when dealing with uncertainties in the real world. In contrast, RL, through extensive trial-and-error learning, can implicitly capture system dynamics and find robust strategies to handle various disturbances.

Mainstream Algorithms and Training Paradigms. In the reviewed literature, policy gradient-based algorithms, particularly Proximal Policy Optimization (PPO), have become the mainstream choice for solving bipedal robot motion control problems. PPO, introduced by Schulman et al. [1], improves stability during learning by introducing a "surrogate" objective function and a clipping mechanism, allowing for larger learning rates and more training steps while achieving a good balance between sample efficiency and stability. Experimental results in Paper 1 show that PPO outperforms other online policy gradient methods in benchmark tasks such as simulated robot locomotion and Atari games, making it the foundation for many subsequent works [1]. For example, the Concurrent Teacher-Student (CTS) architecture proposed in Paper 2 is built on an improved PPO approach, using data samples from both teacher and student policies to update the network, further enhancing policy learning efficiency and performance [2]. Paper 17 also used PPO as a baseline algorithm when studying symmetry constraints, validating the superiority of rigid equivariant constraints [17].

In addition to PPO, other algorithms such as Soft Actor-Critic (SAC) have also been applied to robot learning. Paper 3 combines Generalized State-Dependent Exploration (gSDE) with SAC and successfully conducted direct training on multiple real robots, demonstrating the effectiveness of this combination [3]. Furthermore, the probabilistic reward framework proposed in Paper 12, although not specifying a particular algorithm, offers a design philosophy that is compatible with policy gradient methods like PPO. It provides a new approach for defining reward functions for complex motion tasks [12].

Sim-to-Real Transfer. Due to the safety risks and high time costs associated with large-scale reinforcement learning training on real robots, "simulation-to-reality" (Sim-to-Real) transfer has become a key step in making RL practical. The core challenge of this paradigm lies in the "reality gap," which refers to the differences between the simulated environment and the real world in terms of dynamics, sensor noise, actuator characteristics, and other factors. The reviewed literature proposes various strategies to bridge this gap (Fig. 2).

Fig. 2. Sim-to-Real: A Critical Component for Deploying Reinforcement Learning in Robotics

Domain Randomization is one of the most commonly used and effective methods. By randomizing various physical parameters (such as mass, friction coefficients, joint damping, etc.) and environmental conditions (such as terrain textures, obstacle locations, etc.) during simulated training, it forces the policy to learn a conservative and robust behavior pattern, enabling it to adapt to unseen changes in the real world. Paper 4 explicitly states that guiding the policy to be sufficiently conservative through random modeling is key to achieving zero-shot transfer success [4]. Similarly, the RMA algorithm in Paper 6 also relies on extensive domain randomization to train its adaptation module [6].

The dynamic response of actuators (motors) is a significant source of the reality gap. Paper 4 proposes the use of an unsupervised actuator network (UAN) to learn the actuator's dynamic model, enabling more realistic reproduction of its behavior in simulation and effectively mitigating the "reward hacking" problem [4]. This approach does not rely on torque sensors, making it more practical.

Unlike the approach of trying to perfectly match the real world in simulation, another idea is to equip the policy with real-time adaptation capabilities. The Rapid Motor Adaptation (RMA) algorithm proposed in Paper 6 is a prime example of this concept. RMA consists of a base policy and an adaptation module, the latter of which can adjust the policy in fractions of a second based on proprioceptive feedback to handle new terrain or load changes, achieving outstanding online adaptation performance [6].

These techniques collectively form the mainstream approaches for Sim-to-Real transfer, enabling reinforcement learning policies to be directly deployed onto real robots and achieve success without the need for fine-tuning. This has greatly advanced the practical application of reinforcement learning in the field.

Evolution of Exploration Strategies. Efficient exploration is key to the success of reinforcement learning, especially in continuous action spaces. Traditional Gaussian noise exploration can lead to erratic, unstable movements, increasing the risk of damage. To address this issue, Paper 3 proposes Generalized State-Dependent Exploration (gSDE). gSDE correlates exploration noise with the robot's state (or learned features) and periodically re-samples the noise, generating smoother exploration behavior that aligns better with the current motion state [3]. This exploration strategy not only enhances the safety of the learning process, allowing it to be directly conducted on real robots, but also accelerates convergence. Experimental results show that when combined with SAC, gSDE achieves competitive results on multiple benchmark tasks while significantly reducing wear during training [3]. This advancement paves the way for directly learning complex skills on real systems in the future.

Leveraging Prior Knowledge: Symmetry Constraints. Bipedal robots typically have a left-right symmetrical structure, and their movements should exhibit corresponding symmetry. However, standard deep neural networks do not inherently possess this property, which may lead to asymmetrical gaits that negatively affect both performance and aesthetics. Papers 17, 14, 20, and 32 focus on how to integrate symmetry prior knowledge into reinforcement learning.

Papers 17 and 14 propose constructing neural network layers that are equivariant under reflection or rotation. Paper 14 refers to this as the MDP Homogeneous Network, where weight binding is used to enforce the network to produce symmetric outputs for symmetric state-action pairs, structurally ensuring the symmetry of the policy [14]. Experimental results show that this hard constraint method outperforms soft methods, such as data augmentation, in terms of sample efficiency and task performance [17].

Papers 20 and 32 explore encouraging symmetry through data augmentation (by duplicating and mirroring training data) and designing mirror loss functions. This approach is simple to implement and does not require modifications to the network architecture, but theoretically, it is less rigorous than the hard constraint methods [20, 32].

Leveraging prior knowledge such as symmetry not only enables the generation of more natural and efficient movements, but also effectively reduces the policy search space, enhancing learning efficiency. This is an important strategy for improving the performance of reinforcement learning controllers. In addition to the hard constraints in network architecture, research has also explored methods for incorporating symmetry invariance into goal-conditioned tasks, such as data augmentation and mirror loss functions. These approaches have been shown to significantly improve learning efficiency and policy symmetry in tasks like quadruped climbing and dexterous manipulation [11].

2.2 Learning Dynamic Jumping Skills

Dynamic jumping skills are fundamental to enabling agile movement in bipedal robots. Unlike steady walking, jumping involves both a flight phase and a stance phase, with significant energy transitions, complex ground impacts, and precise landing balance control. This places high demands on the dynamic performance and robustness of the controller. Reinforcement learning-based approaches, through end-to-end learning, can automatically discover and optimize the full-body coordinated movements required for jumping, avoiding the cumbersome manual trajectory planning and parameter tuning typical of traditional methods. We categorize these works under dynamic jumping because their main contribution focuses on generating high-performance jump behaviors, even though some also involve elements of sim-to-real or perception.

Generation of Single Jumping Skills. Achieving a single jump is the first step toward more complex dynamic movements. Researchers have successfully used reinforcement learning to teach bipedal robots the complete process of jumping from a standing position, adjusting their posture in the air, and landing stably.

Paper 10 proposed a method for learning high-jumping skills for humanoid robots and validated it through simulation on the Ranger Max humanoid robot model [10]. The key to this study was the design of a reasonable reward function to encourage the robot to maximize jump height. The reward terms typically included positive rewards for jump height, rewards for landing stability (such as torso posture and foot contact forces), and penalties for energy efficiency. In this way, the policy was able to autonomously learn how to coordinate joint efforts at the hips, knees, and ankles to generate sufficient vertical impulse. The study was also inspired by the concept of "injury recovery" and introduced a curriculum learning strategy: training began on a "stronger" robot with looser motor torque limits, and once the policy had converged to some extent, the limits were gradually tightened, transferring the training to a "weaker" real robot. This progressive training approach effectively accelerated the learning process and improved the performance of the final policy [10].

Paper 16 took the research a step further by proposing an RL controller based on a dual-history architecture, which not only enabled standing long jumps and high jumps but also demonstrated outstanding performance on the real bipedal robot, Cassie [16]. The study achieved a forward jump of up to 1.4 m and a high jump of 0.44 m, representing a significant advancement in bipedal robot jumping capabilities at the time. The key to its success was the controller's ability to effectively utilize both the long-term and short-term input/output (I/O) history of the robot, allowing it to better adapt to time-varying dynamics, such as contact events [16] (Fig. 3).

Fig. 3. Jumping Performance Achieved by the Cassie Bipedal Robot

Multi-Gait Continuous Jumping and Transition. A single jump is just the starting point. Achieving continuous jumps (such as consecutive jumps in place or jumping while moving) and smoothly transitioning between different gaits is a higher standard for measuring a robot's agility and movement flexibility.

Paper 12 made breakthrough progress in this area. The study proposed a reward specification framework based on a simple probabilistic periodic cost composed of basic forces and velocities [12]. The clever aspect of this framework is that it does not require reference motion trajectories, thus avoiding the problem of restricting the policy to a specific action pattern. By adjusting the parameters in the reward function, the policy can be intuitively guided to learn common bipedal gaits, such as standing, walking, running, jumping, and galloping. More importantly, this framework successfully trained a general policy capable of seamlessly transitioning between all two-step gaits (such as walking, running, and jumping), even switching in real-time during motion according to commands [12]. For example, the robot can smoothly accelerate from walking to running, then transition from running to jumping, with the entire process appearing fluid and natural. This achievement greatly enhances the agility and adaptability of bipedal robots and is an important step towards developing a "universal motion controller" [12].

The methods proposed in Papers 20, although focused on quadrupedal robots, offer valuable insights for bipedal robots through their "hierarchical learning and control framework" [20]. This framework includes a learned height map predictor for terrain perception, an RL-based centroid-level motion policy for motion planning, and a low-level leg controller for precise tracking [20]. The hierarchical design decomposes complex jumping tasks, with high-level strategies responsible for deciding "how high and how far to jump," and low-level controllers tasked with precise execution, effectively reducing the learning complexity. This approach enabled the Unitree Go1 robot to achieve continuous jumps across two steps at a time on stairs, completing the challenge of 14 steps in 4.5 s [20]. For bipedal robots, a similar hierarchical framework could be applied to accomplish continuous long-distance jumps across multiple obstacles.

Cross-Paper Insights. Although the surveyed approaches vary in objectives—ranging from single high jumps to continuous long-horizon sequences—they share a common reliance on stabilizing contact transitions and generating coordinated whole-body impulses. Reference-based controllers offer interpretability but limited generality, while end-to-end policies learn richer motion patterns but require careful reward shaping and domain randomization. Methods incorporating historical information (e.g., dual-history architectures) appear particularly promising for handling variable contact conditions, suggesting that temporal abstraction is key for improving robustness in real-world deployment.

Challenges and Future Directions in Dynamic Jumping. Although notable progress has been made in learning dynamic jumping skills, several challenges persist. Reward design remains a central difficulty—poorly structured rewards may lead to "reward hacking," where policies exploit loopholes rather than truly accomplishing the task. For instance, a robot may intentionally crash onto the ground to produce misleading height signals. The Unsupervised Actuator Network (UAN) proposed in [4] addresses this issue by enforcing physically plausible behaviors. Another challenge lies in long-horizon planning for continuous jumping. Robots must accurately predict landing locations across multiple jumps while maintaining stability, yet most current methods focus on single or short-sequence jumps. "Blind" controllers such as the one in [5] further highlight the limitation of relying solely on proprioception, lacking foresight for obstacle-aware jumping. Energy efficiency also remains a concern, as jumping is inherently power-intensive and current approaches often rely on coarse penalty designs, leaving room for more sophisticated energy-aware strategies.

Looking forward, dynamic jumping skill learning is expected to progress toward more intelligent, efficient, and generalizable systems. Future controllers should combine jumping with other dynamic skills—such as mid-air disturbance recovery or post-jump rolling—to achieve parkour-like agility. After deployment, robots will require continual adaptation to new environments, supported by online learning and transfer learning mechanisms. Additionally, incorporating insights from human and animal neuromechanics may inspire biologically grounded RL frameworks that yield more natural and efficient jumping behaviors.

2.3 Vision-Based Terrain-Adaptive Jumping

The jumping capability of bipedal robots is highly constrained in flat, known environments. To achieve true adaptability, robots must perceive the terrain and adjust jump-ing strategies according to step height, gap width, or stepping stone distribution. In recent years, integrating visual perception with reinforcement learning has become a central approach for terrain-adaptive jumping.

From "Blind" Motion to Vision-Guided Movement. Early RL controllers relied solely on proprioception (joint states, IMU), which is easy to model and supports zero-shot transfer [4]. Such controllers show robustness on deformable terrains like mud and rubble [5], yet lack the ability to anticipate future terrain, leading to inefficiency or dangerous failures on cliffs or gaps [5, 9] (Table 1).

Table 1. Transition from "Blind" Motion to Vision-Guided Jumping

Aspect	Blind (Proprioception Only) Control	Vision-Guided Control
Sensor Input	Joint angles, joint velocities, IMU only	RGB / depth camera + proprioception
Terrain Awareness	No long-range perception; reacts only on contact	Long-range terrain prediction before contact
Advantages	Easy to simulate; small sim-to-real gap; high robustness on deformable terrain	Enables proactive planning; suitable for stairs, stepping stones, and gaps
Limitations	Cannot anticipate hazardous terrain (e.g., cliffs, pits); inefficient responses	Sensitive to lighting, weather, occlusion; requires heavier computation
Representative Work	Paper 5: robust blind locomotion on mud, snow, rubble [5]	Paper 9: vision-guided gait selection and stair navigation [9]

To address this, onboard RGB or depth cameras were introduced to provide long-range percep-tion. Vision enables predicting terrain changes and supports navigation on visually dependent terrains such as stairs and stepping stones [9].

End-to-End Vision-Motion Learning. End-to-end frameworks directly map visual inputs (e.g., depth images) to joint com-mands via a single neural network. Paper 7 introduced the first system capable of handling stairs, curbs, stepping stones, and gaps using a depth camera, trained with RL then distilled into a lightweight policy for real-time execution [7]. Paper 8 further extended this paradigm by first learning multiple parkour skills and consolidating them into a unified vision-based controller, enabling low-cost robots to autonomously navigate complex terrain [8].

Hierarchical and Modular Vision Fusion. Due to the instability and opaqueness of end-to-end structures, hierarchical or modular architectures are commonly adopted (Table 2).

Table 2. End-to-End vs. Hierarchical Vision-Motion Learning

Aspect	End-to-End Vision-Motion Learning	Hierarchical / Modular Vision-Control Fusion
System Architecture	Single neural network performing perception → decision → control	Multiple modules: perception, planning, low-level control
Input to Policy	Raw images (RGB / depth)	Processed terrain maps, height maps, semantic info
Strengths	Simple pipeline; strong capability for complex visuomotor mapping	Stable learning; easier to debug; flexible integration with classical algorithms
Weaknesses	Hard to train; perception tightly coupled with control; low interpretability	Requires additional perception modules and training steps
Representative Work	Paper 7: end-to-end traversal of stairs, curbs, stepping stones [7]; Paper 8: multi-skill vision-based parkour [8]	Paper 20: height-map predictor + RL centroid planner + leg controller [20]; Paper 25: 4D terrain reconstruction from point clouds [25]

Paper 20 designed a system combining a learned height-map predictor, an RL-based centroid-level strategy, and a low-level leg controller [20]. Similarly, Paper 25 used a 4D fully convolutional network to reconstruct local terrain from noisy point clouds, producing clean maps for control [25]. These perception–planning–execution frameworks reduce learning difficulty and provide stronger robustness, and have shown success in both bipedal and quadrupedal systems [20, 25].

Cross-Modal Learning and Reality Gap. One of the major challenges when introducing vision into reinforcement learning is the difficulty of simulating RGB images. Real-world effects such as lighting, material properties, textures, and motion blur are extremely difficult to accurately reproduce in simulation. This results in a sharp performance drop of vision-based policies trained in simulation when deployed in the real world. Paper 9 introduces Cross-Modal Supervision (CMS) as an innovative solution to this problem [9]. CMS leverages shifted proprioceptive data (such as impact force upon landing or trunk posture changes) to supervise the learning of the visual module. The key idea is that when the robot steps onto a stair with its rear foot, its proprioception records this event, while the vision module should be able to predict this event from prior images. This allows the visual module to be trained using real-world data supervision, eliminating the need for simulated RGB images, effectively bypassing the issue of visual simulation [9]. This method demonstrated high-performance visual walking on complex terrains such as stairs and slopes with less than 30 min of real-world data, showing significant potential [9].

Cross-Paper Insights. End-to-end visual locomotion offers strong generality but suffers from fragility to perception noise, whereas hierarchical perception–planning–control pipelines trade generality for improved interpretability and robustness. Cross-modal supervision (CMS) provides a compelling solution to the visual-reality gap by anchoring visual predictions to reliable proprioceptive signals. However, existing systems still lack standardized benchmarks for long-range jumping, and few studies compare how perception choices affect robustness under lighting or weather disturbances, leaving an important empirical gap for future research.

Challenges and Future Directions in Vision-Based Terrain-Adaptive Jumping. Vision-based terrain-adaptive jumping continues to face significant challenges, particularly in enabling policies to effectively leverage visual information for long-range planning while simultaneously coordinating short-time-scale motion control. Many existing methods remain limited to short-range, reactive behaviors, and their ability to support long-distance jumps requiring precise landing prediction has yet to be fully validated. Real-world vision systems are also highly susceptible to lighting variation, weather conditions, dust, and occlusions, making the design of a robust perception module crucial for reliable terrain interpretation. Moreover, processing high-resolution visual data demands substantial computational resources, posing difficulties for lightweight models that must maintain perceptual accuracy under real-time onboard constraints.

Looking ahead, vision-based jumping is expected to evolve toward more intelligent and versatile frameworks that integrate multi-modal sensory inputs—from RGB and depth cameras to LiDAR, IMUs, and joint encoders—forming a more comprehensive and resilient environmental model. Beyond geometry, future systems will incorporate semantic understanding to support higher-level planning and decision-making. Building on approaches such as Cross-Modal Supervision (CMS), robots will increasingly rely on online learning from real-world data to continuously improve perception and motion performance, enabling adaptation to new environments and task variations over time.

2.4 Full-Body Coordinated Jumping Control

With the advancement of bipedal robot technology, its applications are no longer limited to simple locomotion but are evolving towards more complex mobility manipulation tasks. Jumping manipulation refers to robots executing dynamic movements like jumping while using their entire body (including arms and torso) to perform specific tasks, such as throwing objects, pushing or pulling heavy loads, or adjusting posture mid-air to counter disturbances. This requires the controller to achieve high coordination between the legs and arms, and between the lower and upper body, making it an important measure of the robot's overall capability. This section focuses on studies where full-body coordination—including arm–leg coupling—plays a central role in task execution, even if the underlying locomotion strategies relate to earlier sections.

Full-Body Coordination in Jumping. During jumping, the robot's center of mass (CoM) trajectory and angular momentum must be precisely controlled to ensure both takeoff power and landing stability. Traditional jumping control usually focuses only on the coordination of the lower limbs. However, the swinging of the upper body (especially the arms) can generate significant reactive forces and torques, which greatly impact jumping performance.

The "Dual History Architecture" proposed in Paper 16 provides an effective framework for achieving full-body coordination. This architecture encodes the robot's long-term and short-term input/output (I/O) history, allowing the policy to learn complex coordination patterns across multiple time steps. In jumping tasks, this architecture allows the policy to autonomously discover and use arm swings to assist with takeoff or stabilize the aerial posture. For example, when performing a high jump, the robot can swing its arms upward, like a human, to increase vertical momentum; during the flight phase, the robot can adjust arm positions to control its pitch angle, preparing for a smooth landing. The universal policy presented in Paper 12 also applies to jumps involving upper-body movements, demonstrating smooth full-body coordination during transitions between different gaits.

Jumping Elements in Motion Manipulation Tasks. Jumping can serve operational tasks, such as reaching high objects or crossing obstacles. In complex parkour scenarios, a fully learning-based approach trains high-level navigation to select and control low-level skills (jumping, climbing), integrating a perception module to reconstruct obstacles from occluded data for agile navigation [21]. This shows that skill-specific learning enables robots to handle highly dynamic motions.

Paper 4 introduces the "Athletic Manipulator," enhancing strength and workspace via full-body coordination [4]. Although focused on ground operations like lifting, throwing, and dragging, its two-stage training—pre-training a full-body controller, then fine-tuning with task rewards—can guide jumping manipulation. Basic jumping skills are first trained, followed by task-specific rewards (e.g., maximizing throw distance) to optimize arm movements during jumps [4].

Paper 24 proposed the Reward Fusion Module (RFM) for 6D end-effector pose tracking in wheeled quadruped robotic arms [24]. RFM nonlinearly integrates rewards from multiple tasks, handling hierarchical and multi-stage features. For jumping control, a composite reward can combine jumping performance (height, distance), manipulation accuracy, and energy efficiency, with RFM balancing these conflicting objectives to guide the policy toward an optimal solution [24].

Hierarchical and Collaborative Control Frameworks. Complex jumping manipulation tasks often require hierarchical or collaborative frameworks to manage task complexity. Hierarchical strategies have proven effective in complex environments; for instance, a two-layer strategy uses high-level 3D spatial understanding to navigate under hanging obstacles, enabling motion control in confined and unstructured spaces [22].

Paper 18's RoboDuet framework exemplifies dual collaborative strategies: one for speed tracking (motion) and one for 6D end-effector pose control (manipulation), achieving full-body control [18]. Though designed for quadrupeds, this concept applies to bipedal robots. In jumping tasks, a high-level planner can handle overall task goals (e.g., "jump and grab a handle"), while two lower-level sub-strategies manage jump stability and precise arm control, collaborating via shared state information [18].

For local manipulation, a low-level RL controller combined with a high-level behavior cloning planner allows trajectory parameterization and synchronization, supporting high-precision tasks during motion [23]. Paper 19's Visual Body Control (VBC) framework similarly uses a low-level controller for body speed and end-effector tracking and a high-level planner to provide commands from visual input, enabling the robot to decide when, where, and how to jump and control its end-effector [19] (Fig. 4).

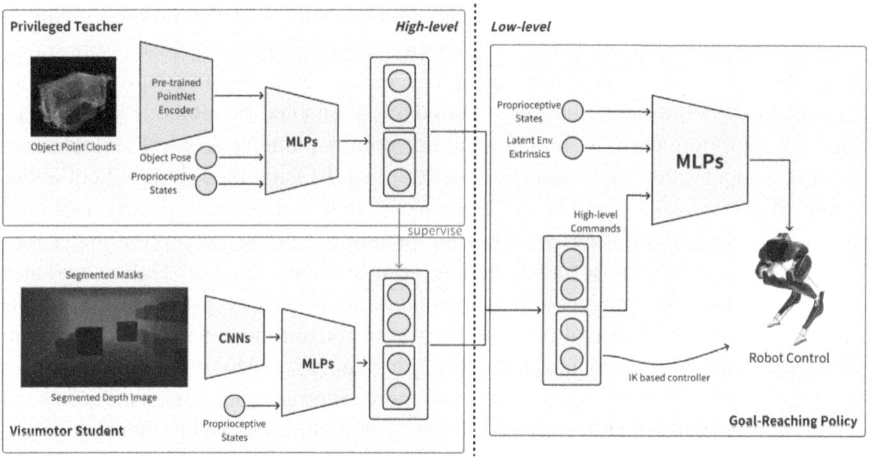

Fig. 4. Visual Whole-Body Control (VBC) Framework

Integrated Challenges and Future Directions in Full-Body Coordinated Jumping Manipulation. Full-body coordinated jumping manipulation presents challenges far beyond those of basic jumping. Bipedal robots have high-dimensional and redundant degrees of freedom, making it difficult to simultaneously satisfy locomotion and manipulation objectives—an NP-hard problem that greatly expands the state–action space explored by reinforcement learning, often resulting in slow and unstable training. Conflicting task goals further complicate control design: while precise manipulation may require steady or slow movement, powerful jumping demands rapid force generation, requiring careful reward shaping and architectural design to resolve these trade-offs. Moreover, performing manipulation during dynamic jumps imposes extreme real-time constraints, as complex neural inference and large-volume sensor processing must be executed with minimal latency on embedded hardware.

Looking forward, full-body coordinated jumping manipulation is expected to evolve toward more integrated and autonomous systems that unify perception, locomotion, and manipulation into a single closed-loop framework. Future robots will autonomously plan composite tasks that combine jumping and manipulation, leveraging multimodal perception and real-time decision-making. Incorporating imitation learning can further accelerate training and produce more intuitive, human-like behaviors. Ultimately, these capabilities are anticipated to support demanding real-world applications, including firefighting, disaster response, and space exploration, enabling robots to perform highly dynamic operations in extreme environments.

3 Conclusion

This survey reviewed recent advances in reinforcement learning–based jumping control for bipedal robots, highlighting key methods, challenges, and emerging trends. Reinforcement learning, especially deep RL, has become a central driver for enabling highly dynamic and agile jumping, supported by policy gradient algorithms such as PPO, advanced training frameworks like CTS, and effective sim-to-real techniques including domain randomization and rapid motor adaptation. Research has progressed from single-jump skills to multi-gait continuous jumping, vision-based terrain-adaptive jumping, and whole-body coordinated jumping manipulation, enabled by methods such as end-to-end visual-motion learning, hierarchical perception–planning–control architectures, cross-modal supervision, dual-history policies, reward-fusion modules, and collaborative control frameworks. Despite these advances, major challenges persist, including robust reward design, long-horizon dynamic planning, reliable perception under real-world noise, efficient semantic understanding, coordination under high DoF redundancy, and strict real-time requirements. The current literature also lacks systematic evaluations of perception robustness in real outdoor environments, and most whole-body jumping manipulation approaches remain validated only in simulation. Moving forward, research would benefit from unified benchmarks and more rigorous comparisons across sensing modalities, control architectures, and sim-to-real strategies, while development is expected to advance toward more general, intelligent, and autonomous systems through multimodal sensory fusion, online adaptation, bio-inspired design, and deeper integration of locomotion and manipulation, ultimately enabling bipedal robots to perform agile, resilient, and versatile tasks in complex real-world environments.

Disclosure of Interests. The authors have no competing interests to declare that are relevant to the content of this article.

References

1. Schulman, J., Wolski, F., Dhariwal, P., et al.: Proximal policy optimization algorithms (2017). arXiv preprint arXiv:1707.06347
2. Wang, H., Luo, H., Zhang, W., et al.: Cts: Concurrent teacher-student reinforcement learning for legged locomotion. IEEE Robotics and Automation Letters (2024)
3. Raffin, A., Kober, J., Stulp, F.: Smooth exploration for robotic reinforcement learning. Conference on robot learning. PMLR, pp. 1634–1644 (2022)
4. Hwangbo, J., Lee, J., Dosovitskiy, A., et al.: Learning agile and dynamic motor skills for legged robots. Science Robotics **4**(26), eaau5872 (2019)

5. Lee, J., Hwangbo, J., Wellhausen, L., et al.: Learning quadrupedal locomotion over challenging terrain. Science Robotics **5**(47), eabc5986 (2020)
6. Kumar, A., Fu, Z., Pathak, D., et al.: Rma: Rapid motor adaptation for legged robots (2021). arXiv preprint arXiv:2107.04034
7. Agarwal, A., Kumar, A., Malik, J., et al.: Legged locomotion in challenging terrains using egocentric vision. Conference on Robot Learning. PMLR, pp. 403–415 (2023)
8. Zhuang, Z., Fu, Z., Wang, J., et al.: Robot parkour learning (2023). arXiv preprint arXiv: 2309.05665
9. Loquercio, A., Kumar, A., Malik, J.: Learning visual locomotion with cross-modal supervision (2022). arXiv preprint arXiv:2211.03785
10. Li, J., Ye, L., Sun, Y., et al.: Recovery from injury: learning bipedal jumping skills with a motor output torque limit curriculum. International Conference on Intelligent Robotics and Applications. Singapore: Springer Nature Singapore, pp. 577–588 (2023)
11. Hoeller, D., Rudin, N., Sako, D., et al.: Anymal parkour: Learning agile navigation for quadrupedal robots. Science Robotics **9**(88), eadi7566 (2024)
12. Siekmann, J., Godse, Y., Fern, A., et al.: Sim-to-real learning of all common bipedal gaits via periodic reward composition. 2021 IEEE International Conference on Robotics and Automation (ICRA), pp. 7309–7315. IEEE (2021)
13. Ding, J., Gan, Z.: Breaking symmetries leads to diverse quadrupedal gaits. IEEE Robo. Auto. Lett. **9**(5), 4782–4789 (2024)
14. Van der Pol, E., Worrall, D., van Hoof, H., et al.: Mdp homomorphic networks: Group symmetries in reinforcement learning. Adv. Neural. Inf. Process. Syst. **33**, 4199–4210 (2020)
15. Sheng, Q., Zhou, Z., Li, J., et al.: A comprehensive review of humanoid robots. SmartBot **1**(1), e12008 (2025)
16. Li, Z., Peng, X.B., Abbeel, P., et al.: Reinforcement learning for versatile, dynamic, and robust bipedal locomotion control. The Int. J. Robo. Res. **44**(5), 840–888 (2025)
17. Su, Z., Huang, X., Apraez, D.O., et al.: Leveraging symmetry in rl-based legged locomotion control. In: 2024 IEEE[C]//RSJ International Conference on Intelligent Robots and Systems (IROS), pp. 6899–6906
18. Pan, G., Ben, Q., Yuan, Z., et al.: RoboDuet: Learning a Cooperative Policy for Whole-body Legged Loco-Manipulation. IEEE Robotics and Automation Letters (2025)
19. Liu, M., Chen, Z., Cheng, X., et al.: Visual whole-body control for legged loco-manipulation (2024). arXiv preprint arXiv:2403.16967
20. Yang, Y., Shi, G., Lin, C., et al.: Agile continuous jumping in discontinuous terrains. 2025 IEEE International Conference on Robotics and Automation (ICRA), pp. 10245–10252. IEEE (2025)
21. Miki, T., Lee, J., Wellhausen, L., et al.: Learning to walk in confined spaces using 3d representation. 2024 IEEE International Conference on Robotics and Automation (ICRA), pp. 8649–8656. IEEE (2024)
22. He, Z., Lei, K., Ze, Y., et al.: Learning visual quadrupedal loco-manipulation from demonstrations. In: 2024 IEEE RSJ International Conference on Intelligent Robots and Systems (IROS), pp. 9102–9109
23. Jiang, K., Fu, Z., Guo, J., et al.: Learning Whole-Body Loco-Manipulation for Omni-Directional Task Space Pose Tracking with a Wheeled-Quadrupedal-Manipulator. IEEE Robotics and Automation Letters (2024)
24. Hoeller, D., Rudin, N., Choy, C., et al.: Neural scene representation for locomotion on structured terrain. IEEE Robo. Auto. Lett. **7**(4), 8667–8674 (2022)
25. Mittal, M., Rudin, N., Klemm, V., et al.: Symmetry considerations for learning task symmetric robot policies. 2024 IEEE International Conference on Robotics and Automation (ICRA), pp. 7433–7439. IEEE (2024)

Research and Application of Cobot Adaptive Machine Tool Loading and Unloading Based on AI Vision

Lu Jie(✉), Zhao Jiangying, and Cai Anxu

Magna Powertrain (Jiangxi) Co., Ltd., Jiangxi, China
jie.lu1@magna.com

Abstract. With the advancement of Industry 4.0, the manufacturing industry is undergoing a profound transformation. As a vital component of smart manufacturing, cobot play a crucial role in enhancing production efficiency and reducing costs. With technological progress, the performance of cobot is continuously improving while costs are gradually decreasing. This paper focuses on integrating cobot with AI vision in production, achieving quality error prevention and precise guidance through AI-enabled visual self-learning. The project resulted in a 30% increase in loading/unloading efficiency and a 15% reduction in the workpiece scrap rate. Research indicates that the application of cobot on machine tools lays a foundation for the realization of intelligent factories.

Keywords: Cobot · AI vision · Quality improvement, Intelligentization

1 Introduction

Industrial robots are multifunctional, multi-degree-of-freedom mechatronic devices based on repeatable programming and automatic control, capable of efficiently performing various operational tasks in the manufacturing process. They feature high precision, high flexibility, and high reliability, making them one of the core pieces of equipment in modern smart manufacturing. They are typically used to perform repetitive, strenuous, and high-precision tasks in industries such as automotive, casting, and furniture manufacturing. Cobots, on the other hand, are intelligent robots capable of working synergistically with humans on production lines. It combines the high efficiency and precision of robots with the flexibility and creativity of humans, enabling seamless human-robot collaboration through safe interaction and intelligent sensing technologies. Such robots are not only cost-effective but also safe and convenient, significantly promoting the development of manufacturing enterprises. The concept of Cobots: compared to traditional industrial robots, they feature a smaller size, higher flexibility, and enhanced safety, allowing them to work directly alongside humans. The differences between the two are shown in Table 1:

Cobots exhibit excellent flexibility and ease of use. Traditional industrial robots typically require complex programming and debugging processes, whereas Cobots employ

H. Wei and J. Niu (Eds.): COBOT 2025, CCIS 2893, pp. 478–487, 2027.
https://doi.org/10.1007/978-981-92-2074-8_28

Table 1. Differences between robots

ROBOT	COBOT
Unresponsive to environmental changes	Senses the surrounding environment and actively stops upon contact
Low safety due to its heavy weight and high speed	High safety (lightweight, safe operating speeds), compliant with ISO 10218 and ISO 15066
Emphasizes precision and simple, repetitive tasks	Emphasizes flexibility and ease of use
Low integration with peripheral accessories	Plug-and-play ecosystem
Requires engineers who have undergone long-term professional training	Can be operated by general engineers

intuitive programming interfaces, such as teach pendants or drag-and-drop programming, enabling even non-professionals to get started quickly. Cobots are simple to operate, easy to move and redeploy, and adapt to multi-variety, small-batch production needs.

Cobots also feature high precision and high repeatability. Although the payload capacity of Cobots is typically lower than that of traditional industrial robots, they excel in precision operations and high-repeatability tasks. Therefore, researching the application of AI vision Cobots in machine tool loading and unloading provides a new technological path for the development of smart manufacturing and industrial automation.

1.1 Overall Solution Design

Manual operation is the most traditional method for machine tool loading and unloading. Workers need to manually place raw materials onto the machine tool and remove the finished products after processing. Although this method is simple and direct, it suffers from issues such as low efficiency, high labor intensity, and being error-prone. Firstly, the efficiency of manual operation is limited by the worker's physical strength and attention, making it difficult to maintain high efficiency for long periods. Secondly, repetitive labor can easily lead to worker fatigue, increasing the risk of operational errors, which in turn affects product quality and production safety. Furthermore, manual operation also faces challenges such as rising labor costs and difficulty in recruitment.

In the overall approach of applying Cobots to machine tool loading and unloading, they achieve high-precision motion control and intelligent decision-making by integrating advanced sensors, control systems, and AI algorithms. The system architecture mainly includes the robot body, flexible fixture, AI vision system, control system, and expert team support system. As shown in Fig. 1:

In the requirements analysis phase, the specific requirements for machine tool loading and unloading must be clarified, including workpiece type, production takt time, and precision requirements. The solution design phase then involves designing the robot body, flexible fixture, and AI vision system based on these requirements, and determining the architecture and algorithms of the control system. The system integration phase integrates the various components to ensure all parts can work synergistically.

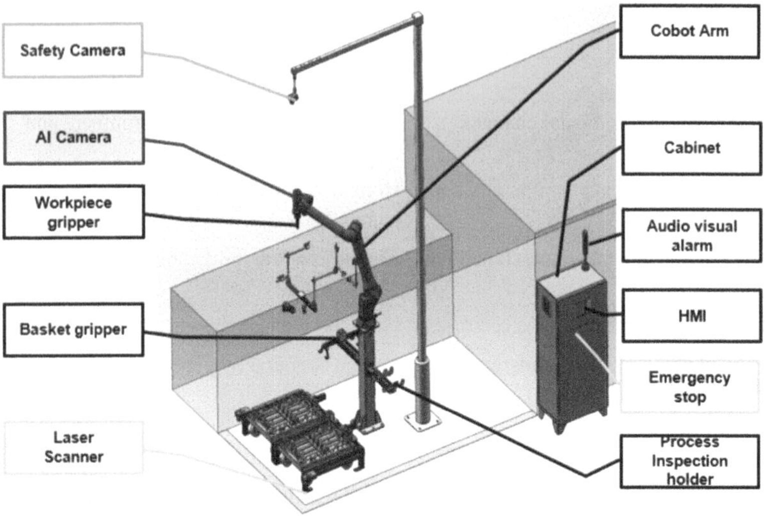

Fig. 1. Overall Layout of the Cobot

The debugging and optimization phase optimizes system performance through repeated testing and adjustment, ensuring its reliability and stability in practical applications. Finally, in the practical application phase, the system is put into production to conduct an effectiveness evaluation.

2 Overall Implementation

2.1 Application of Flexible Fixtures

Flexible fixtures are core equipment for Cobots when handling multi-variety production tasks. Compared to the single adaptability of traditional dedicated fixtures, new flexible fixtures achieve rapid and precise switching for workpieces of different sizes and shapes through innovative modular designs and pneumatic/electric adaptive gripping mechanisms. As shown in Fig. 2:

(a) Modular Combination: The fixture adopts a modular design, facilitating quick combination and replacement according to different workpiece characteristics.
(b) (b) Adaptive Gripping: Pneumatic/electric adaptive gripping mechanisms can flexibly adjust the gripping method to adapt to various workpiece shapes and sizes.
(c) Intuitive and Easy Operation: As shown in Fig. 2, the slide groove structure design makes the adjustment of end fingers intuitive and easy to operate, further simplifying the changeover process.

This solution can greatly enhance the collaborative robot's ability to adapt to multi-variety, small-batch production tasks. The rapid changeover mechanism reduces changeover costs, improves the flexibility and economy of the production line, and variable gripping force control ensures consistency and high quality during the processing of different workpieces.

Fig. 2. Flexible Fixture

2.2 Design of Gear Flipping Mechanism

The gear flipping mechanism is specialized equipment designed for flipping gears of different weights, types, and diameters. Design requires comprehensive consideration of factors such as workpiece characteristics, production takt time, and equipment compatibility to ensure an efficient and stable production process. Adopting a modular design concept based on different workpiece characteristics and production needs facilitates combination and replacement. It can be automatically integrated with other equipment on the production line, achieving automated control and information management. Equipped with safety protection devices to ensure the safety of operators and equipment, as shown in Figure 3

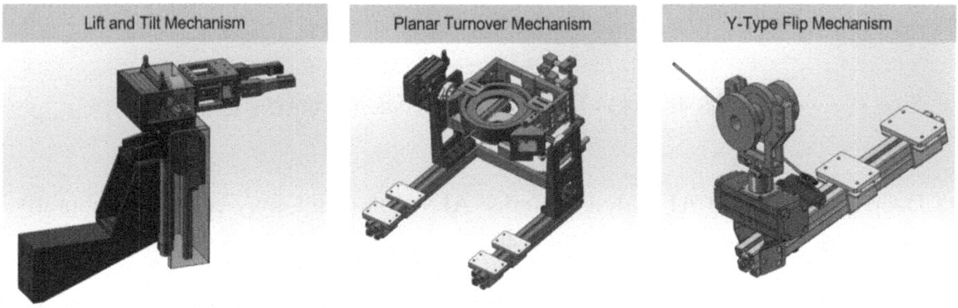

Fig. 3. Flipping Mechanism

2.3 AI Vision Technology

In the implementation of this project, AI vision is a key applied technology. The AI vision application utilizes computer vision and deep learning technologies to achieve precise recognition and positioning of workpieces. In the machine tool loading and unloading process, workpiece recognition and positioning are critical steps, directly impacting the precision and efficiency of loading/unloading. The AI vision system, through high-resolution cameras and the use of advanced mathematical models and image processing algorithms, can rapidly identify workpiece shape, size, and position information. Additionally, the AI vision system can also incorporate 3D vision technology to obtain 3D coordinate information of the workpiece, further improving positioning accuracy. As shown in Figure 4

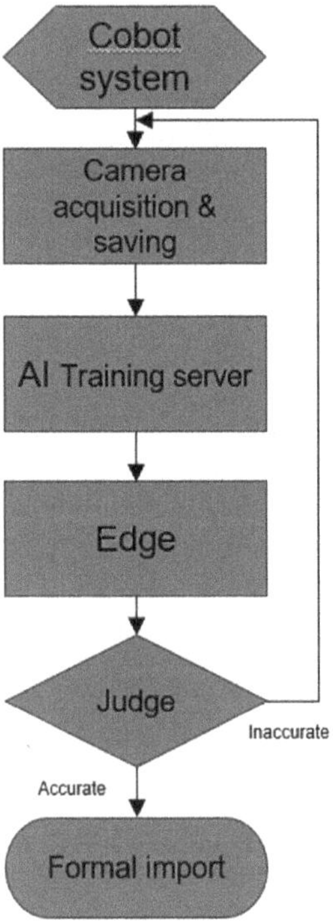

Fig. 4. Robot Vision Decision-Making Flowchart

(1) Deep Integration of AI Vision Systems: AI vision technology equips collaborative robots with autonomous perception and decision-making capabilities, overcoming the limitations of traditional teach-pendant programming.
(2) Dynamic Positioning Compensation: Employs deep learning algorithms (e.g., YOLOv5) to achieve workpiece pose recognition, with a positioning accuracy of ±0.05mm. This addresses the issue of datum shift caused by machine tool vibration.

(3) Real-time Defect Detection: Performs visual quality inspection simultaneously during the loading and unloading process. In one case study, the use of a ResNet model reduced the miss rate from 1.2% (manual inspection) to 0.15%.
(4) Self-learning Path Planning: A reinforcement learning-based obstacle avoidance algorithm enables the robot to dynamically optimize its motion trajectory in complex layouts. The collision warning response time is less than 50ms (Fig. 5).

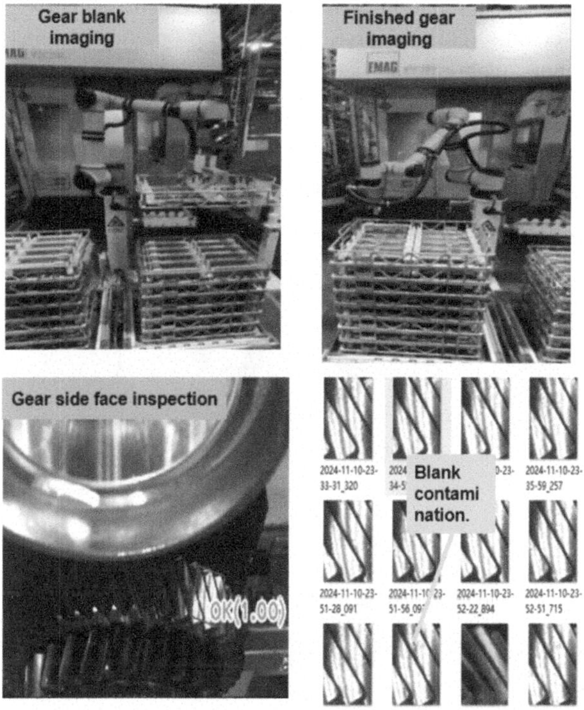

Fig. 5. AI Vision Judgmen

This system has the following advantages:

(1) Multiple judgments from a single capture. Clear imaging is achieved through customized light sources, lenses, parameters, and capture angles.
(2) The AI system employs various neural network algorithms to perform different tasks. Models trained on the server are downloaded into the robot controller for edge computing, eliminating the need for additional controllers.
(3) Perfect integration of the robot arm + vision system + AI algorithms. Simplifying complex tasks. One engineer completes all debugging. Completing what traditionally took two weeks within three days, saving 80% of the time。

2.4 AI Vision Judgment

The main components of the entire automation system. The safety system consists of five layers, as shown in Figure 6:

Layer 1: Visual safety cameras mounted overhead to cover the entire workspace of the robot arm (approx. 3000 x 5500 mm).

Layer 2: LiDAR sensors covering a range of 150-200 mm from ground level.

Layer 3: Emergency stop button on the robot teach pendant.

Layer 4: Emergency stop button on the electrical control cabinet.

Layer 5: The cobot's (collaborative robot's) stop-on-contact functionality, which enables human-robot collaboration.

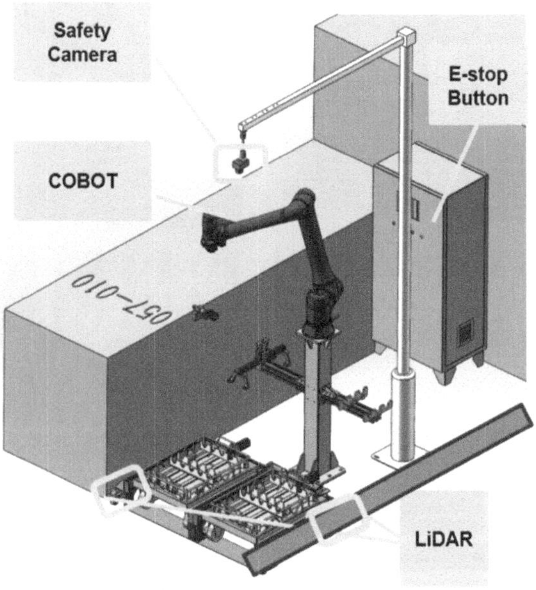

Fig. 6. Schematic Diagram of Protection Scheme

The vision security system can intelligently distinguish between people and objects. When a person or a part of their body enters a designated area, the camera immediately detects the intrusion and triggers a low-level signal through the connection to the robot controller's Safety I/O (SF3) port. Once the SF3 Safety I/O is triggered, it directly cuts power to the arm motors without requiring logical judgment. When the intrusion is cleared, the SF3 signal returns to a high level, and motor output can be resumed.

The LiDAR provides 270-degree static detection. If an additional laser reflection appears within the defined range, it is interpreted as an intrusion trigger. Therefore, during actual production, the presence of stationary objects like material carts or stools within the LiDAR's range can also trigger an alarm, resulting in false positives. This issue is addressed in personnel training. The light source is a 905nm infrared laser, classified as Class 1 (in compliance with IEC 60825-1:2014 and EN 60825-1:2014), making it safe for the human eye within the specified regulations.

The Emergency Stop button on the control cabinet is connected to the robot controller's SF1 (Safety I/O SI0-1, SI0-2). SF1 complies with PL=d, Category 3. The LiDAR and vision security system are connected to the robot controller's SF3 (SI1-1, SI1-2). SF3 complies with PL=d, Category 2. In a worst-case scenario, where a person rushes towards the robot at high speed and is stopped by a material cart, the calculated stopping distance is approximately 413mm.

$$\begin{aligned} Sp &= Sh + Sr + Ss + C + Z \\ &= Vh(Tr + Ts) + Vr(Tr) + Vs(Ts) + C + z \\ &= 1600[(0.015 + 0.124) + (0.05)] + 800(0.015 + 0.124) \\ &= 413 \text{ mm} \end{aligned}$$

Based on the actual on-site conditions, the safety distance has been set accordingly. Practical tests confirm that all safety requirements are met.

2.5 Cobot Monitoring System

(1) Robot Status Monitoring: Displays the status of all robots on the layout diagram using five color-coded icons (① Normal Operation, ② Alarm Pause, ③ Fault Stop, ④ Robot Offline, ⑤ Planned Downtime).

(2) Individual Robot Status Monitoring: Shows detailed information for each robot, including its corresponding equipment information, workpiece name, program name, production count, and current operating status.

(3) Input/Output Statistics: Provides real-time statistics on the input and output quantities for each robot.

(4) Efficiency Statistics: Allows for querying the distribution of the five statuses for all or individual robots over a specific time period (Fig. 7).

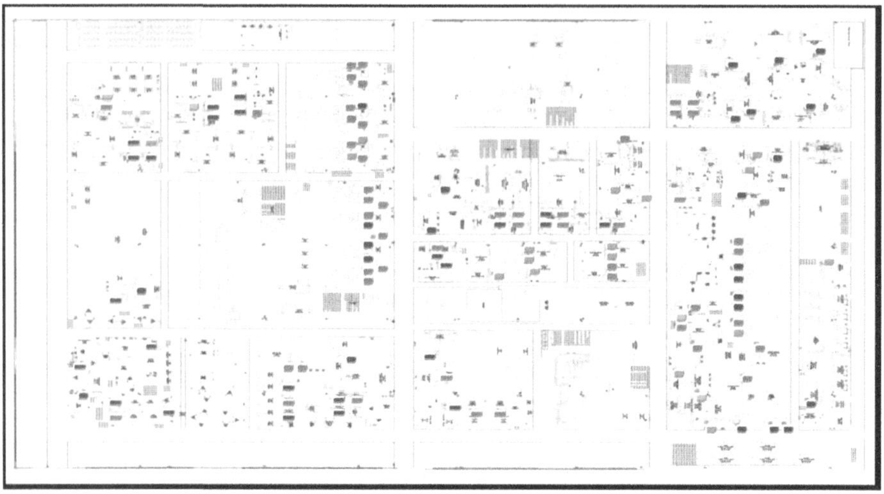

Fig. 7. Cobot Monitoring System

3 Results of Implementation

After the implementation and installation of Cobots, automated processing was achieved for most parts, eliminating the need for manual loading and unloading by personnel and enabling continuous, 24-h uninterrupted production. They have efficiently replaced manual operations, allowing the number of machines each employee can manage to increase from the original 2–3 to 4–5.

The application of Cobots in machine tool loading and unloading, along with the integrated use of flexible fixtures and AI vision systems, provides a reference for automation upgrades in other manufacturing sectors. The project's implementation has also promoted the advancement of robot control technology and AI algorithms, driving the development of intelligent manufacturing technology.

4 Conclusion

The application of AI vision-based Cobots in machine tool loading and unloading has brought significant technological progress and industrial upgrading. Firstly, in terms of production efficiency, with their high-precision, high-speed loading and unloading operations, Cobots have significantly shortened production cycles and boosted capacity, directly generating millions in economic benefits for the company. Secondly, in terms of product quality, through AI cameras, Cobots achieve precise workpiece recognition and positioning, reducing human operational errors and significantly lowering the scrap rate. Finally, the application of Cobots has also reduced labor costs and safety risks, laying a foundation for the further development of intelligent factories.

References

1. Zhang, B., Gao, J., Wang, J., et al.: Design of circular washer dimension measurement system based on machine vision. Tool Engineering **57**(7), 141–145 (2023)
2. Dai, K., Xie, K.: Design of sorting robot fault detection system based on internet of things. Comp. Measur. Cont. **29**(8), 37–41 (2021)
3. Han, Y., Feng, D., Wu, G.: Structural dynamic displacement measurement method based on machine vision and UAV. J. Vibr. Shock **41**(19), 1–7 (2022)
4. Liu, J., Qi, H., Liu, Z., et al.: A precise attention tracking system based on machine vision. J. Sys. Simulat. **35**(10), 2087–2100 (2023)
5. Ren, R., Zhang, S., Ren, J.: Collision detection for manipulator based on nonlinear extended momentum observer. Contr. Eng. China, 1–8 (2023)

A Fringe Projection High-Reflectivity Image Inpainting Method Based on Blender Simulation and Cascaded Network

Xingyu Gao, Rong He, and Weiming Li(✉)

Guangxi Key Laboratory of ZHIYU Humanoid Robots, Guangxi Engineering Research Center for Intelligent Vision and Collaborative Robotics, GuangXi Minzu University, Nanning 530006, China

liaifan@buaa.edu.cn

Abstract. Fringe Projection Profilometry (FPP) is extensively employed in industrial inspection due to its non-contact nature, high precision, and efficiency. However, when measuring highly reflective surfaces such as metals, strong specular reflections cause intensity saturation and clipping in captured images. This results in severe phase retrieval errors and voids in the 3D reconstruction. To address the limitations of intrusive physical coating methods and the scarcity of high-quality ground truth data for traditional deep learning, we propose a simulation-driven deep learning approach for fringe restoration using Blender. First, we constructed a high-fidelity virtual FPP system using the computer graphics software Blender. By precisely controlling Physically Based Rendering (PBR) material properties and employing a "dual-rendering" mechanism, we generated a large-scale dataset of pixel-aligned "saturated specular–ideal diffuse" pairs. This effectively overcomes the bottlenecks associated with acquiring ground truth and ensuring data alignment in physical experiments. Second, we designed a "detect-and-repair" two-stage cascaded network. The first stage employs an Attention U-Net incorporating attention gates to precisely localize saturated regions and generate modulation masks, thereby providing spatial priors for the restoration process. The second stage features a "Repair-Net" based on dilated convolutions and residual connections. Guided by the masks, this network leverages context from unsaturated regions to intelligently infer and reconstruct the missing fringe textures. Furthermore, a combined loss function integrating L1 regularization and the Structural Similarity Index (SSIM) was introduced to ensure high pixel-level accuracy and structural fidelity in the restored images. Finally, comparative experiments and quantitative analyses demonstrate that our method exhibits excellent robustness across objects with varying reflectivities. Compared to the standard U-Net, our method improves the SSIM to 0.9902, effectively eliminating phase nonlinear distortions and ripple errors caused by specular highlights. The restored fringe images enable direct, high-precision phase retrieval, resulting in smooth, complete, and void-free 3D reconstructions. This approach holds significant potential for industrial online inspection applications.

Keywords: Fringe Projection Profilometry (FPP) · High Dynamic Range (HDR) · Deep learning · Blender Simulation · Image Restoration · Phase Retrieval

H. Wei and J. Niu (Eds.): COBOT 2025, CCIS 2893, pp. 488–510, 2027.
https://doi.org/10.1007/978-981-92-2074-8_29

1 Introduction

Fringe Projection Profilometry (FPP), a representative active non-contact 3D optical measurement technique, is widely employed in fields such as industrial manufacturing, reverse engineering, quality inspection, robot vision, and biomedicine, owing to its high precision, speed, and data density [1]. A typical FPP system projects pre-coded sinusoidal fringe patterns onto the object surface using a digital projector and captures the deformed fringe images, which are modulated by the surface topography, from a different viewing angle3. Subsequently, phase information is extracted from these deformed fringes using phase retrieval algorithms—such as Phase Shifting Profilometry (PSP) [2] or Fourier Transform Profilometry (FTP) [3]—to reconstruct the 3D coordinates combined with system calibration parameters.

However, the application of FPP to industrial objects with high reflectivity, such as metal workpieces and smooth molds, presents severe challenges [4]. The smooth surfaces of such objects are prone to strong specular reflections, causing local light intensities to exceed the camera sensor's dynamic range. This results in not only overexposure (saturation) in highlight regions but also potential underexposure in non-highlight areas due to the concentration of reflected light energy. Such saturation severely compromises measurement accuracy by introducing nonlinear distortions to the sinusoidal waveforms and causing intensity truncation in highlight areas, which leads to a loss of valid fringe information. Consequently, this induces errors in wrapped phase calculation and failures in absolute phase retrieval. Ultimately, this results in large-scale voids or geometric distortions in the reconstructed 3D point clouds, failing to meet the stringent requirements for integrity and precision in industrial inspection [5, 6].

To address the challenges of High Dynamic Range (HDR) measurement on highly reflective surfaces, extensive research has been conducted globally. Existing solutions can be primarily categorized into five classes:

First, Physical Surface Treatment [7]. The most direct approach involves coating the object's surface with a thin layer of diffuse developing agent (e.g., titanium dioxide), thereby transforming its surface property from specular to diffuse reflection. While this is a common trade-off in commercial applications, its limitations are significant. Spraying increases operational complexity, and this contact-based method is ill-suited for automated on-machine inspection. Furthermore, precise control over coating thickness and uniformity is difficult, inevitably introducing additional systematic errors.

Second, Polarization-based Methods [8, 9], By adjusting the relative orientation of the polarizers (typically to a cross-polarized configuration), specular reflections with specific polarization directions can be selectively filtered out. This effectively suppresses overexposure in highlight regions, thereby preserving greater surface detail. However, the efficacy of this method relies heavily on the optical properties of the target surface. For complex workpieces made of non-metals or possessing varying surface roughness, the extinction effect often diminishes significantly. Simultaneously, the introduction of polarizers inevitably results in light energy loss, which may compromise fringe contrast in dark regions. Furthermore, achieving optimal suppression often necessitates precise matching and adjustment of polarization angles, thereby increasing the complexity of system configuration

Third, Multi-Exposure Techniques [10–12]. These methods involve capturing a sequence of images of the same scene at varying exposure durations (i.e., using low exposure for highlights and high exposure for dark regions). Subsequently, image fusion algorithms are employed to synthesize a high-quality full-field HDR fringe image by selecting unsaturated pixels that fall within the optimal modulation range from the sequence [13]. However, the requirement to capture and process a large number of image makes data acquisition and processing extremely time-consuming. This severely constrains its potential for dynamic scenes or high-speed online inspection.

Fourth, Adaptive Projection Methods [14, 16]. The core principle here lies in actively regulating the pixel-wise intensity of projected fringes to adapt to the spatially varying reflectivity of the object surface. For instance, Sun et al. [17]. Proposed an intensity adjustment method based on pixel mapping. By pre-calibrating the geometric mapping between camera and projector pixels, the projection intensity is reduced in high-reflection zones while maintaining or boosting brightness in dark areas. Nevertheless, this approach faces distinct bottlenecks: establishing an accurate camera-projector pixel mapping in high-reflection areas is inherently challenging; excessive intensity reduction may compromise the Signal-to-Noise Ratio (SNR) in dark regions; and the cumbersome calibration and iteration processes reduce overall measurement efficiency.

Fifth, Image Processing and Deep Learning Approaches [18]. With advances in computational imaging, research focus has shifted towards algorithmic post-processing of saturated images without altering hardware acquisition workflows (specifically preserving "single-shot" capabilities). On one hand, traditional image enhancement algorithms have been explored. For example, Li et al. [19]. Utilized an improved Retinex algorithm combined with bilateral filtering for edge preservation to restore fringe information in highly reflective regions of blades. On the other hand, deep learning-based methods have demonstrated superior nonlinear restoration capabilities. These typically employ U-Net and its variants (e.g., ResUNet) as backbone networks, training models to learn the end-to-end mapping from "saturated specular fringes" to "ideal sinusoidal fringes." For instance, Yang et al. [20]. Designed a specialized network to detect low-modulation regions and enhance fringe quality. Meanwhile, Wan et al. [21]. Optimized the single-shot measurement process by using ExSNet to intelligently predict optimal exposure times, followed by FrANet for fringe analysis to effectively mitigate specular interference.

While deep learning-based restoration methods offer a viable pathway for "single-shot" measurement of highly reflective objects, their performance hinges critically on the availability of large-scale, high-quality training datasets. In the context of FPP, constructing paired real-world datasets—comprising both "saturated specular images" and "ideal diffuse ground truth images" of the same object under an identical pose—presents a formidable challenge. To acquire Ground Truth (GT), researchers are often compelled to employ laborious physical procedures. For instance, Xi et al. [22] first captured high-reflection images of metal workpieces. Subsequently, while maintaining the workpiece in an absolutely stationary position, they applied a diffuse developing agent to the surface and recaptured the scene to obtain ideal fringes as ground truth. This process is not only operationally complex and time-intensive but also fails to completely eradicate minor pose deviations between the two acquisitions, posing significant obstacles to precise data alignment.

To circumvent the difficulties associated with real-world data acquisition, some research has shifted towards utilizing Computer Graphics (CG) [23, 25] to generate synthetic datasets. For example, Ueda et al. [26] leveraged the Unity engine to construct a virtual FPP system, simulating objects with varying reflective properties by adjusting the material's "Smoothness" parameter to generate paired data. The effectiveness of this approach has been preliminarily validated.

Addressing the aforementioned limitations in industrial on-machine inspection and the critical bottleneck of acquiring high-fidelity paired data, this paper proposes a Blender simulation-driven deep learning approach for fringe restoration on highly reflective surfaces. The primary innovation of this work lies in the construction of a high-fidelity virtual monocular structured light experimental platform using Blender, a professional software suite known for its robust Physically Based Rendering (PBR) capabilities. We demonstrate that by precisely controlling PBR material parameters, it is possible to quantitatively and realistically simulate the physical transition of an object's surface from specular reflection (highly reflective) to ideal diffuse reflection (highlight-free). Leveraging this characteristic, we generated a large-scale, pixel-level strictly aligned paired dataset of "saturated specular images" and "ideal diffuse images" (GT), while maintaining identical pose, lighting, and camera parameters. This fundamentally resolves the issues of ground truth acquisition and alignment inherent in physical experiments.

This study aims to utilize this synthetic dataset to train a deep neural network capable of directly restoring fringe information lost due to saturation from untreated highly reflective images. This enables high-precision 3D reconstruction of specular surfaces without sacrificing measurement efficiency or requiring physical contact. To this end, we propose a Two-stage Cascaded Network comprising two functional modules: the first stage precisely identifies saturated specular regions, while the second stage specifically reconstructs the missing fringe information within these regions. The restored high-quality fringe images serve directly as input data for subsequent N-step Phase Shifting Profilometry (PSP) or multi-frequency heterodyne methods, facilitating high-precision phase calculation.

2 Principle

2.1 Principle of Phase Shifting Profilometry (PSP)

In a typical FPP system, a Digital Light Processing (DLP) projector is employed to project a sequence of sinusoidal fringe patterns with specific phase shifts onto the target surface. Upon projection, these patterns are modulated by the object's surface height distribution, resulting in geometric deformation. The intensity distribution of the $k-$th deformed fringe image captured by the camera can be mathematically expressed as:

$$I_k(x, y) = A(x, y) + B(x, y)\cos\left[\phi(x, y) + \delta_k\right](k = 1, 2, ..., N) \quad (1)$$

Here, (x, y) denotes the pixel coordinates. $A(x, y)$ represents the background intensity, which primarily depends on the surface reflectivity and ambient lighting conditions. $B(x, y)$ is the modulation intensity (or AC component), characterizing the fringe visibility. When $B(x, y)$ approaches zero, the fringe information becomes ambiguous or lost, indicating a low-modulation region. $\phi(x, y)$ is the wrapped phase to be retrieved, typically ranging within $(-\pi, \pi]$. The 3D information of the object is encoded within this phase map. δ_k denotes the known phase shift of the $k-$th projected fringe pattern. Therefore, to accurately reconstruct the object's 3D geometry, the primary objective is to robustly retrieve the wrapped phase for each pixel from the image sequence.

2.2 Phase Analysis Methods

N-step Phase Shifting Algorithm

The N-step Phase Shifting Algorithm stands as one of the most widely adopted and high-precision phase retrieval techniques in the field of FPP. This method involves the precise control of a projector to sequentially cast a minimum of three sinusoidal fringe patterns featuring a constant phase increment. In the standard N-step formulation, the phase shift for the $k-$th pattern is typically defined as:

$$\delta_k = (k-1) \cdot \frac{2\pi}{N} \tag{2}$$

where $k = 1, 2, \ldots, N$. For the N-step phase shifting algorithm, the $A(x, y)$ and $B(x, y)$ at each pixel can be derived using the following analytical formulas:

$$A(x, y) = \frac{1}{N} \sum_{k=1}^{N} I_k(x, y) \tag{3}$$

$$B(x, y) \cos[\phi(x, y)] = \frac{2}{N} \sum_{k=1}^{N} I_k(x, y) \cos(\delta_k) \tag{4}$$

By solving this system of simultaneous equations, the wrapped phase can be derived as:

$$\phi(x, y) = \arctan\left(\frac{\sum_{k=1}^{N} I_k(x, y) \sin(\delta_k)}{\sum_{k=1}^{N} I_k(x, y) \cos(\delta_k)}\right) \tag{5}$$

Multi-frequency Heterodyne Method (Phase Unwrapping).

Due to its inherent periodic discontinuities, the wrapped phase fails to directly represent the true height information of the object. To retrieve continuous and authentic phase information, phase unwrapping is required to convert the wrapped phase into the absolute phase $\Phi(x, y)$, expressed as:

$$\Phi(x, y) = \phi(x, y) + 2\pi k(x, y) \tag{6}$$

Here, $k(x, y)$ is an integer representing the cumulative number of periods at the pixel (x, y) commonly referred to as the fringe order. Ultimately, leveraging the principle of optical triangulation and incorporating system calibration parameters, the absolute phase $\Phi(x, y)$ is mapped to the 3D spatial coordinates (X, Y, Z) of each pixel, culminating in the reconstruction of the 3D point cloud.

Impact of High Reflection on Phase Retrieval

As deduced from Eqs. (1) and (5), a fundamental prerequisite for obtaining high-precision wrapped phases using the N-step algorithm is the accurate acquisition of intensity values for each fringe image. However, when high-reflection regions exist on the object surface, the camera sensor is highly prone to over-saturation. In this scenario, the incident light intensity exceeds the sensor's linear dynamic range, causing the recorded pixel intensities to be "clipped" to the maximum quantization value (e.g., 255 for an 8-bit depth). Consequently, the image intensity suffers severe distortion. The intensity distribution in saturated regions no longer adheres to a sinusoidal waveform but instead exhibits a "peak plateau effect," as illustrated in Fig. 1

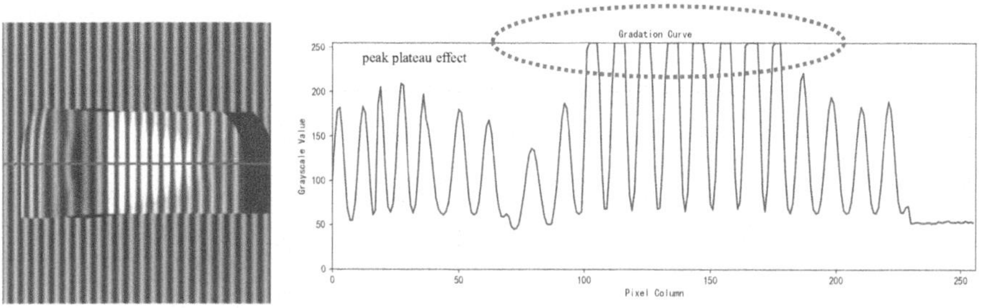

(a) (b)

Fig. 1. Specular fringe image and its intensity profile

Furthermore, this nonlinear intensity clipping violates the orthogonality condition of the phase-shifting algorithm, leading to anomalies in the calculated modulation intensity. Such waveform distortion directly introduces significant wrapped phase errors, which subsequently propagate to the phase unwrapping stage. Therefore, the high-reflection issue fundamentally undermines the image acquisition stage—the bedrock of the FPP measurement chain—rendering the phase retrieval theory invalid and ultimately compromising the accuracy and integrity of the 3D reconstruction.

2.3 Deep Learning-Based Fringe Image Restoration Network

To address the challenges of fringe image distortion and phase retrieval failure induced by specular reflections, we propose a deep learning-based two-stage "detect-and-repair" network model. This model is designed to directly restore fringe images approximating ideal diffuse conditions from saturated specular inputs, thereby providing high-quality input data for subsequent N-step Phase Shifting Profilometry (PSP) and multi-frequency heterodyne phase unwrapping. The overall architecture of the proposed network is illustrated in Fig. 2

In traditional N-step PSP, the calculation of the wrapped phase is highly dependent on the accuracy of pixel intensity values. Once saturation occurs, these values suffer from truncation distortion. The primary objective of the proposed network model is to retrieve and restore the intensity "Ground Truth" (GT) inherent to these saturated pixels.

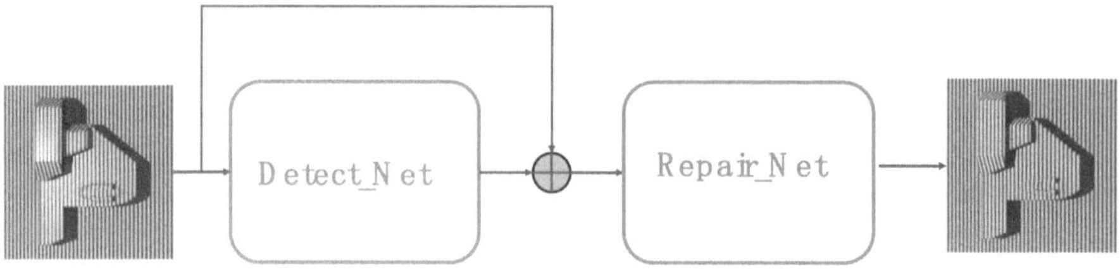

Fig. 2. Architecture of the proposed network model

Specifically, the model aims to learn an end-to-end nonlinear mapping from "saturated specular images" to "ideal diffuse images." The proposed model comprises two cascaded sub-networks: The first-stage network (architecture shown in Fig. 3) is dedicated to "defect localization," specifically identifying and segmenting saturation regions induced by specular reflection. Leveraging the proven efficacy of U-Net in feature extraction and boundary localization for semantic segmentation, we adopted Attention U-Net, which incorporates attention mechanisms, as the backbone architecture. This network retains the classic symmetric "Encoder-Decoder" structure and Skip Connection mechanism of U-Net, effectively fusing multi-scale contextual information to recover spatial details.

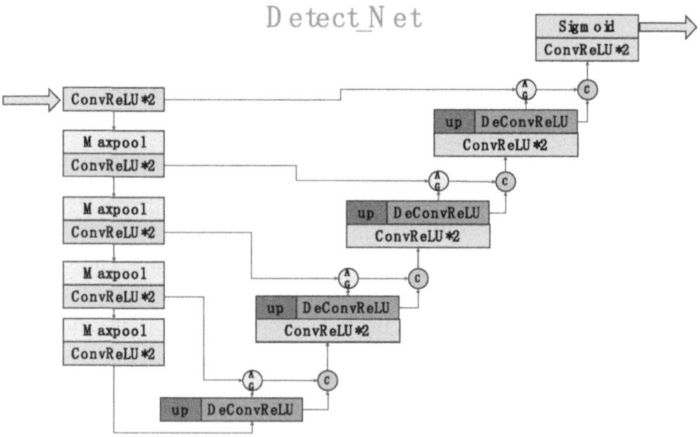

Fig. 3. Architecture of the Detect_Net.

Building on this, we introduced Attention Gate (AG) modules at the skip connections (architecture depicted in Fig. 4).

The AG module utilizes high-level semantic features from the decoder as gating signals to adaptively weight the feature maps transferred from the encoder. This mechanism enables the network to automatically focus on salient regions relevant to the target (i.e., saturated areas) while effectively suppressing background noise. Ultimately, it generates a pixel-level Modulation Mask that not only delineates the spatial location of distortions but also provides crucial spatial priors for the subsequent restoration module.

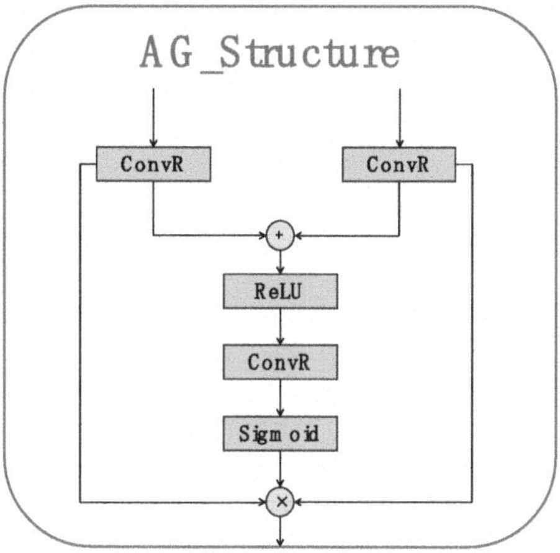

Fig. 4. Architecture of the AG module.

The second-stage network, designated as Repair-Net (architecture shown Fig. 5), is dedicated to the task of "information restoration." Taking the concatenation of the original specular image and the modulation mask along the channel dimension as input, this module aims to intelligently "infer" and reconstruct missing sinusoidal waveforms within saturated regions. It achieves this by leveraging valid fringe textures from surrounding unsaturated areas as context, guided by the mask. To effectively capture global information without increasing network depth or parameter count, we incorporated Dilated Convolution and Residual Connection techniques. By configuring varying dilation rates, dilated convolutions exponentially enlarge the neuron's receptive field to establish long-range pixel dependencies, thereby surmounting the challenge of large-scale information loss in highly reflective areas. This design not only accurately inpaints missing intensity data but also effectively reconstructs the ideal intensity distribution implied in Eq. (1). This ensures that the subsequent N-step PSP operates on accurate intensity values, ultimately yielding correct wrapped and absolute phases for high-precision 3D reconstruction.

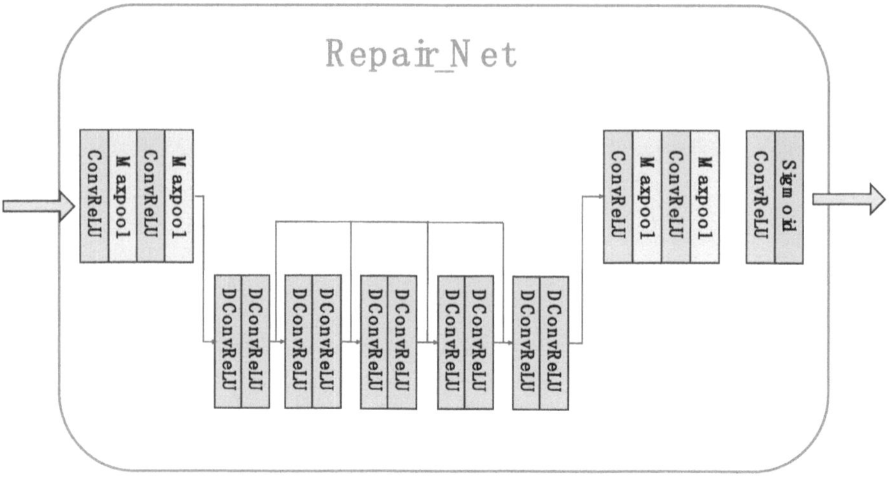

Fig. 5. Architecture of the Repair_Net.

2.4 Loss Function Design

In image restoration tasks, the selection of the loss function plays a decisive role in determining the quality and structural fidelity of the generated images. For the proposed network model, we compared and analyzed several loss functions widely employed in low-level vision tasks and ultimately formulated a combined loss function.

Mean Squared Error (MSE) Loss: MSE stands as one of the most prevalent loss functions in image reconstruction. It quantifies the mean of the squared differences between the predicted values and the Ground Truth (GT). Although MSE imposes heavy penalties on large errors via the squared term, thereby facilitating rapid model convergence, its averaging property often leads to the generation of overly smooth textures, resulting in blurred edges and a loss of high-frequency details in the output images. It is mathematically defined as follows:

$$\mathcal{L}_{MSE} = \frac{1}{N}\sum_{i=1}^{N}(Y_i - \hat{Y}_i)^2 \tag{7}$$

L1 Loss (Mean Absolute Error, MAE) In contrast, L1 loss computes the mean absolute difference between the predicted values and the ground truth. Unlike MSE, L1 loss does not amplify errors via squaring—particularly for outliers such as extreme pixel values caused by specular reflections. Consequently, it exhibits superior robustness to outliers. This characteristic fosters a more stable training process and encourages the generation of images with sharper edges and richer details. The formula is defined as:

$$\mathcal{L}_{\mathcal{L}_1} = \frac{1}{N}\sum_{i=1}^{N}\left|Y_i - \hat{Y}_i\right| \tag{8}$$

Combined Loss Function: To further enhance restoration performance, we introduced the Structural Similarity (SSIM) loss. SSIM comprehensively evaluates perceptual image quality across three dimensions—Luminance, Contrast, and Structure—within a value range of [0, 1]. To synergize pixel-level accuracy with structural integrity—ensuring that the restored image numerically approximates the ground truth while faithfully preserving the periodic structure of sinusoidal fringes—we ultimately formulated a combined loss function. This function is constituted by the weighted sum of the L1 loss and the SSIM loss, defined as follows:

$$L_{total} = w_1 \cdot L_{L1} + w_2 \cdot (1 - L_{SSIM}) \tag{9}$$

Here, w_1 and w_2 denote the weighting coefficients for the L1 and SSIM loss components, respectively. This combined loss is designed to steer the model towards achieving pixel-level reconstruction accuracy while simultaneously prioritizing the overall visual quality and structural fidelity of the fringe patterns. Consequently, it yields optimal, high-quality input data tailored for subsequent phase retrieval.

3 Experiments and Analysis

3.1 Dataset Construction

We established a high-fidelity virtual fringe projection system within the Blender simulation environment. The objective was to acquire strictly pixel-aligned pairs of "specular" and "diffuse" images, serving as the network input and Ground Truth (GT), respectively.

Data Acquisition Process

Regarding test objects, we curated a selection of CAD models exhibiting complex geometric features from the open-source 3D repository Thingiverse [27, 28]. These included various metal mechanical components and smooth plaster casts. Upon importation, all models underwent unified scale normalization and material initialization to accurately simulate target objects in authentic industrial scenarios (as shown in Fig. 6).

Fig. 6. Rendered examples of diverse CAD models exhibiting specular reflections.

Given that the subsequent phase retrieval employs a "4-step phase-shifting" algorithm combined with "three-frequency heterodyne," we pre-generated a total of 12 sinusoidal fringe patterns, covering 3 distinct frequencies with 4 phase shift steps each. The core strategy for data generation lies in the "Dual Rendering" mechanism. For each object at an identical pose, we conducted two rendering acquisition passes while maintaining absolute consistency in camera intrinsic/extrinsic parameters, projector settings, and projected fringes. First Pass: The material was configured with high glossiness (low roughness) to simulate input images exhibiting strong specular reflections. Second Pass:

The material parameters were adjusted to ideal diffuse reflection (high roughness) to simulate reflection-free Ground Truth images. By randomly transforming the object models and their spatial poses, we efficiently generated a large-scale paired dataset. Ultimately, the dataset comprises 3,072 experimental samples, which were randomly partitioned into training and validation sets at a ratio of 8:2.

Analysis of Methodological Advantages
Compared to conventional computer-generated dataset methodologies (e.g., directly simulating fringes via mathematical formulas superimposed with Gaussian noise), the Physically Based Rendering (PBR) simulation approach proposed herein demonstrates significant advantages. Traditional methods are essentially oversimplifications based on ideal mathematical models, struggling to accurately replicate complex real-world optical phenomena…such as the intricate Bidirectional Reflectance Distribution Function (BRDF) properties of surfaces, projector nonlinear response, ambient light interference, and, crucially, the nonlinear saturation effects within high-reflection zones.

Conversely, our Blender-based system realistically simulates the physical transition of surface properties from "strong specular reflection" to "ideal diffuse reflection" by precisely controlling material attributes. This approach not only yields fringe images with high realism but ao ensures that the saturation characteristics and fringe distortions in high-reflection areas are highly consistent with actual physical laws. Consequently, deep learning models trained on this dataset can more effectively learn feature distributions and restoration patterns inherent to real specular scenes, thereby exhibiting superior generalization and robustness in practical industrial inspection.

3.2 Implementation Details and Training Strategy

We adopted an End-to-End joint training strategy, treating the detection and restoration networks as a unified entity for gradient backpropagation and parameter updating to ensure synergistic optimization of the two-stage tasks. In the data preprocessing phase, all input raw specular images and corresponding GT labels were converted to single-channel grayscale maps and normalized, mapping pixel values to the [0, 1] interval. The model was implemented in PyTorch and trained on a workstation equipped with an NVIDIA GeForce RTX 5090 GPU. The network parameters were optimized using the Adam optimizer. Regarding the combined loss function in Eq. (9), the weighting coefficients were empirically set to $w_1 = 0.7$ and $w_2 = 0.3$ to balance the pixel-level and structural constraints. Other hyperparameters were configured as follows: initial learning rate set to 1×10^{-4}. ., Batch Size to 32, and total Epochs to 150. To mitigate potential validation loss oscillation in the later stages of training and further enhance generalization, we introduced a dynamic learning rate scheduling mechanism (Reduce LR On Plateau). The Validation Loss was monitored in real-time; if the loss did not show significant decrease for 10 consecutive Epochs (i.e., Patience = 10), the learning rate was automatically decayed to 0.5 times its current value (Factor = 0.5).

3.3 Validation of Synthetic Data

To validate the effectiveness and physical realism of the synthetic data, we conducted a detailed comparative analysis of intensity distributions between the "input images" (specular samples) and "Ground Truth" (GT) labels within the dataset. As illustrated in Fig. 7, panels (a) and (d) display the ideal diffuse label image and the input image suffering from specular highlight interference, respectively. Panels (b) and (e) indicate the cross-sectional position of the 128th row marked by red lines in the corresponding images, while panels (c) and (f) present the corresponding pixel intensity profiles for that row. As observed in (a) and (c), the label image exhibits distinct fringes with uniform contrast, completely devoid of high-reflection regions. Its intensity profile reveals a perfect sinusoidal waveform, effectively simulating the fringe distribution characteristics typical of an ideal diffuse surface. In contrast, as evident in (d), intense specular reflections appear in the central region of the object in the input image. Examination of the intensity profile in (f) reveals a significant "flat-topped truncation" (or clipping) phenomenon in pixel intensity values within the highlight region (highlighted by red circles). This causes severe distortion of the sinusoidal waveform and blurring of boundaries. This implies a loss of valid fringe information in this region. Conversely, in non-specular regions, the fringes maintain a near-sinusoidal distribution. This characteristic is highly consistent with physical phenomena caused by camera overexposure in real-world measurements, demonstrating that the dataset accurately mirrors the fringe distortion and information loss induced by high reflection in real scenarios.

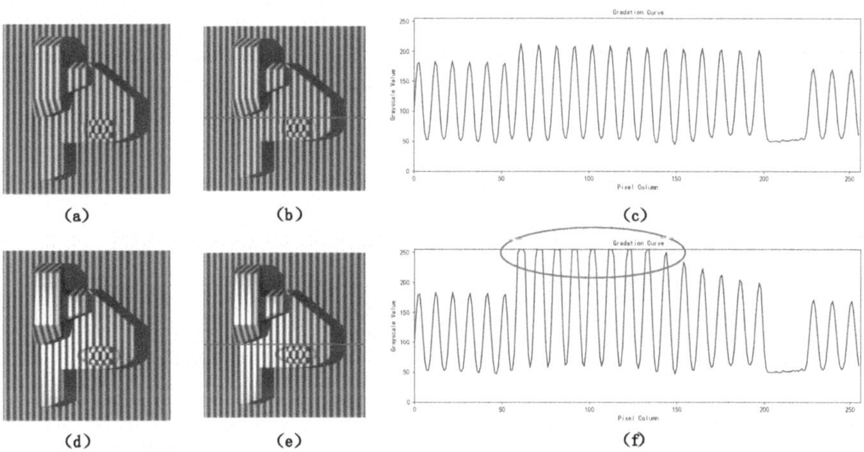

Fig. 7. Comparison of fringe images and their intensity profiles between ideal diffuse (Ground Truth) and specular inputs samples

3.4 Validation of Synthetic Data

To validate the efficacy of the proposed Attention U-Net detection architecture and the combined loss function, we conducted a series of ablation studies on the validation set. Peak Signal-to-Noise Ratio (PSNR) and Structural Similarity (SSIM) were adopted as quantitative evaluation metrics.

Validation of the Attention Mechanism

First, to assess the performance gains attributed to the incorporation of the Attention Gate

(AG) module, we compared the standard U-Net (Baseline) with our proposed network. As presented in **Error! Reference source not found.**, upon integrating the AG mechanism, the SSIM improved from 0.9837 to 0.9955, and the PSNR rose from 36.31 dB to 36.63 dB, compared to the baseline U-Net. These improvements compellingly demonstrate that the Attention Gate effectively enhances the network's ability to focus on saturated specular regions, thereby significantly augmenting the structural fidelity and Signal-to-Noise Ratio (SNR) of the restored images (Table 1).

Table 1. Quantitative results of the ablation study on the Attention Gate mechanism.

DetectNet	Attention gate	LOSS	SSIM	PSNR（dB）
Unet	no	L1+$(1 - L_{SSIM})$	0.9837	36.31
Our network	yes	L1+$(1 - L_{SSIM})$	**0.9955**	**36.63**

Validation of Loss Function Effectiveness

Subsequently, to evaluate the impact of different loss functions on restoration performance, we compared the experimental results of individual loss functions (MSE, L1) against the proposed combined loss function (L1 + SSIM). The results are summarized in Table 2. When relying solely on the MSE loss, the model tended to produce overly smoothed results, yielding the lowest metrics (SSIM: 0.9683, PSNR: 34.74 dB). Employing the L1 loss alone resulted in improved performance (SSIM: 0.9812, PSNR: 36.15 dB). However, by adopting the proposed combined loss function, the model achieved optimal performance, with SSIM and PSNR rising to 0.9955 and 36.63 dB, respectively. These data indicate that the combined loss function substantially enhances structural similarity while ensuring pixel-wise accuracy.

Table 2. Quantitative results of the ablation study on loss functions.

LOSS	SSIM	PSNR
MSE	0.9683	34.74
L1	0.98.12	36.15
L1+$(1 - L_{SSIM})$	**0.9955**	**36.63**

3.5 Training Dynamics and Convergence Analysis

To monitor the model status in real-time and validate the efficacy of the training strategy, we utilized TensorBoard to record loss variations and evaluation metrics throughout the entire training process, as illustrated in Fig. 8. Figure 8(a) and (b) display the convergence curves for Training Loss and Validation Loss, respectively. Overall, the model exhibits exceptional stability, free from gradient explosion or model collapse. The minimal divergence between the final training and validation losses indicates robust generalization capabilities with no distinct signs of overfitting.

Of particular note is the trajectory of the validation loss: Oscillation Phase (Epochs 0–110): During the initial stage, the validation loss exhibits jagged oscillations of substantial magnitude. This reflects the model's search and adjustment process near local minima of the loss function under a relatively high initial learning rate. Transition Phase (Epochs 105–110): A precipitous drop in the curve is observed between epochs 85 and 90, after which it instantaneously smooths out. This critical inflection point provides compelling evidence that the Reduce LR On Plateau dynamic learning rate scheduling mechanism played a decisive role. Upon detecting a plateau in validation performance, the scheduler automatically decayed the learning rate, propelling the model to break through local bottlenecks, navigate into a superior solution space, and achieve fine-grained convergence. Figure 8(c) and (d) illustrate the trends of PSNR and SSIM. During the first 110 epochs, the metric curves fluctuate in tandem with the loss function—a characteristic phenomenon inherent to the optimization process of Stochastic Gradient Descent (SGD). Following the learning rate adjustment, both metrics climbed steadily in the later stages and stabilized at high levels (PSNR > 36 dB, SSIM > 0.99). This validates that, after adequate training, the model possesses stable and high-quality restoration capabilities.

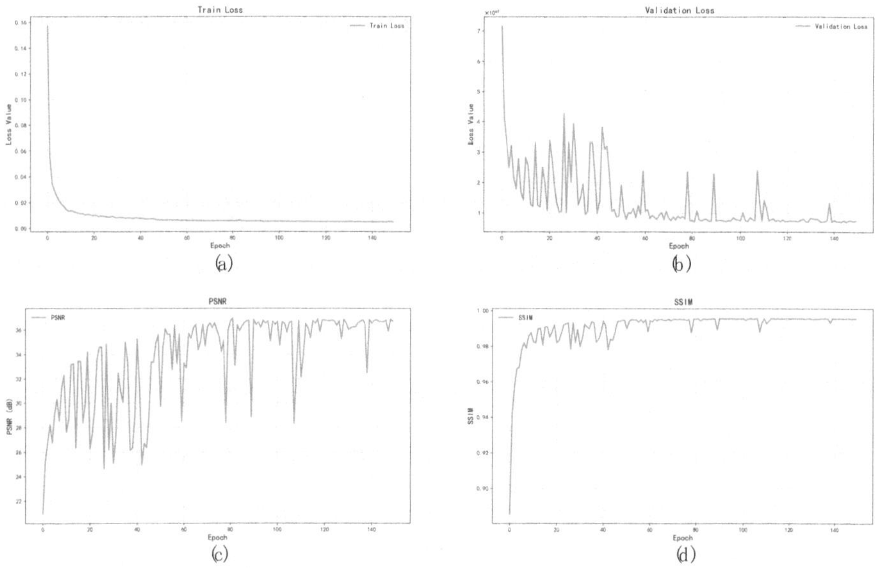

Fig. 8. Evolution curves of loss functions and evaluation metrics during the training process.

3.6 Model Inference and Intensity Analysis

To validate the network's generalization capability across diverse geometric structures, we selected five representative highly reflective objects from the test set for experimentation. Figure 9 presents a visual comparison between the input images, the model outputs, and the Ground Truth (GT).

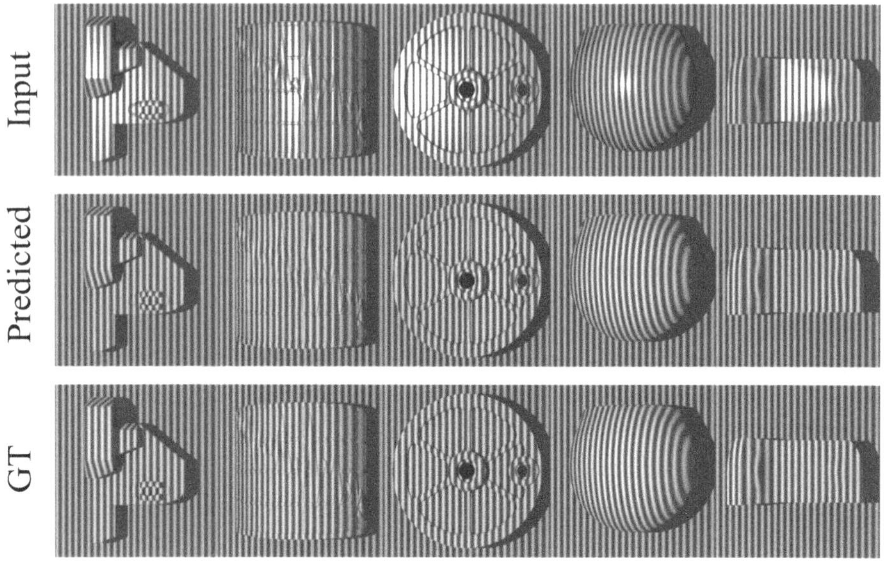

Fig. 9. Visual comparison of fringe restoration results across objects with diverse geometries.

From a visual perspective, our model successfully restores high-reflection regions of varying scales. It not only eliminates overexposure artifacts induced by specular reflection……but also impeccably reconstructs continuous sinusoidal fringe structures, rendering the output images visually approximating the GT to a high degree. This provides compelling evidence of the model's robust capabilities in feature extraction and texture restoration.

Furthermore, we conducted a pixel-level intensity analysis on specific scanlines of last Object, as illustrated in Fig. 10. The comparative profiles clearly reveal that the "Flat-topped Clipping" effect, induced by saturation in the input images, has been completely eliminated in the model's predictions. The restored intensity profiles not only recover the standard sinusoidal waveforms but also align impeccably with the Ground Truth curves in terms of both peak magnitude and phase. This quantitatively validates that the model effectively resolves the issue of nonlinear distortion in specular regions, fundamentally restoring the integrity of the intensity information.

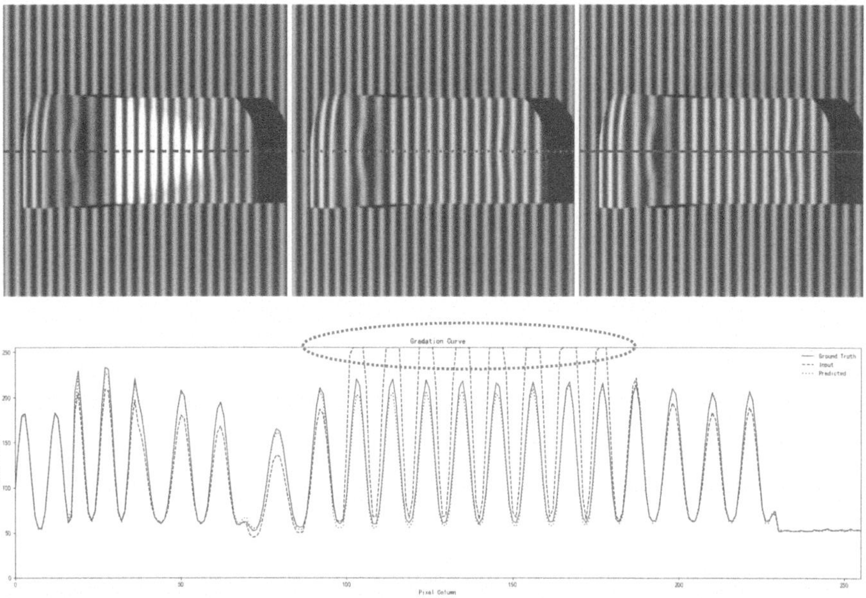

Fig. 10. Fringe restoration results and quantitative analysis of cross-sectional intensity profiles for representative specular objects.

3.7 Analysis of Absolute Phase

The ultimate objective of fringe restoration is to enhance the precision and robustness of subsequent phase retrieval. To this end, using a spherical model as a representative example, we calculated the absolute phases for the Input, Predicted, and Ground Truth (GT) images employing the "4-step phase-shifting" algorithm combined with the "3-frequency heterodyne" method. Figure 11 presents a multi-dimensional visual comparison. Columns 1 through 3 display the captured fringe images, the calculated single-channel wrapped phase maps, and the unwrapped color absolute phase maps, respectively. Column 4 provides magnified close-up views of the central high-reflection region (indicated by red dashed boxes) within the absolute phase maps. Column 5 depicts the error residual maps (Error Map) between the predicted/input phases and the GT. Qualitatively, in the Input row, despite the overall phase contour being discernible, strong specular interference in the center causes visible nonlinear distortions and "ripples" in the magnified view (Column 4), disrupting the smooth gradient transition. Simultaneously, the error map in Column 5 exhibits distinct bright patches (green/yellow regions), indicating significant phase errors. In contrast, the Predicted row demonstrates that fringes in the highlight region have been perfectly reconstructed by the model. The magnified view reveals an extremely smooth and continuous phase distribution, with the original distortion ripples completely eliminated. The error map (last column) transforms into a uniform dark red background, suggesting that the residuals between the predicted and GT phases are minimal and virtually negligible. Quantitatively, the Structural Similarity (SSIM) of the predicted image improved significantly from 0.9259 (Input) to 0.9684. This result compellingly demonstrates that our model not only effectively restores fringe intensity information but also fundamentally corrects phase modulation errors caused by saturation. This ensures the continuity and accuracy of phase data, laying a solid foundation for high-precision 3D reconstruction.

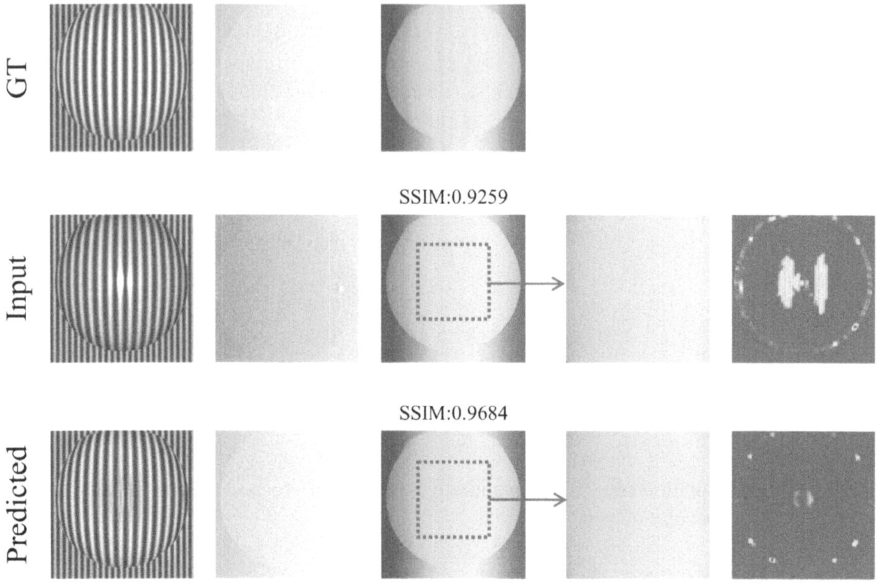

Fig. 11. Comparative analysis of absolute phase retrieval results and error distributions.

3.8 Comparative Experiments Across Varying Reflectivities

To comprehensively assess the model's generalization capability and robustness across diverse surface materials, we constructed three sets of test data with varying reflectivities (Low, Medium, High) by adjusting the specular reflection coefficient, while maintaining constant object pose and projection parameters. Figure 12(a) displays the raw fringe images of the same object under four distinct reflection states. It is evident that as reflectivity increases, the brightness of highlight regions intensifies, whereas the surrounding diffuse dark areas become dimmer due to dynamic range compression. This leads to a significant degradation in fringe modulation. Figure 12(b), (c), and (d) present detailed comparisons between input data and model predictions under Low, Medium, and High reflectivity conditions, respectively.

In the fringe comparison on the left, we focus on two characteristic regions:Highlight Region (Green dashed circle): In the input images, fringe information is completely lost due to overexposure (flat-top effect). Following model inference, the highlights in the predicted images are effectively suppressed, and the sinusoidal fringe structure is clearly restored. Dark Region (Red dashed circle): The input images suffer from extremely low Signal-to-Noise Ratio (SNR) due to underexposure. Conversely, in the predicted images, the intensity is significantly boosted and edge contrast is enhanced, demonstrating the model's dual optimization capability of "highlight suppression" and "dark region enhancement. "Dark Region (Red dashed circle): The input images suffer from extremely

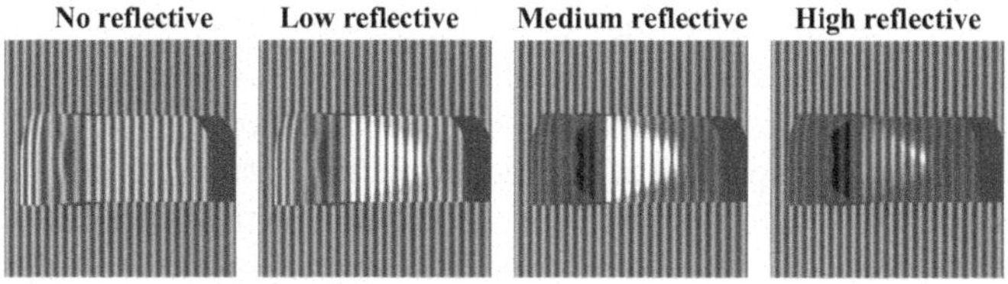

(a) Comparative visualization of fringe patterns under four distinct reflectivity conditions.

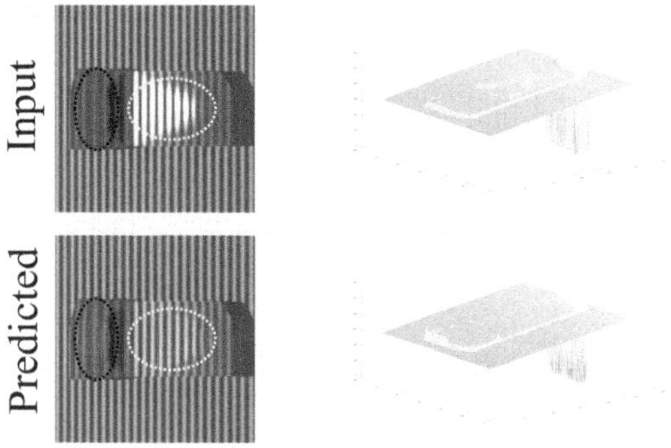

(b) Low reflectivity scenario.Comparison of fringe image and Corresponding 3D phase mapping results.

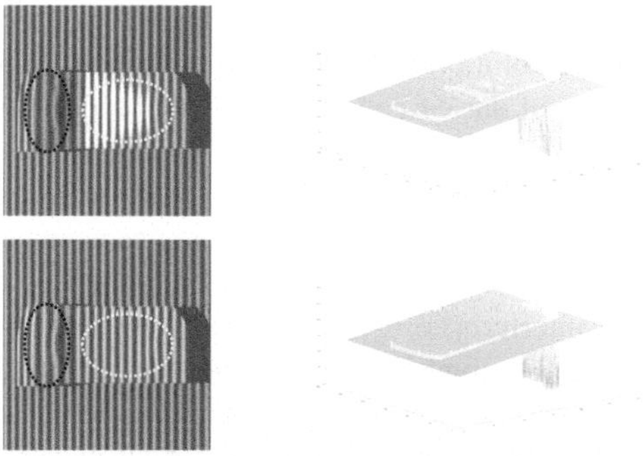

(c) Medium reflectivity scenario.Comparison of fringe image and Corresponding 3D phase mapping results.

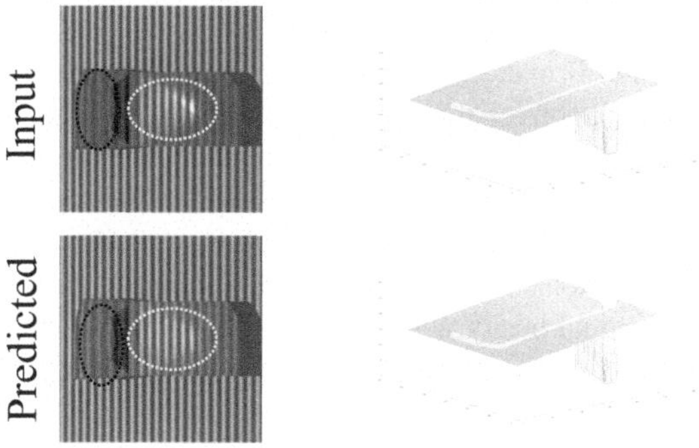

(d) High reflectivity scenario.Comparison of fringe image and Corresponding 3D phase mapping results.

Fig. 12. Comparison of model inference results across objects with varying reflectivities.

low Signal-to-Noise Ratio (SNR) due to underexposure. Conversely, in the predicted images, the intensity is significantly boosted and edge contrast is enhanced, demonstrating the model's dual optimization capability of "highlight suppression" and "dark region enhancement." Right: Comparison of 3D Reconstruction Results. To validate the impact of restoration quality on final measurement, we mapped the fringe patterns to 3D spatial distributions. The reconstruction results from input data exhibit obvious distortions and surface noise in reflective areas, attributed to phase retrieval failure in saturated zones. In contrast, the results from predicted data present a smooth, continuous, and complete 3D surface. Even under extreme high-reflection conditions (Fig. 12d), the model-restored data yields a high-quality object topography. These experiments compellingly demonstrate that the proposed network possesses strong generalization, capable of adaptively handling complex material scenarios ranging from low to high reflectivity. It effectively resolves the issue of reconstruction incompleteness faced by traditional methods in measuring objects with varying reflectivities.

3.9 Comparative Analysis of Inference Results Across Different Models

To further validate the efficacy of the proposed network architecture, we designed a comparative experiment where the detection module in our network was substituted with a standard U-Net to serve as the Baseline. The standard U-Net was selected as the primary benchmark because it represents the most widely adopted backbone architecture for general image restoration tasks, thereby providing a fair standard to evaluate the specific contributions of our cascaded design. Both models were trained and tested on the identical dataset. Figure 13 presents a comprehensive comparison among Input data, U-Net restoration results, Our Network (Our Net) results, and Ground Truth (GT) across three dimensions: fringe images, 2D absolute phase maps, and 3D phase visualizations. In terms of fringe image quality assessment, the input data is plagued by severe highlight and shadow interference (SSIM: 0.9317), resulting in compromised structural integrity. While the baseline U-Net recovers partial fringe information—raising the SSIM to 0.9838—perceptible blurring persists at the boundaries between highlights and shadows. In contrast, the proposed network further elevates the SSIM to 0.9902, achieving a 0.64% performance gain. While numerically seemingly marginal, in the context of precision optical measurement, this signifies a qualitative leap from "usable" to "high-precision" measurement quality.

More pronounced performance disparities are manifested in the quality of phase retrieval, which we analyzed from the perspectives of 2D residual noise and 3D spatial distribution. In the 2D phase maps, the U-Net result (e) exhibits distinct "salt-and-pepper" noise and phase jumps (manifested as black-and-white speckles) in the crescent-shaped shadow area on the right, indicating a failure to differentiate high-frequency fringes from background noise. Conversely, the phase map generated by our network (h) demonstrates exceptional purity with a smooth, uniform distribution, having almost completely eliminated edge noise. This disparity further translates to the 3D spatial distribution characteristics: the U-Net visualization (f) is marred by obvious spikes and outliers at the edges with a rough phase surface, suggesting significant errors in local phase calculation. In contrast, our result (i) is characterized by a smooth, dense surface with natural edge transitions. This compellingly proves that our method not only perfectly restores phase loss induced by highlights but also effectively suppresses mismatches in shadow regions, yielding a high-quality continuous phase distribution. In summary, the incorporation of the Attention Gate mechanism significantly enhances the network's ability to precisely focus on failure regions, while the cascaded restoration strategy ensures the physical consistency of the generated textures. This combined design enables the proposed method to significantly outperform the traditional U-Net architecture in both suppressing phase errors and enhancing phase calculation accuracy.

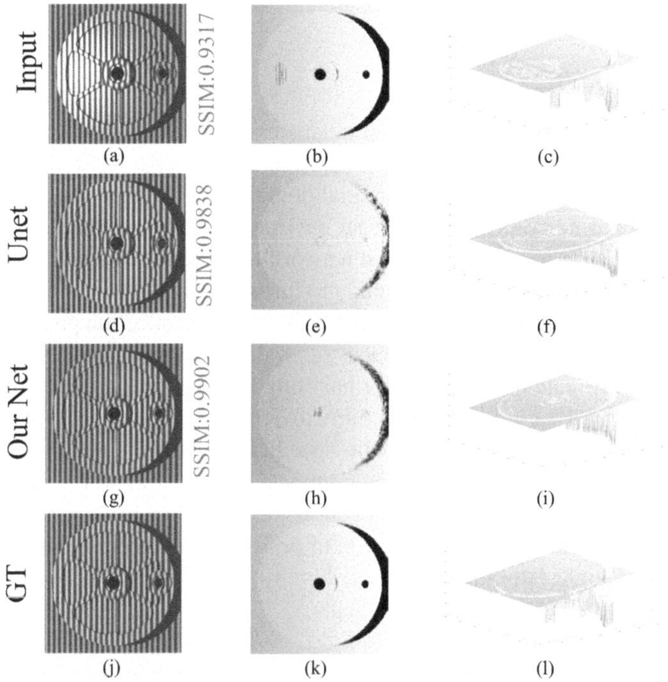

Fig. 13. Comparative analysis of inference results across different models

4 Conclusion and Future Work

4.1 Summary of Contributions

To address the challenges of saturation distortion and phase retrieval failure encountered by Fringe Projection Profilometry (FPP) when measuring highly reflective objects, this paper proposes a deep learning restoration method driven by computer graphics simulation. By constructing a virtual experimental environment and designing a cascaded restoration network, we effectively overcame the bottlenecks of cumbersome physical anti-reflection procedures and the difficulty of acquiring real-world data. The primary contributions of this work are summarized as follows: 1. Construction of a high-fidelity virtual FPP dataset scheme: Leveraging Blender's physical rendering engine, we established a "dual-rendering" simulation mechanism to batch-generate pixel-level strictly aligned "specular-diffuse" paired data. 2.Proposal of a "detect-and-repair" two-stage cascaded network: We designed a cascaded architecture comprising an Attention U-Net detection module and a Repair-Net restoration module. By employing attention mechanisms to precisely localize saturated regions and utilizing dilated convolutions with residual connections to incorporate context, we achieved physical-level restoration of truncated fringe waveforms. 3. Realization of high-precision phase recovery: Experimental results demonstrate the method's adaptability to surfaces with varying reflectivities. Compared to the baseline U-Net, our method improves the Structural Similarity (SSIM) to 0.9902, effectively eliminating phase ripples and edge jumps. The restored images are directly applicable to high-precision phase calculation.

4.2 Future Work

Although the proposed method has yielded satisfactory results in restoring highly reflective fringes, there remains scope for further optimization in complex real-world industrial scenarios. First, considering the potential domain gap between synthetic data and the complex optical noise of real industrial environments, future research will focus on transfer learning or domain adversarial network techniques. The goal is to perform "Sim-to-Real" domain adaptation fine-tuning to further enhance the model's robustness in real-world settings. Second, addressing the multiple inter-reflection phenomena often associated with complex geometric surfaces (e.g., deep holes or grooves), subsequent work plans to incorporate mechanisms for handling inter-reflection interference within the simulation and network design. This aims to achieve comprehensive correction of composite optical distortions involving both "specular saturation" and "inter-reflection." Finally, to meet the stringent real-time requirements of industrial online inspection, we will dedicate efforts to researching lightweight techniques such as network pruning and knowledge distillation. The objective is to drastically reduce computational costs while maintaining high restoration accuracy, thereby propelling this technology towards video-stream-level real-time 3D measurement.

Acknowledgment. This work was supported by the Guangxi Science and Technology Plan Project (2025XTXT04041), the Nanning Major Science and Technology Projects (20231034) and the Youth Science Fund Project of Guangxi Natural Science Foundation (2024JJB160101).

Disclosure of Interest. The authors declare no conflict of interest.

References

1. Pengyu, H., Shuming, Y., Fenghe, Z., et al.: Accurate and dynamic 3D shape measurement with digital image correlation-assisted phase shifting. Measure. Sci. Technol. **32**(7) (2021). https://doi.org/10.1088/1361-6501/ABDDF2
2. Zuo, C., Feng, S., Huang, L., et al.: Phase shifting algorithms for fringe projection profilometry: a review. Opt. Lasers Eng. **109**, 23–59 (2018). https://doi.org/10.1016/j.optlaseng.2018.04.019
3. Yanzhao, L., Yanjun, F., Yuhao, Z., et al.: High dynamic range real-time 3D measurement based on Fourier transform profilometry. Optics and Laser Technol. 138 (2021). https://doi.org/10.1016/J.OPTLASTEC.2020.106833

4. Zhu, J., Wang, K., Liu, H., et al.: Modified model of polarized bidirectional reflectance distribution function for metallic surfaces. Opt. Laser Technol. **99**, 160–166 (2018)
5. Shenzhen, L., Qian, K.: Modeling the measurement precision of Fringe Projection Profilometry. Light, Science & Applications **12**(1), 257–257 (2023). https://doi.org/10.1038/S41377-023-01294-0
6. Burke, J., Pak, A., Höfer, S., et al.: Deflectometry for specular surfaces: an overview. Adv. Opt. Technol. **12**, 1237687 (2023)
7. Palousek, D., Omasta, M., Koutny, D., et al.: Effect of matte coating on 3D optical measurement accuracy. Opt. Mater. **40**, 1–9 (2015)
8. Basel, S., Zhenyue, C.J.J.R., et al.: Multi-polarization fringe projection imaging for high dynamic range objects. Optics Express **22**(8), 10064–71 (2014). https://doi.org/10.1364/OE.22.010064
9. Riviere, J., Reshetouski, I., Filipi, L., et al.: Polarization imaging reflectometry in the wild. ACM Trans. Graph. (TOG) **36**(6), 1–14 (2017). https://doi.org/10.1145/3130800.3130894
10. Yanzhao, L., Yanjun, F., Yuhao, Z., et al.: High dynamic range real-time 3D measurement based on Fourier transform profilometry. Opt. Laser Technol. 138 (2021). https://doi.org/10.1016/J.OPTLASTEC.2020.106833
11. Song, Z., Jiang, H., Lin, H., et al.: A high dynamic range structured light means for the 3D measurement of specular surface. Opt. Lasers Eng. **95**, 8–16 (2017). https://doi.org/10.1016/j.optlaseng.2017.03.008
12. Jiang, H., Zhao, H., Li, X.: High dynamic range fringe acquisition: A novel 3-D scanning technique for high-reflective surfaces. Opt. Lasers Eng. **50**(10), 1484–1493 (2012). https://doi.org/10.1016/j.optlaseng.2011.11.021
13. Wei, F., Shaojing, T., Xiaodong, Z., et al.: Adaptive Fringe Projection for 3D Shape Measurement with Large Reflectivity Variations by Using Image Fusion and Predicted Search. Int. J. Opt. 2020 (2020). https://doi.org/10.1155/2020/4876876
14. Haisen, Y., Yu, L., Jing, Z., et al.: An adaptive fringe projection method for 3D measurement with high-reflective surfaces. Opt. Laser Technol. 170 (2024). https://doi.org/10.1016/J.OPTLASTEC.2023.110062
15. Jianhua, W., Yanxi, Y.: A new method for high dynamic range 3D measurement combining adaptive fringe projection and original-inverse fringe projection. Opt. Lasers in Eng. 163 (2023). https://doi.org/10.1016/J.OPTLASENG.2023.107490
16. Chen, C., Gao, N., Wang, X., et al.: Adaptive projection intensity adjustment for avoiding saturation in three-dimensional shape measurement. Opt. Comm. **410**, 694–702 (2018). https://doi.org/10.1016/j.optcom.2017.11.009
17. Junhua, S., Qiongyi, Z.: A 3D shape measurement method for high-reflective surface based on accurate adaptive fringe projection. Opt. Lasers in Eng. 153 (2022). https://doi.org/10.1016/J.OPTLASENG.2022.106994
18. Liu, H., Yan, N., Shao, B., et al.: Deep learning in fringe projection: A review. Neurocomputing **581**, 127493 (2024). https://doi.org/10.1016/J.NEUCOM.2024.127493
19. Mao-yue, L.I., Ze-long, L.I.U., Wei-xiang, Z.H.A.O., Gui-feng, X.I.A.O.: Blade reflection suppression technology based on surface structured light on-machine detection. Chinese Optics **15**(3), 464–475 (2022). https://doi.org/10.37188/CO.2021-0194
20. Guowei, Y., Min, Y., Nan, Z., et al.: High dynamic range fringe pattern acquisition based on deep neural network. Opt. Comm. 512 (2022). https://doi.org/10.1016/J.OPTCOM.2021.127765
21. MingZhu, W., Lingbao, K.: Single-shot 3D measurement of highly reflective objects with deep learning. Optics express **31**(9), 14965–14985 (2023). https://doi.org/10.1364/OE.487917
22. Xi, D., Hou, L., Wu, F., et al.: Deep learning-based inpainting of high dynamic range fringe pattern for high-speed 3D measurement of industrial metal parts. Adv. Eng. Info. **60**, 102428 (2024). https://doi.org/10.1016/J.AEI.2024.102428

23. Stavroulakis, P., Chen, S., Delorme, C., et al.: Rapid tracking of extrinsic projector parameters in fringe projection using machine learning. Opt. Lasers Eng. **114**, 7–14 (2019). https://doi.org/10.1016/j.optlaseng.2018.08.018
24. Wang, F., Wang, C., Guan, Q.: Single-shot fringe projection profilometry based on deep learning and computer graphics[J]. Opt. Express **29**(6), 8024–8040 (2021)
25. Yi, Z., Shaodong, W., Qing, L., et al.: Fringe projection profilometry by conducting deep learning from its digital twin. Optics Express **28**(24), 36568–36583 (2020). https://doi.org/10.1364/OE.410428
26. Kazumasa, U., Kanami, I., Osanori, K., et al.: Absolute phase retrieval of shiny objects using fringe projection and deep learning with computer-graphics-based images. Applied Optics **61**(10), 2750–2756 (2022). https://doi.org/10.1364/AO.450723
27. Zhou, Q., Jacobson, A.: Thingi10K: A Dataset of 10, 000 3D-Printing Models. CoRR (2016). abs/1605.04797
28. Zhou, Q., Grinspun, E., Zorin, D., et al.: Mesh arrangements for solid geometry. ACM Trans. Graph. (TOG) **35**(4), 1–15 (2016). https://doi.org/10.1145/2897824.2925901

Author Index

H. Wei and J. Niu (Eds.): COBOT 2025, CCIS 2893, pp. 511–512, 2027.
https://doi.org/10.1007/978-981-92-2074-8

The manufacturer's authorised representative in the EU is Springer Nature Customer Service Centre GmbH, Europaplatz 3, 69115 Heidelberg, Germany. If you have any concerns regarding our products, please contact ProductSafety@springernature.com

Printed and bound by CPI Group (UK) Ltd, Croydon, CR0 4YY
07/10/2026
02229353-0019